D0563814

MODERN PRESTRESSED CONCRETE

A prestressed-concrete bridge utilizing precast girders with a composite cast-in-place slab. Owner: Virginia Department of Highways. Consulting engineers: Parsons, Brinckerhoff, Hall, and MacDonald. Prestressed girders designed by the author. (Courtesy Economy Cast Stone Co., Richmond, Virginia.)

MODERN PRESTRESSED CONCRETE

DESIGN PRINCIPLES AND CONSTRUCTION METHODS

James R. Libby

James R. Libby and Associates

San Diego, California

VAN NOSTRAND REINHOLD COMPANY

New York Cincinnati Toronto London Melbourne

VAN NOSTRAND REINHOLD COMPANY REGIONAL OFFICES:
New York Cincinnati Chicago Millbrae Dallas

VAN NOSTRAND REINHOLD COMPANY INTERNATIONAL OFFICES:
London Toronto Melbourne

Copyright © 1971 by LITTON EDUCATIONAL PUBLISHING, INC.

Library of Congress Catalog Card Number: 77–140172

ISBN 0-442-24779-6

All rights reserved. Certain portions of this work copyright © 1961 by Litton Educational Publishing, Inc. No part of this work covered by the copyrights hereon may be reproduced or used in any form or by any means—graphic, electronic, or mechanical, including photocopying, recording, taping, or information storage and retrieval systems—without written permission of the publisher.

Manufactured in the United States of America

Published by VAN NOSTRAND REINHOLD COMPANY
450 West 33rd Street, New York, N.Y. 10001

Published simultaneously in Canada by Van Nostrand Reinhold Ltd.

15 14 13 12 11 10 9 8 7 6 5 4 3 2

Preface

This book has been written with the intention of recording a concise yet comprehensive treatment of the theoretical and practical aspects of prestressed concrete design and construction. The author believes the organization and contents of the book will render it especially useful to practicing engineers engaged in either design or construction. Students of structural engineering should also find this work a valuable textbook.

The first three chapters are devoted to the fundamental principles of prestressed concrete and to the more significant properties of the materials used in this structural form. The theoretical considerations that must be understood in order to design simple and continuous members that will behave in a satisfactory manner, at service loads as well as at design loads, are treated in Chapters 4 through 9.

Practical aspects of prestressed concrete construction which will be of interest to both the designer and builder of prestressed concrete structures and structural elements are covered in Chapters 10 through 13. The last three chapters will be of particular interest to the engineer engaged in the construction of prestressed concrete.

The flexural analysis of prestressed members loaded in the elastic range is treated in detail as is the analysis of these members for ultimate moment capacity. While this may not be necessary for flexural members composed of other materials, each type of analysis is considered necessary with prestressed concrete flexural members if adequate performance at service loads is to be expected and if adequate ultimate flexural strength is to be guaranteed.

The computation of the loss of prestress and long-term deflection of prestressed concrete flexural members, based upon known or assumed shrinkage, creep and relaxation characteristics of the materials, is treated in detail. These methods are now considered practical for general design work, due to the availability and low cost of efficient programable calculators.

Since the first uses of linear prestressing in the United States, which took place in the late 1940's, the methods and materials which have been used in pre-tensioning and post-tensioning have been going through almost constant change. For this reason, no attempt has been made to provide specific details of the hardware that is currently available for use in prestressing. It is felt that the reader desiring specific details will be better served by direct contact with the various commercial firms engaged in marketing these materials.

This book was written during the period of time that the 1963 edition of "Building Code Requirements for Reinforced Concrete" (ACI 318) was being revised for republication as the 1971 edition. The author has worked closely with the members of ACI Committee 318 in order to conform as accurately as possible to the provisions of the revised Code. The prestressed concrete provisions of the 1971 edition of ACI 318 as well as "Standard Specifications for Highway Bridges" (AASHO) have not been reproduced in this book because the author believes the reader who requires this information should have the provisions in their entirety, complete with revisions and addenda, and not extracts which could be misinterpreted.

The author wishes to express his sincere appreciation to his colleagues, Messrs. N. D. Perkins, A. R. McDaniel and F. K. Killman, for their efforts in reviewing and criticizing the manuscript of this book during its preparation.

JAMES R. LIBBY

San Diego, California
February, 1971

Contents

3 | CONCRETE FOR PRESTRESSING

4 | BASIC PRINCIPLES FOR FLEXURAL DESIGN

5 | CRACKING LOAD, ULTIMATE MOMENT, SHEAR AND BOND

1 Prestressing Methods

1-1 Introduction

Prestressing can be defined as the application of a predetermined force or moment to a structural member in such a manner that the combined internal stresses in the member, resulting from this force or moment and from any anticipated condition of external loading, will be confined within specific limits. Prestressing concrete, which is the subject of this book, is the result of applying this principle to concrete structural members, with a view toward eliminating or materially reducing the tensile stresses in the concrete.

The prestressing principle is believed to have been well understood since about 1910, although patent applications relating to types of construction which involved the priciple of prestressing date back to 1888 (Ref. 1).* The early attempts at prestressing were abortive, however, due to the poor quality of the materials that were available in the early days and also to a lack of understanding of the action of creep in concrete. Eugene Freyssinet, the eminent French engineer, is generally regarded as the first to discover the nature of creep in concrete and to realize the necessity of using high-quality

* Numbers in the text refer to the references at the end of each chapter.

concrete and high-tensile steel to ensure that adequate prestress is retained. Freyssinet applied prestressing in structural applications during the early 1930's. The history and evolution of prestressing are controversial and not well documented, and for this reason, they are not discussed further in this book. The reader who is interested may find additional historical details in the references (Refs. 2 and 3).

Many experiments have been conducted to demonstrate that prestressed concrete has properties that differ from those of reinforced concrete. Diving boards and "fishing poles" have been made of prestressed concrete to demonstrate the ability of this material to withstand large deflections without cracking. The more significant fact, however, is that prestressed concrete has proved to be economical in building, bridges, and other structures (under conditions of span and loading) that would not be practical or economical in reinforced concrete.

Prestressed concrete was first used in the U.S. (except in tanks) in the late 1940's. At that time, most U.S. engineers were completely unfamiliar with this mode of construction. Design principles of prestressed concrete were not taught in the universities, and the occasional structure that was constructed with this new material received wide publicity.

The amount of construction utilizing prestressed concrete has become tremendous and certainly will increase in the future. The contemporary structural engineer must be well informed on all facets of prestressing concrete.

1-2 General Design Principles

Prestressing, in its simplest form, can be illustrated by considering a simple, prismatic flexural member (rectangular in cross section) prestressed by a concentric force, as shown in Fig. 1-1. The distribution of the stresses at midspan are as indicated in Fig. 1-2. It is readily seen that if the flexural tensile stresses in the bottom fiber, due to the dead and live loads, are to be eliminated, the uniform compressive stress due to prestressing must be equal in magnitude to the sum of these tensile stresses.

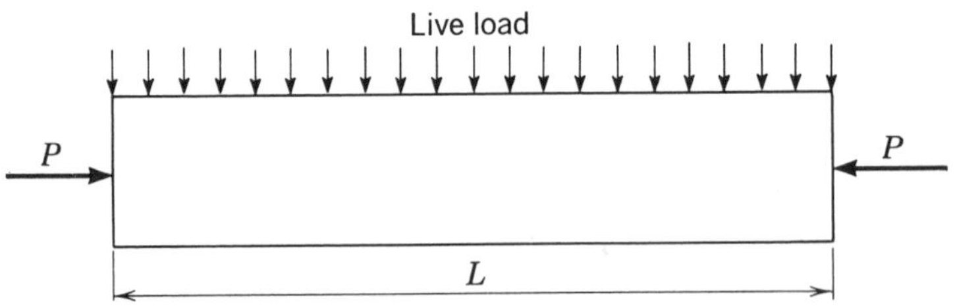

Fig. 1-1 Simple rectangular beam prestressed concentrically.

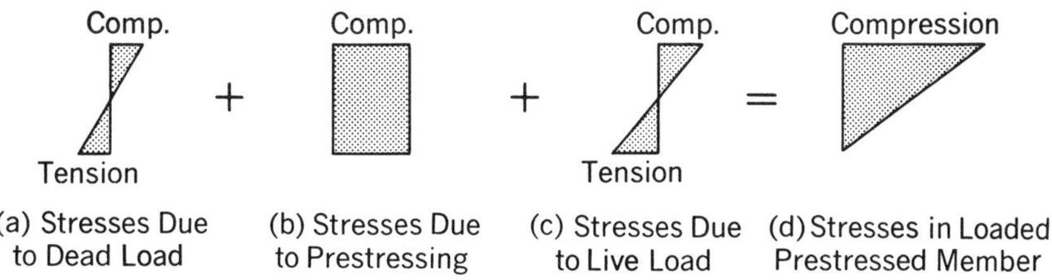

Fig. 1-2 Distribution of stresses at *midspan* of a simple beam concentrically prestressed.

There is a time-dependent reduction in the prestressing force, due to the creep and shrinkage of the concrete and the relaxation of the prestressing steel. If no tensile stresses are to be permitted in the concrete, it is necessary to provide an initial prestressing force that is larger than would be required to compensate for the flexural stresses resulting from the external loads alone. These losses, which are discussed in detail in Sec. 6-2, generally result in a reduction of the initial prestressing force by 10 to 30%. Therefore, if the stress distributions shown in Fig. 1-2 are desired after the loss of stress has taken place (under the effects of the final prestressing force), the distribution of stresses under the initial prestressing force would have to be as shown in Fig. 1-3.

Prestressing with the concentric force just illustrated has the disadvantage that the top fiber is required to withstand the compressive stress due to prestressing in addition to the compressive stresses resulting from the design loads. Furthermore, since sufficient prestressing must be provided to compress the top fibers, as well as the bottom fibers, if sufficient prestressing is to be supplied to eliminate all of the flexural tensile stresses, the average stress due to the prestressing force (P/A) must be equal to the maximum flexural tensile stress resulting from the design loads.

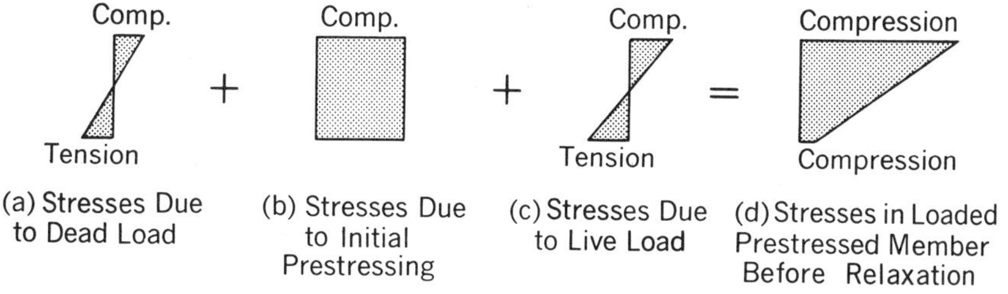

Fig. 1-3 Distribution of stresses at *midspan* of a simple beam under initial concentric prestressing force.

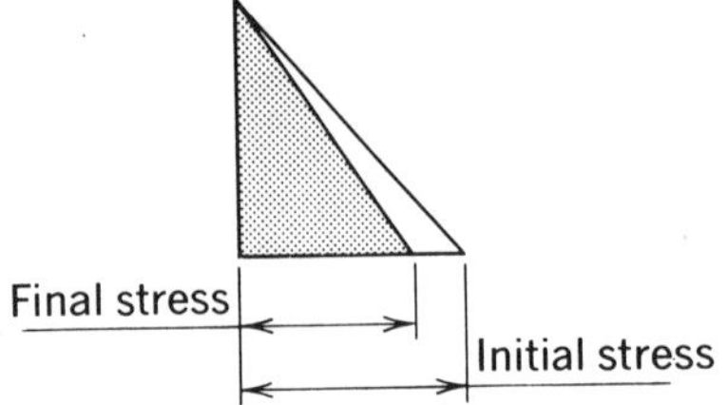

Fig. 1-4 Distribution of stresses due to prestressing force applied at lower third point of rectangular cross section.

If this same rectangular member were prestressed by a force applied at a point one-third of the depth of the beam from the bottom of the beam, the distribution of the stresses due to prestressing would be as shown in Fig. 1-4. In this case, as in the previous example, the final stress in the bottom fiber due to prestressing should be equal in magnitude to the sum of the tensile stresses resulting from the design loads. By inspection of the two stress diagrams for prestressing (Figs. 1-2b and 1-4), it is evident that the average stress in the beam, prestressed with the force at the third point, is only one-half of that required for the beam with concentric prestress. Therefore, the total prestressing force required to develop the desired prestressing of the second example will be only one-half of the amount required in the first example. In addition, the top fiber is not required to carry any compressive stress due to prestressing when the force is applied at the third point.

The economy that results from applying the prestressing force eccentrically is obvious. Further economy can be achieved when small tensile stresses are permissible in the top fibers—these tensile stresses may be due to prestressing alone or to the combined effects of prestressing and any external loads that may be acting at the time of prestressing. This is because the required bottom-fiber prestress can be attained with a smaller prestressing force, which is applied at a greater eccentricity under such conditions. This principle is treated in greater detail in subsequent chapters.

1-3 Prestressing with Jacks

The prestressing force in the above examples could be the result of placing jacks at the ends of the member, if there were abutments at each end of the beam which were sufficiently strong to resist the prestressing force that would be developed by such jacks. Prestressing with jacks, which may or may not remain in the structure, depending upon the circumstances, has been used abroad on dams, dry docks, pavements, and other special structures. This method has been used to a very limited degree in this country, since extremely careful control of the design (including the study of the behavior under overloads), construction planning, and execution of the construction is required if the results obtained are to be satisfactory. Furthermore, the loss of prestress that results from this method is much larger than when other methods are used (*see* Sec. 3-12), unless frequent adjustments of the jacks are made, since the concrete is subjected to constant strain in this method rather than to nearly constant stress as is the case in other methods. For these reasons, and since the type of structures to which this method of prestressing can be applied are very limited and beyond the scope of usual generalities, subsequent consideration of this method is not given in this book (*see* Ref. 5).

1-4 Pre-tensioning

Another method of creating the necessary prestressing force is referred to as pre-tensioning. Pre-tensioning is accomplished by stressing steel wires or strands, called tendons, to a predetermined amount, and then, while the stress is maintained in the tendons, placing concrete around the tendons. After the concrete has hardened, the tendons are released and the concrete, which has become bonded to the tendons, is prestressed as a result of the tendons attempting to regain the length they had before they were stressed. In pre-tensioning, the tendons are usually stressed by the use of hydraulic jacks. The stress is maintained during the placing and curing of the concrete by anchoring the ends of the tendons to abutments that may be as much as 500 ft or more apart. The abutments and appurtenances used in this procedure are referred to as a pre-tensioning bed or bench. In some instances, rather than using pre-tensioning benches, as mentioned above, the steel molds or forms that are used to form the concrete members are designed in such a manner that the tendons can be safely anchored to the mold after they have been stressed. The results obtained with each of these methods is identical, and the factors involved in determining which method should be used are of concern to the fabricator of prestressed concrete, but do not usually affect the designer.

The tendons used in pre-tensioned construction must be relatively small in diameter, since the bond stress between the concrete and the tendon is relied upon to transfer the stress from the tendon to the concrete. It should be recognized that the ratio of bond area to cross-sectional area for a circular wire or bar is

$$\frac{\text{Bond Area}}{\text{Cross-Section Area}} = \frac{4L}{d} \tag{1-1}$$

in which d is the diameter and L is the length. For a unit length, it will be seen from Eq. 1-1 that the ratio, which is also the ratio of the bond area available to the force the tendon can withstand, decreases as the diameter increases. A large number of the small tendons are normally required to develop the required prestressing force.

Pre-tensioning is a major method used in the manufacture of prestressed concrete in this country. The basic principles and some of the methods that are currently used domestically were imported from Europe, but much has been done here to develop and adapt the procedures to the North American market. One of the more recent developments in this country has been the use of pre-tensioned tendons that do not pass straight through the concrete member, but which are deflected or draped into a trajectory that approximates a curve. This procedure was first used on light roof slabs, but it is now used on large structural members. Currently, the use of deflected, pre-tensioned tendons is considered a practical method for the fabrication of precast bridge girders. In other countries, this method is used to a much smaller extent.

Although many of the devices used in pre-tensioned construction are patented, the basic principle is in the public domain. This partially accounts for the very rapid rate of increase in the use of this method in this country. A detailed discussion of the construction procedures and equipment used in pre-tensioned construction is given in Chapter 14.

1-5 Post-tensioning with Tendons

When a member is fabricated in such a manner that the tendons are stressed and each end is anchored to the concrete section after the concrete has been cast and has attained sufficient strength to safely withstand the prestressing force, the member is said to be post-tensioned. In this country, when using post-tensioning, a common method used in preventing the tendon from bonding to the concrete during placing and curing of the concrete is to encase the tendon in a mortar-tight, metal tube (or flexible metal hose) before placing it in the forms. The metal hose or tube is referred to as the sheath or duct and remains in the structure. After the tendon has been stressed, the void between the tendon and the sheath is filled with grout. In this manner, the tendon becomes bonded to the concrete section and corrosion of the steel is prevented.

Rather than use a metal tube or hose, a rubber hose (which may be inflated with air or water or may be stiffened during the placing of the concrete by putting a metal rod in it) has been used to form a hole through the concrete section. After the concrete is sufficiently set, the rubber tube is removed by pulling from one end. The tendon is then inserted in the duct that was formed by the tube. The construction is completed by stressing the tendon, anchoring each end, and grouting it in place.

Another method of preventing post-tensioning tendons from becoming bonded to the concrete is to coat the tendons with grease or a bituminous material, after which, the tendon is wrapped in waterproof paper or plastic.

Tendons of this type are not pressure grouted after stressing. This type of post-tensioning is usually referred to as unbonded construction.

Post-tensioning offers a means of prestressing on the job site. This procedure may be necessary or desirable in some instances. Very large building or bridge girders that cannot be transported from a precasting plant to the job site (due to their weight, size, or the distance between plant and job site) can be made by post-tensioning on the job site. Post-tensioning is used in precast as well as in cast-in-place construction. In addition, fabricators of pre-tensioned concrete will frequently post-tension the members for small projects on which the number of units to be produced does not warrant the expenditures required to set up pre-tensioning faclities. There are other advantages inherent in post-tensioned construction—these will be discussed later.

In post-tensioning, it is necessary to use some type of device to attach or anchor the ends of the tendons to the concrete section. These devices are usually referred to as end anchorages. The end anchorages, together with the special jacking and grouting equipment used in accomplishing the post-tensioning by one of the several available methods, are generally referred to as post-tensioning systems. Many of the systems used in the U.S. were invented and developed in Europe or were modeled after such a system. The various systems are or were patented; this somewhat deterred the early use of the method. Post-tensioning systems and their use are discussed in detail in Chapter 15.

1-6 Pre-tensioning vs Post-tensioning

It is generally considered impractical to use post-tensioning on very short members, because the elongation of a short tendon (during the stressing) is small and would require very precise measurement by the workmen. In addition, many of the post-tensioning systems do not function well with very short tendons. A number of short members can be made in series on a pre-tensioning bench without difficulty and without the necessity of precise measurement of the elongation of the tendons during stressing, since relatively long tendon lengths result from making a number of short members in series.

It has been pointed out that very large members may be more economical when cast-in-place and post-tensioned or when precast and post-tensioned near the job site, rather than attempting to transport and handle large pre-tensioned structural elements.

Post-tensioning allows the tendons to be placed, with little difficulty, through the structural elements on smooth curves of any desired trajectory. Pre-tensioned tendons can be employed on other than straight trajectories, but not without expensive plant facilities and somewhat complicated construction procedures.

The cost of post-tensioned tendons, measured in either cost per pound of prestressing steel or in cost per pound of effective prestressing force, is generally significantly greater than the cost of pre-tensioned tendons. This is due to the larger amount of labor required in placing, stressing, and grouting post-tensioned tendons and to the cost of the special anchorage devices and stressing equipment. A post-tensioned member may require less total prestressing force than an equally strong pre-tensioned member; however, and for this reason, care must be exercised when comparing the relative cost of these modes of prestressing.

The basic shape of an efficient pre-tensioned flexural member may be different from the most economical shape that can be found for a post-tensioned design. This is particularly true of moderate- and long-span members and somewhat complicates any generalizations about which method is best under such conditions.

Post-tensioning is generally regarded as a method of making prestressed concrete at the job site, yet post-tensioned beams are often made in precasting plants and transported to the job site. Pre-tensioning is often thought of as a method of manufacturing that is limited to permanent precasting plants. Yet on very large projects where pre-tensioned elements are to be utilized, it is not uncommon for the general contractor to set up a temporary pre-tensioning plant at or near the job site. Each method of making prestressed concrete has particular theoretical and practical advantages and disadvantages, which will be more apparent after the principles are well understood. The final determination of the mode of prestressing that should be used on any particular project can only be made after careful consideration of the structural requirements and the economic factors that prevail for the particular project.

1-7 Linear vs Circular Prestressing

The subject of prestressed concrete is frequently divided into linear prestressing, which includes the prestressing of elongated structures or elements such as beams, bridges, slabs, piles, etc., and circular prestressing, which includes pipe, tanks, pressure vessels, and domes. There are no generally recognized criteria for the design and construction of circularly prestressed structures. The theory of such construction is relatively simple and is adequately covered in the literature (*see* Refs. 6 through 13). This book has been confined to the structural design and analysis of linear prestressed structures and the methods of prestressing used in this type of construction.

1-8 Application of Prestressed Concrete

Prestressed concrete, when properly designed and fabricated, can be virtually crack-free under normal service loads as well as under moderate overload. This is believed to be an advantage in a structure that is exposed to an especially corrosive atmosphere. Prestressed concrete efficiently utilizes high-strength concretes and steels and is economical even with long spans. Reinforced concrete flexural members cannot be designed to be crack-free, cannot efficiently utilize high-strength materials, and are not economical on long spans.

A number of other statements can be made in favor of prestressed concrete, but there are bona fide objections to the use of this material under specific conditions. An attempt is made to point out these criticisms in subsequent chapters. Among the more significant points to be kept in mind about this material are that, in many structural applications, prestressed concrete is more economical in first cost than other types of construction and, in many cases, if the reduced maintenance costs that are inherent with concrete construction are taken into account, prestressed concrete offers the most economical solution for the structure. This fact has been well confirmed by the very rapid increase in the use of prestressed concrete that has taken place in the United States since 1953. It is well known that the advantage of real economy outweighs all intangible advantages that may be claimed, except for very special conditions. It should also be kept in mind that prestressing is not a panacea for the construction industry, but has definite limitations. Precautions that must be observed in designing and constructing prestressed concrete structures differ from those required for reinforced concrete structures. These precautions are discussed subsequently in Chapter 10.

Illustrations of prestressed structures and structural elements are given in Chapters 11 and 12, where the various types of building and bridge construction are described and compared.

REFERENCES

1. "Proceedings of the Conference Held at the Institution," *The Institution of Civil Engineers*, 5–10 (Feb. 1949).
2. Abeles, P. W., *Principles and Practice of Prestressed Concrete*, pp. 18–20, Crosby Lockwood & Son, Ltd., London, 1949.
3. Dobell, C., "Patents and Code Relating to Prestressed Concrete," *Journal, American Concrete Institute*, **46**, 713–724 (May 1950).
4. Libby, James, R., "The Elastic Design of Simple Prestressed Concrete Beams," *Proc. Western Conf. on Prestressed Concrete*, 71 (1952).
5. Guyon, Y., *Prestressed Concrete*, pp. 4, 49, John Wiley & Sons, Inc., New York, 1953.

6. Dobell, C., "Prestressed Concrete Tanks," *Proc. First U.S. Conference on Prestressed Concrete*, 9 (1951).
7. Hendrickson, J. G., "Prestressed Concrete Pipe," *Proc. First U.S. Conference on Prestressed Concrete*, 21 (1951).
8. Kennison, H. F., "Prestressed Concrete Pipe—Discussion," *Proc. First U.S. Conference on Prestressed Concrete*, 25 (1951).
9. "Proceedings of the Conference Held at the Institution," *The Institution of Civil Engineers*, 52–56 (Feb. 1949).
10. Timoshenko, S., *Theory of Plates and Shells*, McGraw-Hill Book Company, Inc., New York, 1940.
11. "Circular Concrete Tanks Without Prestressing," Bulletin, Portland Cement Association.
12. "Design of Circular Domes," Bulletin, Portland Cement Association.
13. Crom, J. M., "Design of Prestressed Tanks," *Proc. A.S.C.E.*, **37** (Oct. 1950).

2 Steel for Prestressing

2-1 Introduction

It was stated in Sec. 1-2 that the loss in prestress which results from the effects of steel relaxation and the shrinkage and creep of the concrete is generally from 10 to 30% of the initial prestress. The computation of the losses of prestress due to the various causes is discussed in detail in Sec. 6-2, but it is important for the designer of prestressed concrete to be aware that the greater portion of the loss of prestress is normally attributed to be the result of the shrinkage and creep of the concrete. It should also be recognized that this fact accounts the necessity of using high-strength steel, with a relatively high initial stress, in the construction of prestressed concrete.

The shrinkage and creep of concrete produce inelastic volume or strain changes. Because the tendons are anchored to the concrete, either by bond or by end anchorages, the strain changes in the concrete result in strain change in the tendons. Furthermore, since the steel used for prestressing is fundamentally an elastic material at the stress levels employed in normal designs, the reduction of the stress in the tendons, which results from the strain changes in the concrete, is equal to the product of the elastic modulus of the steel and the strain change in the concrete.

It is essential that the loss of prestress be a relatively small portion of the total prestress, in order to attain an economical and practical design. The elastic modulus of steel is a physical property which, for all practical purposes, cannot be altered or adjusted by manufacturing processes. In a similar manner, the inelastic volume changes of concrete of any particular quality are physical properties that cannot be eliminated using practical construction procedures. Therefore, the product of these two factors is normally beyond the control of the designer.

It can be shown that the normal loss of prestress is generally of the order of 15,000 to 50,000 psi. It is apparent that if the loss of prestress is to be a small portion of the initial prestress, the initial stress in the steel must be very high and of the order of 100,000 to 200,000 psi. If a steel having a yield point of 40,000 psi were used to prestress concrete and if this steel were stressed initially to 30,000 psi, the entire prestress could be lost, as was the case indeed in the early attempts at prestressing with low-strength steel and concrete of poor quality.

It should be pointed out that research has been conducted into the use of other materials, such as fiber glass and aluminum alloys, for prestressing concrete. Some of these materials have elastic moduli that are of the order of one-third of that of steel. If such materials could be safely and economically used, the loss of prestress would be reduced to approximately one-third of the loss obtained with steel tendons. Hence, the loss of prestress could possibly be ignored in normal design practice if these materials were employed as tendons. There are, however, many problems to be studied and overcome before these materials can be used safely and with economy. The use of tendons having a low elastic modulus and plastic deformations different from those of steel would result in members having post-cracking deflection and ultimate strength characteristics different from those obtained when steel tendons are used. These problems will be apparent from the subsequent discussion of the desirable physical properties of the steel that is used in prestressing.

There are several basic forms of high-strength steel that are used currently in domestic prestressed work. In general, these can be divided into three groups: uncoated stress-relieved wires; stress-relieved strand, and high-tensile alloy bars. Each of these types of steel is described briefly in the following sections. For a more detailed description of the method of manufacture, chemical composition, and physical properties of these materials, the reader should consult the applicable ASTM Specifications and the references listed at the end of this chapter.

Other types of wire, such as straightened "as-drawn" wire and oil-tempered wire, are used for prestressing in other parts of the world. These materials can be used with satisfactory results, but are not considered here due to the fact that these materials are not normally used in the U.S. The interested reader can find descriptions of these materials in the references listed at the end of this chapter.

2-2 Stress-Relieved Wire

Cold-drawn stress relieved wire, which is commonly used in post-tensioned construction and rarely used in pre-tensioned members, is manufactured to conform to the "Standard Specifications for Uncoated Stress-relieved Wire for Prestressed Concrete" (ASTM Designation A 421). These specifications provide that the wire shall be made in two types (BA and WA), depending upon whether it is to be used with button- or wedge-type anchorages (*see* Chapter 15). Other major requirements in these specifications include the minimum ultimate tensile strength, the minimum yield strength, the minimum elongation at rupture, as well as diameter tolerances. The principal strength requirements of ASTM A 421 are summarized in Table 2.1.

TABLE 2-1 Properties of Stress-Relieved Wire for Prestressed Concrete Required by ASTM A 421.

	Min. Tensile Strength (psi)		Min. Stress at 1% Extension, psi*	
Nominal Diameter, in.	*Type BA*	*Type WA*	*Type BA*	*Type WA*
0.192		250,000		200,000
0.196	240,000	250,000	192,000	200,000
0.250	240,000	240,000	192,000	192,000
0.276		235,000		188,000

* Measured according to procedures specified in ASTM A 421.

Typical stress-strain curves for uncoated, stress-relieved wires are shown in Fig. 2-1. It should be noted that the stress-strain curves for the two wire diameters shown are similar in shape and that the ultimate tensile strength is higher for wires of smaller diameters. It should be pointed out that ASTM A 421 requires a minimum elongation of 4.0% when measured in a gage length of 10 in., which means the steel is quite ductile and has a "plastic range" of considerable magnitude. (The "plastic range" is not shown in Fig. 2.1.)

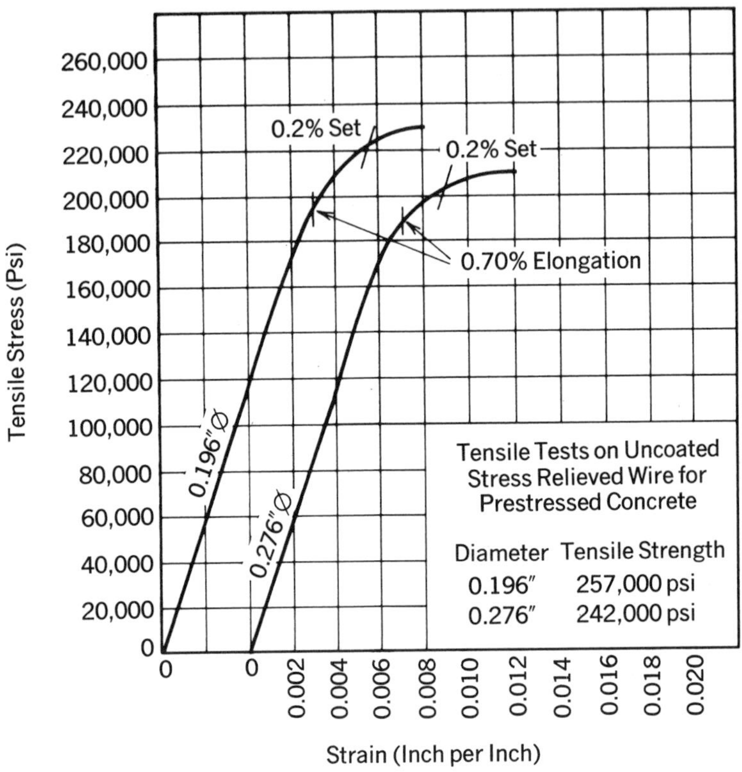

Fig. 2-1 Typical stress-strain curves for wires in elastic range. (Source: C. F. & I. Steel Corp., Trenton, N. J.)

2-3 Stress-Relieved Strand

Stress-relieved strand is made in two forms. The first of these is the seven-wire strand, which is made to conform to the requirements of "Standard Specifications for Uncoated Seven-Wire Stress-Relieved Strand for Prestressed Concrete" (ASTM Designation A 416). Basic strength, area, and weight requirements for seven-wire strands, as provided by ASTM Designation A 416, are given in Table 2-2 for the two grades of strands covered therein. The second form consists of larger strands with diameters up to $1\frac{11}{16}$ in.; these are manufactured with factory-attached end-fittings for use in post-tensioning.

Seven-wire strands are made by twisting six wires, on a pitch of between 12- and 16-wire diameters, around a slightly larger, straight central wire.

TABLE 2-2 Properties of Uncoated, Seven-Wire, Stress-Relieved Strand for Prestressed Concrete.

Nominal Diameter of Strand, in.	*Breaking Strength of Strand, min. lb*	*Nominal Steel Area of Strand, sq in.*	*Nominal Weight of Strands, lb. per 1 000 ft*	*Minimum Load at 1% Extension, lb*
		GRADE 250		
$\frac{1}{4}$ (0.255)	9 000	0.036	122	7 650
$\frac{5}{16}$ (0.313)	14 500	0.058	197	12 300
$\frac{3}{8}$ (0.375)	20 000	0.080	272	17 000
$\frac{7}{16}$ (0.438)	27 000	0.108	367	23 000
$\frac{1}{2}$ (0.500)	36 000	0.144	490	30 600
		GRADE 270		
$\frac{3}{8}$ (0.375)	23 000	0.085	290	19 550
$\frac{7}{16}$ (0.438)	31 000	0.115	390	26 350
$\frac{1}{2}$ (0.500)	41 300	0.153	520	35 100

The strands are stress-relieved after being stranded. Typical stress-strain curves for seven-wire strands commonly used in pre-tensioning and in multi-strand post-tensioning tendons are shown in Figs. 2-2 and 2-3. It is interesting to note that the seven-wire strands are currently available in three grades, all of which meet the minimum requirements of ASTM 416. The original grade has a nominal ultimate tensile strength of 250,000 psi and is frequently referred to as "ASTM Grade" or "250 k grade." The second grade has slightly larger wires than the "ASTM Grade" and a nominal ultimate tensile strength of 270,000 psi—it is referred to as 270 k grade strand. The third grade has a nominal ultimate tensile strength of 270,000 psi similar to 270 k grade strand. However, due to a special process that involves stress-relieving while the strand is under stress, the yield point is raised and the relaxation characteristics of the strand are materially improved. The third grade of strand is marketed under the trade names of "Stabilized," "Lok-Stress," "Thermalized," (etc) strand, depending upon the manufacturer, and will be referred to subsequently as "low-relaxation" strand. It should be noted that seven-wire strand with a nominal diameter of 0.600 in. is also available domestically, but this material is not covered by ASTM A 416. The elastic and plastic characteristics of the seven-wire strand are quite similar to those of stress-relieved wire, as can be seen in comparing Figs. 2-1 and 2-2.

A fourth grade of seven-wire strand, referred to as "Dyform"* strand, is

* "Dyform" is a registered trade name of British Ropes Limited.

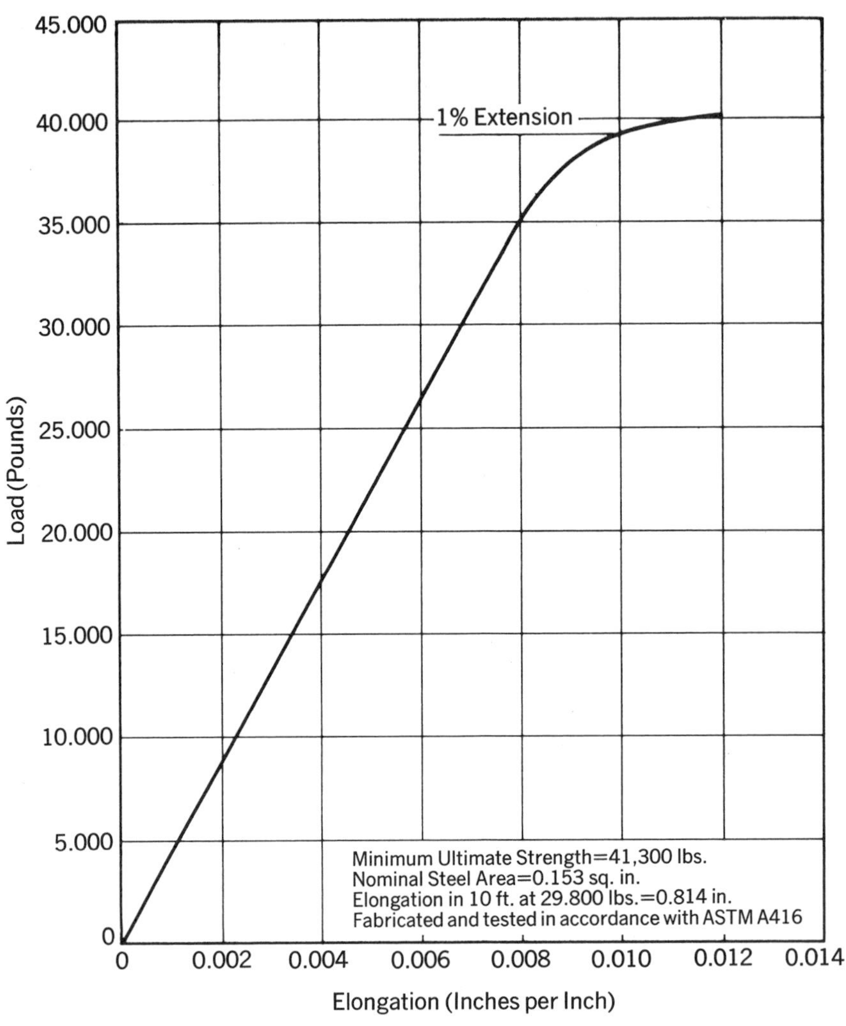

Fig. 2-2 Typical load-elongation curve for ½-in. diameter, seven-wire strands. (*Courtesy C. F. & I. Steel Corp.*)

available from British manufacturers and may be expected to be available from U.S. manufacturers in the future. Dyform strand was originated in Great Britain and is characterized by having been run through a die after having been stranded, thus giving it the cross-sectional shape shown in Fig. 2-4. (Dyform is compared in Fig. 2-4 with a typical seven-wire strand.)

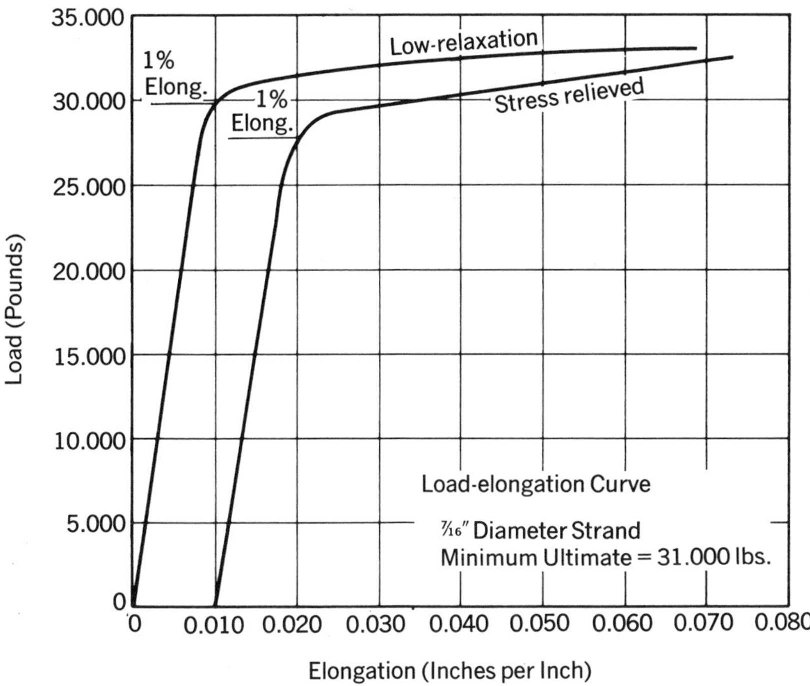

Fig. 2-3 Typical load-elongation curves for 7/16-in. diameter, seven-wire strands. (*Courtesy C. F. & I. Steel Corp.*)

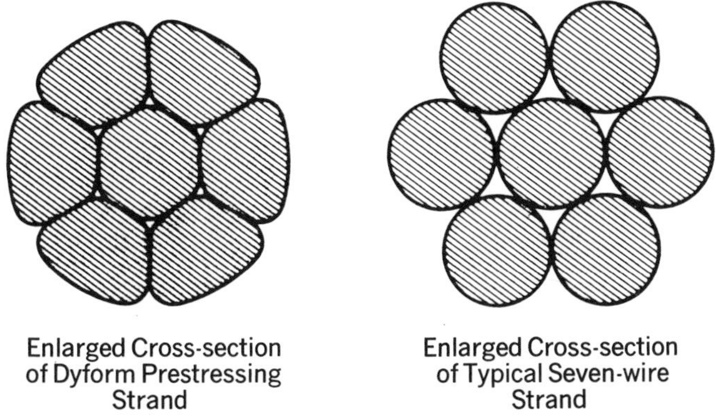

Fig. 2-4 Comparison of Dyform to typical seven-wire strand.

The Dyform strand has the advantage of having greater steel area in any nominal diameter, as compared to a regular strand. With this greater area, the Dyform strand has a larger ultimate tensile strength than a regular strand.

Basic strength, area, and weight properties for Dyform strand are given in Table 2-3.

TABLE 2-3 Properties of Uncoated, Seven-Wire, Stress-Relieved Dyform Strand for Prestressed Concrete

Nominal Diameter in.	*Breaking Strength of Strand, Min. lb*	*Nominal Steel Area in.²*	*Nominal Weight of Strand per 1000 ft lb*
$\frac{5}{16}$	20,000	0.069	230
$\frac{3}{8}$	28,000	0.099	330
$\frac{7}{16}$	38,000	0.134	450
$\frac{1}{2}$	47,000	0.174	600
0.6	65,000	0.253	860

Large diameter strands for post-tensioning were first introduced in the U.S. in the late 1940's. The physical properties of the strands are similar to those for the seven-wire strands, but the large strands would normally be made of larger individual wires and may contain considerably more than seven-wires per strand. The large strands are not covered by ASTM Specifications. Thus, the designer must carefully review all of the physical properties of any strand proposed for use, in order to be assured that satisfactory results will be obtained. The types, sizes, and properties of large-diameter strands that are available should be obtained directly from the manufacturer.

2-4 High-Tensile Strength Bars

Both smooth and deformed high-tensile, alloy steel bars are available in two grades. Nominal diameters from $\frac{1}{2}$ to $1\frac{3}{8}$ in. are standard. Larger bars are available by special arrangement with some bar manufacturers. The bars are made from an alloy steel conforming to "Standard Specification for Hot-Rolled Alloy Steel Bars" (ASTM Designation A 322). This specification is a general specification that covers the chemical composition of many grade designations of alloy steel bars. The sizes and some of the physical properties of the available bars are summarized in Table 2-4.

The bars are cold-stretched in order to raise the yield point and to render the bars more elastic at stress levels below the yield point, after which, they are stress-relieved in order to improve the ductility and stress-strain characteristics. There is no standard ASTM specification covering the minimum requirements for the bars after processing.

TABLE 2-4.

SMOOTH BARS

Nominal Diameter of Bar, in.	Nominal Steel Area of Bar sq. in.	Breaking Strength Minimum: Kips		Yield Strength Minimum at 0.7% Extension: Kips	
		Grade 145	Grade 160	Grade 145	Grade 160
$\frac{1}{2}$	.196	28	31	25	27
$\frac{5}{8}$	.307	45	49	40	43
$\frac{3}{4}$	.442	64	71	58	62
$\frac{7}{8}$	.601	87	96	78	84
1	.785	114	126	102	110
$1\frac{1}{8}$	.994	114	159	129	139
$1\frac{1}{4}$	1.227	178	196	160	172
$1\frac{3}{8}$	1.485	215	238	193	208

DEFORMED BARS

Nominal Bar Diameter in.	Effective Steel Area sq. in.	Minimum Breaking Strength of Bar		Minimum Yield Strength of Bar at 0.7% Extension		Weight (lb/ft)
		Grade 150 Kips	Grade 160 Kips	Grade 150 Kips	Grade 160 Kips	
$\frac{5}{8}$	0.290	43	46	37	39	0.97
$\frac{3}{4}$	0.433	65	69	55	59	1.53
1	0.854	128	136	109	116	3.00
$1\frac{1}{4}$	1.295	194	207	165	176	4.55

Minimum yield strength (0.20% offset): 0.85 f_s'.
Approximate modulus of elasticity (based on nominal area) 30,000,000 psi.
Elongation in twenty (20) diameters after rupture, minimum 4.0%.
Reduction of area (from measured area), minimum 20%.
Permissible variation bar size is +0.030 in., −0.010 in. from nominal specified diameter for smooth bars and +3%, −2% in weight for deformed bars.

A typical load-strain curve for a high-tensile strength bar is given in Fig. 2-5.

2-5 Yield Stress

As will be noted from an examination of the stress-strain curves for the various types of prestressing steels, these steels do not have definite yield points. For this reason, an arbitrary yield stress must be specified in order to

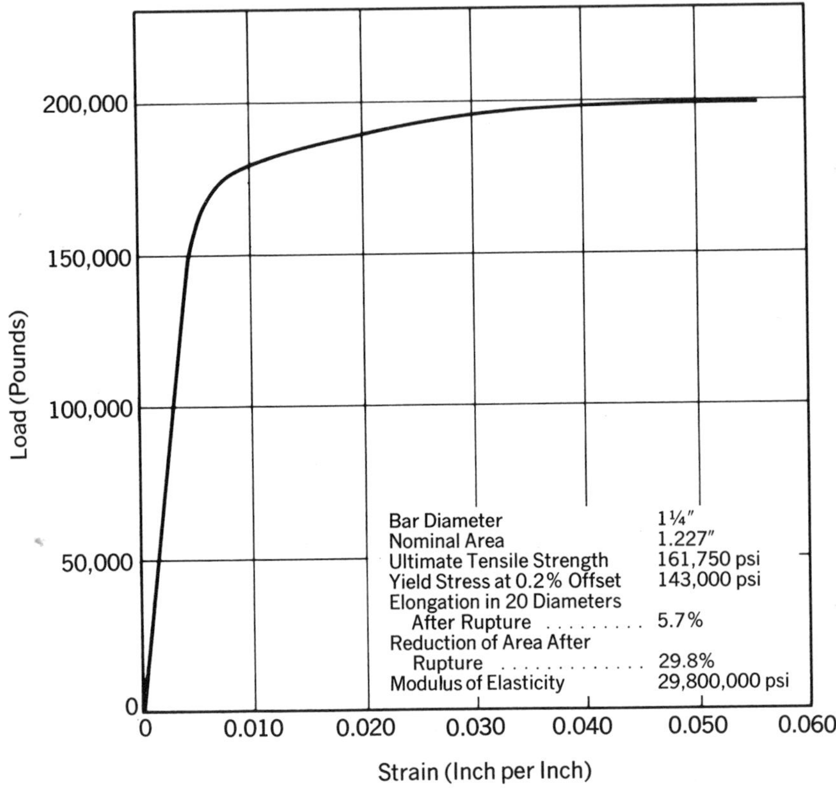

Fig. 2-5 Typical load-strain curve for a prestressing bar. (*Courtesy of Stressteel Corp.*)

define the stress that is taken to be the yield stress. Because there is no definite yield point, the yield stress does not have the importance that it has in a steel with a definite point, such as normal mild steel. Yield stresses taken as being the stress at a 0.20% offset are often used for materials not having a definite yield point. Minimum yield stresses at 1% extension are specified in the standard ASTM specification for wire and strand. Minimum yield strengths at 0.7% extension are specified for high-strength bars. Some research work has been done using the stress of a 0.10% offset as the yield stress. Hence, the term yield stress as related to prestressing materials is not a precise term and the reader is cautioned to use the term with care and to be certain of the definition of yield stress in any discussion or recommendations where the term is used.

2-6 Plasticity

Plasticity at very high stress levels is as essential in prestressing steel as it is in reinforcing steel. The object is to ensure that ultimate bending moments will be reached only after large and very apparent plastic deformations have taken place. The use of brittle steel could result in a sudden failure similar to that which is characteristic of an over-reinforced concrete member. In order to avoid this possibility, the normal practice is to specify that the prestressing steel will have a minimum elongation at rupture of 4 to 8%, depending upon the type of steel used and the method that is used to measure the elongation at rupture (*see* ASTM A 416 and ASTM A 421).

2-7 Relaxation and Creep

Relaxation is defined as the loss of stress in a material that is placed under stress and held at a constant strain, whereas creep is defined as the change in strain for a member held under constant stress. Although tendons are not subjected to constant strain or to constant stress, it is generally agreed the condition more closely approximates a condition of constant strain, and hence, relaxation studies are made to evaluate the loss of prestress that can be attributed to the inelastic behavior of the steel. A typical relaxation curve for stress-relieved prestressing wire is shown in Fig. 2-6.

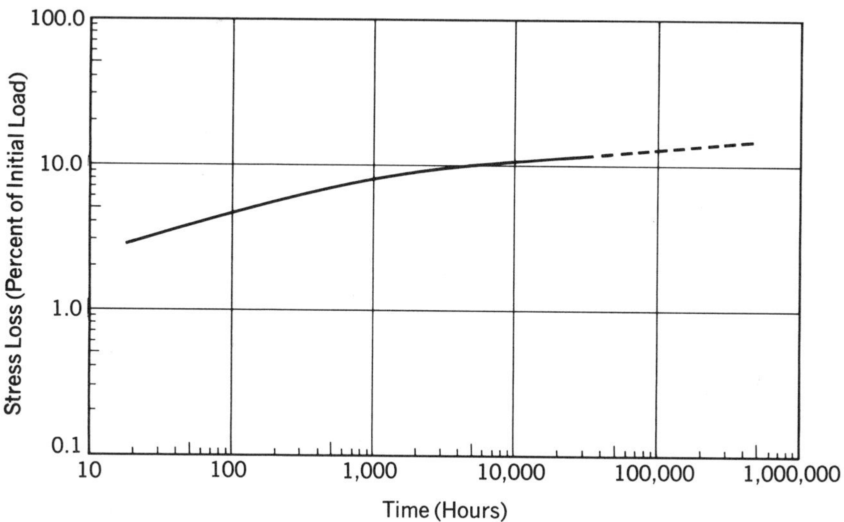

Fig. 2-6 Stress loss vs time for a stress-relieved, prestressed-concrete wire initially loaded at 70% of the guaranteed ultimate tensile strength and held at constant length at 85.0°F. (*Courtesy C. F. & I. Steel Corp.*)

For stress-relieved wire and strand, the loss of stress due to relaxation at normal temperatures can be estimated with sufficient accuracy for design purposes using the following relationship:

$$\Delta f_{sr} = f_{si} \frac{\log t'}{10} \left(\frac{f_{si}}{f_y} - 0.55 \right) \tag{2-1}$$

where Δf_{sr} = the relaxation loss at time t' hours after prestressing, f_{si} = the initial stress and f_y = the 0.10% offset stress for the steel under consideration (Ref. 1). The logarithm of the time t' is to the base 10. This relationship is only applicable when the ratio of f_{si}/f_y is equal to or greater than 0.55.

Using Eq. 2-1 with the value of $f_y = 256{,}000$ psi and $f_{si} = 189{,}000$ psi, that would be the approximate values for the 270 k grade strand illustrated in Fig. 2-2, after t' time of 100,000 hr (11.4 yr) and 400,000 hr (45.6 yr), the relaxation loss would be as follows

$$\Delta f_{sr} = f_{si} \frac{\log t'}{10} \left(\frac{f_{si}}{f_y} - 0.55 \right)$$

$$t' = 100{,}000 \text{ hr}$$

$$\Delta f_{sr} = 189 \times \frac{5.00}{10} (0.74 - 0.55)$$

$$= 189 \times 0.095 = 18.0 \text{ ksi}$$

$$t' = 400{,}000 \text{ hr}$$

$$\Delta f_{sr} = 189 \times \frac{5.60}{10} (0.74 - 0.55)$$

$$= 189 \times 0.106 = 20.0 \text{ ksi}$$

Assuming Eq. 2-1 accurately relates the relaxation of prestressing steel, the efficiency of the steel at various levels of initial stress after 50 yr of service can be studied through the use of Fig. 2-7. From this plot it will be seen that the increase in effective stress f_{se} is nearly equal to the increase in initial stress f_{si} up to the point where $f_{si} = 0.60 f_y$. Above this stress, the efficiency is progressively reduced. For an initial stress ratio of 0.90, virtually no gain in effective stress is realized for increases in initial stress.

In the fabrication of pre-tensioned concrete members, the tendons are stressed to an initial level, held at that elongation for a period of time, and then released. At the time the prestressing force is transferred to the concrete (tendons released), the stress in the tendons is less than the initial stress. This is due to the relaxation that has taken place in the interval between initial stress and release, as well as to the elastic shortening of the concrete

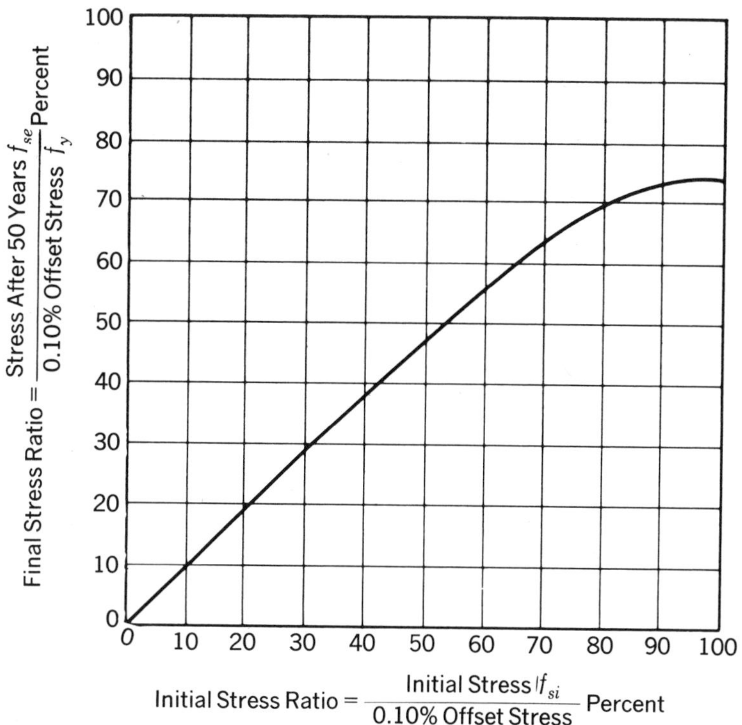

Fig. 2-7 Final stress ratio vs initial stress ratio.

that takes place upon transferring the prestressing force to the concrete. Therefore, in the case of pre-tensioning, the effect of relaxation of the steel can be estimated at time t_n', assuming the tendon was stressed at time zero and released at time t_r'. The following relationship is used

$$\frac{f_{st}}{f_{si}'} = 1 - \left(\frac{f_{si}'}{f_y} - 0.55\right)\left(\frac{\log t_n' - \log t_r'}{10}\right) \tag{2-2}$$

in which f_{si}' is the effective stress in the steel after release of the tendons and f_{st} is the stress in the steel at time t_n'. In Eq. 2-2, the stresses are in ksi and time is in hours. In using Eq. 2-2, one should compute the relaxation loss occurring in the steel from the time of initial stressing until the time of release, using Eq. 2-1. This relaxation loss should be added to the loss due to elastic shortening of the concrete and the sum subtracted from the initial steel stress to determine the value of f_{si}' for use in Eq. 2-2.

Examples of how Eqs. 2-1 and 2-2 can be used in estimating the loss of prestress in pre-tensioned and post-tensioned members are given in Sec. 6-2.

Elevated temperatures have an adverse effect on the relaxation of prestressing steel. For applications where the prestressing tendons will be subjected to temperatures in excess of 100°F for extended periods of time, larger allowances should be made for the relaxation of the steel (Refs. 7 and 8).

Relaxation curves for low-relaxation strand stressed initially to 70% of the guaranteed ultimate tensile strength and held at various temperatures is shown in Fig. 2-8. In Fig. 2-9, the relaxation curves for stress-relieved strand and low-relaxation strand are compared. It should be noted that the 50 yr stress loss is reduced from about 15% to 3% by the special process used in making low-relaxation strand.

From Figs. 2-8 and 2-9, one can conclude that the relaxation of the low-relaxation strand could be taken as 20-25% of that for stress-relieved strands (Eq. 2-1) for applications at normal temperatures. For applications in which the tendons may be exposed to temperatures above 100°F, only the low-relaxation-type strand should be used.

There are little data to be found in the literature relative to the relaxation

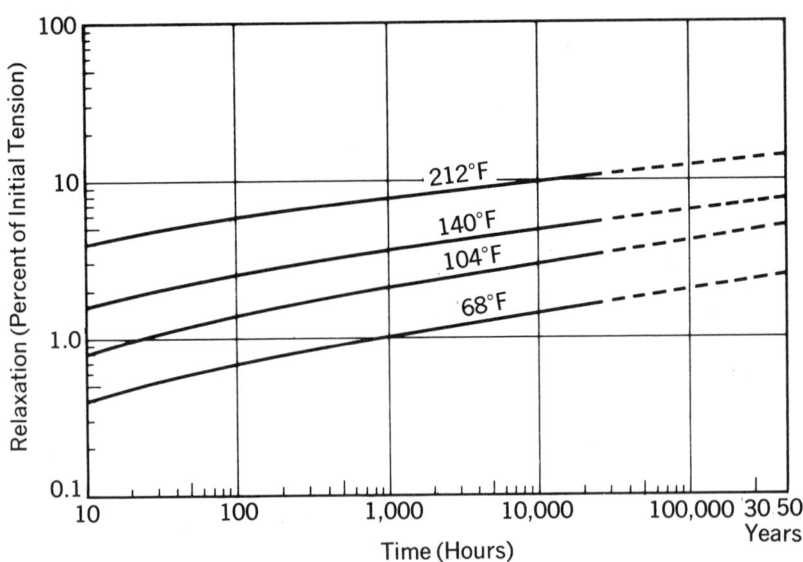

Fig. 2-8 Relaxation vs time curve for low-relaxation strand stressed initially to 70% of the guaranteed ultimate tensile strength and held at various temperatures. (*Courtesy of C. F. & I. Steel Corp.*)

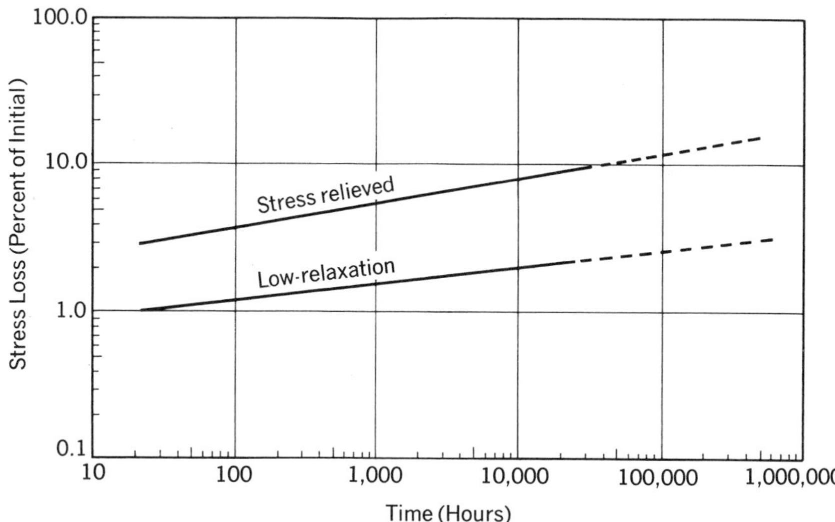

Fig. 2-9 Relaxation vs time curves for stress-relieved and low-relaxation seven-wire strands held at constant length at 85°F. Initial stress was 70% of the guaranteed ultimate tensile strength.

of high-strength bars. It is known that the relaxation characteristics of the bars are affected by the chemical composition and metallurgical history of the bars, as well as the initial stress to which the bars are subjected. It is also known that the relaxation of the bars can be of the same order as that of stress-relieved wire and strand. For this reason, it is recommended that not less than 10% be assumed as the relaxation loss for bars stressed initially to 70% of their guaranteed ultimate tensile strength, and 12% when the initial stress is 80% of the guaranteed ultimate tensile strength, unless test data are available for the particular bars being used (i.e., same chemical composition and metallurgical history). (Refs. 3 and 6.)

2-8 Corrosion

Since the strength of a prestressed-concrete flexural member is dependent upon the adequacy of the tendons, it is essential that the tendons do not deteriorate due to corrosion. Prestressing steels are subject to normal oxidation in approximately the same degree as structural-grade steels. The tendons are normally of small proportions; however, and for this reason, it is essential that they be protected against heavy oxidation.

Protection against corrosion is effected in pre-tensioned construction by the concrete that surrounds the tendons. In post-tensioned construction, the tendons are protected by grout that is injected around the tendons after stressing or by grease or bituminous materials placed on the tendons, as was explained earlier. Steel is not attacked when the pH of the surrounding environment is higher than 8, as is usually the case of concrete made with Portland cement.

It must be emphasized that research has shown that a light, hard oxide on the tendons is desirable in pre-tensioned members, since such an oxide improves the bond characteristics of the tendons (*see* Sec. 5-7). It should also be desirable in bonded, post-tensioned work, because the flexural bond would be improved.

In order to protect prestressing steel between the time it leaves the factory and the time it is finally incorporated in a structure, it has become standard practice (in many areas) to wrap the steel in waterproof paper, with the inclusion of a vapor-phase inhibitor.

A vapor-phase inhibitor is a white, fine-grained powder consisting of an organic compound containing nitrogen. The material vaporizes (sublimes) and, if the vapors are confined, the material will recrystallize on the surface of the steel and will prevent oxidation. The action of vapor-phase inhibitors can be nullified by the following:

(1) Temperature greater than 160°F.
(2) Free-running water over the surface of the steel.
(3) An acidic environment (pH less than 6.5).
(4) Free circulation of fresh air.
(5) Coatings or films on the steel that prevent the vapor from contacting the steel.
(6) The powder not being in the immediate vicinity of the steel (further than 12 in.).
(7) An environment containing a high concentration of chlorides.

The material will work in either air or water, providing the above conditions do not exist to nullify the action.

Post-tensioned construction is occasionally protected against the effects of corrosion by using galvanized tendons. This procedure is not used frequently because galvanized tendons are not as strong as bright tendons of the same size, due to the fact that some of the diameter of a galvanized tendon is composed of low-strength zinc. As a result, galvanized tendons are materially more expensive than bright tendons of equal strength. (For equal diameters, galvanized, seven-wire strands are approximately 15% lower in strength and 10% more in cost.) Furthermore, the various types of anchorage devices used in post-tensioning with the parallel-wire systems either cannot anchor galvanized wire, due to the low coefficient of friction, or cannot be used without damaging the zinc coating. For these reasons, the use of galvanized wire is generally considered to be impractical with parallel-wire systems. The use of galvanized, large-diameter strand is feasible under some conditions. Galvanized, seven-wire strand can be used in some of the more modern anchorage devices, but they rarely are used due to the cost.

A type of corrosion referred to as "pitting corrosion" is the cause of some deterioration (and even failures) of prestressed concrete structures. Calcium chloride or sodium chloride in the concrete or grout is generally considered to be the cause of this type of corrosion. It is for this reason that chlorides must never be permitted, even in the very small amounts, in the concrete or grout used in prestressed-concrete construction. (*See* Ref. 4.)

Prestressing steels, particularly wires and strands, are very susceptible to a type of deterioration that is called "stress corrosion." This type of corrosion has occurred relatively infrequently. Stress corrision is characterized by a breakdown of the cementitious portion of the steel, resulting in fine cracks in the steel. These fine cracks render the steel nearly as brittle as glass. Since little is known about this type of corrosion, there is no way to be certain that it will not occur during construction of a prestressed member. It is known that nitrates (not to be confused with the rust-inhibiting nitrites), chlorides, sulfides, and some other agents can result in stress corrosion under certain conditions. It is also known that the steel is more susceptible to this type of corrosion when highly stressed; this accounts for the name stress corrosion.

Another cause of delayed failure, which can occur in high-strength steels, is called "hydrogen embrittlement." This phenomenon, which apparently results when steel is exposed to hydrogen ions but not to hydrogen molecules, is characterized by a decrease in ductility and tensile strength. Hydrogen embrittlement may be promoted by electroplating steel with cadmium or zinc, as well as from corrosion and electrical currents. Confining the prestressing steel in an environment having a pH greater than 8 is thought to be the best protection against the absorption of hydrogen.

It is interesting to note that aluminum powder, which causes the release of hydrogen gas (molecular hydrogen), has been used for many years as an expansive additive for the grouting of post-tensioned tendons. This practice has apparently not been harmful, because failures have not been reported in structures so constructed. Additives which obtain expansion by the release of nitrogen gas are also being used.

2-9 Application of Steel Types

The same basic steel can be used in pre-tensioning and post-tensioning, but in the former it is necessary that the individual tendons are not so large that they cannot be adequately bonded to the concrete, since this bond is relied upon to transfer the prestressing force from the steel to the concrete. In post-tensioning, as has been explained, end anchorages are used to transfer the prestressing force to the concrete—the grouting (when used) is relied upon to protect the steel against corrosion and to develop flexural bond stress, i.e., bond stresses resulting from changes in the externally applied loads. Bond stresses are discussed in detail in Secs. 5-7 and 5-8. It should be mentioned here that, while in Europe it is customary to use wires up to 0.276 in. in diameter as pre-tensioning tendons, the usual practice in this country has been to use the uncoated, seven-wire strands described in Sec. 2-3. Little or no use of high-tensile alloy bars has been made in pre-tensioning in this country, although favorable results have been obtained experimentally in Europe with bars up to $\frac{5}{8}$ in. in diameter. (*See* Ref. 5.)

2-10 Idealized Tendon Material

One may wish to consider the properties an ideal material for prestressing concrete would have. Some characteristics are desirable from one standpoint and not from another. For example, high tensile strength, coupled with a low elastic modulus, permits a high strain under initial stress, which minimizes the losses of stress due to the inelastic properties of the concrete. On the other hand, a high tensile strength results in a small area of tendon being required and this, coupled with a low modulus of elasticity, could result in very high deflections upon the application of an overload that would cause cracking. In actuality, the steels we currently have available generally result in designs that are efficiently balanced in "serviceability" and "ultimate strength" characteristics. Perhaps steels with somewhat higher strengths could be used efficiently. Steels without any relaxation loss would obviously be advantageous.

It is also possible that materials of very high strength and low elastic modulus will eventually be used in combination with non-prestressed mild reinforcing in order to achieve efficient and economical construction.

The major desirable physical characteristics of the material to be used for prestressing tendons can be summarized as follows:

(1) High strength that allows high prestressing stresses.
(2) Elastic up to high stress levels.
(3) Plastic at very high stress levels.
(4) Low elastic modulus at time of stressing in order to minimize the loss of prestress.
(5) High elastic modulus after bonding in order to contribute to stiffness of the member.
(6) Low creep or relaxation losses at the stress levels normally employed in prestressing and at elevated temperatures.

(7) Resistant to corrosion.
(8) Small diameter or relatively large surface area of the individual tendons to achieve good bond characteristics.
(9) Absence of dirt and lubricants on the surface.
(10) Straightness to facilitate handling and placing.

From the above description one will see there is no known material that has all of these desirable qualities. The high-strength steels that are currently used are reasonable compromises.

2-11 Allowable Steel Stresses

The two most significant design criteria for prestressed concrete in the United States are the "Standard Specifications for Highway Bridges," which is published by the American Association of State Highway Officials (AASHO) and "Building Code Requirements for Reinforced Concrete (ACI 318)" published by the American Concrete Institute.

The stresses permitted in the prestressing steel in the tenth edition of the AASHO Specification (1969) are as follows:

(1) Temporary stress before losses due to creep and shrinkage $0.70f_s'$ (Overstressing to $0.80\,f_s'$ for short periods of time may be permitted provided the stress, after seating of the anchorage, does not exceed $0.70\,f_s'$).
(2) Stress at design load (after losses) $0.60\,f_s'$ or $0.80\,f_{sy}$ (f_{sy} = yield stress) whichever is smaller.

The steel stresses permitted by ACI 318-71 are as follows:

(1) Due to jacking force $0.80f_s'$ but not greater than the maximum value recommended by the manufacturer of the steel or of the anchorages.
(2) Pretensioning tendons immediately after transfer, or post-tensioning tendons immediately after anchoring $0.70f_s'$

From this it will be seen that stresses as high as $0.80\,f_s'$ are permitted by both criteria during jacking. This stress is of a temporary nature. In addition, both criteria limit the initial stress to $0.70\,f_s'$. Initial stress is defined as the stress in pretensioning tendons immediately after transfer (release). In post-tensioned tendons, initial stress is defined as the stress in the tendons immediately after anchoring.

The AASHO Specification also restricts the effective stress (initial stress minus the loss of prestress) to $0.60\,f_s'$. This requirement was in ACI 318-63, but was not included in ACI 318-71. There is no justification for restricting the effective stress to $0.60\,f_s'$. It is the author's opinion that this requirement will eventually be eliminated from the AASHO Specification.

REFERENCES

1. Magura, Sozen and Siess, "A Study of Stress Relaxation in Prestressing Reinforcement" Civil Engineering Studies, Structural Research Series 237, University of Illinois, Urbana, Illinois (Sept. 1962).
2. Cahill, T., "The Development of Stabilized Wire and Strand," *Wire and Wire Products* (Oct. 1964).
3. Bannister, J. L. "Steel Reinforcement and Tendons for Structural Concrete," *Concrete*, **2**, No. 8 (Aug. 1968).
4. Szilard, Rudolph, "Corrosion and Corrosion Protection of Tendons in Prestressed Concrete Bridges," *Journal of the American Concrete Institute, Proceedings*, **66**, No. 1 (Jan. 1969).
5. Base, G. P. "An Investigation of Transmission Length in Pretensioned Concrete," Research Report No. 5, Cement and Concrete Association, London, 1958.
6. Bannister, J. L. Private Communication.
7. de Strycker R., "The Influence of Temperature and Variations of Stress on the Creep of Prestressing Steels," *Revue de Metallurgie*, Paris, **56**, No. 1, 49–54 (Jan. 1959).
8. Papsdorf and Schwier, "Creep and Relaxation of Steel Wire, Particularly at Slightly Elevated Temperatures," *Stahl und Eisen*, **78**, No. 14, 937–947 (July 10, 1958).
9. Monfore, G. E. and Verbeck, G. J. "Corrosion of Prestressed Concrete Wire in Concrete," *Journal of the American Concrete Institute*, **32**, No. 5, 491–515 (Nov. 1960).
10. Podolny, W. Jr. and Melville, T. "Understanding the Relaxation in Prestressing," *Journal of the Prestressed Concrete Institute*, **14**, No. 4, 43–54 (Aug. 1969).

3 Concrete for Prestressing

3-1 Introduction

It is presumed that the reader is familiar with the basic physical properties of portland-cement concrete, which is the principal constituent of prestressed concrete. It is important that a proper concrete be employed in prestressed construction—only the factors that are particularly important in this type of construction are considered here. General data pertaining to the factors affecting the physical properties of concrete can be found in the references (*see* Refs. 1, 2, and 3).

Although concretes having 28-day cylinder compressive strengths of 5000 to 6000 psi are relatively easily obtained in most localities today, such concrete cannot be employed effectively in reinforced-concrete flexural members. This is not true in the case of prestressed concrete where the use of 5000-psi concrete is common and efficient, due to the reduction in the dead weight and cost of the members (which is derived from the use of stronger concretes). Furthermore, as has been explained, volume changes of concrete affect, to a very significant degree, the amount of prestressing that is lost. Since the high-strength concretes generally undergo substantially smaller volume changes than the lower-strength concretes, their use is desirable, if not necessary, in many prestressing applications.

Volume changes in concrete are affected by many variables, but in practice, the control of these variables is generally limited to specifications which govern the amount of water used in the concrete mixture, the types and proportions of the aggregates, the type and amount of cement in the mixture, the use of admixtures, and the method and duration of the curing. The water content of the concrete mixture is kept as low as possible, since by so doing, the shrinkage of the concrete is reduced, the strength is increased, and the creep is reduced. All of these are desirable effects. Care must be taken to ensure that there is sufficient water in the mixture to avoid honeycomb and permit the concrete to be properly placed and compacted.

3-2 Cement Type

Although concrete containing modern, high early-strength (type III) portland cement yields shrinkages that are slightly greater than those obtained with normal portland cement (type I), there is evidence that the combined loss of prestress, due to all changes in concrete volume, is less when high early-strength cement is used (Ref. 4). The very high shrinkages that were associated with the type III cements of 30 to 40 years ago have been substantially reduced as a result of improved cement-manufacturing techniques (Ref. 1). In addition, since the normal prestressed concrete product is precast, the strains due to shrinkage and the high heat of hydration generated with type III cement can take place virtually without restraint, and, as a result, the cracking associated with these phenomena in cast-in-place construction does not occur in well executed precast work.

In applications where type III cement is not used, due to the higher cost or to the unimportance of early strength, type I portland cement is recommended. In structures that are to be subjected to exposure to sea water or moderately reactive aggregates, the use of type II (modified) portland cement is considered a good practice.

Cement used in prestressed-concrete work should conform to ASTM C 150. There is some evidence that high alumina cement should not be used in prestressed concrete, since some of these cements contain significant quantities of sulfides that can undergo changes which may lead to hydrogen embrittlement of the prestressing steel (Ref. 5).

3-3 Admixtures

Admixtures, which make the concrete mixture more plastic, retard the initial set, accelerate the final set, and reduce the amount of water required in the mixture, are often used in prestressed work. Admixtures frequently facilitate placing and handling of the concrete, as well as obtaining the desired high early strengths. Admixtures that entrain air are also used in prestressed concrete to provide resistance to deterioration due to freezing and thawing.

Admixtures used in prestressed concrete construction generally are specified to be either Type A or D, as defined in A.S.T.M. Designation C 494 "Standard Specification for Chemical Admixtures for Concrete." These can be further defined as either lignosulfonates, organic acids, or polymers.

Another admixture that has been proposed for use as an accelerator consists of reground hardened cement paste (Ref. 6). This admixture is made by regrinding a cement paste (water/cement ratio of 0.50) that is seven days old. The material is reground 2 to 7 days before use and is ground to the fineness of the cement being used. Laboratory tests on the materials to be used on any one job should be made before using this technique.

The addition of an air-entraining agent is considered to increase slightly the shrinkage and creep of concrete. Because these admixtures also permit a reduction in the mixing water, without adversely affecting the workability, it is generally accepted that the shrinkage and creep characteristics of air-entrained and non-air-entrained concrete of equal workability are equal.

The use of admixtures is considered a good practice where job conditions can be improved thereby; however, care must be exercised to ensure that admixtures containing calcium chloride or other chlorides are not used, since the chloride ion may result in pitting or stress corrosion of the prestressing tendons. A number of the well known admixtures commonly used in modern concrete practice contain chloride ions and should not be used in prestressed concrete.

3-4 Slump

European literature emphasizes the desirability of using low-slump concrete in the manufacture of prestressed products. Experience in this country would indicate that the use of no-slump concrete should generally be confined to products that can be made on a vibrating table or vibrating pallets and, perhaps, to shallow members that are of such cross-sectional shape that all areas of the member are readily accessible to internal vibrators. For average prestressed members that are too large to be produced on a vibrating table or that have large bottom flanges that cannot be readily vibrated with internal vibrators, it has been found that good results are obtained when the slump of the plastic concrete is between 2 and 4 in. There are established plants in the U.S. that manufacture precast, prestressed members with no slump concrete, but the majority of plants do not.

3-5 Curing

The best method of curing concrete is by keeping the concrete surfaces thoroughly wet as long as possible. This applies to prestressed concrete just as it does to other modes of concrete construction. Because of the necessity of obtaining high early strength in prestressed concrete, in order to permit rapid re-use of the manufacturing facilities, steam curing at atmospheric pressure is often employed to accelerate the hardening of the concrete. Steam curing is discussed in detail in Sec. 3-13.

Hot water has also been successfully used in the manufacture of precast members. In some instances, hot water is circulated through pipes that are close to the concrete elements being cured (while the exposed surfaces are kept wet) and, hence, the system heats concrete in much the same way that hot water is used to heat dwellings or buildings. Hot water is sometimes used in lieu of cold water in much the same way that cold water is used in curing —matts or burlap are placed on the exposed concrete surfaces, thus keeping them wet with hot water ($\pm 150°$F). Each of these methods can be satisfactory and each offers certain advantages.

Heat, applied by any means, can be used to accelerate the curing of concrete, but it is extremely important that the concrete be kept wet during the curing period.

3-6 Concrete Aggregates

The aggregates used in the manufacture of normal concrete members are usually satisfactory for use in prestressed concrete. However, because of the higher strengths required for prestressed concrete, difficulty has been experienced, in some localities, in finding suitable natural aggregates for prestressed construction. Where a choice of aggregates is available, the selection should be made after considering the ease of obtaining the necessary strength, as well as the magnitude of the elastic and inelastic volume changes that might be expected with the different types available. Lightweight aggregates of the expanded shale or clay type have been used with good results in this country. Care must be exercised in employing lightweight aggregates in prestressed concrete, in order to assure that a reasonable estimate of the volume changes that occur with the lightweight concrete are taken into account when estimating the loss of prestress.

The gradation of the concrete aggregates is often not too important and adequate results can frequently be obtained with a relatively fine sand. The sand must be clean, without a large percentage passing the 100 mesh screen. With a sand that has a large amount of fines, it may be wise to reduce the total amount of sand in the concrete as a means of reducing the water required in the concrete. It should be recognized that sand, and particularly fine sand, demands more water than coarse aggregates and coarse sands. In any case, laboratory tests should be performed on concretes with unusual gradations before they are used. If possible, the strength, shrinkage, and creep characteristics of the proposed concrete should be determined to be within acceptable limits before it is used.

3-7 Strength of Concrete

A principal reason concrete with a minimum 28-day cylinder compressive strength of the order of 4000 to 6000 psi is used in prestressed-concrete work is that concrete of this quality exhibits lower volume changes than concretes of lower quality. Another reason for using higher strength concretes is that efficient use can generally be made of such concretes in the flexural design of prestressed concrete (this is not the case in reinforced concrete design).

In some areas it is difficult to consistently produce concrete of high quality

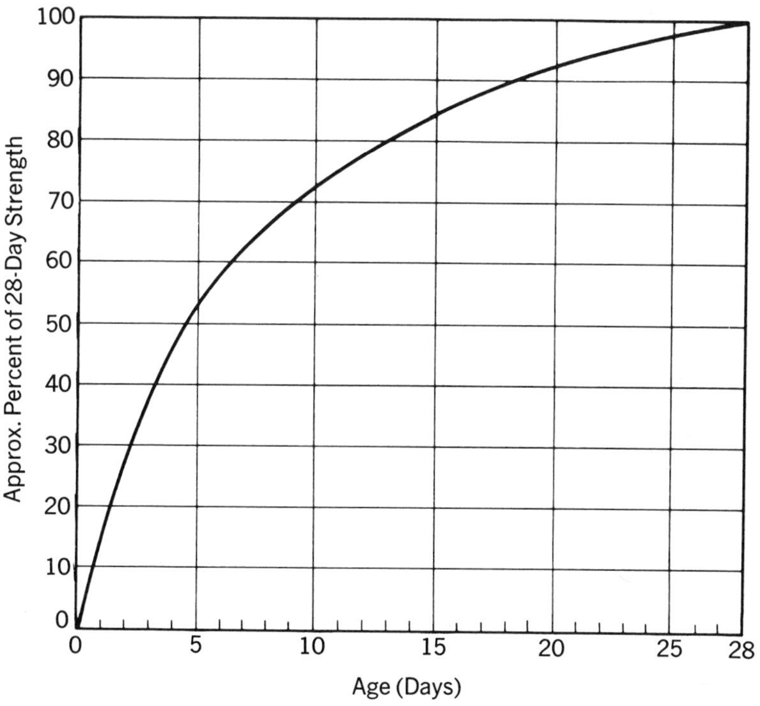

Fig. 3-1 Strength-time curve for concrete.

with local materials. The designer of prestressed-concrete structures should carefully investigate this problem on each project undertaken. It is generally possible to prepare reasonably economical designs with concretes of moderately high strength, and it is better to anticipate this problem and provide for it in the design stage rather than struggle with what may be an almost impossible situation during construction.

Whenever possible, and always on major jobs, the concrete mixes used should be trial batched and laboratory tested before use on the job. The mixes employed in the work should have laboratory strengths 10 to 20% higher than that required by the job specifications.

Although the rate at which concrete gains strength is a function of many variables, it is sometimes advantageous to have a curve that approximates the strength-time relationship for concrete. Such a curve for concrete made with type I cement cured at 70°F is given in Fig. 3-1. The curve is useful in estimating the early strength of concretes of various specified 28-day strengths, as well as in predicting the 28-day strength from known early strengths.

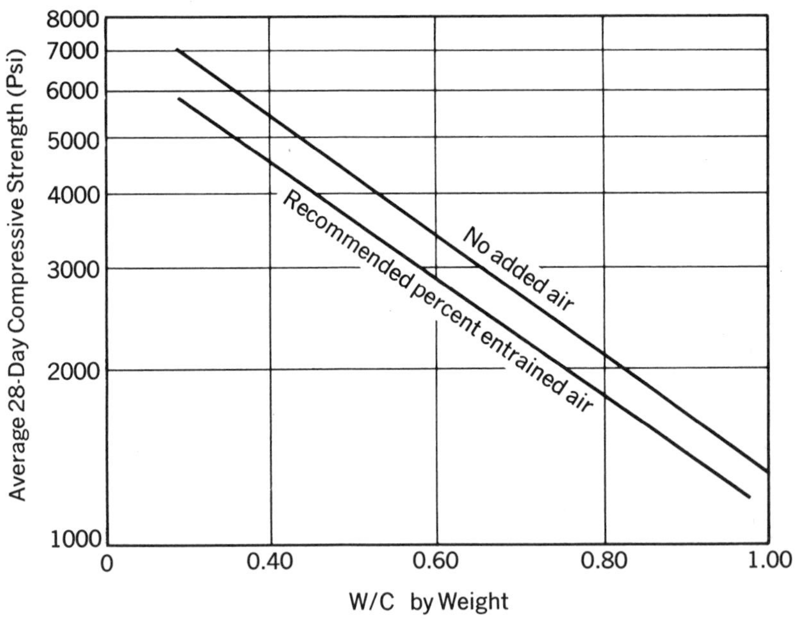

Fig. 3-2 Strength in relation to water-cement ratio for air-entrained and non-air-entrained concrete. Strength decreases with an increase in water-cement ratio; or with the water-cement ratio held constant, use of air entrainment decreases the strength by about 20%. (*See* Ref. 3-2.)

It should be understood that the curve given in Fig. 3-1 is for use when the strength-time relationship for a particular concrete being used is not known. For large projects, the strength-time curve should be developed for the particular concrete being used, in order that the work can be accurately planned in advance and so that any low strength concrete that occurs can be detected at an early age. Considerable variations in the rate of attaining the compressive strength occur between concretes composed of different materials, as well as concretes that are mixed and cured in different manners.

The 28-day compressive strength of concrete as a function of the water-cement ratio for air-entrained and non-air-entrained concrete is illustrated in Fig. 3-2. In Fig. 3-3, the 28-day compressive strength of concrete is shown as a function of the voids-cement ratio. These curves are useful in estimating the quantities of cement and water that must be used to achieve a desired concrete strength.

The compressive strength that will eventually be achieved by concrete can be estimated as being $1.30 f_c'$ and $1.15 f_c'$ for normal and high early strength cements, respectively, in which f_c' is the 28-day cylinder strength. Cube compressive strength of concrete is frequently used in Europe rather than cylinder compressive strength. The cylinder strength is generally considered to be

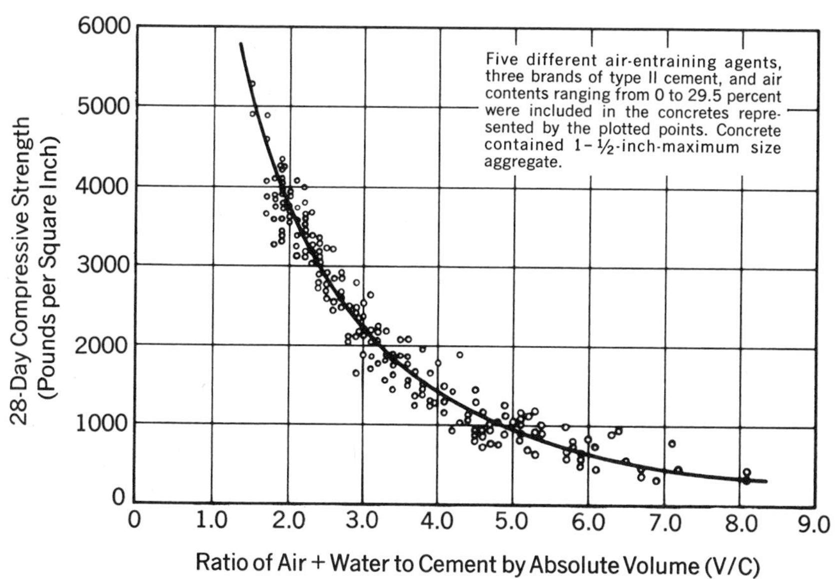

Fig. 3-3 Compressive strength of concrete in relation to voids-cement ratio. (*See* Ref. 3-2.)

80% of the cube strength, and this relationship is recommended for use as a conversion factor between cube and cylinder strengths.

The tensile strength of concrete can be estimated from

$$f_t = \tfrac{1}{3}\sqrt{wf_c'} \tag{3-1}$$

in which w is the unit weight of the concrete in pounds per cubic foot. The modulus of rupture can be taken as follows:

$$f_r = a\sqrt{wf_c'} \tag{3-2}$$

in which a is a constant that normally varies between 0.60 to 0.70. Because of the variation that is found in the modulus of rupture, the upper or lower limit of Eq. 3-2 should be used in computations in such a manner that the result will be conservative.

The strength of lightweight concrete should always be determined by tests. The curves of Figs. 3-1, 3-2 and 3-3 should not be expected to apply to lightweight concrete.

3-8 Elastic Modulus

The magnitude of the elastic modulus of concrete is important to the designer of prestressed concrete, because it must be known when computing the deflections and the losses of prestress. Unfortunately, the elastic modulus of concrete is a function of many variables, including the type and amount of ingredients used in making the concrete (cement, aggegrates and water), as well as the manner and duration of curing the concrete, age at the time of loading, rate of loading and other factors (Ref. 1). When possible, it is recommended that measurements be made in order to determine the magnitude of the elastic modulus on important works where the loss of prestress may be very critical or for structures on which deflections must be computed with the highest possible precision. When it is not possible to predetermine the elastic modulus for the concrete that is to be used in a design, it may be assumed to be

$$E_c = w^{1.5}\, 33\sqrt{f_c'} \tag{3-3}$$

in psi, for values of w (unit weight of the concrete) between 90 and 155 pounds per cubic foot. For normal weight concrete, the relationship may be taken as

$$E_c = 57{,}000\sqrt{f_c'} \tag{3-4}$$

It should be noted that f_c' in this relationship is the compressive strength of the concrete, as determined by tests of standard 6 × 12 in. cylinders made in accordance with A.S.T.M. Designation C 192 and tested in accordance with

A.S.T.M. C 39 at the age of 28 days or such earlier age as the concrete is to receive its full service load or maximum stress.

It should be pointed out that Eq. 3-3 is intended to give a result that approximates the value that would be obtained if the concrete were tested in accordance with A.S.T.M. Designation 469. Because the value of the elastic modulus thus obtained is intended to be the secant modulus at a stress of 40% of the ultimate strength, a higher value for the elastic modulus (as much as 10%) could be anticipated in applications where the concrete is stressed to lower levels.

In prestressing concrete, the prestressing force is often transferred to the concrete at a relatively early age (one to fourteen days, depending upon the materials and method of curing used). Hence, at the time of stressing, the concrete frequently has a strength that is somewhat less than the minimum specified at the age of 28 days. Eq. 3-3 gives a means of approximating the modulus of elasticity of the concrete at a given age by relating it to the cylinder strength, which varies with age.

ILLUSTRATIVE PROBLEM 3-1 A post-tensioned beam is designed to be stressed when the concrete strength is 4000 psi. The specifications also provide that the minimum cylinder compressive strength at the age of 28 days shall be 5000 psi. Compute the elastic modulus that should be used in computing instantaneous deflections and stress losses at the time of stressing and at the age of 28 days if (1) the concrete is normal concrete and (2) the concrete is lightweight —weighing 100 lb per cubic foot.

(1)

$$E_c = 57{,}000\sqrt{f_c'}$$

At time of stressing:

$$E_c = 57{,}000 \times \sqrt{4000} = 3{,}605{,}000 \text{ psi}$$

At 28 days:

$$E_c = 57{,}000 \times \sqrt{5000} = 4{,}030{,}000 \text{ psi}$$

(2)

$$E_c = 100^{1.5} \times 33 \times \sqrt{f_c'} = 33{,}000\sqrt{f_c'}$$

At times of stressing:

$$E_c = 33{,}000 \times \sqrt{4000} = 2{,}090{,}000 \text{ psi}$$

At 28 days:

$$E_c = 33{,}000 \times \sqrt{5000} = 2{,}330{,}000 \text{ psi}$$

Creep of concrete, which is discussed in detail in Sec. 3-11, is defined as the increase in strain that occurs when a concrete member or specimen is subjected to constant stress. Because the elastic modulus of concrete is equal to stress divided by strain, an increase in strain has the effect of decreasing the modulus of elasticity. The elastic modulus that has been corrected to take into account the effect of creep at some particular time is referred to as the "effective modulus" or "reduced modulus." This can be expressed mathematically as follows:

$$\text{Elastic modulus} = E_c = \frac{\text{Stress}}{\text{Elastic Strain}} \tag{3-5}$$

$$\text{Effective modulus} = \text{E}'_c = \frac{\text{Stress}}{\text{Elastic Strain} + \text{Creep Strain}} \tag{3-6}$$

which can be rewritten

$$E'_c = \frac{E_c}{1 + C_t} \tag{3-7}$$

where C_t* is the creep ratio equal to the ratio of the creep strain to the elastic strain, or

$$C_t = \frac{\text{Ultimate Creep Strain}}{\text{Elastic Strain}} \tag{3-8}$$

C_t is a function of many variables, but principally of the relative humidity, concrete quality, duration of applied load, and age of concrete when loaded (Ref. 3-7). Methods of estimating C_t are given in Sec. 3-11.

The reduced modulus is frequently used in computing deflections of reinforced and prestressed concrete, as well as the losses of prestress in prestressed-concrete members. These methods are discussed in greater detail in Secs. 6-2 and 6-3.

3-9 Shrinkage

The shrinkage of concrete is an important factor to the designer of prestressed concrete for several reasons. As has been stated previously, the shrinkage of the concrete contributes to the loss of prestress. The magnitude of the shrinkage must also be known with reasonable accuracy when the deflection of prestressed members are being computed with the more sophisticated methods. The deflection of composite prestressed-concrete members cannot

* Some authors use a creep coefficient that is equal to $\frac{\text{total strain}}{\text{elastic strain}}$. If this coefficient is represented by C_c, we can write $C_c = 1 + C_t$. One must be careful to be sure that these factors are used properly.

be computed without knowing the shrinkage characteristics of each of the concretes involved. In addition, the magnitude of the concrete shrinkage must be estimated in order to evaluate secondary stresses (due to volume changes) that may result.

The effects of concrete shrinkage in prestressed-concrete structures are considerably different than those in reinforced-concrete structures. In reinforced concrete, the shrinkage strains are resisted by compressive stresses in the reinforcing steel, whereas in prestressed concrete, the prestressing steel is always in tension and is causing compressive strains that add to the shrinkage strains in the concrete. In addition, reinforced-concrete structural members are normally cracked with many closely spaced minute cracks that tend to relieve the effect of shrinkage stresses, which may exist due to intended or unintended restraint of the structure. The designer of prestressed-concrete structures must give particular attention to the effects of shrinkage, creep, and temperature variations. If these movements are restrained, forces of very high magnitude can result with the very real possibility of serious structural and non-structural damage. This subject is discussed in greater detail in Chapter 10.

The shrinkage of concrete is known to be related to the loss of moisture. It has been demonstrated by many researchers that concrete will expand if subjected to 100% relative humidity or if submerged in water. The amount of shrinkage obtained in dry storage can also vary widely. Shrinkage of concrete is known to be a function of the following:

A. Composition of the cement.
B. Physical properties of the aggregate.
C. Maximum aggregate size.
D. Quantity of water.
E. Method and duration of the curing.
F. Temperature and humidity of atmosphere during service.
G. Volume to surface ratio of the concrete.
H. Admixtures.

A considerable amount of data is available in the literature relative to the effect of each of these variables. The discussion that follows is of a general nature but is considered sufficiently accurate for most design purposes. The designer of prestressed concrete should recognize that he can often control shrinkage to some degree through careful consideration of the materials and methods specified for each project. The effect of each of the above variables can be summarized as follows.

A. Cement. High early strength portland cement (type III) would normally be expected to have a shrinkage that is 10% higher than normal portland cement (type I) or modified portland cement (type II) (Ref. 1). In addition, a cement which may exhibit a large amount of shrinkage may have a total shrinkage that is 100% greater than that of a cement, which due to its chemical composition, exhibits a small amount of shrinkage. This is an extreme range, however, and it may be beneficial to investigate the cements available in any locality, in order to determine if any of the commonly used cements have exceptionally high or low shrinkage characteristics. As has been previously stated, there is some evidence that the use of a high early strength cement of good quality may result in a concrete that will exhibit somewhat lower total volume changes in prestressed construction than that which would be obtained with normal cement (type I) (Refs. 1, 4 and 11).

B. Aggregates. The physical properties of the larger aggregate particles have considerable influence on the shrinkage of concrete. This is due to the fact that the concrete aggregate reinforces the cement paste and resists its contraction. Aggregates with higher elastic moduli are stiffer and hence restrict the contraction of the paste to a greater degree. Aggregates that have a low volume change in themselves, due to drying, generally result in concrete with lower shrinkage. Concretes containing aggregates of quartz, limestone, dolomite, granite or feldspar are generally low in shrinkage, whereas those containing sandstone, slate, trap rock or basalt may be relatively high in shrinkage. Therefore, if aggregates of the latter type, or gravels containing a large portion of such minerals, are used, an allowance should be made for a relatively high shrinkage value. Concretes made with soft, porous sandstone may shrink 50% more than concretes made with hard dense aggregates (Refs. 1 and 8).

Lightweight concrete aggregates manufactured by expanding clay or shale have been used to a significant extent in prestressed-concrete structures. High-quality expanded shale or clay aggregates that are not crushed after burning, and hence are coated and less absorptive than crushed material, have been reported as having drying shrinkage characteristics that are approximately of the same magnitude and rate as were found with normal aggregates (Ref. 12). Other research has indicated that lightweight aggregates may have shrinkages as much as 50% greater than normal aggregates at the age of 300 days (Ref. 13). When the use of lightweight aggregates is contemplated, the designer should investigate the shrinkage characteristics of the actual concrete mix proposed.

C. Aggregate Size. Aggregate size also has a marked effect in the magnitude of concrete shrinkage. This is due to two reasons. The first of these is that the larger particles offer more restraint to the shrinkage of the mortar. For this reason, increasing the aggregate size from $\frac{3}{4}$ to $1\frac{1}{2}$ in. would be expected to reduce the shrinkage as much as 20%. The second factor is that in changing maximum aggregate size from $\frac{3}{4}$ to $1\frac{1}{2}$ in., about 10% less water is required to achieve the same consistency, and this reduction in water will account for a reduction in shrinkage of as much as 20%. The result is that a change in aggregate size from $\frac{3}{4}$ to $1\frac{1}{2}$ in. may result in a reduction in the shrinkage by about 40% (Ref. 9).

D. Water. The amount of water in a concrete mix is the most important single factor affecting shrinkage of concrete. The shrinkage of concrete has been found to vary directly with the water content of concrete, as is illustrated in Fig. 3-4. Therefore, the amount of water used in prestressed-concrete construction should be kept to the minimum amount required for the consistency necessary for proper placing and compaction. The water required to obtain the necessary plasticity in a concrete mix is a function, among other things, of the amount of mortar (cement + sand) in the mix. For this reason, it is desirable to keep the quantity of mortar as low as practicable.

E. Curing. There is little if any concrete shrinkage during curing, if the concrete is kept sufficiently moist to prevent the loss of moisture. Some investigators report that ultimate shrinkage is unaffected by an increase in the duration of curing time (Ref. 9). There is evidence that curing concrete at an elevated temperature (atmospheric pressure steam curing) will result in a reduction in shrinkage by as much as 30% (Refs. 10 and 17). The acceleration in curing that is obtained from steam curing apparently leads to a more complete hydration of the cement. Hence, less free water remains available for evaporation and the shrinkage is reduced. Atmospheric pressure steam curing has resulted in reductions of shrinking between 10 and 30% for type I cement and 25 to 40% for type III cement, when compared to specimens that were moist cured for 6 days (Ref. 11).

F. Temperature and Humidity. Temperature and humidity each affect the rate and magnitude of concrete shrinkage. Higher temperatures during service would be expected to result in somewhat greater shrinkage, because the higher temperatures would result in greater moisture loss. Variations in temperature and humidity results in higher shrinkage (and creep) than is obtained under constant conditions. Therefore, estimates of shrinkage made on laboratory tests may be low (Ref. 22).

The relative humidity during service has a marked effect on shrinkage, with lower humidities resulting in greater shrinkages. Schorer's formula for calculating ultimate drying shrinkage is

$$\text{Ultimate shrinkage} = 12.5 \times 10^{-6}(90 - H) \qquad (3\text{-}9)$$

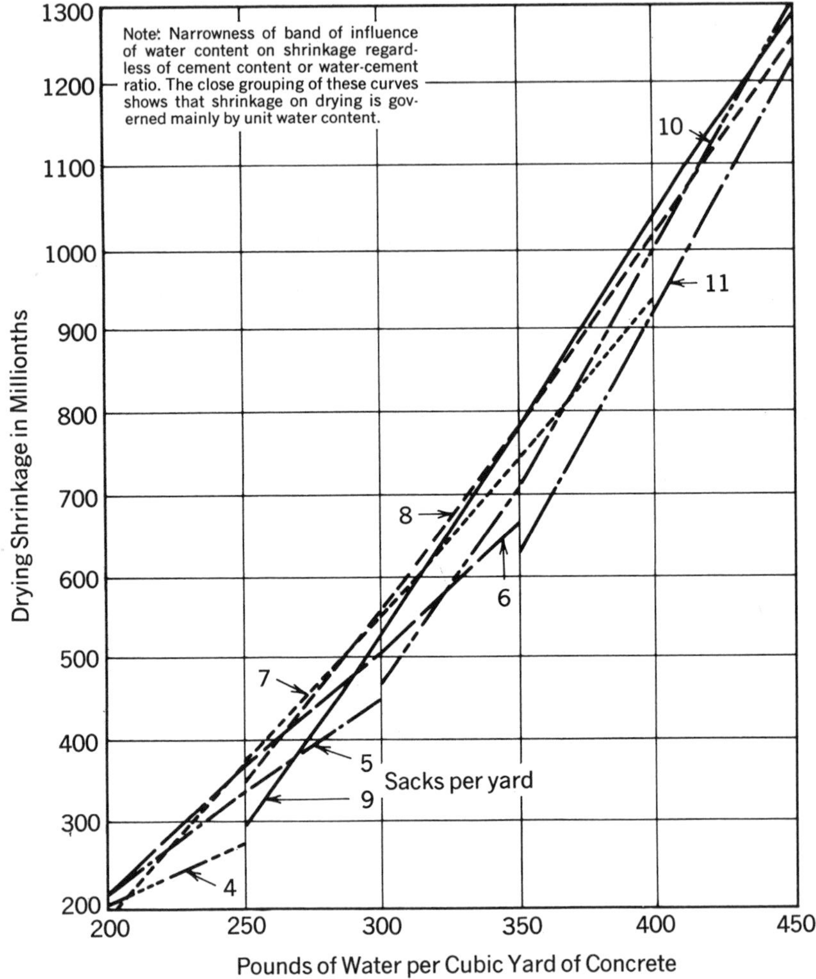

Fig. 3-4 The interrelation of shrinkage and water content.

where H is the relative humidity. A variation of Eq. 3-9 is given in Sec. 3-10 for estimating the effect of humidity on ultimate shrinkage.

G. Size of Member. The size of a member affects the magnitude and rate of shrinkage. Since shrinkage is caused by the evaporation of moisture from the surface, members that have low volume-to-surface ratios will be expected to shrink more, as well as more rapidly, than members having high volume-to-surface ratios. The curve of Fig. 3-5 shows the relationship between the shrinkage size coefficient and the volume-to-surface ratio based on a specimen having a volume-to-surface ratio of 1.5 in., which is the value for a cylindrical specimen having a diameter of 6 in. This curve is useful in estimating the effect of member size when test data for specimens having a diameter of 6 in. are available.

H. Admixtures. Admixtures may increase, decrease or have practically no effect on the amount of concrete shrinkage. The more commonly used admixtures in prestressed work are of the water-reducing and water-reducing retarding type (classified in ASTM 494 as types A and D, respectively). Admixtures of these types can be further classified according to their general chemical composition and, as such, are categorized as lignosulfonates, organic acids or polymers. Test data are available that would indicate that the lignosulfonates tend to increase shrinkage (from 5 to 50%) when compared with the control concrete (a concrete without admixture but having the same slump). In the

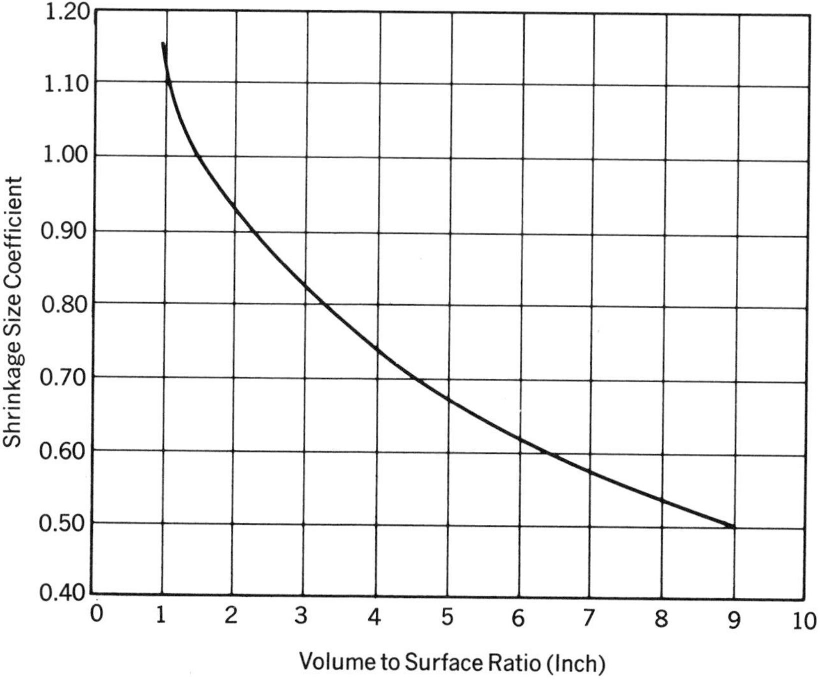

Fig. 3-5 Shrinkage size coefficient vs volume-to-surface ratio.

same tests, organic acid types of admixtures showed shrinkages from 89 to 117% of the control concrete, while the polymer type admixtures revealed shrinkages of from 98 to 112% of the control concrete. Calcium chloride, which should not be used as an admixture in prestressed concrete (*see* Sec. 2-8) increases concrete shrinkage about one third (Ref. 15).

3-10 Estimating Shrinkage

The best method of estimating the amount of shrinkage of concrete that should be used in any structural design is through the use of shrinkage tests. Established precasting plants and firms engaged in supplying ready-mixed concrete, cement or aggregates should have shrinkage test results available for typical concrete mixtures obtainable in the localities they serve. In the event such data are not available, the designer must either make tests or use his judgment in estimating the unrestrained shrinkage of concrete for his particular conditions. A conservative estimate is recommended if tests are not made.

It is recommended that the estimated, ultimate, unrestrained shrinkage for concrete specimens (made with type I cement aggregates of estimated normal or lower shrinkage characteristics) with a volume-to-surface ratio of 1.5 and with a maximum aggregate size of 1.50 in. and no admixtures should be computed as follows, if they are to be used at normal temperatures and at 50% relative humidity:

$$\varepsilon_u = 200 + \frac{1100}{230}(W - 220) \tag{3-10}$$

in which the shrinkage is expressed in micro-inches per inch (inches/inches $\times 10^{-6}$) and W is the total water content in pounds per cubic yard. (*See* Fig. 3-6.) This value should be increased or decreased in order to account for particular characteristics of materials or conditions of service, as described in Sec. 3–9. The effect of relative humidity, if other than 50% during service, can be taken into account by increasing or decreasing the value obtained from Eq. 3-10 by multiplying the ultimate shrinkage value by the Shrinkage Humidity Factor from Fig. 3-7. (*See* Sec. 3-11 for approximate humidities for various service conditions.) The shrinkage can be assumed to take place according to the following relationship:

$$\frac{\text{Shrinkage Strain at Time } t}{\text{Ultimate Shrinkage Strain}} = 0.157 \log_e t - 0.115 \tag{3-11}$$

in which t is the time in days reckoned from the end of the curing period. It is assumed no shrinkage would take place during curing if the concrete is kept

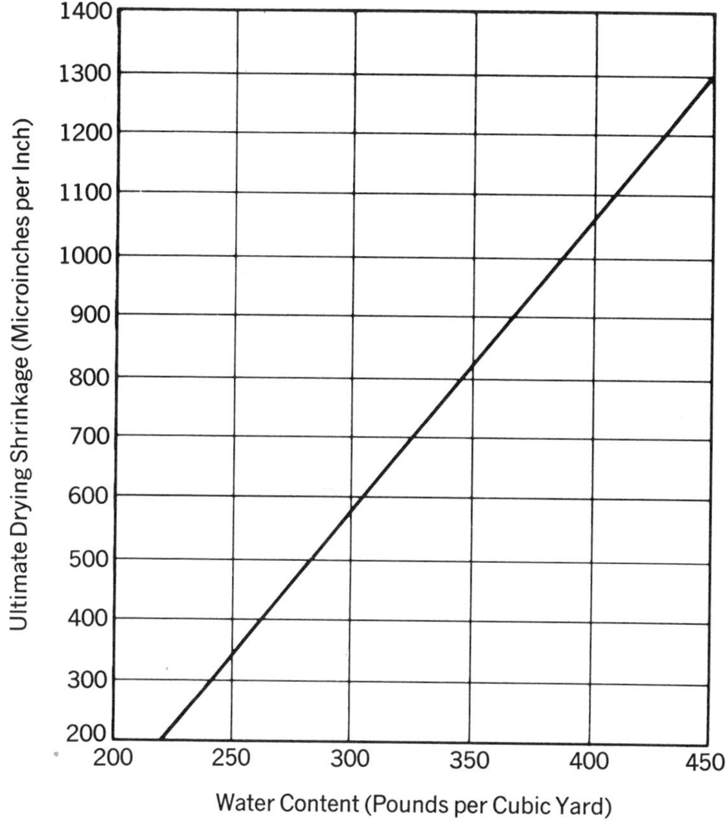

Fig. 3-6 Ultimate shrinkage vs water content.

wet. This relationship is shown plotted in Fig. 3-8. Equation 3-11 can also be used for determining the rate of creep (*see* Sec. 3-11).

An alternate relationship for the shrinkage at time t (days) after drying commences is

$$\frac{\text{Shrinkage Strain at Time } t}{\text{Ultimate Shrinkage Strain}} = \frac{t^a}{b + t^a} \tag{3-12}$$

in which a and b are parameters that can be determined experimentally for any particular concrete. When it is not possible to determine the values of a and b experimentally, the value of a can be assumed to be 1 and the value of b can be assumed to be 35 and 55 for moist and steam cured concretes, respectively.

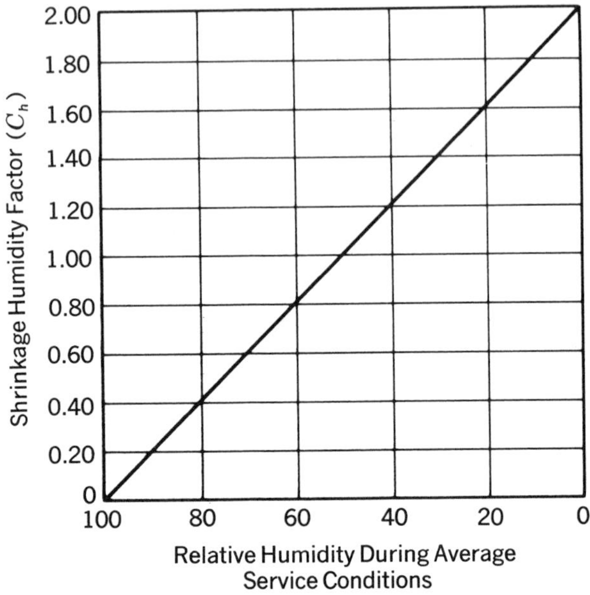

Fig. 3-7 Shrinkage humidity factor (C_h) vs relative humidity.

An inexpensive method of measuring shrinkage under job conditions is shown in Fig. 3-9. The method consists of casting a plain concrete post that is 5 in. by 8 in. in section and 18 ft long. The post can be cast in a horizontal position and set up vertically after curing or can be cast and cured in the vertical position. In either case, the concrete must be water (or steam) cured to simulate normal production methods. The shrinkage measurements should be started at the time curing is stopped. The strain measurements may be measured with embedded electronic strain gages or with the device shown in Fig. 3-9. This device indicates three times the actual shrinkage, due to the lever arrangement. A gage length of 12.5 ft (150 in.) is suggested, since total shrinkage strains of the order of 0.05 to 0.10 in. should be measured over lengths of this order. The specimen should be exposed to ambient temperature and humidity conditions in order to make the specimen as similar to job-site conditions as possible. In order to eliminate solar effects on the specimen, it is recommended that the specimen be shaded at all times and that shrinkage measurements be made only in the very early morning. Using this method, the data required for the shrinkage strain vs time curve can be obtained and plotted for the concrete mix proposed for use in any project.

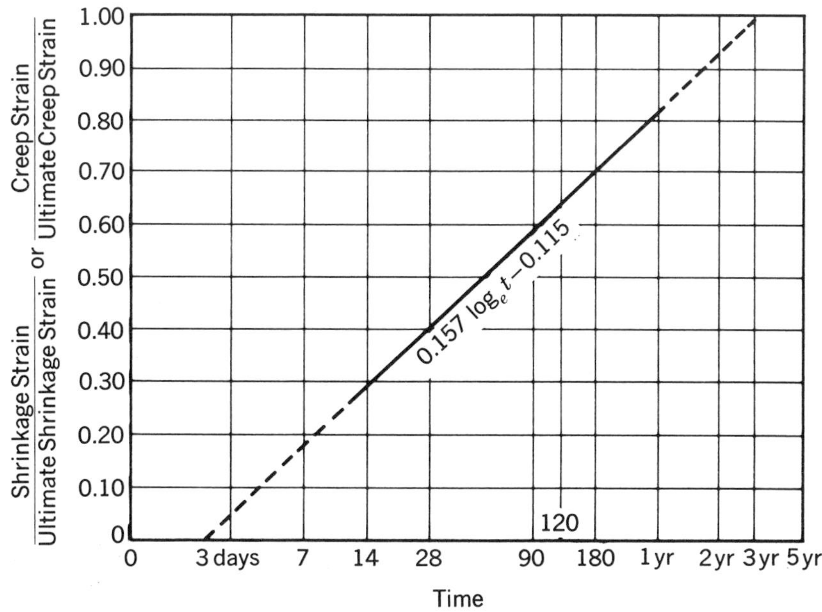

Fig. 3-8 Shrinkage vs time (creep vs time).

ILLUSTRATIVE PROBLEM 3-2 Estimate the shrinkage of a specimen of concrete, having a volume-to-surface ratio of 1.5 in., that is to be made with type III cement, 1.5 in. aggregate of high quality, and which is to have a 28-day compressive strength of 5000 psi. Relative humidity during service will be 70%.

SOLUTION:

From Fig. 3-2 $W/C = 0.44$ for $f_c' = 5000$ psi.

Assume 7 sacks of cement per cubic yard $= 7 \times 94 = 658$ pounds per cubic yard.

Water content $= 0.44 \times 658 = 290$ pounds per cubic yard.

$$\text{Ultimate unit shrinkage at 50\% humidity} = 200 + \frac{1100}{230}(290 - 220)$$

$$= 200 + 335$$

$$= 535 \times 10^{-6} \text{ in./in.}$$

$$\text{Ultimate unit shrinkage at 70\% humidity} = 0.60 \times 535 \times 10^{-6} = 321 \times 10^{-6} \text{ in./in.}$$

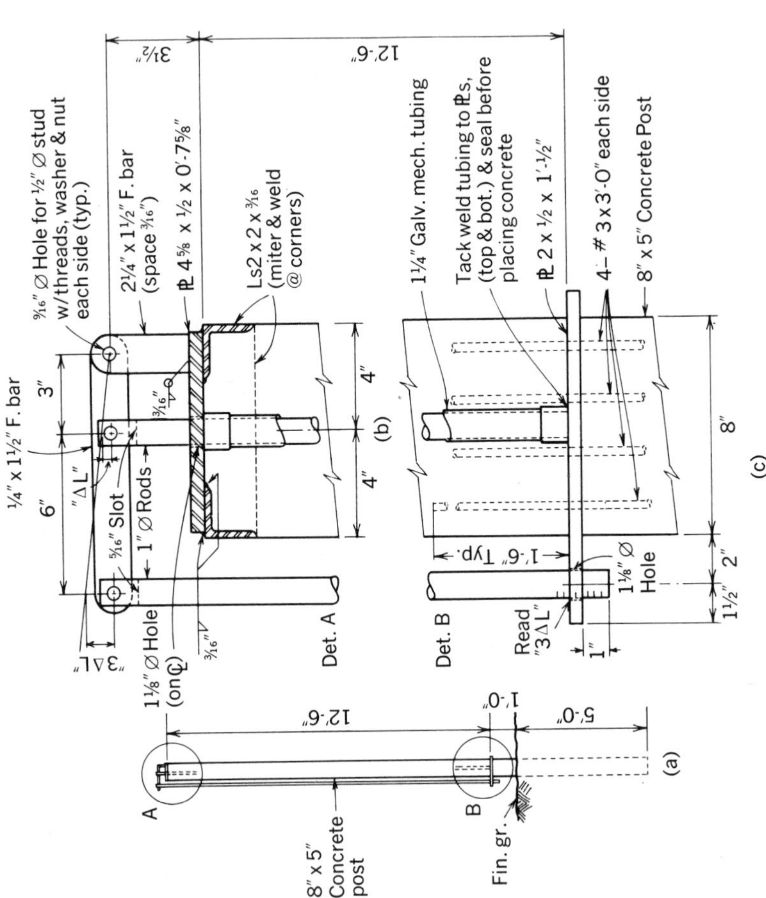

Fig. 3-9 (a–c) Concrete shrinkage test set-up.

3-11 Creep of Concrete

The creep of concrete is defined as the strain change which takes place in concrete that is subjected to constant stress and, as is the case with shrinkage, is associated with the loss of moisture from the concrete. Unlike shrinkage, creep is affected by the stress level in the concrete as well as by the "maturity" of the concrete, which is defined as the age or degree of hydration of the concrete, at the time of loading. Relaxation of concrete is defined as the loss of stress in concrete that is subjected to a constant strain. Prestressed concrete is not subjected to constant stress or strain, but rather to varying conditions as a result of relaxation of the prestressing tendons, shrinkage, and creep of the concrete. However, it is generally considered that for purposes of computing loss of prestress, the loss due to the plasticity of the concrete is more accurately estimated using creep, rather than relaxation curves for the concrete.

The best means of determining the creep characteristic of concrete that is to be used in any one design is by testing the concrete under conditions that approximate the service conditions as closely as possible.

A suggested method for job-site testing concrete for creep characteristics is illustrated in Fig. 3-10. The method consists of loading two reinforced concrete posts, which are 5 in. by 8 in., in section in such a manner that they are subjected to a constant moment over a length of about 12.5 ft. With the bases of the posts rigidly fixed, the deflection of the post top can be measured over a long period of time as a means of determining the creep characteristics of the concrete. The initial deflection of each post should be of the order of 0.20–0.40 in., depending upon the elastic modulus and the magnitude of the applied moment. The ultimate deflection should be from 2 to 5 times this value, depending upon the concrete creep characteristics. The specimens should be shaded if possible to eliminate the effects of the sun, but they should be exposed to temperatures and humidity that approximate service conditions. Deflection measurements should be made in the very early morning to eliminate solar effects.

Progressive prestressing plants should have test data performed on concrete of the type normally used in their products for the information and guidance of engineers contemplating the use of their products. Special tests are not always possible nor are test data always available. The discussion that follows is intended for use in approximating the creep at various ages during varied conditions of service.

The rate at which creep takes place for the purposes of losses of prestress and deflection computations can be assumed to be the same as for shrinkage (*see* Fig. 3-8). Hence, it can be assumed to follow the expression

$$\frac{\text{Creep Strain at Time } t}{\text{Ultimate Creep Strain}} = 0.157 \log_e t - 0.115 \qquad (3\text{-}13)$$

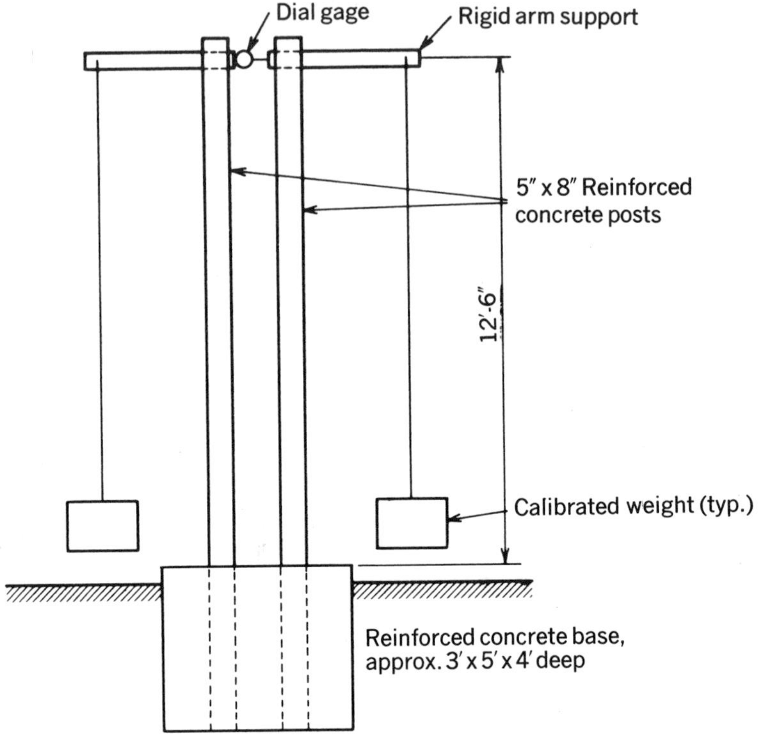

Fig. 3-10 Test set-up for creep of concrete.

where t is in days. Equation 3-13 assumes the creep has reached its limiting value in 1517 days, which is sufficiently accurate for deflection and loss of prestress calculations. Tests have shown creep may not reach its limiting value for from 3 to 4 yr.

The expression of Eq. 3-12 can also be used to predict the creep strain at time t. This expression becomes

$$\frac{\text{Creep Strain at Time } t}{\text{Ultimate Creep Strain}} = \frac{t^{a'}}{b' + t^{a'}} \tag{3-14}$$

in which t is time in days, reckoned from the time the stress is applied, and a' and b' are parameters that can be determined experimentally for each particular concrete. Values of 0.60 and 10 can be used for a' and b', respectively, when test data are not available.

The volume-to-surface ratio affects creep less than it does shrinkage.

Creep is thought to be the effect of two factors. The first of these is basic creep and is independent of moisture movement. The second is drying creep, which results from the concrete losing moisture to its environment. Only the drying creep is affected by member size and shape and it is believed drying creep has no more effect after about 3 months. The relationship between the creep size coefficient and the volume-to-surface ratio is shown in Fig. 3-11, from which it will be seen that the creep size coefficient is equal to 1 for members with a volume-to-surface ratio of 10, which is an indication that drying creep has a negligible effect on more massive members.

Creep strain is generally expressed in terms of the initial elastic strain through the use of the "creep ratio." The creep ratio is defined in Eq. 3-8 as the ultimate creep strain divided by the elastic strain. Recommended curves for use in estimating the creep ratio for normal water-cured concrete are given in Figs. 3-12 and 3-13 as a function of the average humidity conditions to

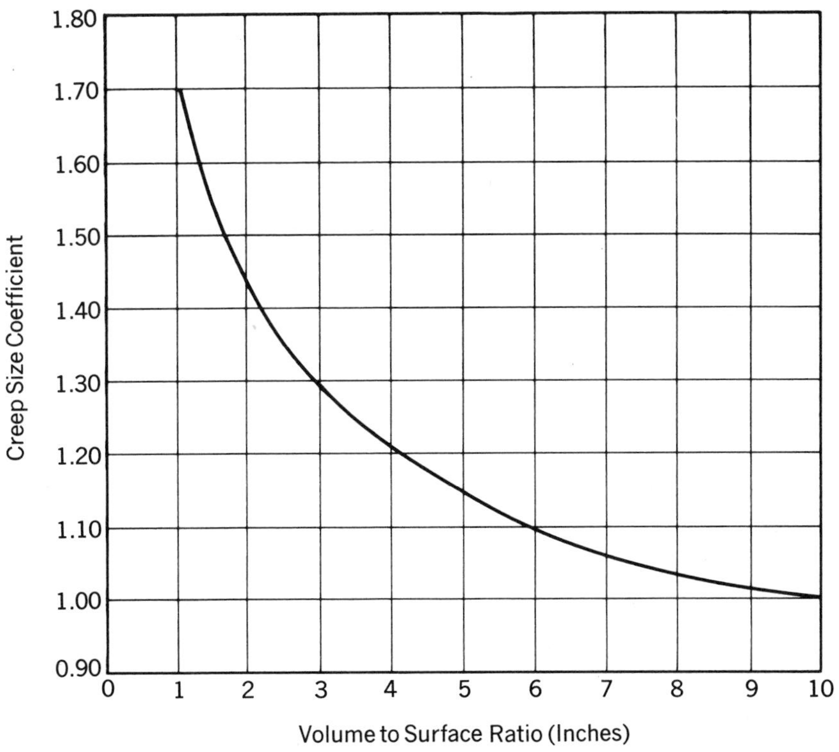

Fig. 3-11 Creep size coefficient vs volume-to-surface ratio.

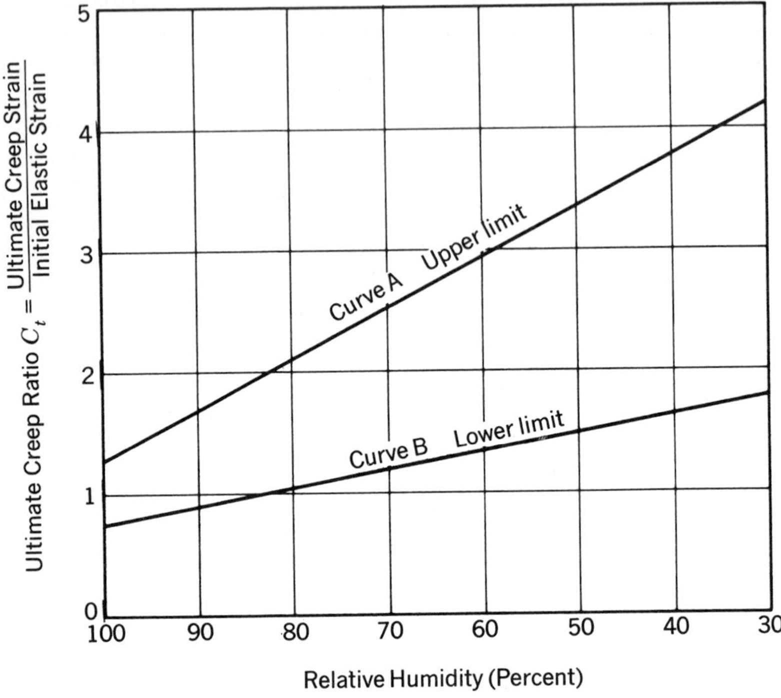

Fig. 3-12 Creep ratio (C_t) vs relative humidity for water-cured concrete of prestressing quality (normal aggregates).

which the concrete will be exposed in service. The limits for the creep ratio for concrete of prestressing quality can be expressed mathematically as follows:

$$\text{Upper Limit of Creep Ratio} = 1.25 + 2.75\,\frac{100 - H}{65} \tag{3-15}$$

$$\text{Lower Limit of Creep Ratio} = 0.75 + 0.75\,\frac{100 - H}{50} \tag{3-16}$$

in which H is the mean humidity to which the concrete is exposed in service.

For concrete of ordinary quality, the relationships for the upper and lower limits of the creep ratio are as follows:

$$\text{Upper Limit } C_t = 2.00 + 2.00\,\frac{100 - H}{50} \tag{3-17}$$

$$\text{Lower Limit } C_t = 1.00 + 1.00\,\frac{100 - H}{50} \tag{3-18}$$

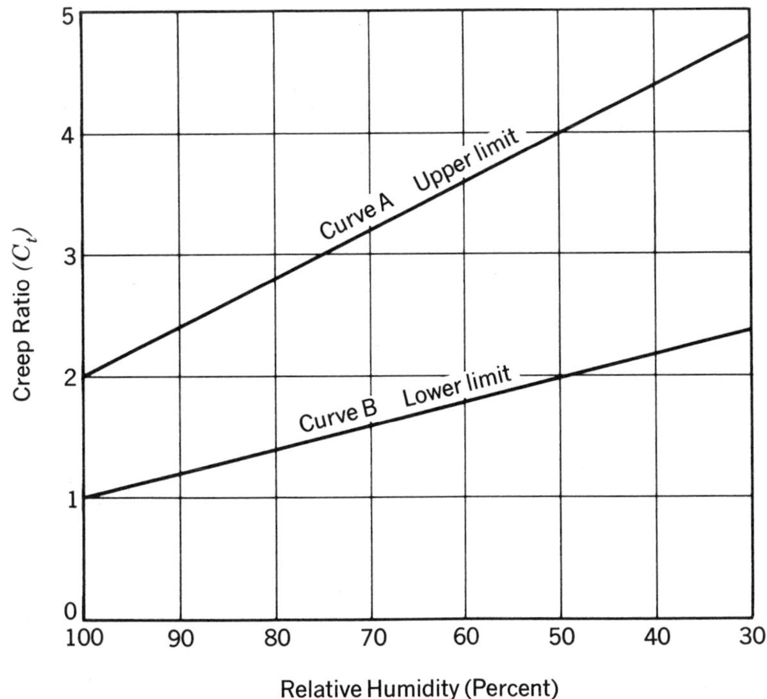

Fig. 3-13 Creep ratio (C_t) vs relative humidity for concrete of ordinary quality.

Humidity during service can be approximated as follows:

Service Conditions	*Approx. Humidity (H)*
In water	100%
Very near water (close to a large body of water)	90%
Near water (valleys with rivers in dense forests)	70%
Normal conditions (relatively dry climate, high mountains)	50%
Enclosed buildings (heated in winter)	35%

Local climatological data for most cities of the United States are available.*

* U.S. Department of Commerce, Weather Bureau, "Local Climatological Data." Available from the Superintendent of Documents, Government Printing Office, Washington, D.C. 20402.

These data include averages and extremes for temperature and relative humidity throughout the year.

The amount of creep strain is also a function of the relative compressive strength of the concrete at the time the concrete is subjected to stress. This factor can be approximated in terms of the 28-day compressive strength, as well as in terms of the age at loading. These approximate relationships are shown in Figs. 3-14(a) and 3-14(b), respectively. Expressed mathematically, Fig. 3-14(a) is as follows:

$$M_c = 1.80 - 1.28\left(\frac{f'_{ci}}{f'_c} - 0.375\right) \qquad (3\text{-}19)$$

in which M_c is the maturity coefficient, f'_{ci} is the concrete strength at the time of stressing, and f'_c is the cylinder strength at the age of 28 days. The relationship of Fig. 3-14(b) can be expressed as:

$$M_c = 1.80 - 0.238\,(\log_e t) \qquad (3\text{-}20)$$

in which t is the age of the concrete in days. It should be recognized that Eqs. 3-19 and 3-20 are applicable to concrete made with type I cement cured at 70°F and 50% relative humidity.

It must be pointed out that wide variations in the amount of creep that is experienced can result from the type of aggregate employed, the design mix

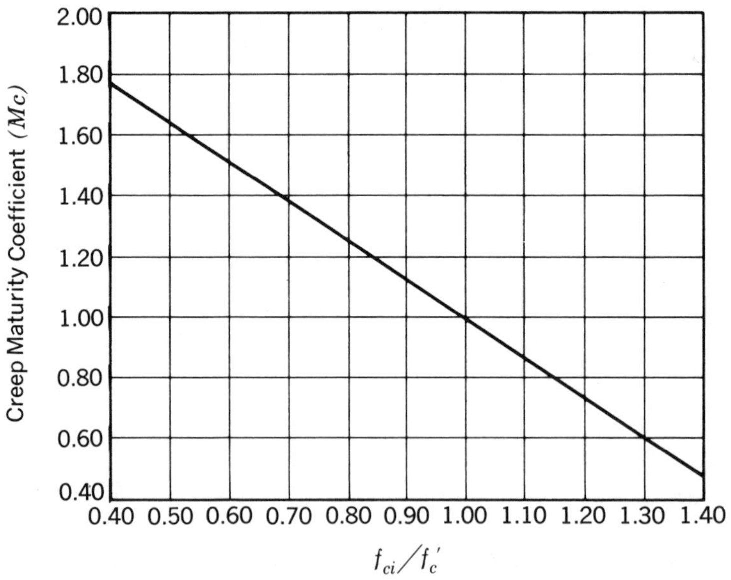

Fig. 3-14 (a) Creep maturity coefficient vs f'_{ci}/f'_c.

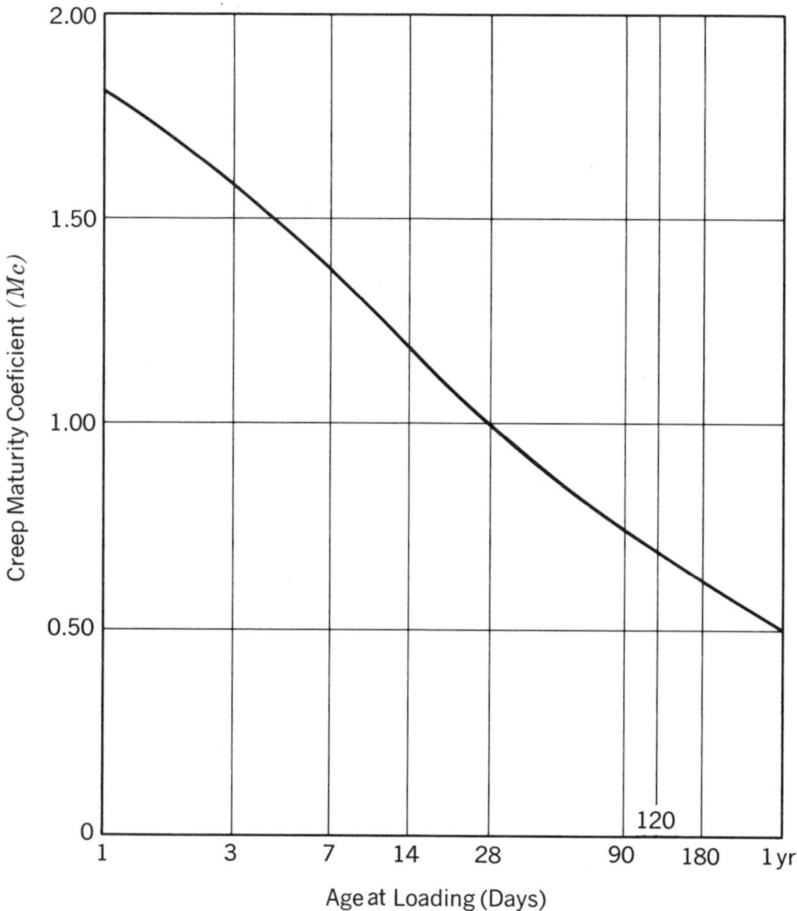

Fig. 3-14 (b) Creep maturity coefficient vs age at loading, in days.

(lower quality concrete would be expected to creep more than the higher quality concrete normally used in prestressed work), and the quality or chemical composition of the cement. The prudent designer will consider these factors in his work.

The total ultimate creep strain can be computed using the above factors as follows:

Ultimate Creep Strain =

$$\begin{bmatrix}\text{Elastic Strain at}\\ \text{time of Stressing}\end{bmatrix}\begin{bmatrix}\text{Creep Ratio}\end{bmatrix}\begin{bmatrix}\text{Creep Maturity}\\ \text{Coefficient}\end{bmatrix}\begin{bmatrix}\text{Creep Size}\\ \text{Coefficient}\end{bmatrix} \qquad (3\text{-}21)$$

There is evidence that steam curing may reduce creep as much as 50% (Refs. 10 and 17). The factors that affect shrinkage of concrete, as described in Sec. 3-9, except as described herein, are considered to affect creep in about the same degree, and the designer should consider them in estimating creep strain.

ILLUSTRATIVE PROBLEM 3-3 Compute the ultimate creep for a natural aggregate concrete (average quality) that contains good quality normal portland cement and has a 28-day strength of 5000 psi and is stressed when the strength is 3500 psi. Mean humidity during service is anticipated to be 35% and average stress is 900 psi. Assume the concrete has the properties described by Curve *A* of Fig. 3-12 and the volume-to-surface ratio is 5.

SOLUTION:

$$f'_{ci} = 3500 \text{ psi} \; f'_c = 5000 \text{ psi} \; \frac{f'_{ci}}{f'_c} = 0.70$$

$$\text{Creep Ratio} = 1.25 + 2.75\,\frac{100 - 35}{65} = 4.00$$

$$\text{Maturity Ratio} = 1.80 - 1.28(0.70 - 0.375) = 1.39$$

$$E_{ci} = 57{,}000\sqrt{3500} = 3.37 \times 10^6 \text{ psi}$$

The creep size coefficient is 1.15 (from Fig. 3-11).

$$\text{Elastic Strain} = \frac{900}{3.37 \times 10^6} = 267 \times 10^{-6} \text{ in./in.}$$

$$\text{Ultimate Creep Strain} = 267 \times 10^{-6} \times 4.00 \times 1.39 \times 1.15 = 1707 \times 10^{-6} \text{ in./in.}$$

3-12 Relaxation of Concrete

When concrete is subjected to a constant strain, as is the case when a concrete member is stressed by jacking and anchoring (shimming) against rigid abutments, a loss of stress takes place with the passage of time. The stress relaxation of concrete is affected by the factors that affect the creep of concrete, in about the same manner.

The rate of stress loss can be estimated from the curve shown in Fig 3-15 (Ref. 16). Because the loss of stress due to relaxation can be very large, in comparison with the final stress, the design of structures that will be affected by this phenomenon must be done with great care. Provision should normally

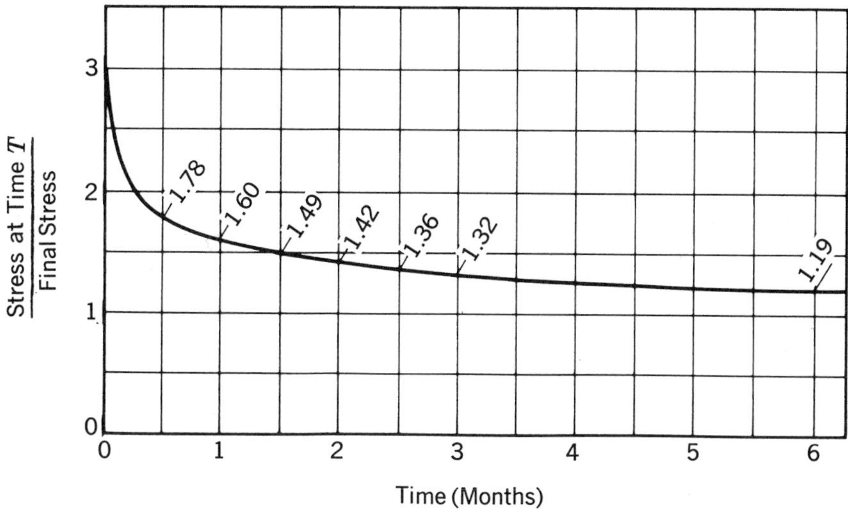

Fig. 3-15 Stress loss curve.

be made for checking the stress retained in the structure, from time to time, and adjusting it when necessary. The relaxation properties of the concrete to be used in important structures should be determined for the conditions of service by experimental means when stressing by constant strain is contemplated.

3-13 Low-Pressure (Atmospheric Pressure) Steam Curing

In the manufacture of structural concrete products, it is often desirable or necessary to accelerate the early hydration of the cement in the products in order that a rapid re-use can be made of the manufacturing facilities. In the case of precast reinforced concrete, it may be necessary to obtain a concrete strength of 1000 to 2000 psi at the age of 24 hr or less so that the products can be safely stripped and moved to storage for further curing and the forms can be re-used. Concrete strengths from 3000 to 4500 psi are required in the manufacture of pre-tensioned concrete before the pre-tensioning tendons can be released, the products removed from the prestressing bench, and the bench re-used. Since forms for structural concrete and pre-tensioning benches represent rather large capital investments that are tied up while the concrete gains these required strengths, it is apparent that a reasonable expenditure for accelerating the hydration of the cement can be justified if the time required for the concrete to attain the adequate strength can be sufficiently reduced.

Low-pressure or atmospheric-pressure steam curing, which is referred to simply as steam curing in this book, is often employed to accelerate the curing of concrete. Well executed steam curing can result in more than 60% of the standard cured 28-day compressive strength being attained in 24 hr. This method consists of confining the concrete products in hot, nearly saturated air at atmospheric pressure by isolating the products from the normal atmosphere in an enclosure into which steam is injected.

Another process that employs steam at elevated pressure and is referred to as high-pressure steam curing (or as autoclaving) is more effective than low-pressure steam curing and is being used rather extensively in concrete-block-manufacturing plants. High-pressure steam curing requires that the concrete products be placed in a steel pressure vessel in which the pressure can be increased above atmospheric. Because of this, this method is considered impracticable for large, structural concrete products, particularly those that are made on pre-tensioning benches.

Hot water and hot oil are being used in some applications, in which case, the hot fluid is pumped through longitudinal cavities in the forms or through pipes in or on the casting beds, thus heating the concrete products. This method can give results similar to those obtained by steaming, if the products are not allowed to dry out, by keeping all exposed surfaces moist during heating.

Chemical admixtures designed to accelerate the hydration are available; however, many of these contain calcium chloride, which cannot be safety used in prestressed concrete, due to the danger of the chloride ion causing corrosion of the high-strength steel. Furthermore, admixtures in themselves generally do not accelerate the hardening of the concrete sufficiently. In the manufacture of precast structural concrete, the concrete is frequently made with high early-strength cement. An admixture that accelerates the set is used and the resulting concrete is cured with low-pressure steam, as described above.

It is generally agreed that the optimum curing cycle to be used in employing steam to increase the early strength of concrete is influenced by the following considerations (Refs. 1 and 17):

(1) *Delay period:* After placing and vibrating the concrete, the concrete must be allowed to attain its initial set before the steam is applied.
(2) *Rate of increasing the concrete temperature:* The temperature of the atmosphere surrounding the concrete and hence the concrete temperature is increased at a specific rate to a maximum temperature.
(3) *Duration of maximum temperature:* The maximum temperature is generally maintained for a specific period of time.
(4) *Rate of cooling:* The temperatures of the concrete and the atmosphere surrounding it are reduced slowly.

The normal American procedure is to employ a delay period of 2 to 6 hr, depending upon the type of cement being used. The longer delay periods are used with the slower setting cements and when higher maximum temperatures are used. Specifications on steam curing usually require that the temperature be increased at a rate less than 1°F per minute to the maximum temperature, which is not greater than 165°F. At this rate the maximum temperature will be reached 1 or 2 hr after steaming is commenced, depending upon the ambient temperature. Most steam-curing facilities do not permit the temperature to be increased at a precise rate, and, as a result, the temperature is usually increased in a few small increments over a period of time. The maximum temperature is maintained between 140°F and 165°F for a period that varies from 10 to 20 hr. The products are then allowed to cool more or less slowly, depending upon the practice at the particular plant.

The procedure described above, which is the general practice in the manufacture of structural concrete products as well as concrete block and pipe, is based upon research that is reviewed in a very brief discussion of steam

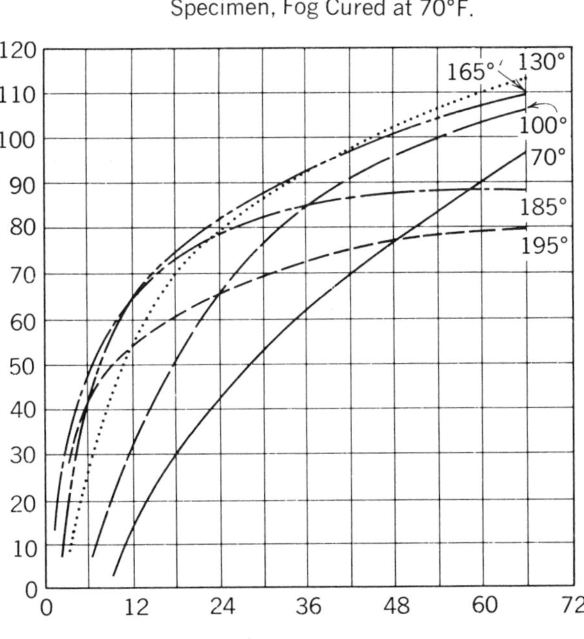

Fig. 3-16 Effect of early steam curing at various temperatures. Steam curing was started immediately after specimens were cast. Compressive strength at 3 days of specimen fog cured at 70°F was 2000 psi. The mix was composed of 1.37 bbl of type II cement per cubic yard and the water-cement ratio was 0.55 (Source: *Concrete Manual*, U. S. Bureau of Reclamation.)

curing in the *Concrete Manual* published by the Bureau of Reclamation. In this discussion it is stated, "A delay of 2 to 6 hr prior to steam curing will result in higher strength at 24 hr than would be obtained if steam curing were commenced immediately after filling of the forms as was the case in the tests from which the data plotted in Fig. 140 were derived." The Fig. 140 referred to is reproduced here as Fig. 3-16, in which it is seen that these tests would lead the casual reader to the conclusion that temperatures of from 100 to 165°F will not adversely affect the 28-day strengths, whereas temperatures of 185 and 195°F seem to give 3-day strengths that are very adversely affected by the increase of the temperature above 165°F. The large adverse effect on the 3-day strength illustrated in the figure is the result of the steam being applied without the desired delay period. More recent research has revealed that the higher temperatures can be used to advantage if the delay period is adequate, without any adverse effects upon the 28-day cylinder strengths (Ref. 18).

It is recommended that tests be performed as a means of determining the optimum curing cycle that should be used under specific conditions. By employing the trial and error method, one can determine the delay period, maximum temperature, and time required at maximum temperature that will yield optimum results.

3-14 Cold Weather Concrete

Frequently, in fabricating precast, prestressed members for state or federal projects during the winter months, manufacturers are required to conform to standard specifications that were originally written for job-site winter concrete. Such specifications frequently provide that concrete cannot be placed when the temperature of the ambient air reaches a particular minimum value and that the aggregates and mixing water shall be heated in such a manner that the plastic concrete does not go below a specific value. Although these specifications may be necessary for concrete that is to be placed and allowed to cure without having the surrounding air artificially heated, such specifications frequently have a deterimental effect on the concrete used in plant-produced products that are to be steam cured. There is no question that the aggregates used in precasting plants should be kept sufficiently warm to prevent ice or frost from being in the plastic concrete and that the plastic concrete should not be allowed to freeze. However, it is known that higher concrete strengths are obtained for concretes mixed and placed at lower temperatures than for concretes that are mixed and placed at higher temperatures. This is attributed to the fact that cool plastic concrete mixes require less water for workability than do warmer mixes. The use of very-hot mixing water can have a very serious detrimental effect on the strength of concrete, and in particular, on the early strength.

3-15 Allowable Concrete Flexural Stresses

The two most significant design criteria for prestressed concrete in the United States are the "Standard Specifications for Highway Bridges," which is published by The American Association of State Highway Officials (AASHO) and "Building Code Requirements for Reinforced Concrete" (ACI 318) published by the American Concrete Institute.

The concrete stresses permitted in the tenth edition of the AASHO Specification (1969) are:

(1) Temporary stresses before losses due to creep and shrinkage:

Compression

Pretensioned members	$0.60 f'_{ci}$
Post-tensioned members	$0.55 f'_{ci}$

Tension

Single elements without auxiliary nonprestressed reinforcement in the tension zone	$3\sqrt{f'_{ci}}$
Segmental elements without auxiliary nonprestressed reinforcement in the tension zone	zero
Where the calculated tension stress exceeds these values, auxiliary nonprestressed reinforcement shall be provided to resist the total tension force in the concrete computed on the assumption of an uncracked section. The maximum tensile stress shall not exceed	$7.5\sqrt{f'_{ci}}$

(2) Stress at design load after losses have occurred:

Compression	$0.40 f'_c$

Tension

In zones initially precompressed with prestressed reinforcement or, in zones with nonprestressed reinforcement that is sufficient to resist the total tension force in the concrete computed on the assumption of an uncracked section	$3\sqrt{f'_c}$ (but not to exceed 250 psi)
In zones without reinforcement	zero

(3) Cracking stress:

Modulus of rupture from tests, or if not available	$7.5\sqrt{f'_c}$

(4) Anchorage bearing stress:
Post-tensioned anchorage (but not to exceed f'_{ci}) $0.6 f'_{ci} \sqrt[3]{A_c/A_b}$

In the above, A_c is the maximum area of the portion of the anchorage surface that is geometrically similar to and concentric with the area of the bearing plate of post-tensioning steel and A_b is the area of the bearing plate. The other notation is standard.

The concrete stresses permitted in ACI 318-71 are:

(1) Flexural stresses immediately after transfer, before losses, shall not exceed the following:
(a) Compression $0.60 f'_{ci}$
(b) Tension stresses in members without bonded auxiliary reinforcement (unprestressed or prestressed) in the tension zone $3\sqrt{f'_{ci}}$

Where the calculated tension stress exceeds this value, reinforcement shall be provided to resist the total tension force in the concrete computed on the assumption of an uncracked section.

(2) Stresses at service loads, after allowance for all prestress losses, shall not exceed the following:

(a) Compression $0.45 f'_c$
(b) Tension in precompressed tensile zone $6\sqrt{f'_c}$
(c) Tension in precompressed tensile zone in members where computations based on the transformed cracked section and on bilinear moment deflection relationships show that immediate and long-term deflections do not exceed the allowable values. $12\sqrt{f'_c}$

(3) The permissible stresses may be exceeded when it is shown experimentally or analytically that performance will not be impaired.

Comparison of these allowable stresses will reveal the requirements of AASHO are much more conservative. This is reasonable because bridge structures are subjected to more severe conditions of service (i.e., fatigue, weather, temporary overloads, etc.).

REFERENCES

1. Troxell, G. E., Davis, H. E. and Kelley, J. W., "Composition and Properties of Concrete." McGraw-Hill Book Company, Inc., New York, 1956.
2. "Concrete Manual,." U.S. Bureau of Reclamation, 1952.
3. Guyon, op. cit. pp. 53–66.
4. Glanville, W. H. "Studies in Reinforced Concrete III—The Creep or Flow Concrete Under Load." Tech. Paper 12, Department of Scientific and Industrial Research, Building Research (England).
5. Szilard, Rudolph. "Corrosion and Corrosion Protection of Tendons in Prestressed Concrete Bridges." *Journal of the American Concrete Institute, Proceedings*, **66**, No. 1 (Jan. 1969).
6. Protopopescu, Dan. Private communication.
7. Scordelis, Branson and Sozen. "Deflection of Prestressed Concrete Members." *Journal of the ACI, Proceedings*, **60**, No. 12 (Dec. 1963).
8. Troxell, Raphael and Davis. "Long-Time Creep and Shrinkage Tests of Plain and Reinforced Concrete." *A.S.T.M. Proceedings*, **59** (1958).
9. Carlson, Roy W. "Drying Shrinkage of Concrete as Affected by Many Factors." *A.S.T.M. Proceedings*, **38**, Part (1938).
10. Klieger, Paul. "Some Aspects of Durability and Volume Change of Concrete for Prestressing." Portland Cement Association, Research Department Bulletin 118 (Nov. 1960).
11. Hanson, J. A. "Prestress Loss As Affected by Type of Curing." *Journal of the Prestressed Concrete Institute*, 9, No. 2, 69–93 (Apr. 1964).
12. "Lightweight Aggregate Concrete." Housing and Home Finance Agency, Washington, D.C. (Aug. 1949).
13. Furr, H. L. and Sinno, R. "Creep in Prestressed Lightweight Concrete." Research Report No. 69–2, Texas Transportation Institute, Texas A. & M. University, College Station, Texas.
14. Leonhardt, Fritz. "Prestressed Concrete Design and Construction." Wilhelm Ernst and Sohn, Berlin, 1964.
15. Ivey, D. L. and Hirsch, T. J. "Effects of Chemical Admixtures in Concrete and Mortar." Research Report 70–3, Texas Transportation Institute, Texas A. & M. University, College Station, Texas.
16. Guyon, Y. "Prestressed Concrete," p. 63, John Wiley & Sons, Inc., New York, 1953.
17. "Recommended Practice for Atmospheric Pressure Steam Curing of Concrete." Reported by ACI Committee 517, *Journal of the American Concrete Institute*, **66**, No. 8 (Aug. 1969).

18. Saul, A. G. A. "Principles Underlying the Steam Curing of Concrete at Atmospheric Pressure." *Magazine of Concrete Research*, **2**, 127 (Mar. 1951).

19. Plowman, J. M. "Maturity and the Strength of Concrete." *Magazine of Concrete Research*, **8**, No. 22, 13 (Mar. 1956).

20. Schorer, H. "Prestressed Concrete, Design Principals and Reinforcing Units." *ACI Journal*, **39**, No. 4, 493–528 (July 1943).

21. Ross, A. D. "Creep of Concrete Under Variable Stress." *ACI Journal Proceedings*, **29**, No. 9, 739–758 (Mar. 1958).

22. Fintel, Mark and Khan, Fazlur R. "Effects of Column Creep and Shrinkage in Tall Structures—Prediction of Inelastic Column Shortening." *Journal of the American Concrete Institute*, **66**, No. 12, 957–67 (Dec. 1969).

4 Basic Principles of Flexural Design

4-1 Introduction

The basic principles and mathematical relationships used in the design and analysis of prestressed-concrete flexural members are not unique to this type of construction. Virtually all of the fundamental relationships are based upon the normal assumptions of elastic design, which forms the basis of the study of the strength of materials. Although the form in which the relationships appear in a discussion of prestressed concrete may be somewhat modified to facilitate their application, the student of engineering should have little difficulty in understanding these modified relationships.

Two major forms of design problems are encountered by the engineer engaged in the design of prestressed concrete flexural members. These are frequently referred to as the *review* of a member and as the *design* of a member.

The review of a member actually consists of the determination of the concrete flexural stresses, under all conditions of loading and prestressing, in order to confirm the compliance of these stresses with the applicable design criteria. In addition, the review of a member must include a study of the ultimate moment that the section can be expected to develop (this is done to ensure adequate safety against a flexural failure). An investigation of the shear stresses must be made and the adequacy of the web reinforcing that is specified for the member must be confirmed. It should be apparent that in order to review a member as described here, the dimensions of the concrete section, the properties of the materials, the amount and eccentricity of the prestressing steel, the amount of non-prestressed reinforcing, as well as the amount of web reinforcing, must be known.

The design of a member consists of selecting and proportioning a concrete section in which the stresses in the concrete do not exceed the permissible values under any condition of loading or prestressing. Design also includes the determination of the amount and eccentricity of the prestressing force that is required for the specific section. The design of a member must include a study of the moment that the section can develop at ultimate load, and the determination of the amount of non-prestressed reinforcing that may be required. Additionally, a study of the shear stresses must be made and the amount of web reinforcing that may be required for adequate shear strength at ultimate load must be determined. It must be emphasized that the design of a flexural member is normally a trial and error procedure. The designer must assume a concrete section and compute the prestressing force and eccentricity that are required to confine the concrete stresses within the allowable limits under all loading conditions. In the design of a member, several adjustments of the trial section are normally required before a satisfactory solution is found.

This chapter is devoted to the consideration of the fundamental principles pertaining to the determination of the concrete stresses due to prestressing, the determination of the prestressing force and eccentricity required for a specific distribution of stresses due to prestressing, the consideration of the pressure line in simple flexural members that are loaded in the elastic range, and other topics related to flexural analysis and design. The problems given in this chapter are confined to the "review" type. The procedures used in preparing preliminary designs by the trial and error procedure are treated in Sec. 7-8.

The elastic analysis and design of prestressed flexural members can be done rapidly and accurately only after the fundamental theorems and axioms have been thoroughly mastered. Many of the operations discussed in this chapter can be done more rapidly by the use of the simple expedients treated in Chapter 7. These "classical" methods should be well understood, however, before attempting the use of the expedients. The design and analysis of continuous prestressed members, which are treated in Chapter 8, also require complete familiarity with the principles presented in this chapter.

4-2 Mathematical Relationships for Prestressing Stresses

The stresses due to prestressing alone are generally combined stresses due to a direct load eccentrically applied. Therefore, these stresses are computed using the following well-known relationship for combined stresses

$$f = \frac{P}{A} \pm \frac{My}{I} \tag{4-1}$$

in which f is the fiber stress at the distance y from the centroidal axis, P is the axial force, A is the area of the cross section, M is the moment acting on the section, and I is the moment of inertia of the cross section. In this book compressive stresses are taken to be positive.

Since the moment due to the prestressing is equal to the prestressing force multiplied by the eccentricity of this force (i.e., $M = Pe$), and since the square of the radius of gyration is equal to the moment of inertia divided by the area of the cross section ($r^2 = I/A$), the above relationship can be rewritten

$$f = \frac{P}{A}\left(1 \pm \frac{ey}{r^2}\right) \tag{4-2}$$

Using y_t and y_b to denote the distances from the centroidal axis to the top and bottom fibers, respectively, and by assuming the eccentricity to be positive when it is on the same side of the center of gravity as the fiber under consideration, the top and bottom fiber stresses for a prestressing force applied eccentrically below the center of gravity are expressed by

$$f_t = \frac{P}{A}\left(1 - \frac{ey_t}{r^2}\right) \tag{4-3}$$

$$f_b = \frac{P}{A}\left(1 + \frac{ey_b}{r^2}\right) \tag{4-4}$$

where f_t and f_b are the stresses in the top and bottom fibers due to the prestressing alone, respectively. A positive stress is compressive in the above relationships.

These relationships are the same for the stresses resulting from the initial and the final prestressing forces (*see* Sec. 1-2). In computing these stresses, one would of course use the initial prestressing force when computing the initial stresses and the final force when computing the final stresses. Frequently, the designer assumes a ratio between the final and the initial prestressing forces for design purposes, since the relaxation of the prestressing force cannot be accurately estimated until the design is nearly complete (*see*

Sec. 6-2). Therefore, if the designer bases his computation on the final prestressing force and has assumed that the total relaxation will be 15% of the initial force, for example, the stresses resulting from the initial prestressing force can be determined by dividing the final stresses by 0.85.

The experienced designer generally prefers to design with the final prestressing force assumed to be from 75% to 90% of the initial force. A comprehensive study of the losses of stress cannot be made until the basic design is finalized. If, when this study is made, it is found the loss will be greater than assumed, the initial prestressing force can be increased so that the final force will be satisfactory. The advantage of this procedure will be apparent after consideration of the data presented in Chapter 7.

ILLUSTRATIVE PROBLEM 4-1 Compute the stresses due to prestressing alone in a beam with a rectangular cross section 10 in. wide and 12 in. high that is prestressed by a final force of 120 k at an eccentricity of 2.5 in. State whether the stresses are compressive or tensile. Compute the stresses due to the initial prestressing force, if the ratio between the final force and the initial force is 0.85.

SOLUTION:

$$A = 120 \text{ sq in. } I = \frac{10 \times 12^3}{12} = 1440 \text{ in.}^4 \; r^2 = \frac{1440}{120} = 12 \text{ sq in.}$$

Final stresses:

$$f_t = \frac{120{,}000}{120}\left(1 - \frac{2.5 \times 6}{12}\right) = -250 \text{ psi (tension)}$$

$$f_b = \frac{120{,}000}{120}\left(1 + \frac{2.5 \times 6}{12}\right) = +2250 \text{ psi (compression)}$$

Initial stresses:

$$f_t = \frac{-250}{0.85} = -294 \text{ psi (tension)}$$

$$f_b = \frac{+2250}{0.85} = +2650 \text{ psi compression)}$$

ILLUSTRATIVE PROBLEM 4-2 Compute the prestressing force and eccentricity that would be necessary in the beam of Prob. 4-1, in order to obtain a bottom-fiber compression of 2400 psi and a top-fiber tension of 350 psi, by equating the relationships for stresses due to prestressing in the top and bottom fibers.

SOLUTION:

$$\frac{P}{A}\left(1 - \frac{ey_t}{r^2}\right) = -350 = \frac{P}{120}\left(1 - \frac{e \times 6}{12}\right)$$

$$\frac{P}{A}\left(1 + \frac{ey_b}{r^2}\right) = +2400 = \frac{P}{120}\left(1 + \frac{e \times 6}{12}\right)$$

$$2400 - \frac{2400\,e \times 6}{12} = -350 - \frac{350\,e \times 6}{12}$$

$$1025\,e = 2750$$

$$e = 2.68 \text{ in.}$$

$$P = \frac{120 \times 2400}{1 + \dfrac{2.68 \times 6}{12}} = 123{,}000 \text{ lb}$$

The familiar principle of superposition is used to determine the combined effect of the prestressing and the other loads that may be acting simultaneously on a prestressed beam. Although it is possible to write a single equation that will accurately define the stress at any particular point in a beam, it is normally less confusing if the effect of each load (or prestressing) is computed separately and the net effect is determined by algebraically adding the effects of the several loads.

ILLUSTRATIVE PROBLEM 4-3 Compute the net initial and final concrete stresses in the extreme top and bottom fibers at the center line of a beam that is 10 in. wide, 12 in. deep and on a span of 25 ft. The beam is to support an intermittent, uniformly distributed live load of 0.45 k/ft and is to be prestressed with a final force of 120 k positioned with an eccentricity of 2.50 in. The ratio between the final and initial prestressing forces is assumed to be 0.85.

SOLUTION:

$$A = 120 \text{ in.}^2 \qquad I = \frac{10 \times 12^3}{12} = 1440 \text{ in.}^4$$

$$S = \frac{1440}{6} = 240 \text{ in.}^3 \qquad r^2 = \frac{I}{A} = \frac{1440}{120} = 12 \text{ in.}^2$$

Stresses due to final prestress:

$$f_t = \frac{120{,}000}{120}\left(1 - \frac{2.5 \times 6}{12.0}\right) = -250 \text{ psi (tension)}$$

$$f_b = \frac{120{,}000}{120}\left(1 + \frac{2.5 \times 6}{12.0}\right) = +2250 \text{ psi (compression)}$$

Stresses due to initial prestress:

$$f_t = \frac{-250 \text{ psi}}{0.85} = -294 \text{ psi (tension)}$$

$$f_b = \frac{+2250}{0.85} = +2650 \text{ psi (compression)}$$

Stresses due to the dead load of the beam alone:

$$w_{DL} = \frac{120}{144} \times 0.150 = 0.125 \text{ k/ft}, \qquad M_{DL} = 0.125 \times \frac{(25)^2}{8} = 9.78 \text{ k-ft}$$

$$f_t = +\frac{9.78 \times 12{,}000}{240} = +488 \text{ psi (compression)}$$

$$f_b = \frac{9.78 \times 12{,}000}{240} = -488 \text{ psi (tension)}$$

Stresses due to live load alone:

$$w_{LL} = 0.45 \text{ k/ft}, \qquad M_{LL} = 0.45 \times \frac{(25)^2}{8} = 35.2 \text{ k-ft}$$

$$f_t = \frac{+35.2 \times 12{,}000}{240} = +1760 \text{ psi (compression)}$$

$$f_b = \frac{-35.2 \times 12{,}000}{240} = -1760 \text{ psi (tension)}$$

Combined stresses:

	Top Fiber	*Bottom Fiber*
Initial prestress	− 294 psi	+2650 psi
Beam dead load	+ 488 psi	− 488 psi
Initial prestress plus dead load	+ 194 psi	+2162 psi
Live load	+1760 psi	−1760 psi
Initial prestress plus total load	+1954 psi	+ 402 psi
Final prestress	− 250 psi	+2250 psi
Dead load of beam	+ 488 psi	− 488 psi
Final prestress plus dead load	+ 238 psi	+1762 psi
Live load	+1760 psi	−1760 psi
Final prestress plus total load	+1998 psi	+ 2 psi

4-3 Pressure Line in a Beam with a Straight Tendon

At any section of a beam, the combined effect of the prestressing force and the externally applied load will result in a distribution of concrete stresses that can be resolved into a single force. The locus of the points of application of this force in any beam or structure is called the pressure line.

This can be illustrated by considering a rectangular beam prestressed by an eccentric, straight tendon, as is shown in Fig. 4-1. Such a beam would have a distribution of stresses due to prestressing alone at every cross section, as is shown in Fig. 4-2(a). It is readily seen that the force resulting from the distribution of internal prestressing stresses (C) is equal in magnitude to the prestressing force. In addition, it is applied at the same point as the prestressing force at every section, since the prestressing force and the eccentricity are both constant throughout the length of the beam.

If a uniform load of such magnitude, that results in the bottom-fiber prestress being nullified at midspan, is applied to the beam, the resulting stress distribution would be as indicated in Fig. 4-2(b) and the pressure line at this point would then be applied at a point $d/6$ above the centroidal axis of the beam. At the quarter point of this beam, under the same loading conditions,

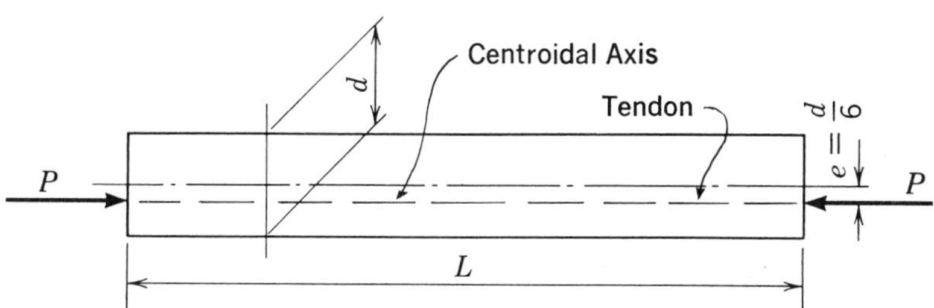

Fig. 4-1 Simple rectangular beam prestressed by an eccentric straight tendon.

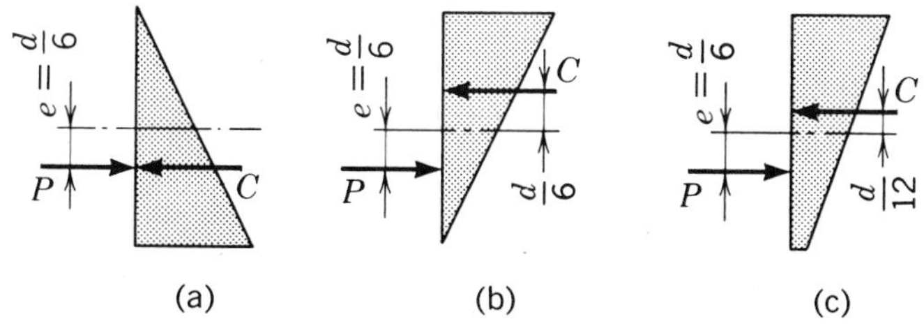

Fig. 4-2 Stress distributions and pressure-line locations for a simple rectangular beam prestressed with a straight eccentric tendon (a) due to prestressing alone, (b) at midspan under full design load, and (c) at quarter point under full design load.

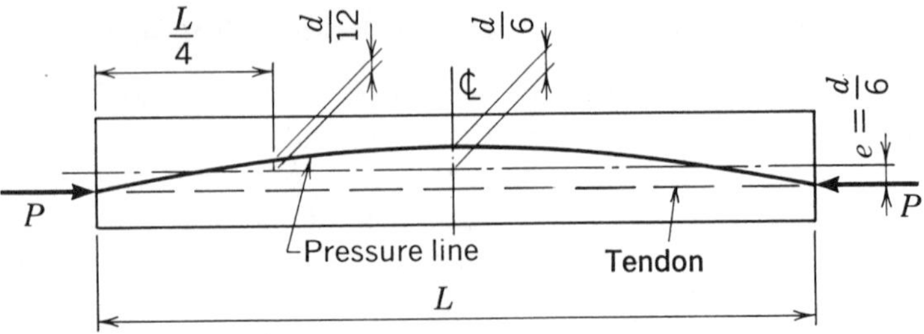

Fig. 4-3 Location of pressure line in a simple beam of rectangular cross section, prestressed by a force at $e = d/6$ and under uniform load resulting in zero bottom-fiber stress at the midspan.

the stresses due to the external load are only 75% as much as those at midspan. The stress distribution resulting from the combination of prestressing and the flexural stresses due to the external load would be as shown in Fig. 4-2(c). At this point the pressure line is located at a distance of $d/12$ above the centroidal axis. At the support, since there are no flexural stresses resulting from the external load, the pressure line remains at the level of the steel. Plotting the location of the pressure line for this loading reveals that it is a parabola with its vertex at the center of the beam, as shown in Fig. 4-3.

In a similar manner, it can be shown that a larger uniform load would result in the pressure line being moved up even higher, and for a uniform load applied upward rather than downward, the result would be a downward movement of the pressure line. Therefore, it is apparent that the location of the pressure line in simple prestressed beams is dependent upon the magnitude and direction of the moments applied at any cross section and the magnitude and distribution of stress due to prestressing: *A change in the external moments in the elastic range of a prestressed beam results in a shift of the pressure line rather than an increase in the resultant force in the beam as, is the case in beams composed of other materials.*

Due to the change in the strain in the concrete at the level of the steel (assuming the flexural bond strength between the steel and concrete is adequate, as it is in pre-tensioned and bonded post-tensioned beams), there is an increase in the stress in the prestressing steel as a result of applying an external load. This is rarely of importance and the effect is normally disregarded (*see* Sec. 4-11).

ILLUSTRATIVE PROBLEM 4-4 Compute and draw to scale the location of the pressure line for a rectangular beam 10 in. wide and 12 in. deep that is prestressed with a force of 120 k at a constant eccentricity of 2.5 in. and that is supporting a 15 k concentrated force at midspan of a span of 10 ft. Use an exaggerated vertical scale in a sketch, and dimension the location of the pressure line at the midspan, quarter point, and end of the beam. Neglect the dead weight of the beam.

$$A = 120 \text{ sq in.} \qquad \frac{r^2}{y} = \frac{I}{A_y} = 2 \text{ in.} \qquad S = \frac{I}{y} = 240 \text{ in.}^3$$

SOLUTION:

Stresses due to prestressing:

$$f_t = \frac{120{,}000}{120}\left(1 - \frac{2.5 \text{ in.}}{2.0 \text{ in.}}\right) = -250 \text{ psi}$$

$$f_b = \frac{120{,}000}{120}\left(1 + \frac{2.5 \text{ in.}}{2.0 \text{ in.}}\right) = +2250 \text{ psi}$$

At the end of the beam, moment = zero. Therefore, the pressure line is at $e = 2.50$ in.

At the midspan:

$$M = \frac{PL}{4} = \frac{15 \text{ k} \times 10 \text{ ft}}{4} = 37.5 \text{ k ft}$$

$$f = \pm\frac{37.5 \times 12{,}000}{240} = \pm 1880 \text{ psi}$$

Stress distribution at midspan:

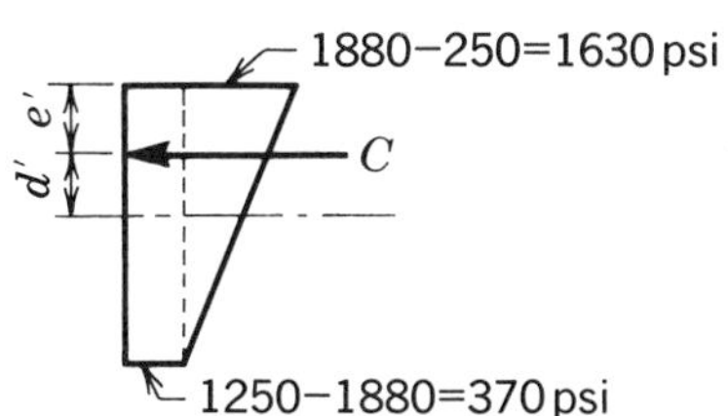

$$d' = \frac{6 \times 370 \times 120 + (1260/2) \times 120 \times 4}{370 \times 120 + (1260/2) \times 120}$$

$$d' = 4.73 \text{ in.}$$

$$e' = 6.00 \text{ in.} - 4.73 \text{ in.} = 1.27 \text{ in.}$$

At the quarter point, the moment due to the external load is only one-half that at the midspan. Therefore, the flexural stresses due to the applied load are only one-half of those at the center line, or ± 940 psi.

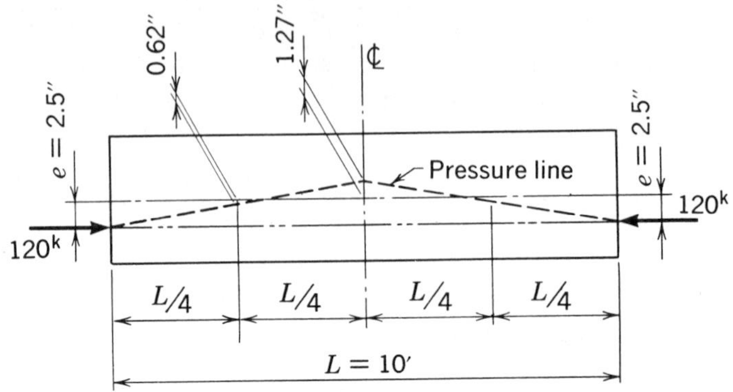

Fig. 4-4 Location of pressure line, Prob. 4-4.

Stress distribution at quarter point:

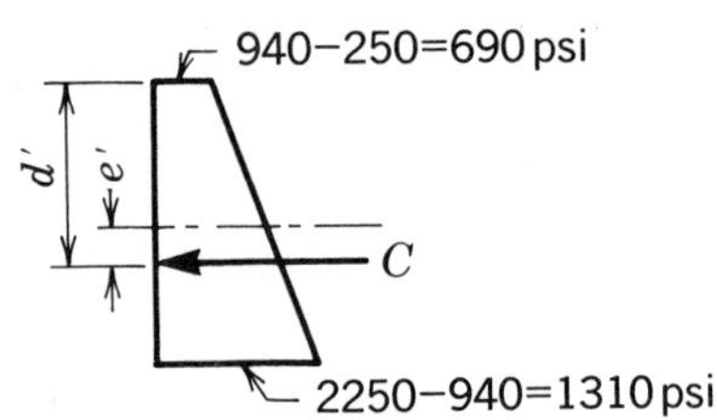

$$d' = \frac{6 \times 690 \times 120 + (620/2) \times 120 \times 8}{690 \times 120 + (620/2) \times 120} = 6.62 \text{ in.}$$

$$e' = 6.62 \text{ in.} - 6.00 \text{ in.} = 0.62 \text{ in. below the centroid.}$$

The results are shown plotted in Fig. 4-4.

4-4 Variation in Pressure-Line Location

If tensile stresses are not permitted in the bottom fibers of a simple prestressed concrete beam, when it is subjected to service loads, the distribution of stresses will be as shown in Fig. 4-5. Also shown in Fig. 4-5 is the cross section of the beam. The force C is the resultant of the stresses in the concrete (pressure line) and it obviously must be equal in magnitude and opposite in direction to the prestressing force P, since the horizontal forces acting on the cross section must be in equilibrium. In addition, from the relationship for

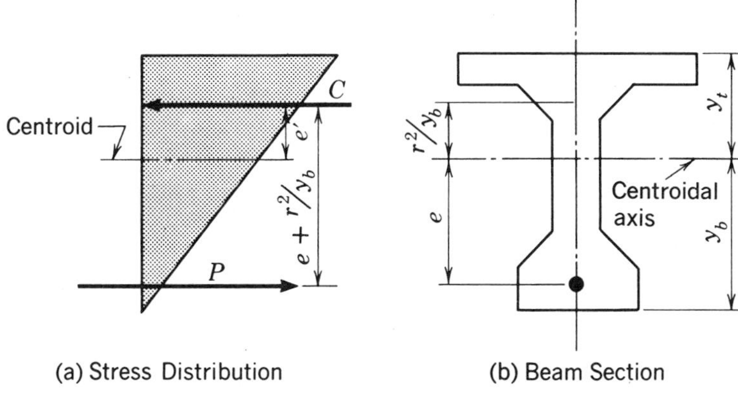

Fig. 4-5 Relationship between prestressing force, pressure line, and section properties of a beam having zero stress in bottom fiber under design load.

combined stresses developed in Sec. 4-2, we can write the relationship for the stress in the bottom fibers as follows:

$$\frac{C}{A}\left(1 - \frac{e' y_b}{r^2}\right) = 0$$

from which we obtain

$$e' = r^2/y_b$$

The eccentricity e' of the resultant C should not be confused with the eccentricity of the prestressing force.

Another requirement of equilibrium is that the internal and external moments are equal in magnitude and opposite in direction at every section. It follows, then, that the total external moment that the beam is resisting at this section is numerically equal to

$$M_T = M_{DL} + M_{LL} = C(e + r^2/y_b) = P(e + r^2/y_b) \tag{4-5}$$

in which e is the normal eccentricity of the prestressing force.

The above example further illustrates that prestressed beams, functioning in the elastic range, resist the moment due to externally applied loads by the movement of the resultant of the stresses in the concrete, rather than by an increase in the prestressing stress, as was brought out in Sec. 4-3. From Eq. 4-5, it is apparent that if $M_T = 0$, the product of C multiplied by the quantity $(e + r^2/y_b)$ must also be equal to zero and the concrete stresses would be distributed as shown in Fig. 4-6. If the external moment (M_T) were some value less than that which nullifies the precompression of the

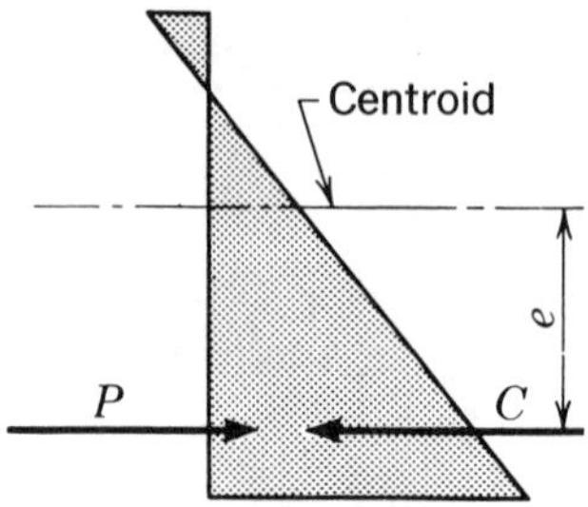

Fig. 4-6 Distribution of stress and location of C when external moment $= 0$ (prestress alone).

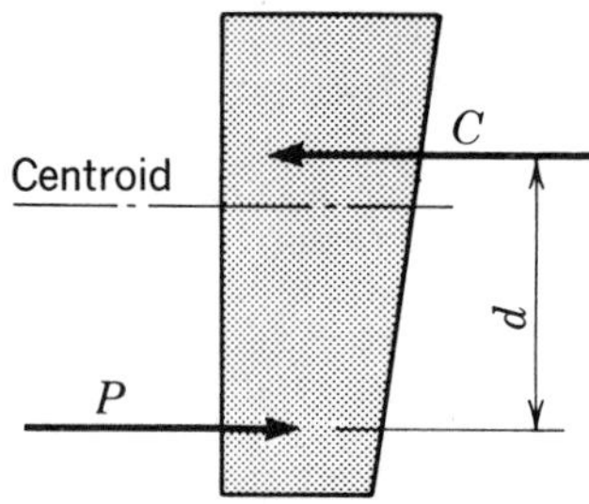

Fig. 4-7 Distribution of stress and location of resultant C when external moment is of nominal magnitude.

bottom fibers, the force C would be applied above the location of the prestressing steel at a distance (d) equal to

$$d = \frac{M_T}{P} \tag{4-6}$$

This condition is illustrated in Fig. 4-7.

The relationship given by Eq. 4-5 is extremely useful in the preliminary design of beams as well as in checking the final design. Since the value of $(e + r^2/y_b)$ is normally of the order of 65% of the depth of the beam section (it varies between the approximate limits of 33 to 80% for different cross sections) for a given superimposed moment, the designer can assume a dead weight for the beam and estimate the prestressing force required for different depths of construction. The use of this relationship is illustrated in Prob. 4-5 and in Sec. 7-8, in the discussion of preliminary design.

ILLUSTRATIVE PROBLEM 4-5 Compute the maximum concentrated load that can be applied at the midspan of a beam that is 10 in. wide, 12 in. deep, prestressed with 120 k at an eccentricity of 2.5 in., and is to be used on a span of 10.0 ft center to center of bearings without tensile stresses resulting in the bottom fibers.

SOLUTION: (Using the basic relationships for flexural design)

$$A = 120 \text{ in.}^2 \qquad \frac{r^2}{y} = \frac{I}{Ay} = 2.0 \text{ in.} \qquad S = \frac{I}{y} = 240 \text{ in.}^3$$

$$P = 120 \text{ k} \qquad e = 2.5 \text{ in.} \qquad f_b = \frac{120{,}000}{120}\left(1 + \frac{2.5}{2.0}\right) = 2250 \text{ psi}$$

$$\text{Maximum allowable moment} = \frac{2250 \times 240}{12{,}000} = 45.0 \text{ k-ft}$$

$$\text{Moment due to dead load} = \frac{wl^2}{8} = \frac{120}{144} \times 0.15 \times \frac{(10)^2}{8} = 1.56 \text{ k-ft}$$

$$\text{Moment due to concentrated load} = \frac{PL}{4} = 45.0 \text{ k-ft} - 1.6 \text{ k-ft} = 43.4 \text{ k-ft}$$

$$P = \frac{43.4 \times 4}{10} = 17.4 \text{ k}$$

Using Eq. 4-5 the computation of the maximum permissible moment becomes

$$M_T = 120 \text{ k} \, \frac{2.5 \text{ in.} + 2.0 \text{ in.}}{12} = 45.0 \text{ k-ft}$$

4-5 Pressure-Line Location in a Beam with a Curved Tendon

It has been shown in Sec 4-3 that the pressure line for prestressing alone in a prismatic beam is coincident with the prestressing force when the beam is prestressed with a straight tendon. This can also be demonstrated for a beam prestressed with a tendon that changes slope, as shown in Fig. 4-8. By

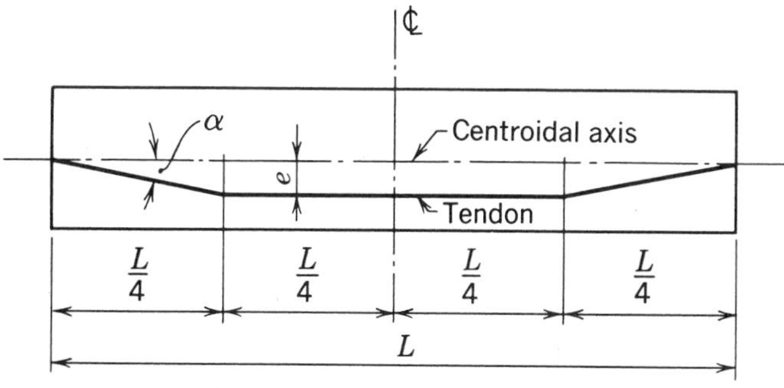

Fig. 4-8 Beam prestressed with tendon that slopes at ends.

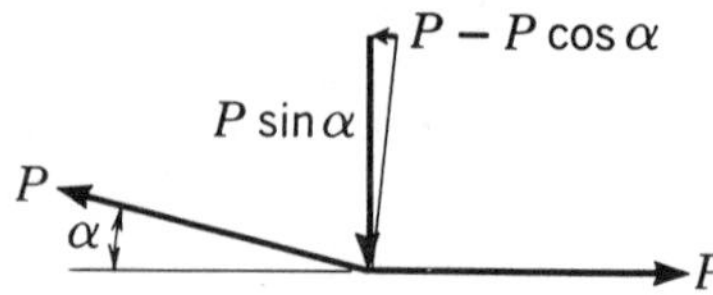

Fig. 4-9 Free-body diagram of Fig. 4-8 tendon at quarter point.

inspection, the forces acting on the concrete at the point where the tendon changes slope are determined to be as indicated in Fig. 4-9. The forces acting on the concrete are shown at their respective points of application in the free-body diagram of Fig. 4-10. In order to determine where the pressure line is acting at the center of the beam, the conditions of statics at point A are investigated. The sum of the vertical forces are equal to zero, since $P \sin \alpha$ is acting downward at the end of the beam and upward at the quarter point. The sum of the horizontal forces indicates the force R must be equal to P, since

$$\Sigma H = P \cos \alpha + (P - P \cos \alpha) - R = 0 \therefore\ R = P$$

To determine the distance from the centroidal axis to the point of application of the force R (and hence the location of the pressure line at the center of the beam), moments are taken about point A as follows:

$$\Sigma M_A = (P \sin \alpha)\frac{L}{2} + (P - P \cos \alpha)e - (P \sin \alpha)\frac{L}{4} - Px = 0$$

and

$$\frac{PL \sin \alpha}{4} + Pe - Pe \cos \alpha - Px = 0$$

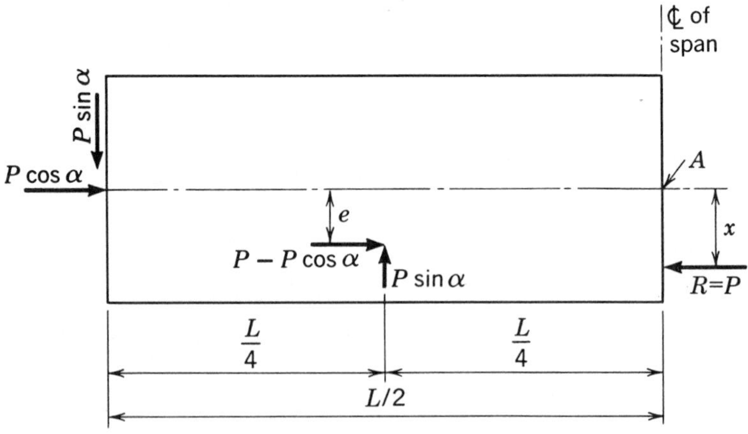

Fig. 4-10 Free-body diagram for half of the beam shown in Fig. 4-8.

but

$$\tan \alpha = \frac{4e}{L} = \frac{\sin \alpha}{\cos \alpha}$$

and

$$\sin \alpha = \frac{4e \cos \alpha}{L}$$

therefore,

$$Pe \cos \alpha + Pe - Pe \cos \alpha - Px = 0$$

hence,

$$x = e$$

and the pressure line is coincident with the location of the tendon.

If a beam with a curved tendon, as shown in Fig. 4-11, is considered, it is readily seen that in stressing the tendon, the natural tendency for the tendon to straighten out is resisted by the concrete. If a short segment of the tendon is studied as a free body, as shown in Fig. 4-12, forces must be present normal to the tendon (neglecting friction) in order to prevent this straightening. If

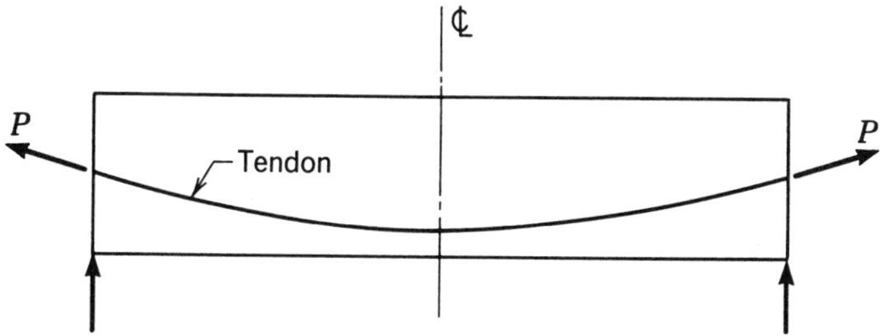

Fig. 4-11 Simple beam with curved tendon.

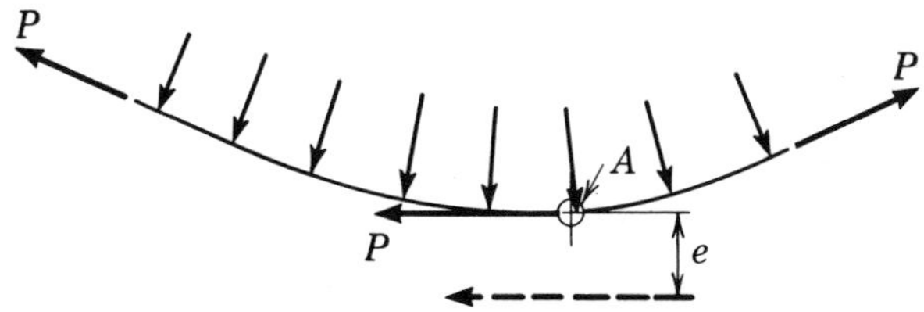

Fig. 4-12 Free-body diagram of portion of curved prestressing tendon.

friction is neglected, the force acting throughout the tendon is uniform, and since the tendon is flexible, it cannot support any bending moments. Therefore, at every point such as point A, the force in the tendon is equal to P and is located on the trajectory of the tendon. If the force were not coincident with the tendon at A, but were located at some distance from A (as shown by the dashed vector), the tendon would have to withstand the moment Pe caused by this eccentric force.

From this analysis then, it can be concluded that the pressure line for prestressing alone in a simple beam, prestressed with a curved tendon, is coincident with the trajectory of the tendon, since the forces in the concrete must be equal and opposite to those in the steel in order to maintain equilibrium. Furthermore, it can be shown that the pressure line moves when an external load is applied to a beam with a curved tendon, just as it does in a beam with a straight tendon.

ILLUSTRATIVE PROBLEM 4-6 Compute and plot to scale the location of the pressure line for the 10 × 12 in. rectangular beam, if prestressed with a force of 120 k, which is on a parabolic curve and which has an eccentricity of 2.5 in. and zero, at the midspan, and end, respectively, if the beam is spanning 10 ft and subjected to a uniformly distributed load of 3.5 k/ft. Neglect dead load of the beam.

SOLUTION:
At midspan:

$$M = 3.5 \text{ k} \times \frac{(10)^2}{8} = 43.8 \text{ k-ft}$$

$$\text{Movement} = \frac{43.8 \text{ k-ft} \times 12 \text{ in./ft}}{120 \text{ k}} = 4.38 \text{ in.}$$

$$\text{Location} = 4.38 \text{ in.} - 2.50 \text{ in.} = 1.88 \text{ in. above c.g.s.}$$

At quarter point:

$$M = 0.75 \times 43.8 \text{ k-ft} = 32.8 \text{ k-ft}$$

$$\text{Movement} = \frac{32.8 \times 12 \text{ in./ft}}{120 \text{ k}} = 3.28 \text{ in.}$$

$$\text{Location} = 3.28 \text{ in.} - 0.75 \times 2.50 \text{ in.} = 1.40 \text{ in. above c.g.s.}$$

At end:

$$M = 0 \qquad e = 0$$

(*See* Fig. 4-13.)

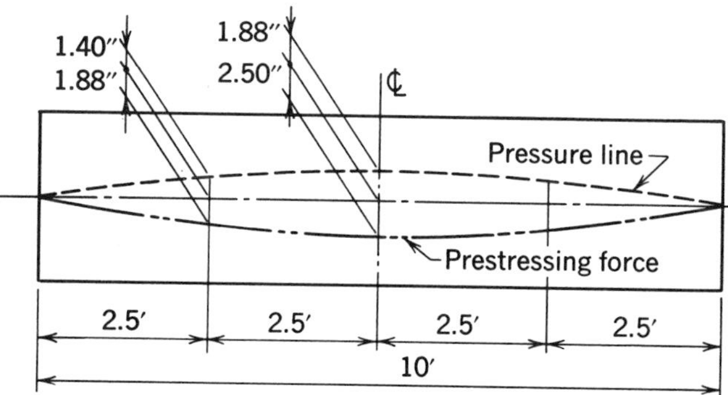

Fig. 4-13 Location of pressure line for Prob. 4-6.

ILLUSTRATIVE PROBLEM 4-7 Calculate the maximum, uniformly distributed load that can be applied to the beam of Prob. 4-6, if the span is 10 ft and the bottom-fiber stress is zero at midspan.

SOLUTION: Under the loaded condition, the pressure line will be at r^2/y_b above the centroid (the upper limit of the Kern zone).

$$r^2/y_b = 2.0 \text{ in. } e = 2.5 \text{ in.}$$

$$M_T = 120 \text{ k} \frac{4.5 \text{ in.}}{12} = 45.0 \text{ k-ft}$$

$$w_{max} = \frac{45 \times 8}{(10)^2} = 3.60 \text{ k/ft}$$

4-6 Advantages of Curved or Draped Tendons

When a beam, such as is shown in Fig. 4-14, is prestressed by a straight tendon, it deflects upward or cambers. It is apparent that the dead weight of the beam itself is acting at the time of the prestressing, since, as the beam cambers, the soffit of the beam is no longer in contact with the soffit form, except at the extremities of the beam. From this consideration, it can be concluded that the actual stresses existing in the concrete at any point in the beam, at the time of prestressing, is equal to the algebraic sum of the stresses caused by the prestressing and the dead weight of the beam itself.

The variation in the stresses, along the length of the beam in the extreme top and bottom fibers, for a beam prestressed with straight tendons is also illustrated in Fig. 4-14. If it is assumed that, for the concrete in the beam under consideration, the maximum permissible bottom-fiber compressive

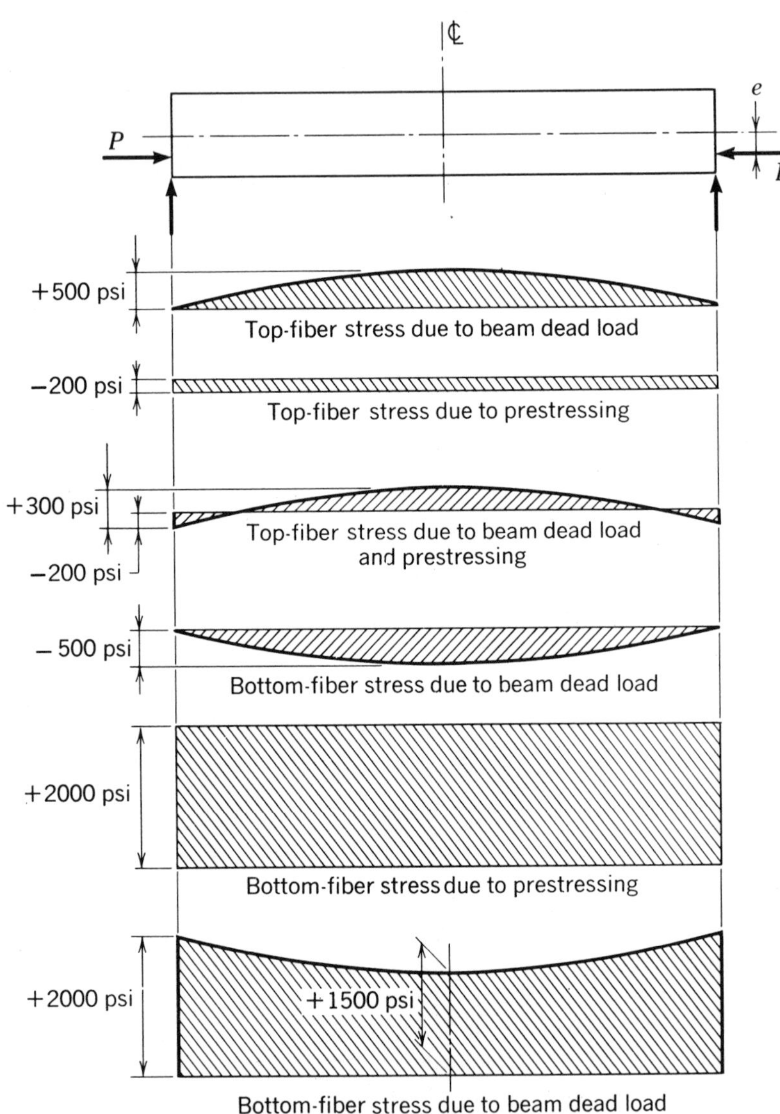

Fig. 4-14 Stress distribution of top and bottom fibers of simple prismatic beam prestressed with a straight tendon.

stress is +2000 psi and the maximum permissible top-fiber tensile stress is −200 psi—the beam as illustrated is prestressed as highly as possible. Assuming no tensile stresses are to be allowed in the bottom fiber under the total load, it will be seen that 1500 psi or 75% of the total prestressing stresses in the concrete at the midspan of the beam are "reserved" for the superimposed loads and 25% are used in carrying the dead weight of the beam alone. Furthermore, the maximum stresses that limit the capacity of the beam occur at the end of the beam, where there are no flexural stresses, rather than near the center, where the flexural stresses are maximum.

If the tendon were placed in the member on a parabolic curve such that the eccentricity were maximum at midspan of the beam and minimum at the ends of the beam, the stresses in the top and bottom fibers would vary along the length of the beam, as illustrated in Fig. 4-15. It will be seen, from an examination of these stress distributions, that the maximum stresses resulting from prestressing in both the top and bottom fibers occur at midspan of the beam. Furthermore, it is apparent that, by careful selection of the magnitude and eccentricity of the prestressing force, it is possible to eliminate the reduction in the capacity of the beam to withstand a superimposed load due to the dead weight of the beam itself, as was the case in the previous example. This can be explained in terms of the pressure line as follows: The prestressing force can be applied lower at the center of the beam than at the ends, without exceeding the permissible stresses, since the dead-load moment of the beam is acting in a direction opposite to that of the prestressing moment. The increase in eccentricity that can be used is equal to M_{DL}/P.

The advantage to be gained from curving the tendons is obviously more important in members in which the external moment that exists at the time of stressing is a large percentage of the total moment. Conversely, if the dead-load moment acting at the time of prestressing is very small, there is little or no advantage (from a standpoint of flexural stresses) in having the prestressing force at a greater eccentricity at the center of the span than it is near the ends.

It is axiomatic in structural engineering that the dead load of structures becomes progressively more important and greater, in respect to the total load, as the span lengths are increased. This is one of the important considerations influencing the normal practice of using straight tendons for short members and using tendons having variable eccentricity, either pre-tensioned or post-tensioned, for longer members. As is discussed in Sec. 4-9, this fact is also important in determining the proper cross-sectional shape of a flexural member.

It should be recognized (*see* Sec. 14-6) that deflected or draped pre-tensioned tendons cannot be placed on smooth curves. They are often placed on trajectories consisting of a series of straight lines that approximate a parabolic

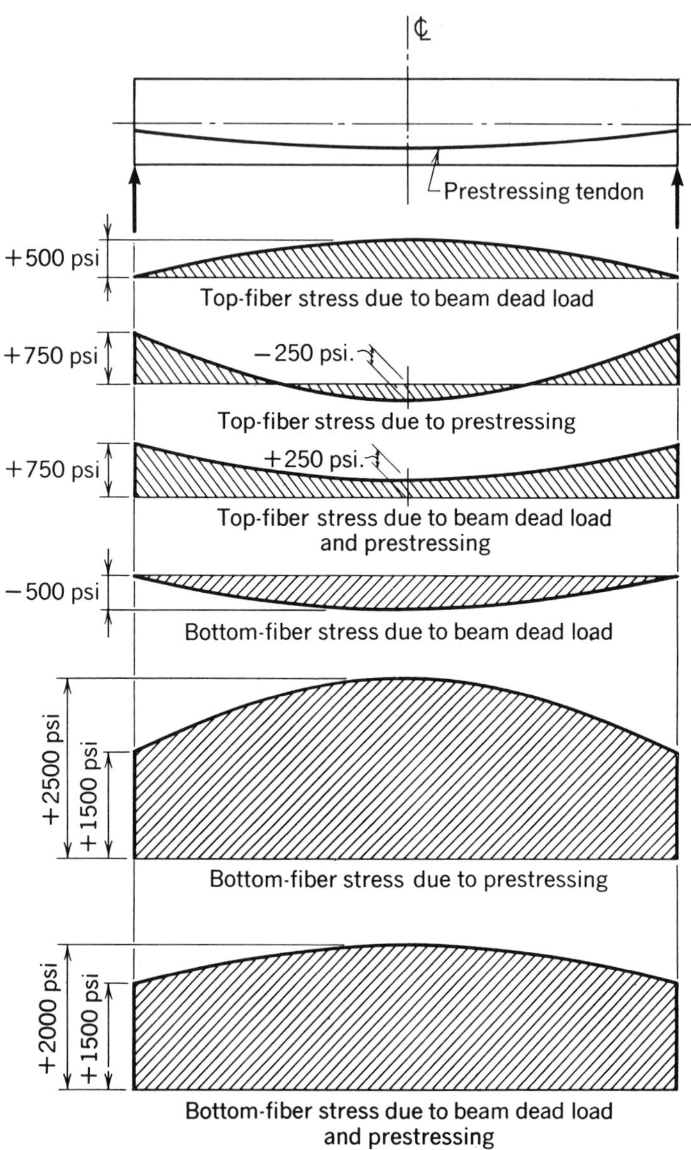

Fig. 4-15 Stress distribution in the top and bottom fibers of a simple prismatic beam prestressed with a curved tendon.

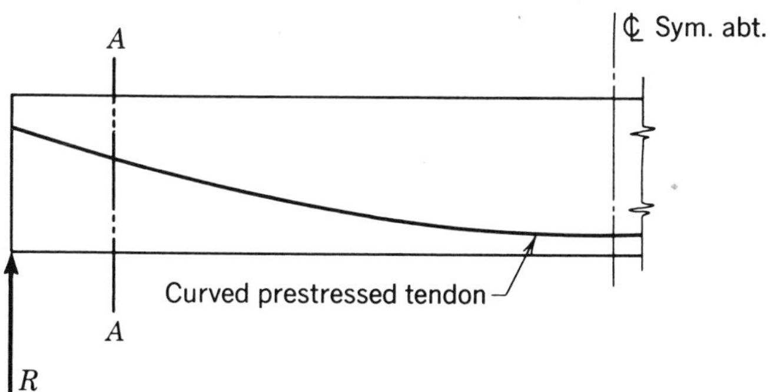

Fig. 4-16 Half-elevation of a simple beam with curved tendon.

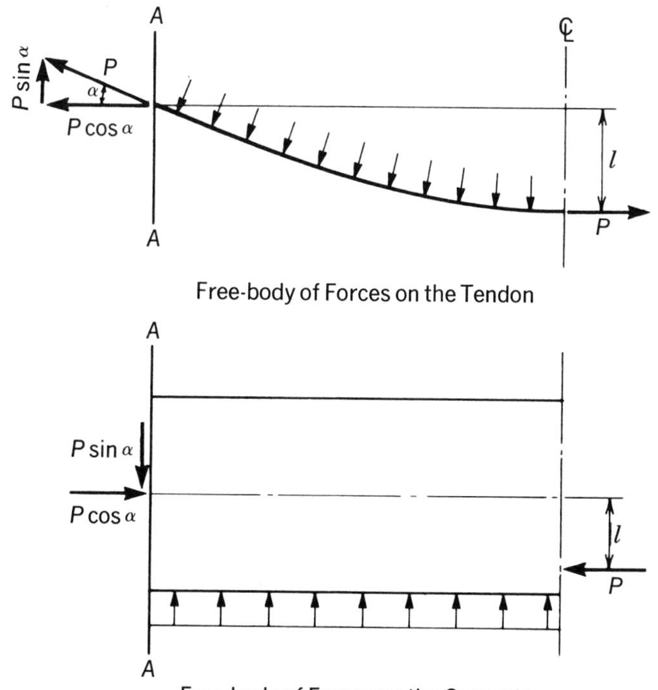

Fig. 4-17 Free-body diagrams for curved tendon and concrete section.

or other curve form. When the term "curved tendon" is used in this book, it is not necessarily meant to infer that the tendon must be post-tensioned.

Another very important beneficial effect of curving the prestressing tendons is the reduction of the shear force that must be carried by the concrete section. This can be illustrated by considering a beam having a curved prestressing tendon that is sloped at an angle of α to the horizontal at the point under consideration, as is illustrated in Fig. 4-16. Inspection of a free-body diagram for this condition, as illustrated in Fig. 4-17, will reveal that the prestressing force P can be resolved into two components: $P \sin \alpha$, which acts vertically upward, and $P \cos \alpha$, which acts horizontally. If the total shear force at the end of the free body due to external loads is V, the concrete must resist the amount $V - P \sin \alpha$, since the tendon is exerting an upward force equal to $P \sin \alpha$ between the center of the span and the end. If the tendon were not curved, the entire shear force V would have to be carried by the concrete section alone.

ILLUSTRATIVE PROBLEM 4-8 Determine the prestressing force and eccentricity required to prestress a slab, 4 ft wide, 8 in. deep, that is to be used on a span of 30 ft and is to be simply supported. Maximum final compression in the bottom fibers is 2000 psi and maximum allowable, final, top-fiber tensile stress is 300 psi. The slab is solid sand and gravel concrete, and the superimposed load is 45 psf. Minimum cover for the tendons is $1\frac{1}{2}$ in. Assume the tendons have a $\frac{3}{8}$ in. nominal diameter and have a final prestressing force of 11 k each. If the tendons are straight and in one row, how many are required and how thick is the concrete cover? If the tendons can be placed in such a manner that the cover at midspan is $1\frac{1}{2}$ in., how many tendons are required?

SOLUTION:

$$\begin{aligned} \text{Slab dead load} &= 4 \text{ ft} \times 100 \text{ psf} = 400 \text{ plf} \\ \text{Superimposed load} &= 4 \text{ ft} \times 45 \text{ psf} = \underline{180 \text{ plf}} \\ & \qquad\qquad\qquad\qquad\quad 580 \text{ plf} \end{aligned}$$

$$M_t = 0.58 \times \frac{(30)^2}{8} = 65.4 \text{ k-ft} \qquad S = \frac{I}{y} = \frac{bd^2}{6} = \frac{48 \times (8)^2}{6} = 512 \text{ in.}^3$$

$$f = \frac{65.4 \times 12{,}000}{512} = 1530 \text{ psi} \qquad \frac{r^2}{y} = \frac{d}{6} = 1.33 \text{ in.}$$

Desired pre-tension (final):

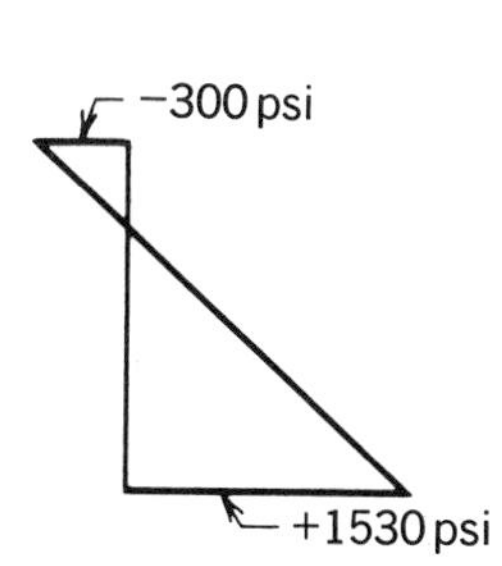

$$\frac{P}{A}\left(1 - \frac{e}{1.33}\right) = -300 \text{ psi}$$

$$\frac{P}{A}\left(1 + \frac{e}{1.33}\right) = 1530 \text{ psi}$$

$$1530 - (1150)e = -300 - 225\,e$$

$$925\,e = 1830$$

$$e = 1.98 \text{ in.}$$

$$P = \frac{1530 \times 48 \times 8}{2.49} = 236{,}000 \text{ lb} \quad \text{Cover} = 1.98 \text{ in.} - \frac{0.375}{2} = 1.79 \text{ in.}$$

$$= 22\text{—}\tfrac{3}{8} \text{ in. diameter tendons} \qquad \cong 1\tfrac{13}{16} \text{ in.}$$

If the tendons can be curved parabolically, the top-fiber stress will not limit the eccentricity

$$e = 2.31 \text{ in.} \qquad \frac{P}{384}\left(1 + \frac{2.31}{1.33}\right) = 1530 \text{ psi}$$

$$P = 214 \text{ k} \cong 20\text{—}\tfrac{3}{8} \text{ in. diameter}$$

ILLUSTRATIVE PROBLEM 4-9 Assuming the maximum final, top- and bottom-fiber stresses allowable are −170 psi and +2000 psi, respectively, determine the maximum superimposed load that can be carried by a 12 in. × 18 in. beam on a simple span of 30 ft. Determine the minimum prestressing force and corresponding eccentricity of the force, if the member is pre-tensioned with straight tendons. Determine the minimum, curved-tendon prestressing force that could be used to carry the same superimposed load if the maximum eccentricity is 6 in. (3 in. from the bottom of the beam to the center of the tendon). What is the ratio of the forces?

SOLUTION:

$$A = 216 \text{ in.}^2 \qquad S = \frac{bd^2}{6} = \frac{12 \times (18)^2}{6} = 648 \text{ in.}^3 \qquad \frac{r^2}{y} = \frac{d}{6} = 3.00 \text{ in.}$$

Stress distribution due to final pre-tension:

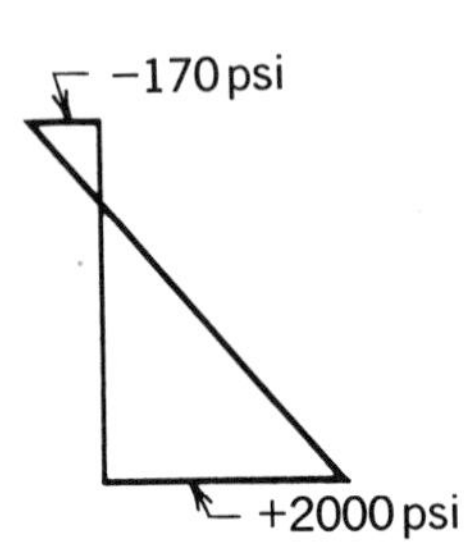

$$\frac{P}{A}\left(1+\frac{e}{3.0}\right)=2000$$

$$\frac{P}{A}\left(1-\frac{e}{3.0}\right)=-170$$

$$-170-56.7e=2000-667\,e$$

$$610.3e=2170$$

$$e=3.54\text{ in.}$$

$$P=\frac{2000\times 216}{1+(3.54/3.00)}=198{,}000\text{ lb. (Straight tendons)}$$

$$M_t=198\text{ k}\times\frac{3.0+3.54}{12}=108\text{ k-ft}$$

$$w_t=\frac{108\times 8}{(30)^2}=0.960\text{ k/ft}$$

$$w_d=0.225\text{ k/ft}$$

$$w_{SL}=0.735\text{ k/ft}$$

With curved tendon:

$$\text{Stress due to dead load}=\frac{0.225\times(30)^2\times 12{,}000}{8\times 648}=\ 470\text{ psi}$$

$$\text{Stress due to superimposed load}=\frac{0.735\times(30)^2\times 12{,}000}{8\times 648}=\frac{1530\text{ psi}}{2000\text{ psi}}$$

Stress distribution due to prestress:

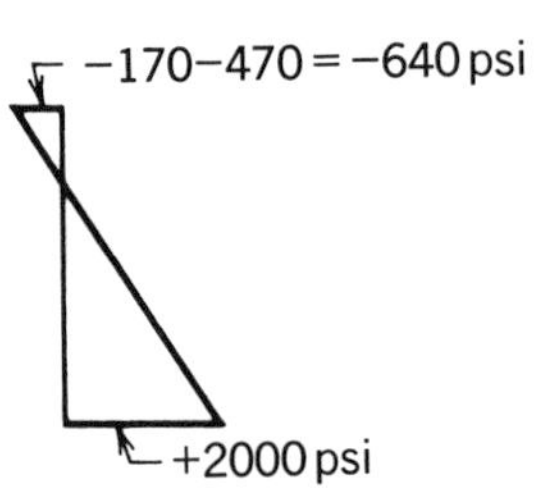

$$\frac{P}{A}\left(1+\frac{e}{3.00}\right)=2000\text{ psi}$$

$$\frac{P}{A}\left(1-\frac{e}{3.00}\right)=-640\text{ psi}$$

$$-640-213\,e=2000-667\,e$$

$$e=5.82\text{ in.}$$

$$P=\frac{2000\times 216}{2.94}=147\text{ k}$$

$$\text{Ratio}=1.35$$

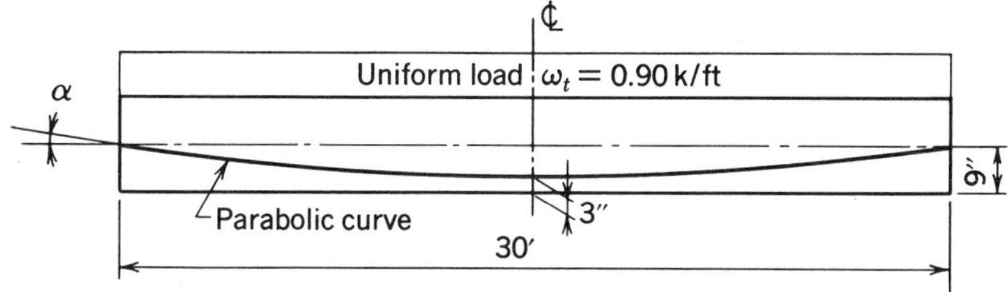

Fig. 4-18 Beam of Prob. 4-10.

ILLUSTRATIVE PROBLEM 4-10 Compute the shear force carried by the prestressing tendon of 120 k and by the concrete section at the ends of the beam for the beam and condition of loading shown in Fig. 4-18.

SOLUTION:

At the ends:

$$\sin \alpha \cong \tan = \frac{4e}{L} = \frac{4 \times 6 \text{ in.}}{12 \times 30 \text{ ft}} = 0.0667$$

$$P \sin \alpha \cong 120 \text{ k} \times 0.0667 \cong 8.00 \text{ k}$$

$$V_t = 0.90 \text{ k/ft} \times 15 \text{ ft} = 13.5 \text{ k}$$

$$V_c = 13.5 \text{ k} - 8.0 \text{ k} = 5.5 \text{ k}$$

4-7 Limiting Eccentricities

It was explained in Sec. 4-6 that a greater eccentricity of the prestressing force can frequently be allowed at the midspan of a beam than at the ends, without exceeding the permissible stresses, due to the dead weight of the beam itself which acts at the time of stressing. The permissible stresses that must be satisfied in a normal, simple beam include a maximum compressive stress in the bottom fiber and a maximum tensile stress in the top fiber under the combined action of the initial prestressing force (before relaxation) and the dead weight of the beam. In addition, a maximum top-fiber compressive stress and a maximum bottom-fiber tensile stress under the combined effects of the total external load and the final prestressing force (after relaxation) are normally specified. For most beams, a number of combinations of prestressing force and eccentricity can be found that will satisfy these conditions of stress. In the interest of economy, however, the minimum force that satisfies the above conditions of stress at the most highly stressed section is usually selected.

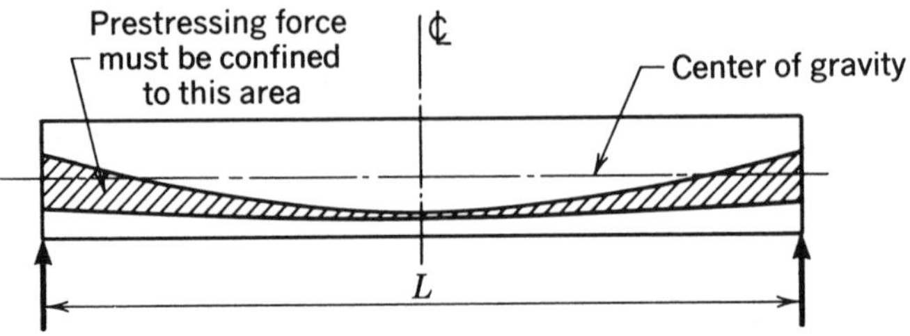

Fig. 4-19 Schematic diagram showing area in which prestressing force must be confined in order to satisfy initial and final stress requirements.

For the force that is selected, one can compute maximum and minimum eccentricities that can be used at various locations along the length of the beam without exceeding the permissible stresses enumerated above. Plotting these eccentricities in a schematic elevation of the beam, which has an exaggerated vertical scale, reveals the limiting dimensions in which the center of gravity of the prestressing force must remain in order to satisfy the conditions of allowable stress. Such a schematic diagram is shown in Fig. 4-19, where the area in which the selected prestressing force must be confined is cross-hatched. It is generally not necessary to make such a diagram in designing beams subjected to normal loading, since by placing the tendons on parabolic (or near parabolic) curves, the stress conditions can generally be satisfied without difficulty. However, when non-prismatic beams, continuous beams, or beams that have acute and unusual stress conditions are encountered, such diagrams facilitate the design.

ILLUSTRATIVE PROBLEM 4-11 Compute the limits of the eccentricity of the prestressing force of 550 k at the midspan, quarter point, and end, for a simple beam, if the allowable tensile stress is 200 psi and zero in the top and bottom fibers, respectively, and if the maximum allowable compressive stress is 2000 psi and 2200 psi in the bottom and top fibers, respectively. The maximum and minimum external-load stresses (total load and beam dead load alone, respectively, are as follows:

Location	*Max. top (psi)*	*Min. top (psi)*	*Max. bottom (psi)*	*Min. bottom (psi)*
Quarter point	+1350	+453	−1530	−328
Midspan	+1800	+605	−2038	−438

The area of the beam is 445 in.^{2} and r^2/y_b and r^2/y_t are equal to 8.99 and 6.50 in., respectively.

SOLUTION:

$$\frac{P}{A} = \frac{550{,}000}{445} = 1235 \text{ psi}$$

At midspan:
Maximum allowable stress due to prestress in bottom fiber:

$$f_b = 2438 = 1235\left(1 + \frac{e}{8.99}\right), \qquad e = 8.70 \text{ in.}$$

For

$$e = 8.70 \text{ in.}, \qquad f_t = 1235\left(1 - \frac{8.70}{6.50}\right) = -420 \text{ psi}$$

Net top-fiber stress $= -420 + 605 = +185$ psi > -200 psi O.K.

Minimum allowable stress due to prestress in bottom fiber:

$$f_b = 2038 = 1235\left(1 + \frac{e}{8.99}\right), \qquad e = 5.83 \text{ in.}$$

For

$$e = 5.83 \text{ in.}, \qquad f_t = 1235\left(1 - \frac{5.83}{6.50}\right) = +123 \text{ psi}$$

Net maximum top-fiber stress $= +123 + 1800 = +1923 < +2200$ psi O.K.

Summary for center line:

$$e =_{\text{max}} 8.70 \text{ in.}, \ e_{\text{min}} = 5.83 \text{ in.}$$

At quarter point:
Maximum allowable stress due to prestress in bottom fiber:

$$f_b = 2328 = 1235\left(1 + \frac{e}{8.99}\right), \qquad e = 8.00 \text{ in.}$$

For

$$e = 8.00 \text{ in.}, \qquad f_t = 1235\left(1 - \frac{8.00}{6.50}\right) = -284 \text{ psi}$$

Net top-fiber stress $= -284 + 453 = +169$ psi > -200 psi O.K.

Minimum allowable stress due to prestress in bottom fiber:

$$f_b = 1530 = 1235\left(1 + \frac{e}{8.99}\right), \qquad e = 2.16 \text{ in.}$$

For

$$e = 2.16 \text{ in.,} \qquad f_t = 1235\left(1 - \frac{2.16}{6.50}\right) = +825 \text{ psi}$$

Net maximum top-fiber stress = $+825 + 1350 = +2175 < +2200$ psi O.K.
Summary for quarter point:

$$e_{max} = 8.00 \text{ in., } e_{min} = 2.16 \text{ in.}$$

At the support:
Maximum allowable stress due to prestress in bottom fiber:

$$f_b = 2000 = 1235\left(1 + \frac{e}{8.99}\right), \qquad e = 5.57 \text{ in.}$$

For

$$e = 5.57 \text{ in.,} \qquad f_t = 1235\left(1 - \frac{5.57}{6.50}\right) = +173 \text{ psi} > -200 \text{ psi O.K.}$$

Minimum allowable stress due to prestress in bottom fiber:

$$f_b = 0 = 1235\left(1 + \frac{e}{8.99}\right), \qquad e = -8.99 \text{ in.}$$

For

$$e = -8.99 \text{ in.,} \qquad f_t = 1235\left(1 + \frac{8.99}{6.50}\right) = +2940 \text{ psi} > +2200 \text{ psi N.G.}$$

For

$$f_t = +2200 \text{ psi} \qquad e = \left(\frac{2200}{1235} - 1\right) 6.50 = -5.07 \text{ in.}$$

For

$$e = -5.07 \text{ in.,} \qquad f_b = 1235\left(1 - \frac{5.07}{8.99}\right) = +537 \text{ psi}$$

Summary at center line of support: $e_{max} = 5.57$ in., $e_{min} = -5.07$ in.

The limits in which the center of gravity of the tendons must fall are shown in Fig. 4-20.

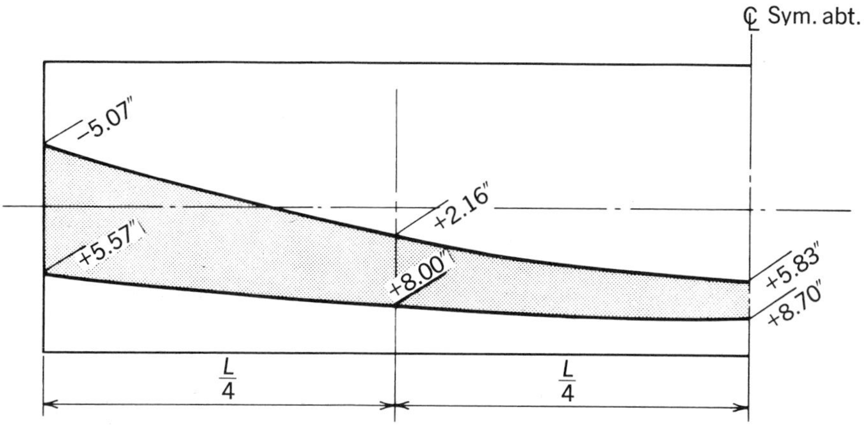

Fig. 4-20 Plot of the limits of the prestressing force for Prob. 4-11.

4-8 Cross-Section Efficiency

In a rectangular beam the distribution of the unit flexural stresses in the concrete under prestress alone and under total load at the midspan may be as is illustrated in Fig. 4-21. The distribution of the forces in this beam will have the identical shape as the distribution of the unit stresses, and the conversion of the unit stresses to forces can be made by multiplying the unit stresses by the width of the cross section. As has been explained, the total moment to which this member is subjected can be computed by determining the distance between the points of application of the resultant forces in the concrete, under the conditions of prestressing alone and when under full load, and by multiplying this distance by the prestressing force.

Analysis of a beam with an I-shaped cross section, such as is illustrated in

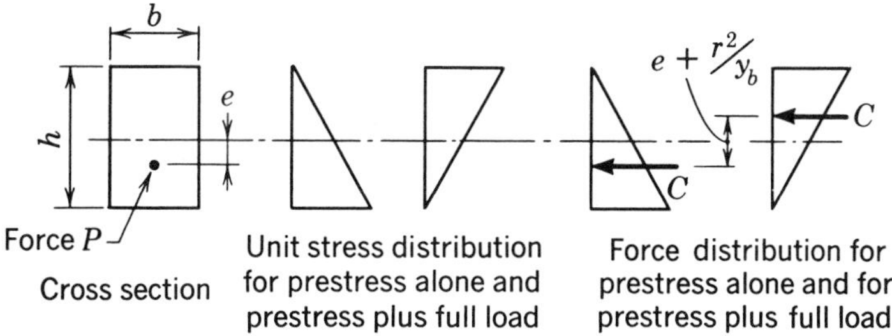

Fig. 4-21 Distribution of unit stresses and forces in a rectangular beam under prestress alone and under prestress plus full load.

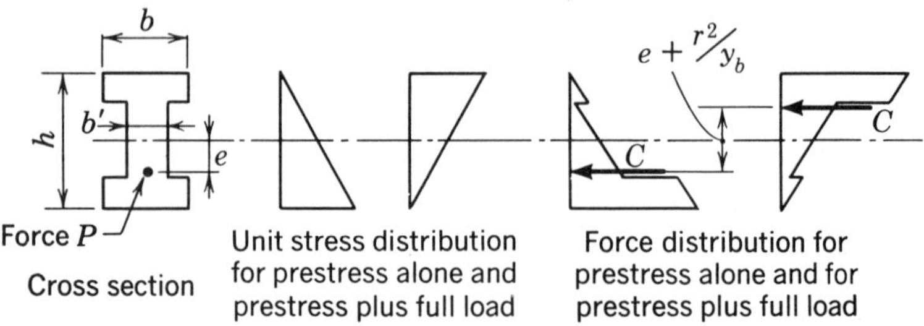

Fig. 4-22 Distribution of unit stresses and forces in a beam with I-shaped cross section, under prestress alone and under prestress plus full load.

Fig. 4-22, will reveal that the distribution of unit stresses varies linearly, as in the case of the rectangular cross section; however due to the variable width of the cross section, the distribution of forces is variable as illustrated. It is apparent that the resultants of the force diagrams for the I-shaped member will be nearer the extreme fibers of the cross section. For this reason, the resultant force in the I-shaped concrete section moves through a greater distance when external load is applied, to nullify the bottom-fiber prestress, than is the case for a rectangular cross section of equal depth. From this consideration, it is obvious that the I-shape will be more efficient and is capable of withstanding a greater load than a rectangular section of equal depth, providing each section is prestressed with a force of equal magnitude and tensile stresses are not allowed in the section.

This consideration is the primary reason for using I, T, and hollow shapes in prestressed flexural members in which major tensile stresses must be avoided and in which construction depth is of importance and must be minimized. Solid slabs and rectangular beams are economical under some conditions of span, loading, and design criteria, but the more complicated shapes generally result in the minimum quantities of prestressing steel and concrete required to carry a particular load condition, and as a result, they are frequently more economical.

The effect of allowing tensile stresses in the top and bottom fibers is discussed subsequently in Secs. 6-10 and 6-11. Selection of an efficient beam cross section for various loading conditions is discussed in Sec. 4-9.

ILLUSTRATIVE PROBLEM 4-12 Determine the maximum total moments that can be imposed upon the I-shaped and rectangular cross sections in Fig. 4-23 if each is prestressed with a straight tendon having an effective force of 200 k and if tensile stresses are not allowed under any condition of loading. (Neglect the effect of initial prestressing force.)

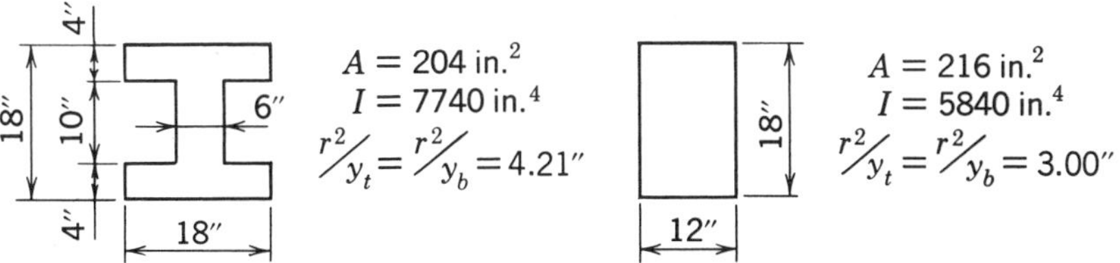

Fig. 4-23 Cross sections compared in Prob. 4-12.

SOLUTION: Since tensile stresses are not allowed under any conditions, the maximum eccentricity of the prestressing force is r^2/y_t. (*See* Eq. 4-5.) Therefore,

For the I-shape:

$$M_T = 200 \text{ k} \left(\frac{4.21 + 4.21}{12}\right) = 140 \text{ k-ft}$$

For the rectangular shape:

$$M_T = 200 \text{ k} \left(\frac{3.00 + 3.00}{12}\right) = 100 \text{ k-ft}$$

4-9 Selection of Beam Cross Section

It has been shown that the location of the pressure line in a prestressed-concrete flexural member changes upon the application of external load. At the end of a member where no moment exists, the pressure line in a simple prestressed-concrete beam is always coincident with the location of the center of gravity of the prestressing force. At the center of the beam, the distance from the center of gravity of the prestressing to the pressure line is equal to the total moment acting at that point divided by the prestressing force. (From Eq. 4-6.)

In order to illustrate the effect of this action on the shape of the optimum concrete section, consider a simple pre-tensioned beam that is prismatic, has straight tendons, and is subjected to a load of such magnitude that the bottom-fiber stress is zero at the midspan. At the end of the beam, the pressure line is coincident with the center of gravity of the prestressing, which is a condition that remains unchanged despite variations in the external load. Therefore, the optimum section at the end would be a shape that is concentric about the prestressing force, since this shape will result in minimum concrete stresses. At midspan, the pressure line acts above the center of gravity of the section, and, therefore, a top flange is necessary to resist this force. Since the stress in the bottom fibers is zero, no bottom flange is required to resist stress under this condition of loading.

The above example illustrates that, as would be expected, the optimum concrete section is materially influenced by the prestressing force and the loading. If in the above example the prestressing tendons were draped in such a manner that there was little or no eccentricity at the ends of the beam, there would be no need for any shape other than a rectangular section, which is easy to form and is efficient in resisting large, concentric, compressive forces. If the load causing zero stress in the bottom fibers at the midspan of the beam were always present, there would be no need for a large bottom flange near midspan, since the pressure line would always be acting near the top of the section and the concrete in the bottom flange would only serve to protect the prestressing steel from fire and corrosion; therefore, a T section would be efficient. On the other hand, if the load that causes zero stress in the bottom fibers at midspan is an intermittent load, and if this intermittent load is very large in comparison to the dead load of the beam itself, a large bottom flange would be required at the center of the beam to resist or "store" the prestressing force until the beam is again required to carry the intermittent load. An I shape is better for this purpose than a rectangular shape, since with an I shape the distance the pressure line can move without tensile stresses resulting in the section is greater than with a rectangular shape of equal depth (*see* Sec. 4-8).

These are the basic principles the designer of prestressed concrete must keep in mind: Bottom flanges are primarily for resisting and retaining the prestressing force until it is needed to resist the external load, at which time the pressure line moves upward; top flanges are needed for fully loaded, flexural members, since the pressure line is in the vicinity of the top flange when the beam is fully loaded (in addition, amply proportioned top flanges ensure that ultimate load failures of the brittle type cannot occur, as is discussed in Chapter 5); flanged shapes permit greater distance between the pressure line and the center of gravity of the prestressing force than is allowed by rectangular shapes, and hence smaller prestressing forces are required; and finally, the webs are primarily effective in resisting shear stresses. A complete understanding and appreciation of these functions will assist the designer in the rapid preliminary design of beams, as well as in obtaining economical and efficient designs.

Due to the fact that the dead load of a prestressed member constitutes a small portion of the total load to which it is subjected for short spans and a large portion of the total load for long spans, the use of I-shaped, hollow-rectangular, and solid-rectangular beams is more common for the short-span members, while T-shaped beams are more often used on long spans.

When straight pre-tensioned tendons are used in applications in which the dead load of the member is large in comparison with the total moment, it is often necessary to supply a large bottom flange to resist the prestressing stresses at the end. In addition, the large bottom flange may be required to assure that the concrete cover for the tendons will be adequate to protect the tendons against corrosion throughout the length of the beam. In such applications, the stress level in the bottom flange at the center of the beam, due to the combined effects of prestressing and dead load, may be relatively low. Due to the smaller area required for post-tensioned tendons, as well as the ease of placing post-tensioned tendons on curved trajectories, the size of the bottom flange of post-tensioned beams are not frequently dictated by the stresses due to prestressing at the ends or by the amount of concrete required to provide adequate concrete cover.

The designer who is experienced in field supervision as well as in the theoretical aspects of prestressed concrete will bear in mind that, although thin webs of 4 or 5 in. in thickness are often theoretically satisfactory with minimum web reinforcement, they often lead to a member in which it is difficult to place and vibrate the concrete. Therefore, honeycomb becomes a real danger. Under normal conditions, 6 in. should be regarded as the minimum web thickness for an I-shaped beam and 7 in. is the preferred minimum thickness if post-tensioning is used.

Extremely narrow, top flanges are dangerous in prestressed concrete, just as they are in structural steel. A top flange can buckle in the same manner as a column if it is of narrow dimensions, unsupported laterally, and too highly stressed. Field experience demonstrates the desirability of a reasonable width of the top flange in order to reduce the transverse flexibility of the girder during handling. This subject is treated in further detail in Sec. 10-5.

The usual ratio between depth of beam to span for simple prestressed-concrete beams varies between from 1 in 16 to 1 in 22, depending upon the conditions of loading, allowable vertical clearance and the type of construction. In lightly loaded, simple T-shaped roof members, the depth-to-span ratio may be as high as 1 in 40. Simple, supported cored slabs of prestressed concrete have been successfully used with depth to span ratios of as high as 1 in 40. Solid, continuous, post-tensioned roof slabs with depth to span ratios as high as 1 in 45 have given good performance.

4-10 Effective Beam Cross Section

In the past, the most commonly used procedure in prestressed-concrete design has been to base the flexural computations in the elastic range upon the section properties of the gross concrete section. The gross section is defined as the concrete section from which the area of steel, or ducts in the case of post-tensioning, has not been deducted and to which the transformed area of the steel has not been added. This procedure has been considered to render sufficiently accurate results in the usual applications of prestressed concrete. The change in the stresses that would result by basing the computations on the net or transformed section properties is not normally significant, in view of the fact that concrete is not a completely elastic material. Furthermore, the modulus of elasticity of concrete is not generally known precisely and a value must be assumed in computing transformed section properties. It is important, however, that the designer of prestressed concrete be aware of the nature of the section that is theoretically involved in the various types of construction, since it can be important under special conditions.

In the case of pre-tensioning, when the prestress is applied the deformation of the concrete is a function of the net section, since the concrete is compressed by the steel, which does not assist the concrete in resisting the prestressing force. The net section is defined as the section that results when the area occupied by the tendons (or ducts in the case of post-tensioning) is deducted from the gross section. Since the pre-tensioned tendons are bonded to the concrete, when there is a change of strain in the concrete at the level of the steel, there must be a corresponding and equal change of strain in the steel. Therefore, when external loads, other than the dead load of the beam, which is acting at the time of prestressing, are applied, the deformation of the member is a function of the transformed section. The transformed section can be defined as the section that results when the area of steel is transformed into an elastically equivalent area of concrete, by multiplying the steel area by the modular ratio and adding this transformed area to the net section at the proper location. Because in normal pre-tensioned practice, the tendons are straight and spaced out in order to achieve adequate bond, the effect of the transformed section is small and little is normally gained by taking these effects into account. The effect of the transformed section will normally be greater in large members with bundled pre-tensioned tendons (*see* Sec. 6-12). However, little is to be gained under normal conditions by including these refinements in the computations.

In the case of post-tensioning, the deformation of a member is a function of the net section under all conditions of prestressing and external load, until such time as grout is injected into the ducts and allowed to harden and bond the tendons to the concrete section. After bond is established, the deformation of the member is a function of the transformed section. As in the case of pre-tensioning, under normal conditions little is gained by including these effects in the computations.

The use of the net section for the computation of stresses that occur before the bonding of the tendons is required by ACI 318-71, whereas the use of the transformed section is optional for stresses that occur after bonding.

The net and transformed sections should be used in computing stresses in long-span, post-tensioned girders that have large concentrations of ducts in relatively small bottom flanges. In such a case, the ducts can have a significant influence on the compressive stresses, in the bottom flange, which result from the prestressing, since the area occupied by the ducts may be a large portion of the total bottom flange area. Additionally, the area of the prestressing steel is generally large and has a significant effect upon the stresses due to superimposed loads under such conditions.

ILLUSTRATIVE PROBLEM 4-13 For the pre-tensioned girder illustrated in Fig. 4-24, compute the stresses in the concrete due to prestressing, based upon the gross- and net-section properties. In addition, compute the concrete stresses, based upon the gross and transformed sections at the center of a 40 ft span, when the externally applied load is 3.13 k/ft. The section properties of the gross section are:

$A = 419$ in.^{2}	$I = 44{,}670$ in.^{4}	$w = 0.44$ k/ft.
$A_s = 3.20$ in.^{2}	$e = 14.60$ in. $-$ 5.20 in. $=$ 9.40 in.	$n = 6$
$y_t = 15.40$ in.	$r^2/y_t = 6.94$ in.	$S_t = 2900$ in.^{3}
$y_b = 14.60$ in.	$r^2/y_b = 7.30$ in.	$S_b = 3060$ in.^{3}
$P = 440$ k		

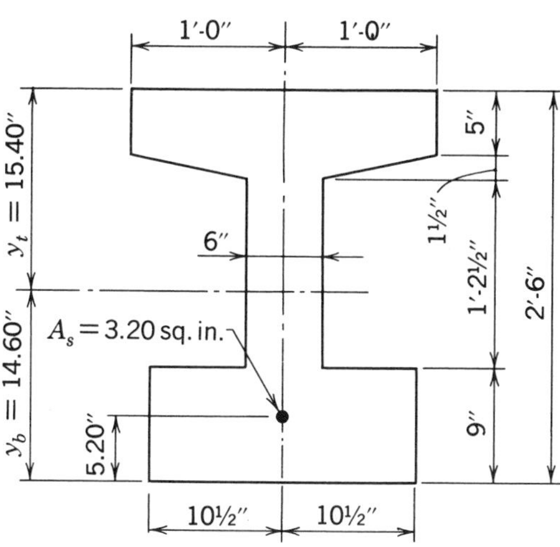

Fig. 4-24 Cross section used to demonstrate the effect of the transformed and net beam cross sections as compared to gross cross section in Prob. 4-13.

SOLUTION:

Stresses due to prestressing based upon gross section:

$$f_t = \frac{440{,}000}{419}\left(1 - \frac{9.40}{6.94}\right) = -373 \text{ psi}$$

$$f_b = \frac{440{,}000}{419}\left(1 + \frac{9.40}{7.30}\right) = +2400 \text{ psi}$$

Compute the section properties for the net section (moment about the top):

A	$\bar{y}$	$A_{\bar{y}}$	$\bar{y}'$	$\bar{y}'^2$	$A_y'^2$	I_0	$I_0 + A_y'^2$
419.0 ×	15.4 =	6450	0.10 in.	0.01	4	44,670	44,674
−3.2 ×	24.8 =	−79	9.50 in.	90.0	−288	0	−288
415.8		6371					44,386

$$y_t = 15.3 \text{ in.} \qquad r^2/y_t = 6.97 \text{ in.} \qquad S_t = 2900 \text{ in.}^3$$

$$y_b = 14.7 \text{ in.} \qquad r^2/y_b = 7.25 \text{ in.} \qquad S_b = 3020 \text{ in.}^3$$

$$e = 14.7 - 5.20 = 9.50 \text{ in.}$$

Stresses due to prestressing based upon the net section:

$$f_t = \frac{440{,}000}{416}\left(1 - \frac{9.50}{6.97}\right) = -380 \text{ psi}$$

$$f_b = \frac{440{,}000}{416}\left(1 + \frac{9.50}{7.25}\right) = +2440 \text{ psi}$$

Stresses due to prestressing, dead and superimposed loads, based upon the gross section, are computed as follows:

$$w_d + w_{SL} = 0.44 + 3.13 = 3.57 \text{ k/ft}$$

$$M_D + M_{SL} = 0.44 \times \frac{(40)^2}{8} + 3.13 \times \frac{(40)^2}{8} = 88 + 627 = 715 \text{ k-ft}$$

$$f_t = \frac{715 \times 12{,}000}{2900} - 373 = +2577 \text{ psi}$$

$$f_b = -\frac{715 \times 12{,}000}{3060} + 2400 = -400 \text{ psi}$$

Compute the transformed section properties (moments about the top):

	A	$\bar{y}$	$A_{\bar{y}}$	$\bar{y}'$	$\bar{y}'^2$	$A_y'^2$	I_0	$I_0 + A_y'^2$
	416 ×	15.3 =	6371	0.40	0.16	66	44,386	44,452
6 × 3.2 =	19 ×	24.8 =	471	9.10	83	1575	0	1,575
	435		6842					46,027

$$y_t = 15.7 \text{ in.}, \; S_t = 2930 \text{ in.}^3$$
$$y_b = 14.3 \text{ in.}, \; S_b = 3220 \text{ in.}^3$$

Stresses in the top and bottom fibers, respectively, due to prestressing, dead and superimposed loads, based upon the net and transformed sections are

$$f_t = -380 + \frac{88 \times 12{,}000}{2900} + \frac{627 \times 12{,}000}{2930} = +2554 \text{ psi}$$

$$f_b = 2440 - \frac{88 \times 12{,}000}{3020} - \frac{627 \times 12{,}000}{3220} = -250 \text{ psi}$$

ILLUSTRATIVE PROBLEM 4-14 Compute the stresses due to prestressing in the top and bottom fibers for the girder of Fig. 4-25 based upon an effective prestressing force of 2380 k located 5.3 in. above the soffit based upon: (1) the gross section properties and (2) the net section properties if the area of the ducts is 39.0 sq in. Also, determine the allowable superimposed live load on the girder based upon: (3) the gross section properties and (4) the transformed section properties if $nA_s = 83.5$ in.^{2}. The design span is 200 ft. What is the ratio between the allowable superimposed live loads of (3) and (4)?

Gross section properties

$A = 2051$ in.^{2} $\quad I = 3.735{,}950$ in.^{4}

$y_t = 53.7$ in. $\quad S_t = 69{,}700$ in.^{3} $\quad r^2/y_t = 34.0$ in.

$y_b = 69.3$ in. $\quad S_b = 54{,}000$ in.^{3} $\quad r^2/y_b = 26.3$ in.

Net section properties

$A = 2012$ in.^{2} $\quad I = 3{,}572{,}900$ in.^{4}

$y_t = 52.5$ in. $\quad S_t = 68{,}000$ in.^{3} $\quad r^2/y_t = 33.8$ in.

$y_b = 70.5$ in. $\quad S_b = 50{,}750$ in.^{3} $\quad r^2/y_b = 25.2$ in.

Transformed section properties

$A = 2096$ in.^{2} $\quad I = 3{,}915{,}500$ in.^{4}

$y_t = 55.0$ in. $\quad S_t = 71{,}300$ in.^{3}

$y_s = 68.0$ in. $\quad S_b = 57{,}600$ in.^{3}

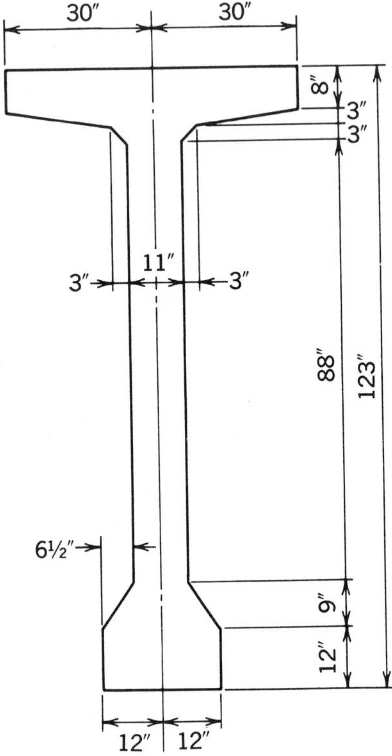

Fig. 4-25 Girder cross section for Prob. 4-14.

SOLUTION:

(1)
$$f_t = \frac{2380}{2051}\left(1 - \frac{64.0}{34.0}\right) = -1021 \text{ psi}$$

$$f_b = \frac{2380}{2051}\left(1 + \frac{64.0}{26.3}\right) = +3980 \text{ psi}$$

(2)
$$f = \frac{2380}{2012}\left(1 - \frac{65.2}{33.8}\right) = -1100 \text{ psi}$$

$$f_b = \frac{2380}{2012}\left(1 + \frac{65.2}{25.2}\right) = +4250 \text{ psi}$$

(3) $$w_d = 2.14^{k/1} M_d = 2.14 \times 5000 = 10{,}700'^{k}$$

$$\frac{l^2}{8} = 5000 \text{ ft}^2$$

$$f_{dt} = \frac{10{,}700 \times 12{,}000}{69{,}700} = +1{,}840 \text{ psi}$$

$$f_{db} = \frac{10{,}700 \times 12{,}000}{54{,}000} = -2{,}380 \text{ psi}$$

Final top fiber stress = +1840 − 1020 psi = +820 psi

Final bottom fiber stress = +3980 − 2380 = 1600 psi

$$w_a = \frac{1600 \times 54{,}000}{12{,}000 \times 5000} = 1.44^{k/1}$$

(4) $$f_{dt} = \frac{10{,}700 \times 12{,}000}{68{,}000} = +1890 \text{ psi}$$

$$f_{db} = \frac{10{,}700 \times 12{,}000}{50{,}750} = -2530 \text{ psi}$$

Final top fiber stress = +1890 − 1100 = +790 psi

Final bottom fiber stress = +4250 − 2530 = +1720 psi

$$w_a = \frac{1720 \times 57{,}600}{12{,}000 \times 5000} = 1.65^{k/1}$$

(5) Ratio = 1.15 (more for 4).

4-11 Variation in Steel Stress

Since the prestressing steel is never located at the extreme fiber of a prestressed beam, but is at some distance from the surface of the concrete, the maximum change in concrete stress that can normally be expected to occur at the level of the center of gravity of the steel is approximately 70 to 80% of the bottom-fiber stress that results from superimposed loads. With concrete that has a cylinder strength of 5000 psi, the stress change in the concrete at the level of the steel could be expected to be of the order of 1500 psi. The modular ratio between the prestressing steel and the concrete can be assumed to be 6 for loads of short duration. As a result, the application of the short-duration, superimposed load would cause an increase in steel stress of approximately 9000 psi, providing the steel and the concrete were adequately bonded. If the steel is not bonded to the concrete, but is anchored at the ends of the member only, the increase in steel stress resulting from the application of the superimposed load would be less than 9000 psi, since the steel can slip in the ducts. The steel increase in stress in unbonded tendons tends to be proportional to the average change in the concrete stress at the level of the steel.

It should be noted that the increase in stress of 9000 psi, which results from the application of the superimposed load, is only about 7% of the final stress normally employed in wire or strand tendons and about 11% of the final stress normally employed in bar tendons. The reduction in the stress in the prestressing steel due to the relaxation of the steel, shrinkage of the concrete, and the plastic flow of the concrete is of the order of 10 to 30% under average conditions (*see* Sec. 6-2). Hence, the stress that exists in the tendon under the superimposed load after all of the losses of prestress have taken place is not as high as the initial stress in the steel.

The small variation in steel stress that occurs in a normal prestressed member subjected to frequent applications of the design load is responsible for the high resistance to fatigue failure that is associated with this material (*see* Sec. 9-6).

ILLUSTRATIVE PROBLEM 4-15 Compute the increase in the stress in the steel at the midspan of the beam in Prob. 4-13.

SOLUTION: The concrete stress at the level of the steel due to the external load of 3.13 k/ft is

$$y_{c.g.s.} = 14.30 \text{ in.} - 5.20 \text{ in.} = 9.10 \text{ in.}$$

$$f_c = \frac{627 \times 12{,}000}{46{,}027} \times 9.1 = 1490 \text{ psi}$$

The increase in steel stress due to the superimposed load is

$$\Delta f_s = nf_c = 6 \times 1490 = 8940 \text{ psi}$$

---

5 Cracking Load, Ultimate Moment, Shear, and Bond

5-1 Action Under Overloads—Cracking Load

It has been shown that a variation in the external load acting on a prestressed beam results in a change in the location of the pressure line for beams in the elastic range. This is a fundamental principle of prestressed construction. In a normal prestressed beam, this shift in the location of the pressure line continues at a relatively uniform rate, as the external load is increased, to the point where cracks develop in the tension fiber. After the cracking load has been exceeded, the rate of movement in the pressure line decreases as additional load is applied, and a significant increase in the stress in the prestressing tendon and the result concrete force begins to take place. This change in the action of the internal moment continues until all movement of the pressure line ceases. The moment caused by loads that are applied thereafter is offset entirely by a corresponding and proportional change in the internal forces, just as in reinforced-concrete construction. The range of loading that is characterized by these different actions is illustrated in the load deflection curve of Fig. 5-1. This fact, that the load in the elastic range and the plastic range is carried by actions that are fundamentally different, is very significant and renders ultimate-moment computations essential for all designs in order

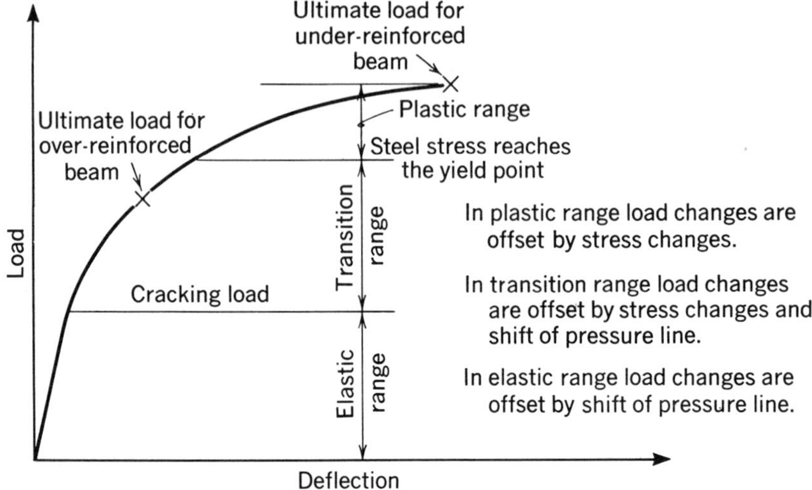

Fig. 5-1 Load-deflection curve for the prestressed beam.

to ensure that adequate safety factors exist. This is true even though the stresses in the elastic range may conform to a recognized elastic design criterion.

It should be noted that the load deflection curve in Fig. 5-1 is close to a straight line up to the cracking load and that the curve becomes progressively more curved as the load is increased above the cracking load.* The curvature of the load-deflection curve for loads over the cracking load is due to the change in the basic internal resisting moment action that counteracts the applied loads, as described above, as well as to plastic strains that begin to take place in the steel and the concrete when stressed to high levels.

In some structures it may be essential that the flexural members remain crack free even under significant overloads. This may be due to the structures being exposed to exceptionally corrosive atmospheres during their useful life. In designing prestressed members to be used in special structures of this type, it may be necessary to compute the load that causes cracking of the tensile flange, in order to ensure that adequate safety against cracking is provided by the design.

Many tests have demonstrated that the load-deflection curves of prestressed beams are virtually linear up to and slightly in excess of the load that causes the first cracks in the tensile flange. For this reason, normal elastic-design

* The presence of non-prestressed reinforcing in the tensile flange will tend to make the cracking load more difficult to detect from a load-deflection curve as well, as from observations of a beam during loading.

relationships can be used in computing the cracking load by simply determining the load that results in a net tensile stress in the tensile flange (prestress minus the effects of the applied loads) that is equal to the tensile strength of the concrete. It is customary to assume that the tensile strength of the concrete is equal to the modulus of rupture of the concrete when computing the cracking load. The modulus of rupture can be estimated from Eq. 3-2.

It should be recognized that the performance of bonded prestressed members is actually a function of the transformed section rather than the gross concrete section (*see* Sec. 4-10). If it is desirable to make a precise estimate of the cracking load, such as is required in some research work, this effect should be considered.

ILLUSTRATIVE PROBLEM 5-1 Compute the total uniformly distributed load required to cause cracking in a beam that is 10 in. wide, 12 in. deep and is supported on a simple span of 25 ft, if the final prestressing force is 120,000 lb applied at an eccentricity of 2.50 in. Assume the modulus of rupture equals $7.2\sqrt{f'_c}$ psi and $f'_c = 5000$ psi.

SOLUTION:

$$A = 120 \text{ in.}^2, \qquad I = 1440 \text{ in.}^4, \qquad r^2/y = 2.00 \text{ in.}$$

$$S = \frac{1440}{6} = 240 \text{ in.}^3$$

$$\text{Modulus of rupture} = 7.2\sqrt{5000} = 509 \text{ psi}$$

$$f_b = \frac{120{,}000}{120}\left(1 + \frac{2.50}{2.00}\right) = 2250 \text{ psi}$$

Therefore, the moment that causes cracking must result in a bottom-fiber stress equal to: 509 psi + 2250 psi = 2759 psi.

$$M_t = \frac{w_t L^2}{8} = f_b S = \frac{2759 \times 240}{12{,}000} = 55.18 \text{ k-ft}$$

$$w_t = \frac{55.18 \times 8}{25.0^2} = 0.706 \text{ k/ft}$$

5-2 Principles of Ultimate Moment Capacity for Bonded Members

When prestressed flexural members that are stronger in shear and bond than in bending are loaded to failure, they fail in one of the following modes:

(1) *Failure at cracking load* In very lightly prestressed members, the

cracking moment may be greater than the moment the member can withstand in the cracked condition and, hence, the cracking moment is the ultimate moment. This condition is rare and is most likely to occur in members that are prestressed concentrically with small amounts of steel. Determination of the possibility of the occurrence of this type of failure is accomplished by comparing the estimated moment that would cause cracking to the estimated ultimate moment, computed as described below. When the estimated cracking load is larger than the computed ultimate load, this type of failure would take place if the member were subjected to the required load. Because this type of failure is a brittle failure, it occurs without warning—designs that would yield this mode of failure should be avoided.

(2) *Failure due to rupture of steel* In lightly reinforced members subjected to ultimate load, the ultimate strength of the steel may be attained before the concrete has reached a highly plastic state. This type of failure is occasionally encountered in the design of structures with very large compression flanges in comparison to the amount of prestressing steel, such as a composite bridge stringer. Computation of the ultimate moment of a member subject to this type of failure can be done with a high precision. The method of computation, as well as the determination of which members are subject to this mode of failure, is described below.

(3) *Failure due to strain* The usual underreinforced, prestressed structures that are encountered in practice are of such proportions that, if loaded to ultimate, the steel would be stressed well into the plastic range and the member would evidence large deflection. Failure of the member will occur when the concrete attains the maximum strain that it is capable of withstanding. It is important to understand that research into the ultimate bending strength of reinforced and prestressed concrete has lead most investigators to the conclusion that concrete, of the quality normally encountered in prestressed work, fails when the limiting strain of 0.0034 is attained in the concrete. Since the ultimate bending capacity is limited by strain rather than stress in the concrete, it is a function of the elastic moduli of the concrete and steel. The magnitude of the ultimate moment for members of this category can also be predicted, as a rule, within the normal tolerances expected in structural design. The ultimate moment of underreinforced sections cannot be predicted with the same precision as the lightly reinforced members described above, since the ultimate moments of underreinforced members are a function of the elastic properties of the steel and the effective stresses in the prestressing steel, whereas the ultimate moment capacities of lightly reinforced members are not.

(4) *Failure due to crushing of the concrete* Flexural members that have relatively large amounts of prestressing steel or relatively small compressive flanges are referred to as being overreinforced. Overreinforced members,

when loaded to destruction, do not attain the large deflections associated with underreinforced members—the steel stresses do not exceed the yield point and failure is the result of the concrete being crushed. Computation of the ultimate moments of overreinforced members is done by a trial and error procedure, involving assumed strain patterns, as well as by empirical relationships. Both methods are discussed below (*see* Ref. 1).

It must be emphasized that there is no clear distinction between the different classifications of failure listed above. For convenience of design, certain parameters, which are a function of the percentage of steel, are used by different authorities to distinguish between the different types of failure that would be anticipated. These parameters, for rectangular sections, are defined as follows:

$$\text{Percentage of steel} = p^* = \frac{A_s^*}{bd} \tag{5-1}$$

In Eq. 5-1, A_s^* is the area of the prestressed reinforcement, b is the width of the compression flange of the member, and d is the distance from the extreme compression fiber to the centroid of the prestressing reinforcing. A dimensionless factor, called the steel index (q''), is used by some authorities. The factor is defined as

$$q'' = \frac{A_s^* f'_s}{bdf'_c} = p^* \frac{f'_s}{f'_c} \tag{5-2}$$

where f'_s and f'_c are the ultimate tensile strength of the prestressing steel and the 28-day cylinder strength of the concrete, respectively. The reinforcement index (q^*) is currently used in certain building codes. It is defined as

$$q^* = \frac{A_s^* f_{su}}{bdf'_c} = p^* \frac{f_{su}}{f'_c} \tag{5-3}$$

In Eq. 5-3, f_{su} is the calculated stress in the prestressing steel under the load resulting in the ultimate moment.

Each of these factors is used in such a manner that the results obtained from them are virtually identical, as will be seen.

In order to simplify the explanation of the theory related to the computation of the ultimate moments, a rectangular section will be assumed throughout the derivation, in order to eliminate the variable of flange width which is frequently encountered with I or T sections. In addition, the following assumptions are made:

(1) Plane sections are assumed to remain plane.
(2) The stress-strain properties of the steel are smooth curves without a definite yield point.

(3) The limiting strain of the concrete is equal to 0.0034, regardless of the strength of the concrete.
(4) The steel and concrete are completely bonded.
(5) The stress diagram of the concrete at failure is such that the average concrete stress is $0.80f'_c$ and the resultant of the stress in the concrete acts at a distance from the extreme fiber equal to 0.42 of the depth of the compression block, as is illustrated in Fig. 5-2.
(6) The strain in the top fiber under prestress alone is equal to zero.
(7) The section is subject to pure bending.
(8) The analysis is for the condition of static loads of short duration.

The definition of the strains illustrated in Fig. 5-2 and used in the derivation are as follows:

ε_c = concrete strain due to prestressing (assumed = 0)
ε_u = concrete strain at ultimate (assumed = 0.0034)
ε_{ce} = concrete strain at the level of the steel due to prestressing
ε_{cu} = concrete strain at the level of the steel at ultimate
ε_{se} = steel strain due to the effective prestress
ε_{su} = steel strain at ultimate

Since equilibrium of the section requires that the forces in the steel and concrete be equal, we can write:

$$T = C$$

or

$$A_s^* f_{su} = 0.80 f'_c b k_u d$$

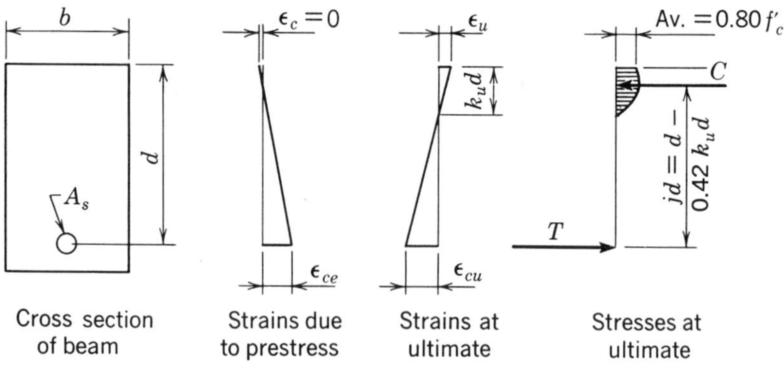

Fig. 5-2 Strain and stress distributions assumed in ultimate moment computations.

and

$$f_{su} = \frac{0.80\, f'_c b k_u d}{A_s^*}$$

or

$$f_{su} = \frac{0.80\, f'_c k_u}{p^*} \tag{5-4}$$

Since the steel index is equal to $q'' = p^* f'_s / f'_c$, Eq. 5-4 can be written

$$f_{su} = \frac{0.80\, f'_s k_u}{q''} \tag{5-5}$$

Comparing the similar triangles of the concrete strains at ultimate, the following relationship is seen

$$\frac{\varepsilon_{cu}}{d - k_u d} = \frac{\varepsilon_u}{k_u d}$$

or

$$\varepsilon_{cu} = \varepsilon_u \left(\frac{1 - k_u}{k_u} \right) \tag{5-6}$$

Using the above value for the concrete strain at the level of the steel at ultimate, the strain in the steel at ultimate (which consists of the sum of the strains due to the effective prestress, the strain in the concrete at the level of the steel resulting from prestressing, and the strain in the concrete at the level of the steel at ultimate) can be expressed as

$$\varepsilon_{su} = \varepsilon_{se} + \varepsilon_{ce} + \varepsilon_{cu} \tag{5-7}$$

or

$$\varepsilon_{su} = \varepsilon_{se} + \varepsilon_{ce} + \varepsilon_u \left(\frac{1 - k_u}{k_u} \right) \tag{5-8}$$

Equation 5-8 can be rearranged to

$$k_u = \frac{\varepsilon_u}{\varepsilon_u + \varepsilon_{su} - \varepsilon_{se} - \varepsilon_{ce}} \tag{5-9}$$

Substituting the value of k_u given in Eq. 5-9 into the relationship of Eq. 5-5, the general equation of the ultimate steel stress is obtained

$$f_{su} = \frac{0.80\, f'_s}{q''} \times \frac{\varepsilon_u}{\varepsilon_u + \varepsilon_{su} - \varepsilon_{se} - \varepsilon_{ce}} \tag{5-10}$$

All of the terms in this relationship are known or assumed, except the steel strain at ultimate ε_{su} and the steel stress at ultimate (f_{su}). Therefore, by employing a trial and error procedure and solving Eq. 5-10 for various assumed values of q'' and ε_{su}, the values of f_{su} can be determined and the results can be then plotted on the stress-strain diagram for the particular prestressing steel to be used. The intersection of the curves thus obtained with the stress-strain curve reveals the values of f_{su} and ε_{su} for different values of q''. This is illustrated in the curve of Fig. 5-3, which is representative of one type of tendon commonly used domestically in pre-tensioning. The data used in plotting the curves are

$f'_s = 275{,}000$ psi
$E_s = 26.7 \times 10^6$ psi
$\varepsilon_{se} = 0.0051$
$\varepsilon_{ce} = 0.0004$, which corresponds to a concrete stress at the level of the steel due to prestressing of 2000 psi, if $E_c = 5 \times 10^6$, and 1600 psi, if $E_c = 4 \times 10^6$ psi

The values of f_{su} obtained through the analysis of the steel stress-strain curve and the values obtained from the frequently used approximate relationship:

$$f_{su} = f'_s\left(1 - 0.5p^*\frac{f'_s}{f'_c}\right) \tag{5-11}$$

are compared in Fig. 5-4. The values obtained from the approximate relationship are conservative, as would be expected. The variation of j, the ratio

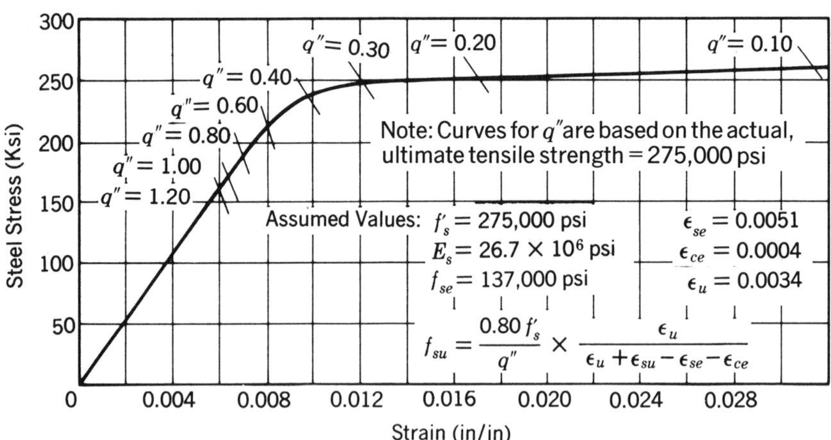

Fig. 5-3 Stress-strain diagram with f_{su}–ε_{su} curves for various values of q'' superimposed.

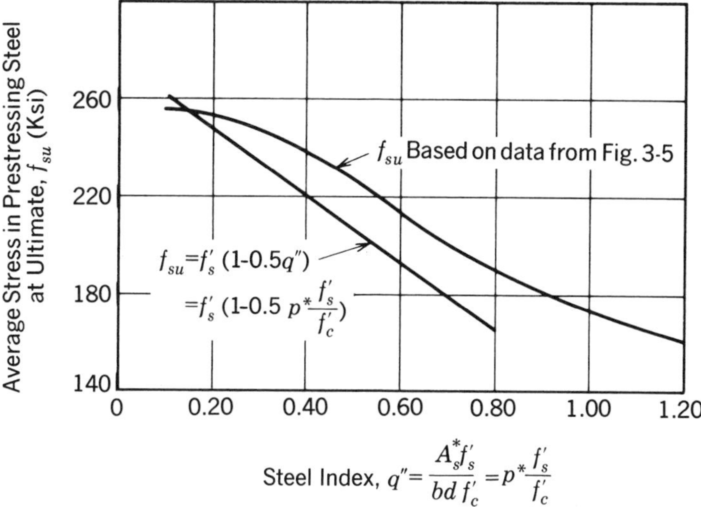

Fig. 5-4 Variation of f_{su} with the steel index. The actual values of f_{su} are shown, based on the stress-strain curve of Fig. 5-3 and the approximate value specified by ACI 318.

between the lever arm and the depth of the section, is shown in Fig. 5-5 as a function of the steel index. In a similar manner, the ratio of the ultimate moment to $f_s' A_s^* d$ as a function of the steel index is shown in Fig. 5-6.

Tests have shown that for lightly reinforced members with very low values of q'' (0.00–0.08), the moment at ultimate can be calculated by the relationship

$$M_u = 0.95 f_s' A_s^* d \tag{5-12}$$

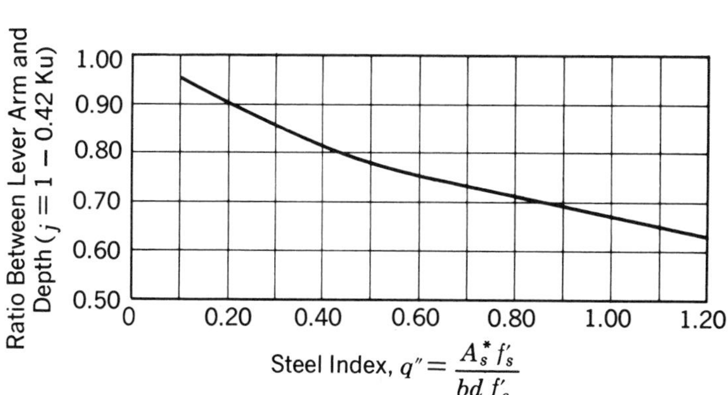

Fig. 5-5 Variation of lever-arm-depth ratio, j, with steel index.

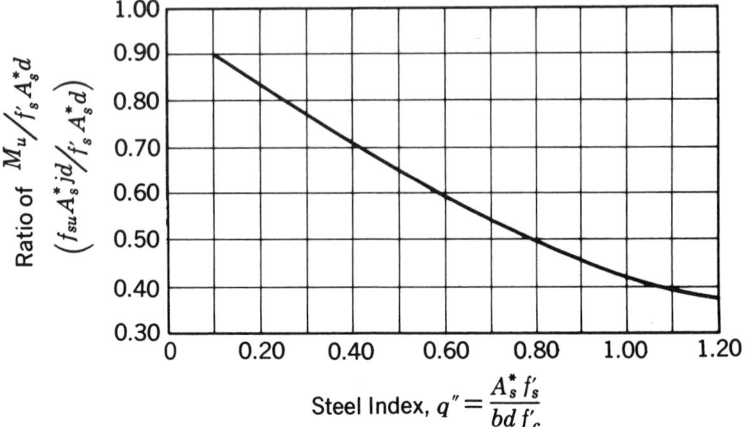

Fig. 5-6 Variation in the factor $M_u/f'_s A^*_s d$ with the steel index.

Under such conditions the member fails as the result of the failure of the steel. The value of k_u is of the order of 0.10 for the very low values of q'' (*see* Fig. 5-5). The concrete is not stressed in the plastic region for these conditions because the steel fails before such concrete stresses can develop.

For the steel studied here, the curve of $M_u/f'_s A^*_s d$ is nearly linear from 0.10 to 0.40 (Fig. 5-6) and is approximately equal to the following relationship

$$M_u = (1 - 0.60q'')A^*_s f'_s d \tag{5.13}$$

In this range of steel indices, the steel would evidence large deformations, if a member were loaded to failure, and the action would be as is described above as underreinforced.

Finally, for values of q'' in excess of 0.40, the steel stress would be below the yield point at failure and the failure would be initiated in the concrete without being the result of excessive elongation of the steel and, hence, would be of the overreinforced type.

As was stated above, the relationships that were developed are applicable to rectangular sections. These relationships are equally accurate for flanged sections, provided the neutral axis of the section at ultimate is within the limits of the flange. If the neutral axis falls outside of the flange area, the same strain distribution applies as in the case of rectangular sections, but due to the variable width of the section, the distance to the resultant of the compressive block is no longer equal to $0.42k_u d$ and must be calculated. To facilitate the calculation of the location of the resultant, the compression block can be assumed to be rectangular rather than curved, as shown in Fig. 5-2, without introducing significant error.

When small quantities of non-prestressed reinforcement are used in combination with small quantities of prestressed reinforcement, the additional ultimate moment due to the non-prestressed reinforcement can be calculated by

$$M_u = 0.90 A_s f_y d \tag{5-14}$$

where A_s and f_y are the area and stress at the yield point (max. value $\cong$ 60,000 psi) of the non-prestressed reinforcement, respectively. For larger amounts of non-prestressed reinforcement or for members with high steel indices, the moment should be determined by trial and error from the basic strain patterns.

The variation that can be expected in the ultimate moment as a result of a variation in the effective steel stress is shown in Fig. 5-7. Examination of this curve will show that small variations in the effective prestress have no significant effect on the ultimate strength of prestressed members. It is important to note that even if errors are made in estimating the losses of prestress, in estimating the stressing friction, or even if the stressing is not carried out to a high precision in the field due to poor workmanship, the effect on the ultimate moment is generally small for flexural members with bonded tendons.

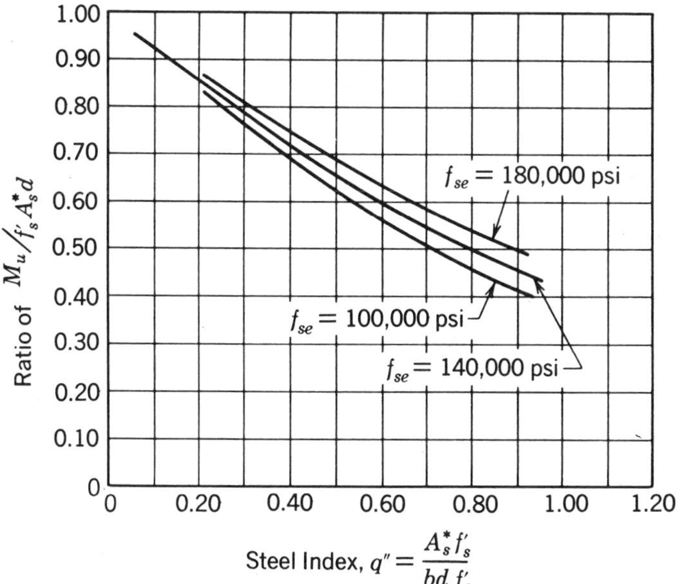

Fig. 5-7 Effect of effective prestressing stress f_{se} on the ratio $M_u/A^*_s f'_s d$ for various values of steel index q''. (After J. Muller.)

ILLUSTRATIVE PROBLEM 5-2 Compute the ultimate moment for the composite section of Fig. 5-8, which consists of the AASHO-PCI type III bridge stringer with a 6.50 in. cast-in-place slab. The steel area of 4.00 sq in. is located with its centroid 5.85 in. above the bottom of the beam. The steel has the characteristics indicated in Fig. 5-3. The concrete strengths are 5000 psi and 3000 psi for the stringer and the cast-in-place deck, respectively.

SOLUTION:

$$\text{Steel index} = q'' = \frac{4.00 \times 275{,}000}{72.0 \times 45.65 \times 3000} = 0.111$$

From Fig. 5-5 it will be seen that $j = 0.95$ (approximately) for $q'' = 0.11$. Therefore, the value of k_u is computed as follows

$$j = 0.95 = 1 - 0.42k_u$$

$$k_u = \frac{0.05}{0.42} = 0.119$$

$$k_u d = 5.45 \text{ in.}$$

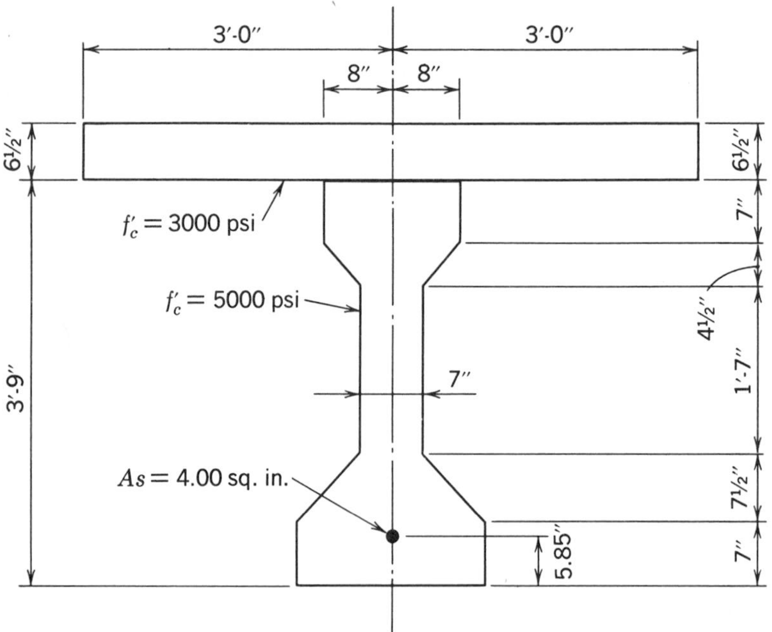

Fig. 5-8 AASHO-PCI type III bridge stringer with composite deck.

The value of $k_u d$ is approximately $5\frac{1}{2}$ in. and the neutral axis falls within the top flange. Hence, the relationships for the ultimate moments of rectangular sections are valid

$$M_u = \frac{4.00 \times 275{,}000}{1000} \times \frac{45.65}{12} (1 - 0.6 \times 0.111) = 3906 \text{ k-ft}$$

ILLUSTRATIVE PROBLEM 5-3 Compute the ultimate moment for the stringer of Prob. 5-2, neglecting the composite action of the deck.

SOLUTION:

Average concrete stress $= 0.80 f'_c = 4000$ psi.

Assume

$$\varepsilon_{se} = 0.0050 \qquad d = 45.00 \text{ in.} - 5.85 \text{ in.} = 39.15 \text{ in.}$$

$$\varepsilon_{ce} = 0.0004$$

$$\varepsilon_u = 0.0034 \qquad \varepsilon_{su} = \varepsilon_{se} + \varepsilon_{ce} + \varepsilon_u \left(\frac{d - k_u d}{k_u d}\right)$$

Try

$$k_u d = 15 \text{ in.}$$

$$C = 7 \text{ in.} \times 15 \text{ in.} \times 4 \text{ k/in.}^2 + 9 \text{ in.} \times 7 \text{ in.} \times 4 \text{ k/in.}^2 + 9 \times 4.5/2 \times 4 \text{ k/in.}^2 = 753 \text{ k}$$

$$\varepsilon_{su} = 0.0054 + 0.0034 \frac{39.15 - 15.00}{15.00} = 0.0109$$

From Fig. 5-3

$$f_{su} = 245{,}000 \text{ psi} \qquad T = 980 \text{ k}$$

$$C < T \text{ try larger } k_u d$$

Assume

$$k_u d = 20 \text{ in.}$$

$$C = 753 \text{ k} + 5 \text{ in.} \times 7 \text{ in.} \times 4 \text{ k/in.}^2 = 893 \text{ k}$$

$$\varepsilon_{su} = 0.0054 + 0.0034 \left(\frac{19.15}{20.00}\right) = 0.0087$$

From Fig. 5-3

$$f_{su} = 223{,}000 \text{ psi}$$

$$T = 893 \text{ k}$$

Compute location of force C from the top of the section:

$$7 \text{ in.} \times 20 \text{ in.} \times 4 \text{ ksi} = 560 \text{ k} \times 10 \text{ in.} = 5600 \text{ k-in.}$$

$$9 \text{ in.} \times 7 \text{ in.} \times 4 \text{ ksi} = 252 \text{ k} \times 3.5 \text{ in.} = 882 \text{ k-in.}$$

$$0.5 \times 9 \text{ in.} \times 4.5 \times 4 \text{ ksi} = 81 \text{ k} \times 8.5 \text{ in.} = 688 \text{ k-in.}$$

$$893 \text{ k} \qquad\qquad 7170 \text{ k-in.}$$

$$d_t = \frac{7170}{893} = 8.03 \text{ in.}$$

$$M_u = 893 \text{ k} \left(\frac{39.15 \text{ in.} - 8.03 \text{ in.}}{12}\right) = 2315 \text{ k/ft}$$

5-3 Principles of Ultimate Moment Capacity for Unbonded Members

Because the prestressing tendons can slip (with respect to the concrete) during loading of an unbonded member, the relationships for ultimate moment capacity developed in Sec. 5-2 do not apply to unbonded beams. The reader will recall that one of the basic assumptions made before the derivation of the relationships of Sec. 5-2 was that the concrete and steel are completely bonded. Because the tendons can slip with respect to the concrete, other variables affect the ultimate moment capacity of unbonded prestressed concrete members. Since 1958, when the "Tentative Recommendations for Prestressed Concrete" (Ref. 4) appeared, normal American practice has been to consider the stress in unbonded prestressing steel under ultimate load to be as follows:

$$f_{su} = f_{se} + 15{,}000 \tag{5-15}$$

(in psi) with the requirements that the effective stress in the prestressing steel be between 0.50 and $0.60 f'_s$ and that the steel index (q'') not exceed 0.30.

Variables that affect the ultimate moment capacity of an unbonded beam, but which do not affect bonded beams in the same manner or not at all, include the following:

(1) Magnitude of the effective stress in the tendons.
(2) Span to depth ratio.
(3) Characteristics of the materials.
(4) Form of loading (shape of the bending moment diagram).
(5) Profile of the prestressing tendon.
(6) Friction coefficient between the prestressing steel and duct.
(7) Amount of bonded non-prestressed reinforcing.

Another relationship that has been suggested for the value of f_{su} in unbonded members (to be used in lieu of Eq. 5-15) is as follows:

$$f_{su} = f_{se} + \left(30{,}000 - \frac{p^*}{f'_c} \times 10^{10}\right) \tag{5-16}$$

in which f_{se} is limited to $0.60 f'_s$, p^* is the percentage of steel, and f_{su}, f_{se} and f'_c are all in psi. Still another relationship that has been more recently proposed is:

$$f_{su} = f_{se} + \frac{1.4 f_c'}{100 p^*} + 10{,}000 \text{ psi} \tag{5-17}$$

The results of tests of members with unbonded tendons as well as Eqs. 5-15, 5-16 and 5-17 are shown in Fig. 5-9 (Ref. 21). Equations 5-16 and 5-17 have

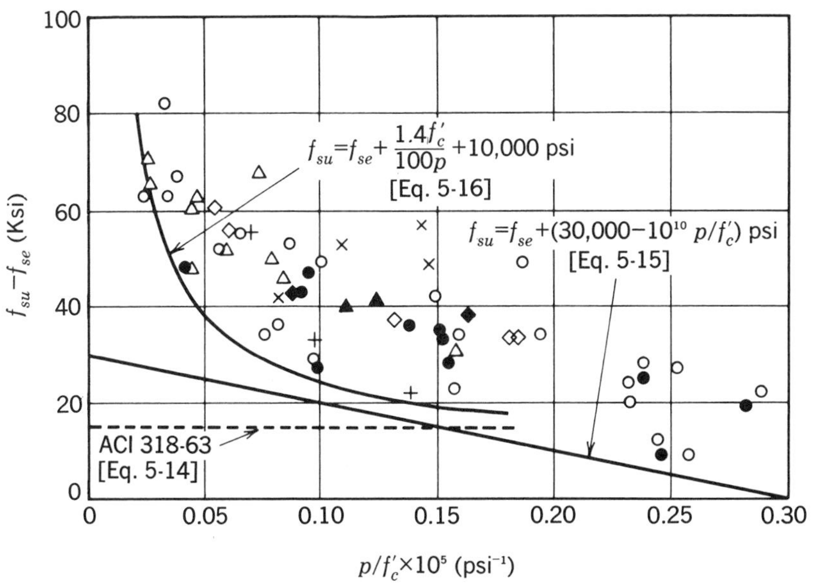

Fig. 5-9 Comparison of values of f_{su}–f_{se} for unbonded beams. Test data and suggested mathematical relationships are shown.

not been widely used, since they have not been included in any of the American codes or standards.

A method of computing the ultimate strength of prestressed members (with unbonded tendons) that takes into account the variables listed above has been proposed by Pannell (Ref. 6). This method is based upon experimental data and is considered slightly conservative. The method provides the value of f_{su} to be as follows:

$$f_{su} = \frac{q_u f'_c}{p} \tag{5-18}$$

in which

$$q_u = \frac{q_e + \lambda}{1 + 1.6\lambda} \tag{5-19}$$

with

$$q_e = \frac{p^* f_{se}}{f'_c} \tag{5-20}$$

and

$$\lambda = \frac{10^6 p^* d}{f'_c L} \tag{5-21}$$

The relationship for the ultimate moment is:

$$M_u = q_u(1 - 0.89q_u)f'_c bd^2 \tag{5-22}$$

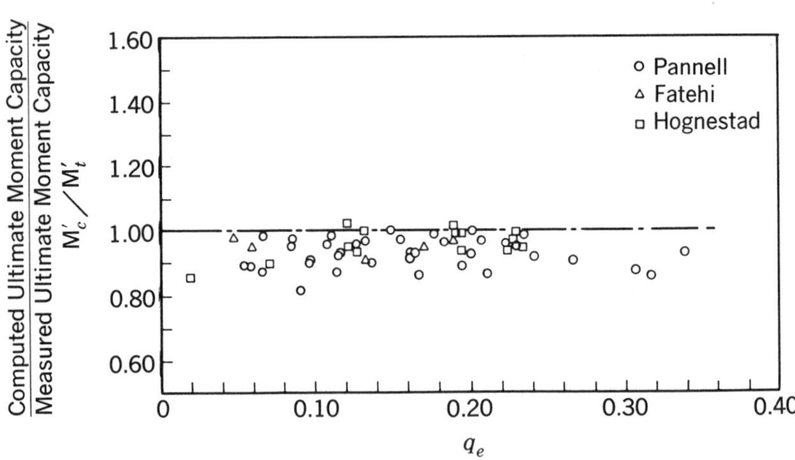

Fig. 5-10 Plot showing the ratio of the computed ultimate moment to that found by tests for various effective steel indices (From Ref. 6).

In Eqs. 5-18 through 5-22 the notation is standard and it should be recognized that the depth of the member d and the span length L must be in the same units.

A plot showing the accuracy of Eq. 5-22 is given in Fig. 5-10, in which the ordinate is the ratio of the calculated ultimate moment to the ultimate moment measured in tests conducted by various investigators.

It should be recognized that the ultimate moment capacity of a member stressed with unbonded tendons, unlike members with bonded tendons, (*see* Fig. 5-7), may be adversely affected by unintentional variations in the effective prestress. Hence, it is considered prudent to exert more care in estimating the losses of prestress and in supervising the stressing of unbonded members than would be considered necessary for bonded members, in order to assure the desired results are obtained.

5-4 Ultimate Moment Code Requirements—Bonded Members

The provisions of "Building Code Requirements for Reinforced Concrete" (ACI 318) (Ref. 5-3) are most commonly used in computing the ultimate moment capacity of prestressed concrete in actual design computations. The accepted relationship for the computation of the ultimate moment capacity of rectangular sections is as follows:

$$M_u = \phi[A_s^* f_{su} d(1 - 0.59q^*)] = \phi\left[A_s^* f_{su}\left(d - \frac{a}{2}\right)\right] \qquad (5\text{-}23)$$

in which $q^* = p^*(f_{su}/f_c')$ (in contrast to q'' which has been taken to be equal to $p^*(f_s'/f_c')$, $a = A_s^* f_{su}/0.85 f_c' b$, and the other terms are as have been previously defined. It should be noted that Eq. 5-23 is almost identical to Eq. 5-13 with the exception that f_{su} is used in lieu of f_s' and the factor ϕ, which is explained below, is introduced.

The Code permits the use of computed values of f_{su} when the properties of the steel are known. Computed values are obtained through an analysis as was done in Sec. 5-2, the result of which is a relationship between the steel index and f_{su}, as shown in Fig. 5-4. When the values of f_{su} are not computed from basic principles, the Code provides the approximate relationship that was given as Eq. 5-11. The relationship

$$f_{su} = f_s'\left(1 - 0.5\, p^* \frac{f_s'}{f_c'}\right)$$

is also shown in Fig. 5-4. It should be realized that the approximate relationship for f_{su} is sufficiently accurate for design purposes, but not for strength evaluation or research purposes.

The relationship for ultimate moment given in Eq. 5-23 is limited by the Code for use in applications where f_{se} is not less than $0.50f'_s$.

Another obvious difference between Eq. 5-13 and 5-23 is the factor ϕ which appears in Eq. 5-23. The factor ϕ is specified by ACI 318 to be equal to 0.90 for flexural computations and is introduced to provide for the possibility that a number of unintended defects in materials and construction may compound and result in an undercapacity.

Equation 5-23 is intended for use in computing the ultimate moment capacity of underreinforced rectangular sections without compressive reinforcing. Underreinforced sections are defined as those in which quality and proportions of the materials used in the member result in the following:

$$\text{Prestressing Reinforcement Index} = q^* = p^* \frac{f_{su}}{f'_c} \leq 0.30 \tag{5-24}$$

$$\text{Combined reinforcement index, rectangular sections} = q^* + q - q' \leq 0.30 \tag{5-25}$$

in which

$$q = \frac{A_s f_y}{bd}, \qquad q' = \frac{A'_s f_y}{bd}$$

A_s and A'_s are the areas of the non-prestressed reinforcement in the tensile and compressive flanges, respectively, f_y is the specified yield strength of the non-prestressed reinforcement, and the other terms are standard.

For flange sections the following requirements must be met:

$$\text{Combined reinforcement index flanged sections} = q_w + q_w - q_w \leq 0.30 \tag{5-26}$$

in which q^*_w, q_w and q'_w are the reinforcement indices computed as for q^*, q and q', respectively, except using b', the width of the web in lieu of b. The tensile force used in computing $q^*_w + q_w - q'_w$ is that which remains from the total tensile force after the flange has been developed.

The effect of compressive reinforcement is taken into account by basic principles.

In order to illustrate the manner in which these requirements are used, consider a prestressed concrete T beam having non-prestressed tensile and compressive reinforcement. The principal dimensions and properties of materials are:

$$
\begin{aligned}
b &= 60.0 \text{ in. (flange width)} & f'_c &= 4.0 \text{ ksi} \\
b' &= 16.0 \text{ in. (web width)} & f'_s &= 270 \text{ ksi} \\
h &= 32.5 \text{ in. (overall height)} & f_y &= 60 \text{ ksi} \\
d^* &= 27.75 \text{ in. (depth to } A_s^*) & A_s^* &= 3.50 \text{ in.}^2 \\
d &= 30.0 \text{ in. (depth to } A_s) & A_s &= 4.00 \text{ in.}^2 \\
d' &= 1.5 \text{ in. (depth to } A'_s) & A'_s &= 1.00 \text{ in.}^2 \\
t &= 4.5 \text{ in. (flange thickness)}
\end{aligned}
$$

From these data, the percentage of steel and f_{su} are computed as follows:

$$p^* = \frac{3.50}{60 \times 27.25} = 0.00214$$

$$f_{su} = 270\left(1 - \frac{0.50 \times 0.00214 \times 270}{4}\right) = 250.5 \text{ ksi}$$

The average depth to the tensile reinforcement is computed by:

$$
\begin{aligned}
A_s^* f_{su} &= 3.50 \times 250.5 = 876.75 \times 27.25 = 23891 \\
A_s f_y &= 4.00 \times 60 \quad = 240.00 \times 30.00 = 7200 \\
& \qquad\qquad\qquad\quad \overline{1116.75} \qquad\qquad \overline{31091}
\end{aligned}
$$

$$\text{average } d = 27.84 \text{ in.}$$

Using the average value for d, the reinforcement indices are computed and compared to the allowable value of 0.30 as follows:

$$q^* = \frac{3.50 \times 250.5}{60 \times 27.84 \times 4} = 0.131 < 0.30 \text{ as required by the Code.}$$

$$q = \frac{4.0 \times 60}{60 \times 27.84 \times 4} = 0.036$$

$$q' = \frac{1.0 \times 60}{60 \times 27.84 \times 4} = 0.009$$

$$q^* + q - q' = 0.158 < 0.30 \text{ as required by the Code.}$$

The total force required to develop the total compressive flange is:

$$0.85 f'_c bt + A'_s f_y = 978 \text{ k} < A_s^* f_{su} + A_s f_y$$

Therefore, the beam must be analyzed as a T beam and not as a rectangular beam. The force required to develop flange overhangs and the compressive reinforcement is computed from:

$$F_{sf} + A'_s f_y = 0.85 f'_c (b - b')t + A'_s f_y = 733.2 \text{ k}$$

The force to be developed by the web is:

$$F_{sw} = A_{s}^{*}f_{su} + A_s f_y - F_{sf} = 383.55$$

The combined reinforcement index for the web is computed as follows:

$$q_w^* + q_w - q_w' = \frac{383.55}{16 \times 27.84 \times 4} = 0.215 < 0.30 \text{ O.K.}$$

Finally, the ultimate moment capacity is computed as follows:

$$M_u = \phi\left[F_{sw}\, d\left(1 - \frac{0.59F_{sw}}{b'df_c'}\right) + 0.85f_c'(b - b')t(d - 0.5t) + A_s'f_y(d - d')\right]$$

$$= \frac{0.90}{12}\left[383.55 \times 27.84\left(1 - \frac{0.59 \times 383.55}{16 \times 27.84 \times 4}\right)\right.$$

$$\left. + 0.85 \times 4(44)(4.5)(25.59) + 1 \times 60 \times 26.34\right]$$

$$= 2110 \text{ ft-kips.}$$

Sections having reinforcement indices exceeding 0.30 are considered to be overreinforced and their capacity may be computed from basic principles (*see* Prob. 5-3) or from:

$$M_u = \phi[0.25f_c' bd^2]$$

which is an approximate, and hence, conservative relationship.

Another important Code requirement, which is equally applicable to bonded and unbonded tendons, provides that the minimum ultimate moment a section is capable of developing must be at least 1.2 times the cracking moment, based upon a modulus of rupture of $7.5\sqrt{f_c'}$. This is to guard against failure at cracking load, which was described in Sec. 5-2.

5-5 Ultimate Moment Code Requirements—Unbonded Tendons

The Code requirements for members prestressed with unbonded tendons are the same as those for members with bonded tendons, with the exception that f_{su} is specified to be as follows:

$$f_{su} = f_{se} + 10{,}000 + \frac{f_c'}{100p^*} \tag{5-28}$$

but not more than f_{sy} (yield strength of the prestressing steel) or $f_{se} + 60{,}000$. In Eq. 5-28 f_{su}, f_{se}, and f_c' are in psi. When information is available for determining a more accurate value of f_{su}, it may be used. Like Eq. 5-23, Eq. 5-28 is limited for use in applications where f_{se} is not less than $0.50f_s'$.

The relationships given in Eqs. 5-24 through 5-26 for determining if the members are to be analyzed as under-or overreinforced are also applicable to members with unbonded tendons. Members with unbonded tendons must have ultimate moment capacities equal to or greater than 1.2 times the cracking moment with the modulus of rupture equal to $7.5\sqrt{f'_c}$, as do bonded members.

Unbonded members that do not have bonded, non-stressed reinforcement could be subject to sudden brittle failure. For this reason the Code requires a minimum amount of non-prestressed reinforcement in the tensile zone of members stressed with unbonded tendons. The minimum amount of bonded reinforcing is computed by

$$A_s = \frac{T_c}{0.5f_y} \tag{5-29}$$

in which T_c is the tensile force in the concrete under dead load plus 1.2 times live load and f_y is limited to 60,000 psi, but the area of steel may not be less than

$$A_s = 0.004A_t \tag{5-30}$$

where A_t is the area of the concrete section between the flexural tension face and the center of gravity of the gross section. The Code permits the use of less bonded reinforcement than would be required by Eqs. 5-29 and 5-30 in two-way slabs if tension does not exist in the precompressed tensile zone under service loads.

ILLUSTRATIVE PROBLEM 5-4 Compute the ultimate moment for the member shown in Fig. 5-11. The member is stressed with an unbonded tendon, $f_{se} = 144$ ksi, $f'_s = 240$ ksi, $f_{sy} = 192$ ksi, $f'_c = 6.0$ ksi and the span is 40.0 ft. Use the Eq. 5-28 for f_{su} and the method proposed by Pannell.

SOLUTION:

By Eq. 5-28:

$$p^* = \frac{3.20}{24 \times 24.8} = 0.00538$$

$$f_{su} = 144 + 10 + \frac{6.0}{0.538} = 165 \text{ ksi} \quad \begin{matrix} < f_{se} + 60 = 204 \text{ ksi} \\ < f_{sy} = 192 \text{ ksi} \end{matrix}$$

$$P_{su} = 3.20 \times 165 = 528 \text{ kips}$$

$$a = \frac{528}{0.85 \times 6 \times 24} = 4.31 \text{ in.}$$

$$M_u = \frac{528(24.8 - 2.16)}{12} = 996 \text{ ft-kips}$$

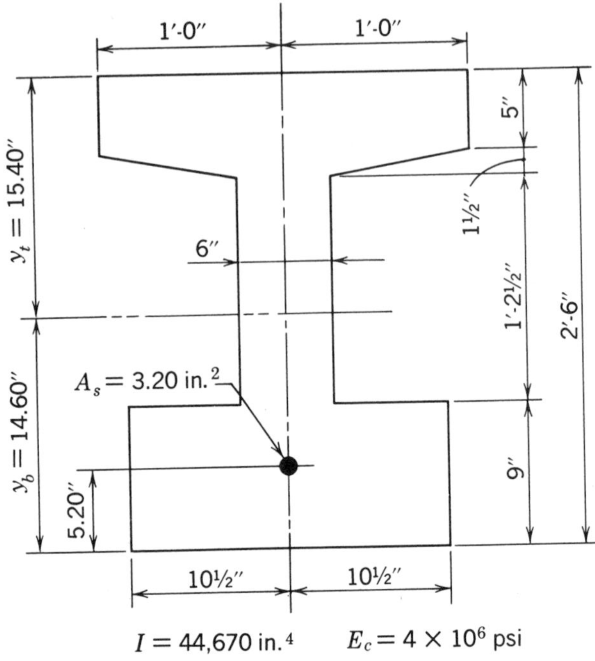

Fig. 5-11 Cross section of a beam used in Prob. 5-4.

By Pannell's method:

$$p^* = \frac{3.20}{24 \times 24.8} = 0.00537, \qquad q_e = \frac{0.00537 \times 144}{6} = 0.129$$

$$\lambda = \frac{10^6 \times 0.00537 \times 2.5}{6000 \times 40.0} = 0.056$$

$$q_u = \frac{0.129 + 0.056}{1 + 0.0895} = 0.170$$

$$M_u = \frac{0.170(1 - 0.80 \times 0.170)6 \times 24 \times 24.8^2}{12} = 1085 \text{ ft-kips}$$

5-6 Shear and Principal Tensile Stresses

The principal tensile stresses, which are the stresses that require "shear reinforcing" near the end of a beam, are not as severe if the beam is prestressed rather than constructed of reinforced concrete. This reduction in intensity of the principal tensile stresses is the result of the compressive stresses induced by the prestressing and, in the case of members with curved or draped tendons, a reduction in the shear forces acting upon the concrete section.

In designing prestressed beams for shear, European practice has been to determine the principal tensile stress in the concrete, and if the principal tensile stress exceeds the permissible value, provide sufficient reinforcing steel to carry the entire shear force (Ref. 8). This condition is most critical under the desired ultimate load, in which case, the shear steel is permitted a stress equal to the yield point of the steel. This procedure is based upon the assumptions that 5000 psi concrete can carry from 300 to 400 psi pure tension at ultimate load and, if this value is exceeded, the entire shear force must be carried by the reinforcing steel, since the concrete and steel cannot work together due to the difference in strain that occurs in each material when under reasonable unit stress.

In this country, the practice has been to assume that the concrete is capable of carrying a specific unit shear stress and the excess shear must be carried by reinforcing steel. Both design procedures are discussed in this article.

In 1954, the "Criteria for Prestressed Concrete Bridges" was published by the Bureau of Public Roads (Ref. 9). This was the first design criteria for prestressed concrete published in the United States and it provided that principal tensile stresses under service load conditions (dead + live + impact) should not exceed $0.03f'_c$. In addition, under ultimate design load, the principal tensile stresses were not permitted to exceed $0.08f'_c$. If the principal tensile stress exceeded these values, stirrups were to be provided to carry the excess principal tensile stress (that above $0.03f'_c$) in the first case (service loads) or the entire principal tensile stress in the second case (ultimate design load). The load factors for the determination of minimum ultimate loads were as follows:

$$\text{Dead Load} + 3(\text{Live Load} + \text{Impact})$$

or

$$2(\text{Dead Load} + \text{Live Load} + \text{Impact})$$

whichever was greater.

The classical relationship for computing principal tensile stress (T_p) is:

$$T_p = \sqrt{v^2 + \frac{C^2}{2}} - \frac{C}{2} \qquad (5\text{-}31)$$

in which v is the shear stress and C is the compressive stress (in psi) which act upon an infinitesimal particle. A complete treatment of the method of computing principal tensile stresses and reinforcing will be found in Ref. 23.

When using the "Criteria for Prestressed Concrete," the point at which maximum shears were considered to exist was specified to be at a distance of 1.5 times the depth of the member from the nearest support.

Subsequently, a method of designing prestressed concrete beams for shear was introduced in the "Tentative Recommendations for Prestressed Concrete" by ACI-ASCE Joint Committee 323 (Ref. 4). Using this method of analysis, it is only necessary to determine the shear force that the concrete is assumed able to carry at ultimate, which is computed by

$$V_c = 0.06 f'_c b' jd \tag{5-32}$$

in which $0.06f'_c$ cannot exceed 180 psi. Reinforcing steel must be provided to carry the shear force over and above that which the concrete can carry. The amount of web reinforcing required is computed by the relationship:

$$A_v = \frac{(V_u - V_c)s}{2 f'_y jd} \tag{5-33}$$

in which V_u is the shear force at the desired ultimate load, s is the stirrup spacing, f'_y is the yield strength of the steel, jd is the lever arm of the section, and the factor 2 was included to take into account the reduction in web reinforcing required in prestressed concrete as a result of the prestressing. For low values of average prestressing stress, the Recommendations state that this factor of 2 may be too high. The reduction in the shear force that is carried by the concrete, due to the curvature of the tendons, should be taken into account in determining V_u.

The Recommendations state that the critical sections for shear are not usually near the ends of the span, but at some point away from the ends in a region of high moment, because the formation of inclined cracks reduces the flexural capacity of the member. Furthermore, it was recommended that the shear be investigated only within the middle half of simple beams that are designed for moving loads and that the amount of web reinforcement required at the quarter point be supplied throughout the end quarters of the spans. For simple beams designed for uniformly distributed loads, the most critical section is assumed to be located at a distance that is equal to the depth of the beam from the support, and it was stipulated that the middle third of the beam should not have less web reinforcing than that required at the third point.

The procedure specified in the 1963 edition of ACI 318 for the shear design of prestressed-concrete members provides that the members are designed only for ultimate load conditions. No option for designing under a working stress criteria is offered. In addition, two types of possible shear failure are recognized. These are illustrated in Fig. 5-12 and are described as follows:

Type I This type of cracking is associated with flexural cracking. Some authorities believe that for this type of crack to adversely affect the capacity of a member, it must be cracked in such a manner that the horizontal projection of the crack is equal in length to the depth of the member, and for this reason, it is believed a flexural crack that occurs at a distance equal to the depth of the member away (in the direction of lesser moment) from the section being investigated may lead to a critical crack (Ref. 10). In addition, principal tensile stresses along a potential crack may be aggravated by flexural cracks that may occur in the vicinity of the potential type I crack, and since principal tensile stresses are normally maximum at the center of gravity of a beam (can be taken to be approximately equal to one-half of the depth of the beam), a flexural crack at one-half the beam depth from the section under consideration can be considered to cause a type I crack. Shear cracking in reinforced- and prestressed-concrete beams are generally type I cracks. These *flexural-shear* cracks begin as flexural cracks extending approximately vertically into the beam. When a critical combination of flexural and shear stresses develop near the top of a flexural crack, the inclined crack forms.

Type II This type of cracking is associated with principal tensile stresses, in areas where there are no flexural cracks, and originates in the web of the member near the centroid where the principal tensile stresses are the greatest; it subsequently extends towards the flanges (Ref. 10). Type II cracking is fairly unusual. It may appear near the supports of highly prestressed beams that have thin webs. It may also appear near inflection points and bar cut-off points of continuous, reinforced-concrete members under axial tension (Ref. 11).

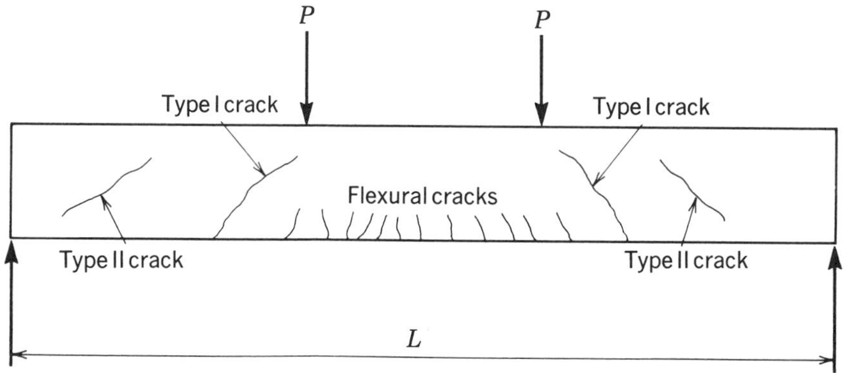

Fig. 5-12 Illustration of flexural cracks and type I and type II shear cracks.

With this design procedure, it is necessary to determine the portions of a beam that may be subject to type I and type II cracking, and then to determine the total force the concrete can withstand in these areas. Steel reinforcing must be provided for the shear force equal to the difference between the ultimate shear capacity that must be provided and the amount of shear required to produce the first type I or type II crack in the concrete section.

The shear force that normal weight concrete can carry in the vicinity of type I cracking is given by the following relationship:

$$V_{ci} = 0.6b'd\sqrt{f'_c} + \frac{M_{cr}}{\dfrac{M}{V} - \dfrac{d}{2}} + V_d \tag{5-34}$$

with a minimum value of $1.7b'd\sqrt{f'_c}$. The notation for Eq. 5-34 is standard, except M and V are to be taken as the moment and shear due to "externally applied loads," which are defined as external ultimate loads acting upon the member, excepting those due to the prestressing, and V_d is defined as the shear due to dead load. In Eq. 5-34 the relationship for the cracking moment is specified to be

$$M_{cr} = \frac{I}{y}(6\sqrt{f'_c} + f_{pe} - f_d) \tag{5.35}$$

in which f_{pe} is the compressive stress in the concrete due to the effective prestress only and f_d is the stress due to dead load. Both f_{pe} and f_d are the stresses that occur at the extreme fiber of a section at which tension stresses are caused by the applied loads. The term

$$\frac{M_{cr}}{\dfrac{M}{V} - \dfrac{d}{2}} + V_d \tag{5-36}$$

is the shear at the section under consideration when flexural cracking occurs at a distance equal to one-half the depth of the beam away. The term

$$0.6b'd\sqrt{f'_c} \tag{5-37}$$

is the shear force required to cause a diagonal shear crack to form after flexural cracking has occurred. The lower limit of $1.7b'd\sqrt{f'_c}$ has been specified since tests have shown this value is only exceeded in beams with extremely low levels of prestress.

The relationship specified to be the shear force a normal weight concrete beam can withstand in the area where type II cracking could occur is as follows:

$$V_{cw} = b'd(3.5\sqrt{f'_c} + 0.3f_{pc}) + V_p \tag{5-38}$$

in which f_{pc} is defined as the compressive stress in the concrete, after all prestress losses have occurred, at the centroid of the cross section resisting the applied loads or at the junction of the web and flange when the centroid lies in the flange. (In a composite member f_{pc} will be the resultant compressive stress at the centroid of the composite section or at the junction of the web and flange when the centroid lies within the flange, due to both prestress and to the bending moments resisted by the precast member acting alone.) V_p is the vertical component of the effective prestress force at the section being considered.

Equation 5-38 is intended to facilitate computations and approximate the shear force a normal weight concrete beam can withstand under dead load plus live load shear that results in a principal tensile stress of $4\sqrt{f_c'}$ at the center of gravity of the section. The magnitude of V_{cw} can be computed using the classical relationship described earlier.

The amount of shear reinforcing placed perpendicular to the axis of a member is specified to be

$$A_v = \frac{(V_u - \phi V_c)}{\phi d f_y} s \tag{5-39}$$

in which V_c is equal to the lesser of V_{ci} (Eq. 5-34) or V_{cw} (Eq. 5-38), ϕ is a factor (equal to 0.85 for shear computations) introduced to account for unintended variations in materials and construction tolerances that may compound and result in an undercapacity, s is the spacing of the web reinforcing, and the remaining notation is standard. However, the minimum amount of shear reinforcement is specified to be

$$A_v = \frac{A_s^*}{80} \cdot \frac{f_s'}{f_y} \cdot \frac{s}{d} \sqrt{\frac{d}{b'}} \tag{5-40}$$

Equations 5-39 and 5-40 apply equally to members made of lightweight concrete, whereas the relationship for v_{ci} and v_{cw}, which are applicable to members of lightweight concrete, are specified to be

$$V_{ci} = 0.1 F_{sp} b'd\sqrt{f_c'} + \frac{M_{cr}}{\dfrac{M}{V} - \dfrac{d}{2}} \tag{5-41}$$

with a minimum value of $0.25 F_{sp} b'd\sqrt{f_c'}$. F_{sp} is the ratio of the splitting tensile strength to the square root of the compressive strength. In this relationship $M_{cr} = (I/y)\,(0.9 F_{sp}\sqrt{f_c'} + f_{pc} - f_d)$ and

$$V_{cw} = b'd\left[0.5 F_{sp}\sqrt{f_c'} + f_{pc}\left(0.2 + \frac{F_{sp}}{67}\right)\right] + V_p \tag{5-42}$$

For lightweight concrete members, V_{cw} can be taken as the shear force due to dead load plus live load which corresponds to a principal tensile stress of $0.6F_{sp}\sqrt{f_c'}$ at the centroidal axis of the section resisting live load.

It should be recognized that ACI 318-63 requires a specific amount of minimum shear reinforcing in prestressed members, unless tests are made to demonstrate it is not required. This provision, if strictly enforced, would prohibit the use of prestressed-concrete slabs which (without shear reinforcing) have proved to be economical and structurally safe in many installations.

The complete shear design requirements of ACI 318-63 should be read and understood, with the several qualifications and definitions of terms contained therein, before attempting to apply these equations on actual problems.

Prestressed-concrete flexural members are generally regarded to be substantially superior to reinforced concrete beams as far as resistance to shear stress is concerned. With respect to flexural-shear cracking (type I cracks) at ultimate load, it can be shown that prestressed members may not be superior to reinforced concrete by considering Fig. 5-13 (from MacGregor). This figure shows the results of an analytical study of the development of

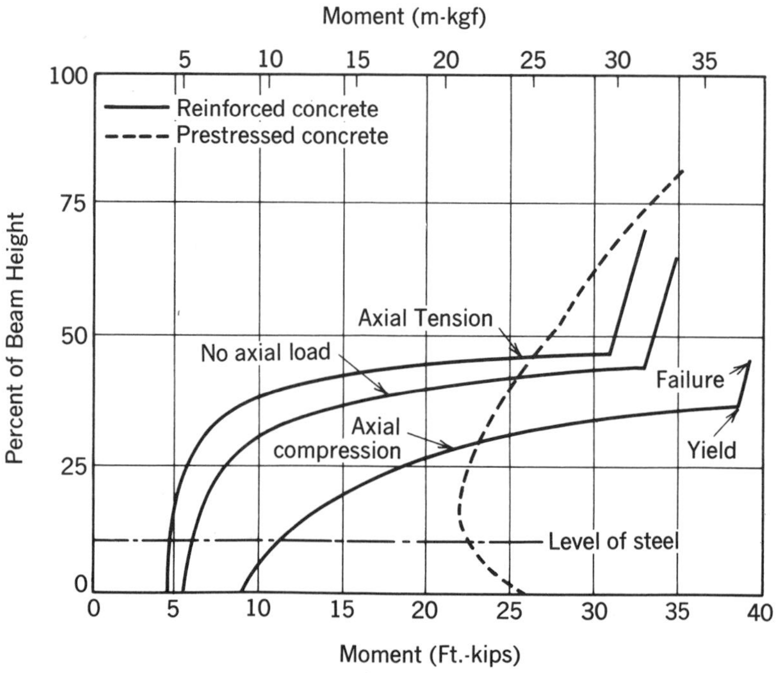

Fig. 5-13. Relationship between moment and flexural crack height (From Ref. 11).

cracks in three reinforced concrete beams and one prestressed concrete beam that had the properties shown in Table 5-1.

TABLE 5-1 Properties of Beams Plotted in Fig. 5-13.

*Type of Beam**	*Reinforcement Ratio* p	*Concrete Cylinder Strength* f'_c, psi	*Steel Yield Stress* f_y, ksi	*Axial Force* N kips	*Effective Prestress Force* F_{SP}, kips	*Calculated Ultimate Moment* M_{fu}, ft-kips
Reinforced concrete	0.015	3000	45	0	—	34.8
Reinforced concrete—Axial tension	0.015	3000	45	−7.2	—	33.0
Reinforced concrete—Axial compression	0.015	3000	45	+27.0	—	39.3
Prestressed concrete	0.0028	5000	250†	—	27.0	34.8

* The effective depth d of all beams was 10.8 in. (27.4 cm)
† Ultimate stress f_{su} rather than yield stress.

From Fig. 5-13 it will be seen that the reinforced-concrete beam that is without axial load cracks at a relatively low moment and the cracks immediately grow to a depth of about 15% of the beam height. The cracks continue to grow in height with a decreasing rate as the moment is increased. The cracks increase in depth with little increase in moment after the reinforcement yields. The reinforced concrete beams with axial loads are affected as one would expect by the axial loads.

The effect of the prestressing on the prestressed beam is to require a considerable increase in the moment required to cause cracking. When the first crack does form, it extends almost to middepth of the beam immediately before equilibrium under the cracking moment is achieved. An increase in the moment results in a significant increase in the height of the crack.

The difference in action between the reinforced- and prestressed-concrete beam can be understood by considering the mechanism of cracking. Immediately prior to cracking, the concrete is sustaining a tensile force in the bottom fibers and this force must be transferred to the steel when cracking occurs. In a typical reinforced-concrete beam, the percentage of steel is relatively high and the additional force can be assumed by the steel, without a large strain, with the result that the height of the crack is not great. In the prestressed beam, the percentage of steel is low with the result that a large crack depth is formed due to the relatively large strain the steel must undergo in resisting the additional tensile force resulting from cracking.

In addition, the magnitude of the shear stresses acting near the top of the flexural crack affects the rate at which the initiating flexural crack develops into an inclined crack. The shear stresses are relatively high if the shear force is high when flexural cracking first occurs. Small shear spans and thin webs are conducive to this situation. The shear stresses also tend to be high in beams with low steel percentages when the flexural cracks extend high in the beam, leaving little uncracked concrete above the crack.

Because of these actions, initiating flexural cracks in prestressed beams become critical rapidly after cracking first occurs and inclined cracks develop shortly after flexural cracking first takes place (Ref. 11).

The shear provisions of ACI 318-71 convert the relationships for V_{ci} and V_{cw} of ACI 318-63 to *unit* shear stress formulas from shear *force* formulas. The current relationship for v_{ci} is

$$v_{ci} = 0.6\sqrt{f_c'} + \frac{V_d + (V_l M_{cr}/M_l)}{b'd} \tag{5-43}$$

in which M_l and V_l are defined as the maximum bending moment due to externally applied design loads and the shear force at the section occurring simultaneously with M_l, respectively, and $M_{cr} = (I/y_t)(6\sqrt{f_c'} + f_{pe} - f_d)$. V_d is the shear force at the section due to dead load (not increased by load factors). The minimum value of v_{ci} is specified to be $1.7\sqrt{f_c'}$. This relationship is similar to Eq. 5-34 in which the term $d/2$ has been removed. It should be noted that y_t is defined as the distance from the centroidal axis of gross section, neglecting the reinforcement, to the extreme fiber in tension. This definition of y_t should not be confused with the definition (distance from the centroidal axis to the top fiber) that is used elsewhere in this book in the discussions of flexural stresses. The current relationship for v_{cw} is

$$v_{cw} = 3.5\sqrt{f_c'} + 0.3f_{pc} + \frac{V_p}{b'd} \tag{5-44}$$

In lieu of Eq. 5-44, v_{cw} can be taken as the shear stress corresponding to a multiple of dead load plus live load, which results in a computed principal tensile stress of $4\sqrt{f_c'}$ at the centroidal axis of the member, or at the intersection of the flange and the web when the centroidal axis is in the flange.

The nominal shear stress is computed by

$$v_u = \frac{V_u}{\phi b'd} \tag{5-45}$$

in which V_u is the design (ultimate) shear force and ϕ is 0.85. If v_u exceeds the lesser of v_{ci} or v_{cw}, reinforcing computed by

$$A_v = \frac{(v_u - v_c)b's}{f_y} \tag{5-46}$$

must be provided. The maximum value permitted for the quantity $(v_u - v_c)$ is $8\sqrt{f'_c}$ and when it exceeds $4\sqrt{f'_c}$, the stirrup spacing must be reduced. The minimum amount of reinforcing permitted is given below in Eqs. 5-48 and 5-49.

An alternate method for computing v_c is permitted for members with an effective prestress which is equal to at least 40% of the tensile strength of the reinforcement. This relationship is

$$v_c = 0.6\sqrt{f'_c} + 700\,\frac{V_u d}{M_u} \tag{5-47}$$

in which v_c is not less than $2\sqrt{f'_c}$ nor greater than $5\sqrt{f'_c}$. M_u is the applied design (ultimate) load moment at the section and V_u is the total applied design (ultimate) load shear force at the section. The term $V_u d/M_u$ shall not be taken greater than 1.0 in Eq. 5-47.

When applying Eq. 5-47, d is the distance from the extreme compression fiber to the centroid of the prestressing tendons. When applying Eqs. 5-43 and 5-44, d is the distance from the extreme compression fiber to the centroid of the tendons or 0.80 times the total depth of the section, whichever is greater.

ACI 318-71 provides for minimum shear reinforcing in reinforced and prestressed members alike. Slabs, footings and concrete joist construction are exempt from this requirement. The minimum amount of reinforcing is

$$A_v = 50b's/f_y \tag{5-48}$$

for reinforced or prestressed members, but for prestressed members the relationship

$$A_v = \frac{A_s^*}{80}\,\frac{f'_s}{f_y}\,\frac{s}{d}\sqrt{\frac{d}{b'}} \tag{5-49}$$

may be used. The shear stress is taken to be maximum at a distance equal to one-half the depth of the member from the support. When the nominal torsion stress τ_u is greater than $1.5\sqrt{f'_c}$ and when web reinforcement is required, the minimum area of closed stirrups shall be:

$$A_v + 2A_0 = \frac{50b's}{f_y}$$

in which A_0 is the area of one leg of a closed stirrup resisting torsion within a distance of s.

If the value of f'_{sp} is specified, f'_{sp} is defined as the average splitting tensile strength of lightweight aggregate concrete in psi; the equations given above for normal concrete apply for members made with lightweight concrete. In such a case, $f'_{sp}/6.7$ is to be substituted for $\sqrt{f'_c}$ and the value of $f'_{sp}/6.7$ may not exceed $\sqrt{f'_c}$. If f'_{sp} is not specified, all values of $\sqrt{f'_c}$ affecting v_c, τ_c and M_{cr} shall be multiplied by 0.75 for all-lightweight concrete and 0.85 the values for sand-lightweight concrete. A linear interpolation is to be used when the sand is partially lightweight.

An interesting and logical theory, relative to the manner in which shear forces are transmitted in reinforced concrete beams and which should be equally applicable to prestressed concrete beams under ultimate loading conditions, has been offered by Kani (Ref. 12). This theory is basically that reinforced concrete beams act like a family of tied arches in transferring the applied loads to the supports. This concept is illustrated in Fig. 5-14. The load on the uncracked beam and the family of arches is shown in Fig. 5-14(a). Arches I and II are shown as freebody diagrams in Fig. 5-14(b) and (c) with the forces acting upon them as they exist before cracking. The freebody diagram for Arch II, after cracking, is shown in Fig. 5-14(d), where it will be noted that the tensile stresses on the top of the arch near the ends have changed in magnitude and distribution as a result of cracking. The tensile stresses on top of the arch function as supports for the hanging of Arch II from Arch I. Arch II is still stable immediately after cracking and the beam will not fail unless additional load is applied, causing the crack to extend and hence reducing the area available for the very important hanging forces. The provision of stirrups or bent bars in the area of the crack between Arches I and II permits the loads on Arch II to be transferred to Arch I in a manner similar to hangers in a suspended span.

Punching shear stresses, as might be caused by a concentrated wheel load on a prestressed slab or by the supporting column on a flat plate structure, are a source of concern to the prestressed concrete designer. Provisions are made for punching shear in reinforced concrete slabs and footings in ACI 318-71. There has not been a great deal of research related to the problem of evaluating prestressed plates under such loading, but what has been done would indicate that the strength of prestressed flat slabs is very high in punching shear. It would be expected that the capacity of a slab to resist punching shear stresses would be a function of the concrete strength, the effective prestress, the magnitude of the perimeter of the loaded area, and the amount of moment at the loaded point. No mathematical relationship that accurately predicts the ultimate strength of a prestressed slab is known to the author, but some test data are available that give an indication of the punching shear capacity of prestressed slabs.

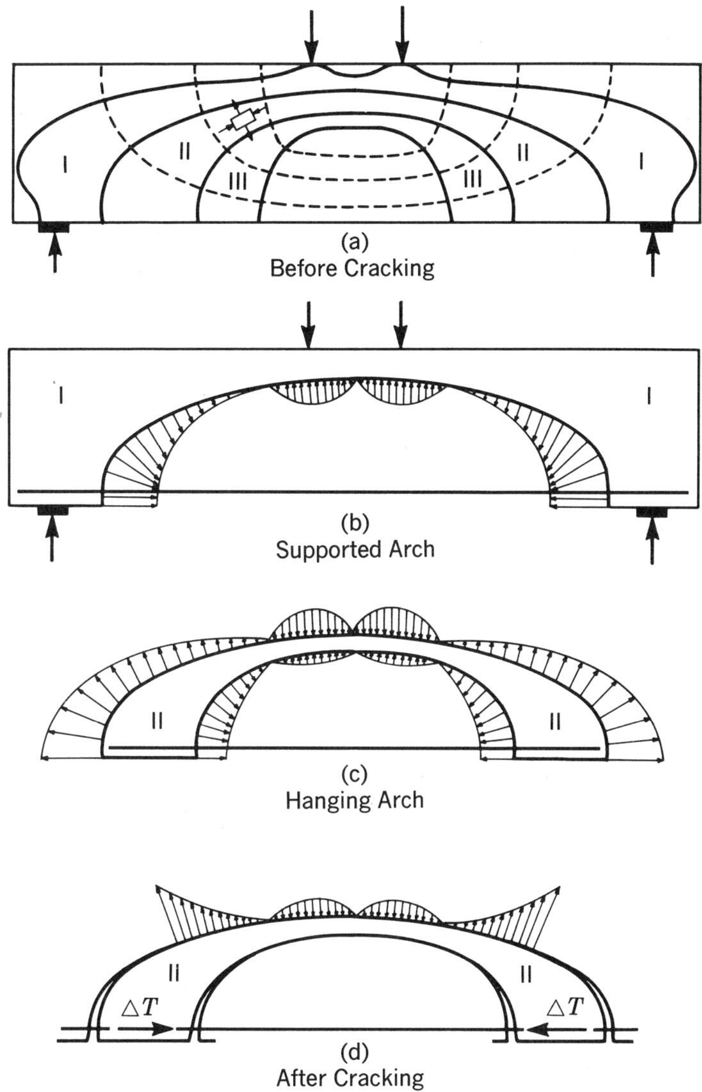

Fig. 5-14 Internal arches in a reinforced-concrete beam (*See* Ref. 12).

A one-way, prestressed-concrete slab with a depth of 9 in. at the point of loading was loaded with a 150 ton ram. The load was applied on a diameter of 1 ft at the top of the slab. The concrete strength was 5000 psi (nominal) and the slab was post-tensioned to a value of 150 psi in the transverse direction and prestressed to approximately 400 psi in the longitudinal direction.

When loaded to ultimate, the punching shear failure occurred at a load of 250,000 lb and the cone of concrete that was punched out varied in diameter from 1 ft at the top to 5 ft at the bottom.

A series of 7-in. thick post-tensioned specimens that were designed to simulate the column-slab connection of a two-way flat slab were constructed and tested in order to evaluate the effectiveness of various reinforcing configurations. The slabs consisted of concrete having a nominal cylinder strength of 5000 psi and were post-tensioned in each direction to 250 psi. Some of the specimens were designed to represent cast-in-place construction with 12-in. square columns while others were provided with lifting collars and were intended to simulate lift slab construction. The cast-in-place specimens failed at loads as follows:

Specimen	*Ultimate Load*	*Type of Supplementary Reinforcing Steel*
1C1	165 k	None
1C2	180 k	4-No. 5 bars 8′-0″ long each way in top of slab
1C3	190 k	Special welded wire fabric in top of slab
1C4	170 k	2-No. 9 bars 11′-6″ long each way in bottom of the slab passing over the column.

The conical section that was punched out at failure had dimensions equal to the column at the bottom and was approximately 6 ft in diameter at the top. The lift-slab specimens generally showed ultimate strengths somewhat higher than that of the cast-in-place specimens, since the lift-slab collars result in a larger loaded area than that offered by the specimens simulating cast-in-place construction.

ILLUSTRATIVE PROBLEM 5-5 Using the provisions of ACI 318-71, investigate the double T slab shown in Fig. 5-15 if it is designed for the following loads:

Double T Slab	49 psf
Roofing	6 psf
Insulation	2 psf
Ceiling	3 psf
Total Dead Load	60 psf
Live Load	20 psf
Total Load	80 psf

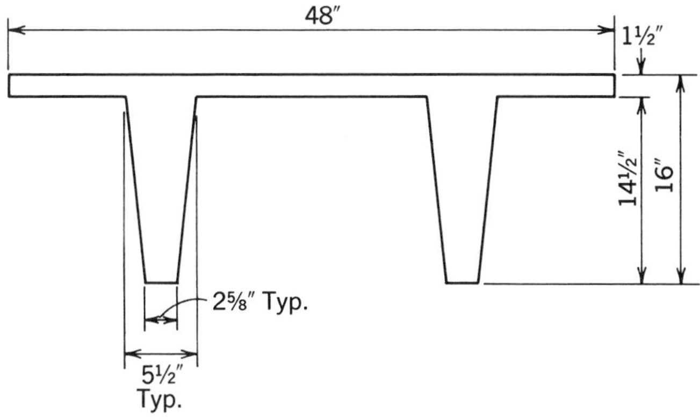

Fig. 5-15 Double T slab.

Assume the member stressed with 4-3/8-in. diameter straight strands in each leg. Each strand has a breaking strength of 20.0 k and an effective prestressing force of 11.2 k. The section properties for the slab are as follows:

$$\text{Area} = 189.5 \text{ in.}^2 \qquad y_t = 5.17 \text{ in.} \qquad S_t = 823 \text{ in.}^3 \qquad r^2/y_t = 4.34 \text{ in.}$$

$$I = 4256 \text{ in.}^4 \qquad y_b = 10.83 \text{ in.} \qquad S_b = 393 \text{ in.}^3 \qquad r^2/y_b = 2.07 \text{ in.}$$

$$e = 10.83 - 4.25 = 6.58 \text{ in.}$$

$$d = 11.75 \text{ in.} \qquad 0.80t = 12.8 \text{ in.}$$

$$A_s^* = \frac{8 \times 20}{250} = 0.64 \text{ in.}^2 \qquad f_y = 40.0 \text{ ksi}$$

$$f_s' = 250 \text{ ksi} \qquad b' \cong 8.0 \text{ in.} \qquad \text{load factor} = 1.4D + 1.7L$$

The concrete cylinder strength at 28 days is specified to be not less than 5000 psi. The design span is 40.0 ft.

SOLUTION:

$$M_{cr} = \frac{I}{y}(6\sqrt{f_c'} + f_{pe} - f_d)$$

$$= \frac{393}{12}(6\sqrt{f_c'} + f_{pe} - f_d)$$

	Service Loads	Design Loads	Max. Design Shear
$w_d = 4 \times 60 =$	$240 \text{ plf} \times 1.4 =$	336 plf	6720 lb
$w_a = 4 \times 20 =$	$80 \text{ plf} \times 1.7 =$	136 plf	2720 lb
		472 plf	9440 lb

At midspan:

$$M_d = 240 \times \frac{40^2}{8} = 48{,}000 \text{ ft-lb}$$

$$f_d = \frac{48{,}000 \times 12}{393} = 1466 \text{ psi}$$

The computations are summarized in Table 5-2. The following factors were used:

$$0.6\sqrt{f'_c} = 42.4 \text{ psi}$$

$$v_{cw} = 3.5\sqrt{f'_c} + \frac{0.3 \times 8 \times 11{,}200}{189.5} = 389 \text{ psi}$$

$$\text{Min. Reinf.} = \frac{0.64}{80} \times \frac{250}{40} \times \frac{12}{11.75}\sqrt{\frac{11.75}{8}} = 0.062 \text{ sq in. per ft}$$

$$\text{Max. stirrup spacing} = 0.75 \times 16 = 12.0 \text{ in.}$$

$$f_{pe} = \frac{8 \times 11{,}200}{189.5}\left(1 + \frac{6.58}{2.07}\right) = +1970$$

$$6\sqrt{f'_c} + f_{pe} = 2394 \text{ psi}$$

A summary of the shear stresses are shown in Fig. 5-16.

ILLUSTRATIVE PROBLEM 5-6 Investigate the beam shown in Fig. 5-17 for shear reinforcing if the beam has the following section properties:

$A = 876 \text{ in.}^2$ $y_t = 25.0 \text{ in.}$ $S_t = 17{,}300 \text{ in.}^3$ $r^2/y_t = 19.8 \text{ in.}$

$I = 433{,}350 \text{ in.}^4$ $y_b = 38.0 \text{ in.}$ $S_b = 11{,}400 \text{ in.}^3$ $r^2/y_b = 13.0 \text{ in.}$

$e = 32.1$ in. at center of span load factor $= 1.4D + 1.7L$

TABLE 5-2

Point	Moment Factor	f_d psi	$6\sqrt{f'_c} + f_{pe} - f_d$ psi	M_{cr} ft-lb	M_l/V_l^* ft	V_d lb	v_u psi	$\frac{V_l M_{cr}}{M_l}$	v_{ci} psi	$1.7\sqrt{f'_c}$ psi	v_{cw} psi	v_c psi
℄	1.00	1466	928	30,400	∞	0	0	0	42.4	120	389	120
0.9	0.99	1319	1075	35,200	98.8	480	10.8	356	50.6	120	389	120
0.8	0.96	1173	1221	40,000	48.0	960	21.6	833	59.9	120	389	120
0.7	0.91	1026	1368	44,800	30.3	1440	32.4	1478	70.9	120	389	120
0.6	0.84	880	1514	49,600	21.5	1920	43.2	2306	83.7	120	389	120
0.5	0.75	733	1661	54,400	15.0	2400	54.0	3627	101.2	120	389	120
0.4	0.64	586	1808	59,200	10.7	2880	64.8	5533	124.6	120	389	125
0.3	0.51	440	1954	64,000	7.28	3360	75.6	8791	161.1	120	389	161
0.2	0.36	293	2101	68,800	4.52	3840	86.4	15,221	228.6	120	389	229
0.1	0.19	147	2247	73,600	2.12	4320	97.2	34,717	423.6	120	389	389
0.0	0.00	0	2394	78,400	0	4800	108	∞	∞	120	389	389

* $\frac{M_l}{V_l} = \frac{Lx - x^2}{L - 2x}$ for a uniformly loaded beam in which L is the span and x is the distance from the support to the point under consideration.

$$v_u = \frac{V_u}{0.85b'd}$$

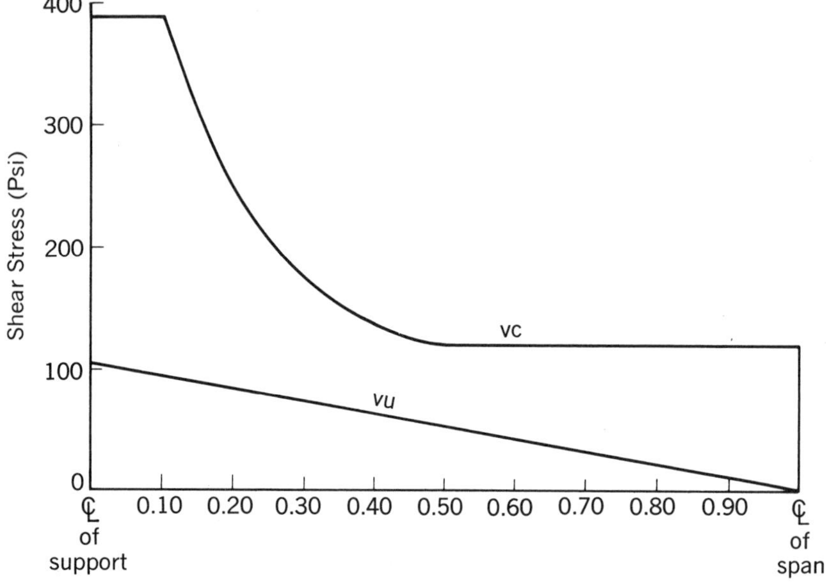

Fig. 5-16 Ultimate shear stress and shear capacity for Prob. 5-5.

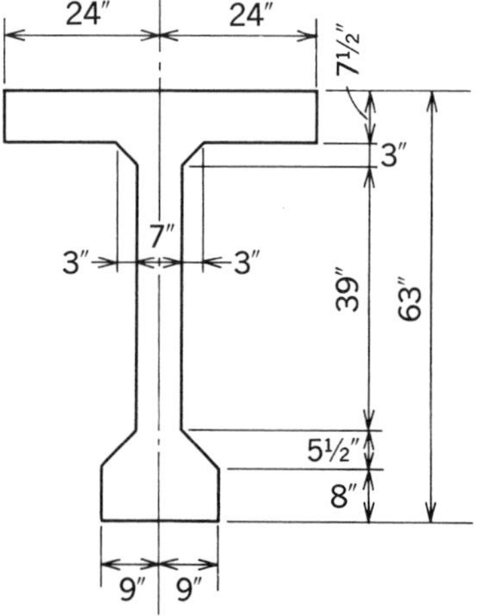

Fig. 5-17 Beam section.

The design dead loads are as follows:

Girder dead load	= 0.911 k-ft
Superimposed dead load	= 0.500 k-ft
Total Dead Load	1.411 k-ft

The design span is 80.0 ft and the design live load is one concentrated load of 70.0 kips applied at the quarter point. The beam is stressed with an effective force of 670 k and the ultimate strength of the tendon is 1115 k, $f'_c = 5000$ psi. The tendon is on a parabolic curve with $e = 0$ at the support and $e = 32.1$ in. at midspan.

$$f_y = 60{,}000 \text{ psi} \qquad f'_s = 270{,}000 \text{ psi}$$

SOLUTION:

	Shear at Service Load	Shear at Design Load
$w_d = 1.411$ k/ft × 40′ =	56.4 kips	79.0 kips

At midspan:

$$M_d = \frac{1.411 \times 80^2}{8} = 1129 \text{ ft-kips}$$

$$f_d = \frac{1129 \times 12000}{11{,}400} = 1188 \text{ psi}$$

$$f_{pe} = \frac{670}{876}\left(1 + \frac{32.1}{13.0}\right) = 2650 \text{ psi}$$

$$d = 25.0 + 32.1 = 57.1 \text{ in. at midspan}$$

$$0.80t = 0.8 \times 63 = 50.4 \text{ in.}$$

At support:

$$f_{pe} = \frac{670}{876}\left(1 + \frac{0}{13.0}\right) = 765 \text{ psi}$$

$$0.80t = 50.4 \text{ in.}$$

$$V_d = 1.411 \times 40 = 56.44 \text{ k}$$

$$V_l = 79.0 + 1.7 \times 52.5$$

$$= 168 \text{ k}$$

$$V_p = 670 \times \frac{4 \times 32.1}{12 \times 80} = 89.6 \text{ k}$$

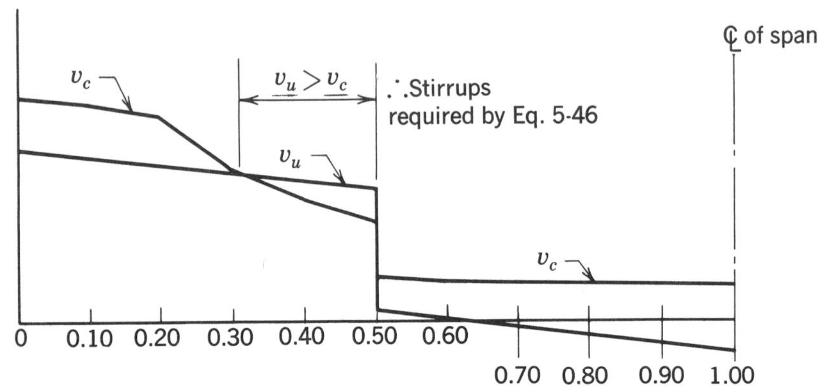

Fig. 5-18 Shear diagram for beam of Prob. 5-6.

At intermediate points:

$$f_d = 1188 \text{ (parabolic factor)}$$

$$f_{pe} = 765 + 1885 \text{ (parabolic factor)}$$

$$d = 25.0 + 32.1 \text{ (parabolic factor), min} = 50.4 \text{ in.}$$

$$V_d = 56.44 - 1.411x \text{ (from 0.0 to 0.5 of span/2)}$$

Where x = the distance from the support to the section under consideration in feet

$$V_l = V_d + 89.25 = 168.3 - 1.975x \text{ (from 0.0 to 0.5 of span/2)}$$

$$M_l = 1129 \times 1.4 \text{ (parabolic factor)} + (1.7)(52.5)x$$

$$= 1580 \text{ (parabolic factor)} + 89.25x \text{ (from 0.0 to 0.5 of span/2)}$$

The computations for the shear stresses and areas of shear reinforcing required are summarized in Table 5-3 and are shown in Fig. 5-18.

5-7 Bond of Prestressing Tendons

Two types of bond stress must be considered in the case of prestressed concrete. The first of these is referred to as "transfer bond stress" and has the function of transferring the force in a pre-tensioned tendon to the concrete. Transfer bond stresses come into existence when the prestressing force in the

TABLE 5-3

Pt.	f_d psi	f_{pe} psi	$6\sqrt{f_c'} + f_{pe} - f_d$	M_{cr} ft-kips	V_l kips	M_l ft-kips	V_d kips	*Actual* d in.	*Eff.* d in.	V_p kips	v_{ci} psi	v_{cw} psi	v_c psi	v_u psi
0.6	998	2350	1776	1687	1.85	2993	22.8	51.9	51.9	35.8	108	578	120*	6
0.5	891	2180	1713	1628	9.75/1288	2970	28.2	49.1	50.4	44.8	138/330	604	138/330	33/430
0.4	760	1970	1634	1552	136.7	2439	33.9	45.5	50.4	53.7	385	629	385	456
0.3	606	1730	1548	1471	144.6	1877	39.5	41.4	50.4	62.7	476	655	476	482
0.2	428	1440	1436	1364	152.5	1283	45.2	36.6	50.4	71.7	629	680	629	509
0.1	226	1120	1318	1252	160.4	657	50.8	31.1	50.4	80.6	1053	705	705	535
0.0	0	765	1189	1130	168.3	0	56.4	25.0	50.4	89.6	∞	731	731	561

* v_c minimum $= 1.7\sqrt{f_c'} = 120$ psi

At point 5, $A_v = \dfrac{(430 - 335)\ 7 \text{ in.} \times 12 \text{ in.}}{60{,}000} = 0.13$ sq in. per foot

At point 4, $A_v = \dfrac{(456 - 401)\ 7 \text{ in.} \times 12 \text{ in.}}{60{,}000} = 0.08$ sq in. per foot

By Eq. 5-48 Min. $A_v = \dfrac{50 \times 7 \times 12}{60{,}000} = 0.070$ sq in. per foot

By Eq. 5-49 Min. $A_v = \dfrac{4.13}{80} \cdot \dfrac{270}{60} \cdot \dfrac{12}{49.1}\sqrt{\dfrac{49.1}{7}} = 0.150$ sq in. per foot.

tendons is transferred from the prestressing beds to the concrete section. The second type of bond is termed "flexural bond stress" and comes into existence in pre-tensioned and bonded, post-tensioned members when the members are subjected to external loads. Flexural bond stress does not exist in unbonded, post-tensioned construction, which accounts for the term "unbonded."

When a prestressing tendon is stressed, the elongation of the tendon is accompanied by a reduction in the diameter due to Poisson's effect. When the tendon is released, the diameter increases to its original diameter at the ends of the prestressed member where it is not restrained. This phenomenon is generally regarded as the primary factor that influences the bonding of pre-tensioned wires to the concrete. The stress in the wire is zero at the extreme end and is at a maximum value at some distance from the end of the member. Therefore, in the length of the tendon from the extreme end to the point where it attains maximum stress, called the "transmission length," there is a gradual decrease in the diameter of the tendon, giving the tendon a slight wedge shape over this length. This wedge shape is often referred to as the "Hoyer Effect" after the German engineer E. Hoyer, who was one of the early engineers to develop this theory. Hoyer, and others more recently, derived elastic theory to compute the transmission length as a function of Poisson's ratio for steel and concrete, the moduli of elasticity of steel and concrete, the diameter of the tendon, the coefficient of friction between the tendon and the concrete, and the initial and effective stresses in the steel (Ref. 13). Laboratory studies of the transmission lengths have indicated a relatively close agreement between the theoretical and actual values. There can be wide variation, however, due to the different properties of concrete and steel and due to surface conditions of the tendons, which affect the coefficient of friction.

There is reason to believe that the configuration of a seven-wire strand (i.e., 6 small wires twisted about a slightly larger center wire) results in very good bond characteristics. It is believed the Hoyer Effect is partially responsible for this, but the relatively large surface area and twisted configuration is believed to result in a significant mechanical bond.

Although these theoretical relationships are of academic interest, they have little practical application, due to the inability of designers and fabricators of prestressed concrete to control the several factors that influence the transmission length. Fortunately, there has been sufficient research into the magnitude of transmission lengths under both laboratory (Ref. 14) and production (Ref. 17) conditions for the following significant conclusions to be drawn.

(1) The bond of clean three and seven-wire prestressing strands and concrete is adequate for the majority of pre-tensioned concrete elements.

(2) Members that are of such a nature that high moments may occur near

the ends of the members, such as short cantilevers, may require special consideration.

(3) Clean smooth wires of small diameter are also adequate for use in pre-tensioning, but the transmission length for tendons of this type should be expected to be approximately double that for seven-wire strands (expressed as a multiple of the diameter).

(4) Under good conditions, the initial transmission length for clean seven-wire strands can be assumed to be 50 to 75 times the diameter of the strand.

(5) The transmission length of tendons can be expected to increase from 5 to 20% within one year after release as a result of relaxation.

(6) The transmission length of tendons released by flame cutting or with an abrasive wheel can be expected to be from 20 to 30% greater than tendons that are released gradually.

(7) Hard non-flaky surface rust and surface indentations effectively reduce the transmission lengths required for strand and some forms of wire tendons.

(8) Concrete compressive strengths between 1500 and 5000 psi at the time of release result in transmission lengths of the same order, except for strand tendons larger than 1/2 in., in which case strengths, less than 3000 psi result in larger transmission lengths.

(9) It would seem prudent to use 3000 psi as a minimum release strength in pre-tensioned tendons, except for very unusual cases. Higher strengths may be required for tendons larger than 1/2 in.

(10) Because of relaxation, a small length of tendon ($\pm 3''$) at the end of a member can be expected to become completely unstressed.

(11) The degree of compaction of the concrete at the ends of pre-tensioned members is extremely important if good bond and short transmission lengths are to be obtained. Honeycombing must be avoided at the ends of the beams.

(12) There is little if any reason to believe that the use of end blocks improves the transfer bond of pre-tensioned tendons, other than to facilitate the placing and compacting of the concrete at the ends. Hence, the use of end blocks is considered unnecessary in pre-tensioned beams, if sufficient care is given to this consideration.

(13) Tensile stresses and strains develop in the ends of pre-tensioned members along the transmission length as a result of the wedge effect of the tendons. Little if any beneficial results can be gained in attempting to reduce these stresses and strains by providing mild reinforcing steel around the ends of these tendons, since the concrete must undergo large deformations and would probably crack before such reinforcing steel could become effective.

(14) Lubricants and dirt on the surface of tendons has a detrimental effect on the bond characteristics of the tendons.

A curve showing the typical variation of stress along the length of a pre-tensioned tendon near the end of a beam is given in Fig. 5-19. It will be seen that this curve is approximately hyperbolic. The stress is zero at the extreme end and for a distance of 4 in., as is assumed to be the case in most applications. This should be considered in the design.

The consensus of many persons who are active in the prestressing industry in the United States is that the transmission length of the seven-wire strands, which are commonly used in this country, is of much smaller magnitude than those encountered when smooth wires are employed, as in the European practice. This is believed to be the result of the misunderstanding of the fact that the seven-wire strands require a short transmission length in comparison to the effective prestress they develop. Although there have been a few reports of very small transmission lengths for seven-wire strands (Ref. 18), most experiments would indicate that the transmission length required for 3/8-in. strands is of the same order as required for 0.20-in. smooth wire, although the effective prestressing forces for the 3/8-in. strand and the 0.20-in. wire would normally be about 11,000 lb and 4300 lb, respectively.

Bond stresses also occur between the tendons and the concrete in both pre-tensioned and bonded, post-tensioned members, as a result of changes in the external load. There are of course no transfer bond stresses in post-tensioned members, since the end anchorage device accomplishes the transfer of stress. Although it is known that flexural-bond stresses are relatively low in prestressed members for loads less than the cracking load, there is an abrupt and significant increase in these bond stresses after the cracking load is

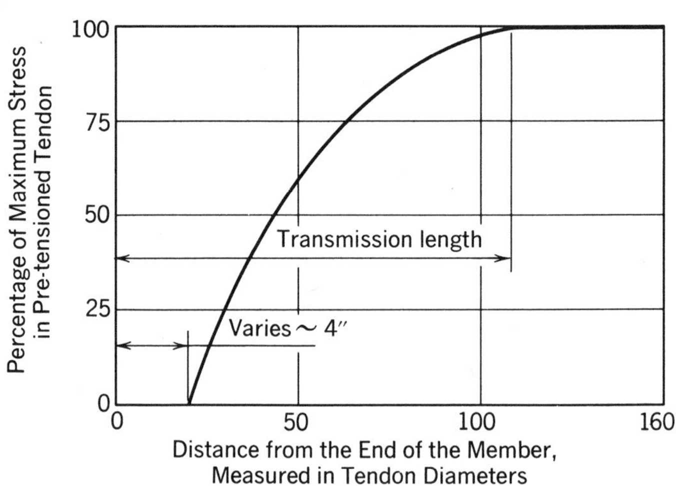

Fig. 5-19 Variation in stress in pre-tensioned tendon near end of beam after relaxation.

exceeded. Because of the indeterminancy which results from the plasticity of the concrete for loads exceeding the cracking load, accurate computation of the flexural-bond stresses cannot be made under such conditions. Again, tests must be relied upon as a guide for design.

The effect of flexural bond is most evident when two identical post-tensioned members, one grouted and one unbonded, are tested to destruction and the results are compared. The load-deflection curve for such tests, when plotted together, would appear as in Fig. 5-20. From these curves, it will be seen that the grouted beam does not deflect as much under a specific load as the unbonded beam. The explanation for this is that the tendon in the bonded beam must undergo changes in strain equal to the strain changes in the concrete to which it is bonded, whereas the unbonded tendon can slip in the duct and the strain changes are averaged. Hence, the grouted beam deforms and deflects as a function of a transformed section. This difference generally results in the cracking load of the grouted beam being from 10 to 15% higher than that of the unbonded beam, while the ultimate load may be as much as 50% higher. The presence of flexural bond results in many very fine cracks in a bonded member in which the cracking load is exceeded, while, in an identical unbonded member subjected to the same load, only a few wide cracks occur. This is significant because removal of the load from the bonded member will result in the fine cracks closing completely, while in the unbonded member, the large cracks do not completely close.*

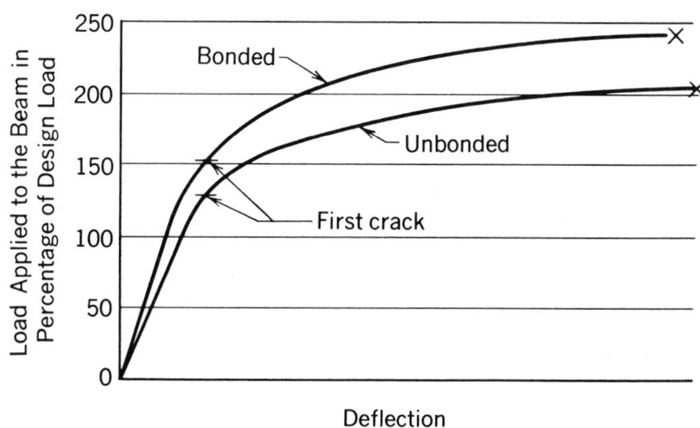

Fig. 5-20 Comparison of load-deflection curves for bonded and unbonded post-tensioned construction.

* The provision of non-prestressed reinforcement, if in sufficient quantity, will result in a non-bonded beam having deflection and cracking characteristics similar to a bonded beam.

It is generally believed that once a bonded prestressed member is cracked, a significant increase in flexural-bond stress occurs at the point of cracking. As the load on the beam is increased, the flexural bond stresses at the crack increase until slip occurs at the cracked section. Further increase in the external loads will be accompanied by additional slip in the tendon. This action will continue until the member fails either by flexure or, (in the case of a pre-tensioned member), when the flexural bond stress is destroyed over a length of a tendon that reaches the zone in which the pre-tension is developed by transfer bond by lack of bond (Refs. 19 and 20). Research has shown that the embedment length, the length from the free end of the beam to the point of maximum steel stress, required to develop a specific steel stress for 1/4, 3/8, and 1/2 in. strands is of the order given in Table 5-4. The data in the table are

TABLE 5-4 Maximum Stresses (psi) that can be Developed at the Section of Maximum Moment for Various Sizes of Seven-wire Strands and Embedment Lengths.

Embedment Length (in.)	1/4-in. *Strand*	3/8-in. *Strand*	1/2-in. *Strand*
20	194,000	160,000	—
30	218,000	187,000	166,000
40	234,000	201,000	180,000
50	250,000	211,000	192,000
60	264,000	220,000	200,000
70	—	229,000	206,000
80	—	238,000	213,000
90	—	247,000	219,000
100	—	257,000	226,000
120	—	—	244,000
140	—	—	272,000

From Ref. 20, p. 798.

applicable to concrete with a cylinder strength of 5500 psi and steel stresses of the order of 150,000 psi. If the distance from the section at which the critical, flexural stresses occur is less than the embedment length required to develop the computed stress in the steel at ultimate, the ultimate load may be controlled by bond rather than by flexure. In such instances, the design should be revised, since it is generally more desirable that the controlling mode of failure be flexural rather than by bond.

The bond provisions of ACI 318-71 apply only to three- and seven-wire strands. It is stipulated that strands of these types shall be extended a distance beyond the critical section (for moment) equal to

$$(f_{su} - \tfrac{2}{3}f_{se})D \tag{5-50}$$

in which D is the nominal diameter of the strand, f_{su} and f_{se} are as defined elsewhere in this chapter and have the units of ksi. The quantity within the parentheses is considered to be dimensionless. If the bonding of the tendons does not extend to the end of the member, the length given in Eq. 5-50 must be doubled.

5-8 Bonded vs Unbonded Post-Tensioning

The structural advantages of bonding post-tensioning tendons by grouting them in place should be apparent from the preceding article. In spite of these advantages, unbonded tendons are widely used in the United States.

Most suppliers of post-tensioning tendons, as well as some prestressed-concrete fabricators and subcontractors, report that the use of tendons coated with a bituminous rust inhibitor and wrapped with paper is significantly lower in cost than the use of tendons placed in a steel sheath or preformed hole and grouted in place after stressing. The proponents of unbonded tendons point out that the lower cracking and ultimate loads that are characteristic of unbonded construction, as well as the fact that the few widely spaced

Fig. 5-21 Unbonded post-tensioned beam under overload. Note the wide cracks in the bottom flange as well as the relatively large spacing of the cracks. (*Courtesy U.S. Naval Civil Engineering Research and Evaluation Laboratory, Port Hueneme, California.*)

cracks which would appear in the tensile flange at loads that exceed the cracking load, can be adequately controlled by including normal reinforcing steel in the tensile flanges to supplement the prestressing. It is claimed that sufficient quantity of the supplementary reinforcing steel can be supplied at less cost than would be required for grouted tendons.

The difference in the spacing of the cracks that appear at overloads in unbonded and bonded construction is clearly illustrated in Figs. 5-21 and 5-22. In Fig. 5-21 an overloaded, unbonded beam is shown and the wide cracks that are spaced 2 to 3 ft on centers are clearly visible. The portion of bonded beam shown in Fig. 5-22 is immediately adjacent to a section of the beam that was demolished when the beam collapsed during testing. The cracks that were open in the bottom flange and web of the beam near ultimate load lie between the easily seen pencil lines. The cracks in the bonded beam were only faintly visible to the unaided eye after the failure of the beam. The effectiveness of

Fig. 5-22 A portion of a bonded post-tensioned beam after testing to destruction. Note the close spacing of cracks located between the pencil lines. The effectiveness of the grouting is illustrated by the fact that the cracks are closed and virtually invisible to the unaided eye. (*Courtesy U.S. Naval Civil Engineering Research and Evaluation Laboratory, Port Hueneme, California.*)

the grouting in the beam of Fig. 5-22 is demonstrated by the fact that the cracks closed so completely after ultimate flexural failure of the beam.

During the testing of the beam shown in Fig. 5-22, the effectiveness of the grouting was also clearly evidenced by the location of the neutral axis of the beam. The location of the neutral axis was determined by measuring flexural strains and was located where it would be expected for the transformed concrete section, which is lower than would be expected for the net or gross concrete section.

The sections of the grouted post-tensioning tendons shown in Fig. 5-23 were taken from the beam shown in Fig. 5-22. The fact that the metal sheath is very well filled with grout and virtually without voids should be noted. In addition, friction tape, which was used to seize the wires when they were being inserted in the metal sheath, is clearly seen in two sections of the tendon. The friction tape did not seriously restrict the flow of grout.

The use of unbonded tendons will certainly result in satisfactory construction if properly designed and constructed. This has been demonstrated

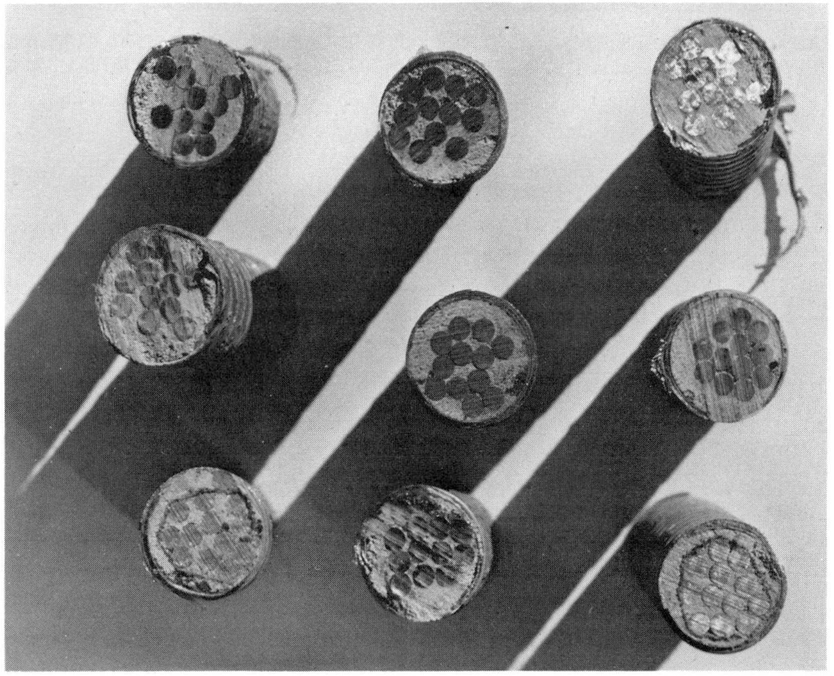

Fig. 5-23 Sections of grouted post-tensioned tendons removed from the test beam in Fig. 5-21. (*Courtesy U.S. Naval Civil Engineering Research and Evaluation Laboratory, Port Hueneme, California.*)

by the large amount of building construction that has been done successfully with this method in the United States. Members that are designed with unbonded tendons should be made to conform to the provisions of ACI 318-71 " Building Code Requirements for Reinforced Concrete " as well as to " Tentative Recommendations for Concrete Members Prestressed with Unbonded Tendons," (Ref. 24).

REFERENCES

1. Muller, Jean. "Flexural Strength of Prestressed Concrete Continuous Structures." Paper presented at the Knoxville Convention of the A.S.C.E., pp. 9–19 (Jan. 1956).
2. Janney, Jr. R., Hognestad, E. and McHenry, D. "Ultimate Flexural Strength of Prestressed and Conventionally Reinforced Concrete Beams," *Journal of the American Concrete Institute*, **27**, 601–620 (Feb. 1956).
3. Building Code Requirements for Reinforced Concrete, ACI Committee 318–63, American Concrete Institute, 1963.
4. ACI-ASCE Joint Committee 323, "Tentative Recommendations for Prestressed Concrete," *Journal of the American Concrete Institute*, **29**, No. 7, 545–578 (Jan. 1958).
5. Warwarvk, J., Sozen, M. and Siess, C. P., "Strength and Behavior in Flexure of Prestressed Concrete Beams," Private Communication.
6. Pannell, F. N. "The Ultimate Moment Resistance of Unbonded Prestressed Concrete Beam," *Magazine of Concrete Research*, **21**, No. 66, 43–54 (Mar. 1969).
7. Rozvany, G. I. N. and Woods, J. F., "Sudden Collapse of Unbonded Underprestressed Structures," *ACI Journal*, **66**, No. 2, 129–135 (Feb. 1969).
8. Guyon, Y. *Prestressed Concrete*, pp. 270, 271, John Wiley & Sons, Inc., New York, 1953.
9. Bureau of Public Roads, "Criteria for Prestressed Concrete Bridges," U.S. Department of Commerce, Superintendent of Documents, Washington, D.C., 1954.
10. Portland Cement Association, Notes from the Building Code Seminar, Chapter 63S-28, pp. 8–31 (1963).
11. MacGregor, J. G. and Hanson, J. M. "Proposed Changes in Shear Provisions for Reinforced and Prestressed Concrete Beams," *ACI Journal*, **66**, No. 4, 276–288 (April 1969).
12. Kani, G. N. J., "A Rational Theory for the Function of Web Reinforcement," *ACI Journal*, **66**, No. 3, 185–197 (Mar. 1969).
13. Janney, Jack R. "Nature of Bond in Pretensioned Concrete," *Journal of the American Concrete Institute*, **25**, 717–736 (May 1954).
14. Hanson, N. W. "Influence of Surface Roughness of Prestressing Strand on Bond Performance," *Journal of the Prestressed Concrete Institute*, **14**, No. 1, 32–45 (Feb. 1969).

15. Stocker, M. F., Sozen, M. A., "Investigation of Prestressed Reinforced Concrete for Highway Bridges, Part VI: Bond Characteristics of Prestressing Strand." Private communication.

16. Kaar, P. H., La Fraugh, R. W. and Mass, M. A., "Influence of Concrete Strength on Strand Transfer Length." *Journal of the Prestressed Concrete Institute*, **8**, No. 5, 47–67 (Oct. 1963).

17. Base, G. D. "An Investigation of Transmission Length in Pretensioned Concrete," Research Report No. 5, Cement and Concrete Association, London, 1958.

18. "Applications." (Panel Discussion), *Journal of the Prestressed Concrete Institute*, **1**, 11 (Sept. 1956).

19. Nordby, G. M. "Fatigue of Concrete—A Review of Research," *Journal of the American Concrete Institute*, **31**, No. 2, 210–215 (Aug. 1958).

20. Hanson, N. W. and Kaar, P. H. "Flexural Bond Tests of Pre-tensioned Prestressed Beams," *Journal of the American Concrete Institute*, **30**, No. 7, 783–802 (Jan. 1959).

21. Yamazaki, Jun, Kattula, Basil T. and Mattock, Alan H. "A Comparison of the Behavior of Post-Tensioned Prestressed Concrete Beams With and Without Bond." Structures and Mechanics Report SM69-3. Department of Civil Engineering University of Washington, Seattle, Washington (Dec. 1969).

22. ACI Committee 318. "Proposed Revision of ACI 318-63: Building Code Requirements for Reinforced Concrete," *Journal of the American Concrete Institute*, **67**, No. 2, 77–186 (Feb. 1970).

23. Libby, James R. *Prestressed Concrete Design and Construction.* The Ronald Press Company, New York, 1961.

24. "Tentative Recommendations for Concrete Members Prestressed With Unbonded Tendons," *Journal of the American Concrete Institute*, **66,** No. 2 (Feb. 1969).

6 Additional Design Considerations

6-1 Introduction

Included in this Chapter is a discussion of a number of principles, pertaining to the elastic design of simple prestressed flexural members, which may not at first be apparent to the student of prestressed concrete. The topics treated in this chapter are often important in problems encountered in practice, and it is well for the designer and student to be familiar with these principles.

The engineer who is often engaged in the design of prestressed structures will become familiar with these relationships as a result of his design experience. On the other hand, the engineer who is only occasionally involved in the design or review of prestressed members will find that this chapter contains valuable, concise reference material presented in a manner intended to facilitate its use.

6-2 Losses of Prestress

The final stress that is required in the prestressing steel at all of the critical sections in a prestressed member should be specified by the designer. If the system of prestressing to be used is specified, the complete stressing schedule and initial stresses required should be computed and indicated on the drawings or in the specifications. In order to do this, it is necessary to either compute or assume a value for the loss of stress, in the prestressing tendons, which results from the several contributing phenomena. The losses of prestress must be included in the computation of gauge pressure and elongation for prestressing, which is discussed in Chapter 15, as is the friction loss that occurs during post-tensioning.

The recommendations of the Bureau of Public Roads, which appeared in 1954 (Ref. 1), stipulated that the losses of prestress in normal concrete members would be computed as follows for pretensioned concrete

$$\Delta f_{si} = 6000 + 16 f_{cs} + 0.04 f_{si} \tag{6-1}$$

and for post-tensioned concrete

$$\Delta f_{si} = 3000 + 11 f_{cs} + 0.04 f_{si} \tag{6-2}$$

where f_{si} is the initial stress in the prestressing steel and f_{cs} is the concrete stress at the level of the steel. For lightweight concrete members, the losses were to be determined by tests. Eqs. 6-1 and 6-2 are each in three parts. The first of these is intended to compensate for shrinkage, which for pre-tensioning was taken to be a strain of approximately 214×10^{-6} in./in. and for post-tensioning one-half of this value, since some shrinkage of the concrete will have occurred before stressing. The second term is intended to account for the effect of concrete creep and, in the case of pre-tensioning, elastic shortening of the concrete. The creep ratio was assumed to be 2.25. Elastic shortening was not included in Eq. 6-2 for post-tensioning, since it is possible to over stress to compensate for this loss in post-tensioning. (Stressing computations rarely provide for elastic shortening.) The third term was to account for the relaxation loss of the prestressing steel, which at that time (1954) was believed to be of the order of 4%.

The Tentative Recommendations of the Joint ACI-ASCE Committee 323, which appeared in 1958 (Ref. 16), permitted the losses of prestress to be assumed to be 35,000 psi and 25,000 psi for pre-tensioned and post-tensioned concrete, respectively, when specific data relative to the materials was not known. These values were also adopted by the American Association of State Highway Officials and have appeared in their Standard Specifications for Highway Bridges since 1961. Structures built with the losses assumed to be as above have given satisfactory service in spite of the fact that the losses may be considerably more or less than those assumed.

As an alternative to using assumed losses, the loss of steel stress can be computed, but it must be emphasized that the results should be regarded as estimates rather than computations, unless the elastic, plastic, and shrinkage properties of the concrete to be used in the construction are known. The various phenomena that contribute to the loss of prestress as well as the method of calculation are discussed below.

Elastic Shortening of the Concrete

When the prestress is applied to the concrete, an elastic shortening of the concrete takes place. This results in an equal and simultaneous shortening of the prestressing steel. The loss in stress in the steel from this shortening is equal to the product of the stress in the concrete at the level of the steel and the ratio of the moduli of elasticity of the steel and concrete. In a simple beam, the critial section for flexural stress, and hence the section at which the losses of prestress should be considered, is normally at the midspan of the beam. When pre-tensioning is used, the concrete stress that should be used in computing the reduction in prestress (due to elastic shortening) is equal to the net, initial concrete stress that results from the algebraic sum of the stress due to initial prestressing and the stress due to the dead load of the beam at the level of the steel. In post-tensioning, the first tendon that is stressed is shortened by the subsequent stressing of all other tendons, and the last tendon is not shortened by any subsequent stressing. Therefore, in post-tensioning, an average value of stress change can be computed and applied equally to all tendons.

Creep of Concrete

The loss in steel stress resulting from the creep of the concrete should also be computed on the basis of stresses that occur at the critical section for flexure rather than average values, since the greatest factor of safety against cracking is generally required at the section of maximum moment. In bonded construction, since creep is time dependent and does not take place to any significant degree until after bond has been established between the steel and the concrete, the strains along the tendons are not averaged in their effect upon the loss of steel stress. In the case of post-tensioned construction. in which the tendons are not bonded to the concrete after stressing, the creep strains become averaged, since the tendon can slip in the member, and for this reason, the average concrete stress at the center of gravity of the steel should be used in the computation of this stress loss.

The magnitude of the creep of concrete, as well as the rate at which it occurs, can be estimated using the data in Sec. 3-11. When possible, the concrete stress at the level of the steel, taking into account the stress-history of the member, should be used in computing the loss of stress due to this phenomenon.

Shrinkage of the Concrete

The rate at which concrete shrinks as well as the magnitude of the ultimate shrinkage can be estimated using the data of Secs. 3-9 and 3-10. The entire shrinkage strain is effective in reducing the steel stress in pre-tensioned construction, while only the amount of shrinkage that occurs after the stressing is of significance in this respect in the case of post-tensioning.

Relaxation of the Prestressing Steel

Relaxation of the prestressing steel should be estimated on the basis of the data presented in Sec. 2-7.

The computation of the losses of prestress due to the shrinkage of the concrete, elastic deformation of the concrete and relaxation of the prestressing steel is straight forward, provided that the parameters governing these phenomena are known. The loss due to creep is not made accurately as easily, because the creep of concrete is a function of both time and the level of stress in the concrete. Since the stress is constantly changing, as a result of the losses in prestress that are occurring, the most accurate method of computing stress loss is to employ a numerical integration that takes into account the several variables. This procedure consists of employing a unit creep curve, the assumed shrinkage curve, and the relaxation curve for the materials under consideration and computing the changes in stress over a large number of increments of relatively short duration. The curves are treated as step functions to facilitate the computations (Ref. 2).

The creep strain in the concrete can be computed using either the rate-of-creep method or the superposition method. These are illustrated in Figs. 6-1 and 6-2, respectively. The rate-of-creep method assumes the creep strain

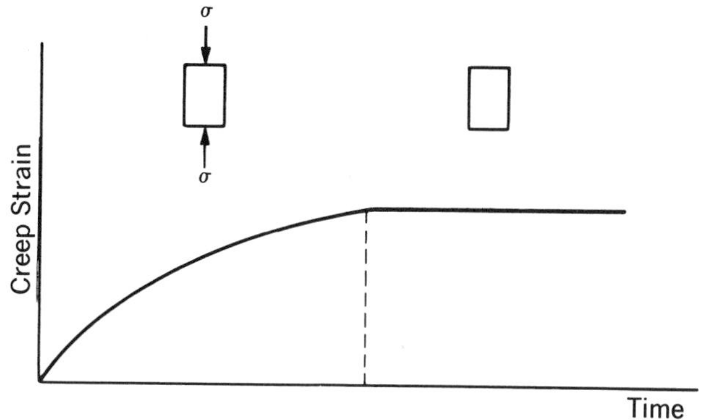

Fig. 6-1 Creep strains by the rate of creep method.

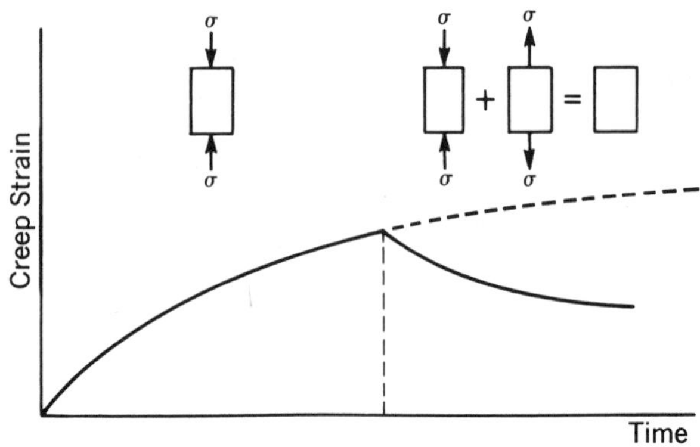

Fig. 6-2 Creep strains by the superposition method.

at time t is equal to the product of the stress and the ordinate of the unit creep curve corresponding to time t. Once the stress is removed, there is no further change in the creep strain. The superposition method predicts the same initial strains as the rate-of-creep method, but assumes the member is subjected to a tensile stress upon removal of the stress and creeps under two opposing fictitious stresses. The creep characteristics in tension and compression are assumed to be equal.

The use of the superposition method is believed to be the more accurate method, but its use is laborious and is considered to be worth using only if the creep curves and other properties of the materials are well known and high precision is required.

The following basic assumptions are made in employing the numerical integration method for predicting prestress losses (as well as in computing deflections).

(1) The initial stresses in the member under consideration are known.

(2) The unit creep vs time curve for the concrete under constant stress is known and can be considered a step function to approximate the curve.

(3) The shrinkage vs time curve is known for the concrete under consideration, and it too can be treated as a step function. The shrinkage characteristics are uniform over the section.

(4) The relaxation vs time relationship for the prestressing steel is known and can be treated as a step function.

(5) Creep strains are proportional to the concrete stress up to 50% of the concrete strength.

(6) Strains vary linearly over the depth of the section.

(7) The stress-strain relationship for the concrete is linear up to 50% of the ultimate strength for loads of short duration.

(8) The stress-strain relationship for steel is linear under short duration loads.

This method lends itself to solution by programmable calculator or computer, since the accuracy will improve by using many increments of short duration. Examples of the use of this method are given in Illustrative Problems 6-1 and 6-2.

ILLUSTRATIVE PROBLEM 6-1 Compute the loss of prestress for the post-tensioned girder shown in Fig. 6-3, assuming the concrete shrinkage takes

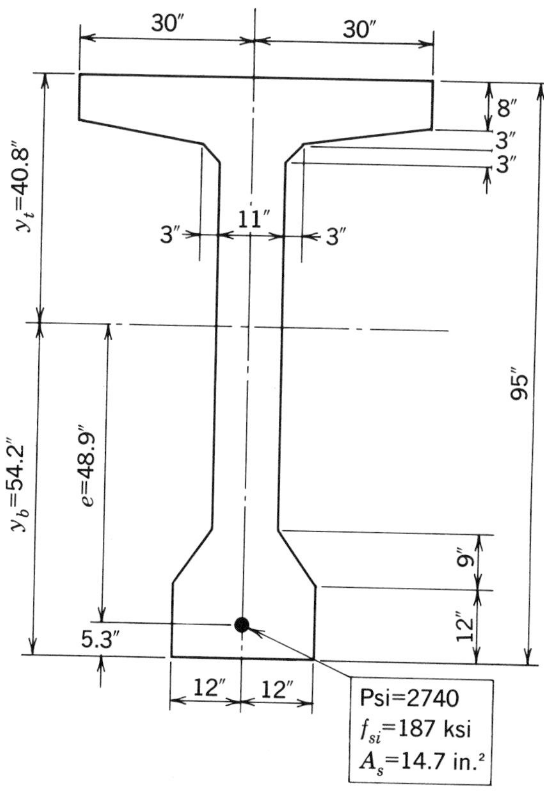

$A = 1743$ in.^{2} $I = 1{,}908{,}000$ in.^{4} $r^2 = 1095$ in.^{2}
$S_t = 46{,}700$ in.^{3} $r^2/y_t = 26.8$ in. $V/s \cong 5.5$
$S_b = 35{,}200$ in.^{3} $r^2/y_b = 20.2$ in.
$S_e = 39{,}100$ in.^{3} $r^2/e = 22.4$ in.

Fig. 6-3 Girder and section properties for Problems 6-1 and 6-2.

place at the rate given by Eq. 3-11, the creep ratio is 2.50, the rate of creep is as given by Eq. 3-13, and the relaxation of the steel is as defined by the Eq. 2-1. Assume $E_c = 4.0 \times 10^6$ psi, $E_s = 28 \times 10^6$ psi, the girder concrete has air cured 30 days before stressing, the ultimate shrinkage is 300×10^{-6} in./in., $f_s' = 270$ ksi, $f_y = 256$ ksi (0.10% offset), and $f_{si} = 187$ ksi. The girder is subject to prestress and girder dead load (1.82 k/ft) for the first 30 days after stressing, after which, a superimposed dead load of 0.50 k/ft is applied. The span is 180 ft. (The effect of elastic shortening is accounted for in the stressing procedure (see Sec. 15-7).)

SOLUTION:

$$\text{Initial stress at c.g.s. due to prestressing} = \frac{2740}{1743}\left(1 + \frac{48.9}{22.4}\right) = +5000 \text{ psi}$$

$$\text{Stress at c.g.s. due to girder dead load} = \frac{1.82 \times 180^2 \times 12{,}000}{8 \times 39{,}100} = -2260 \text{ psi}$$

Stress at c.g.s due to superimposed dead load

$$= \frac{0.50 \times 180^2 \times 12{,}000}{8 \times 39{,}100} = -622 \text{ psi}$$

The equation for shrinkage strain is

$$\epsilon_s = 300 \times 10^{-6}[0.157 \log_e t - 0.115]$$

and the loss of steel stress due to shrinkage is

$$\begin{aligned} \Delta f_{ss} = E_s \epsilon_s &= 300 \times 10^{-6}[0.157 \log_e t - 0.115]28 \times 10^3 \\ &= 1.32 \log_e t - 0.966 \end{aligned}$$

in which Δf_{ss} is in kips per square inch and t is in days.

The incremental stress loss due to shrinkage during time interval $t - 1$ to t is

$$f_{ss} = f_{sst} - f_{ss(t-1)} = 1.32 [\log_e t - \log_e(t-1)]$$

Because the beam is not stressed until the concrete is 30 days old, only shrinkage occurring after 30 days affects the loss of stress. Therefore, for the first increment, $t = 45$ days, $t - 1 = 30$ days ($\Delta t = 15$ days).

The equation for relaxation loss is

$$\begin{aligned} f_{sr} = f_{si} \frac{\log t'}{10}\left[\frac{f_{si}}{f_y} - 0.55\right] &= 187 \frac{\log t'}{10}\left[\frac{187}{256} - 0.55\right] \\ &= 3.37 \log t' = 1.46 \log_e t' \end{aligned}$$

where f_{sr} is in kips per square inch and t' is time in hours. The incremental relaxation loss during time $\Delta t'$ is

$$\Delta f_{sr} = 3.37[\log t' - \log (t' - 1)] = 1.46 [\log_e t' - \log_e (t' - 1)]$$

in which $t' = 0$ and $\Delta t' = 360$ hr.

The creep loss at the level of the steel equals

$$f_{sc} = \frac{(f_{\text{cgs}})\,(\text{Creep Ratio})\,(\text{Creep Coefficient})\,E_s}{E_c}$$

$$f_{sc} = \frac{(f_{\text{cgs}})(2.50)(0.157 \log_e t - 0.115)28 \times 10^3}{4 \times 10}$$

$$f_{sc} = f_{\text{cgs}}\,[2.75 \log_e t - 2.01]$$

in which t is in days, f_{sc} is in kips per square inch and f_{cgs} is the stress in the concrete at the level of the steel due to prestressing. The incremental creep during time Δt is

$$\Delta f_{sc} = 2.75 f_{\text{cgs}}\,[\log_e t - \log_e (t - 1)]$$

The value of Δf_{cgs} for each step is computed using the value of f_{cgs} computed in the previous step. The total change in steel stress is

$$\Delta f_s = \Delta f_{ss} + \Delta f_{sr} + \Delta f_{sc}$$

and the change in f_{cgs} that results from the reduction in the prestressing force can be computed from

$$\Delta f_{\text{cgs}} = -\frac{\Delta f_s}{f_{si}} \begin{bmatrix} \text{Initial stress due to prestressing} \\ \text{at the level of the steel} \end{bmatrix}$$

$$= -\frac{\Delta f_s}{187} \times 5000 = -26.74\,\Delta f_s$$

Subtracting Δf_{cgs} from the previous value of f_{cgs} results in the new value of f_{cgs}. The calculations are summarized in Table 6-1.

Summing the incremental losses of stress gives a total loss of 46.07 ksi, which is of the order of 25% of the initial steel stress.

Reviewing the data in Table 6-1 will reveal that the incremental loss of stress due to relaxation (Δf_{sr}) accounts for from one-fourth to one-third of the total incremental loss of steel stress (Δf_s). One would expect the relaxation loss to be less when the strain in the steel is reducing with the passage of time. This could be accounted for by using the stress remaining in the steel at the end of each time increment (i.e., $f_{sn} = f_{s(n-1)} - \Delta f_s$) in lieu of f_{si} in the relationship for f_{sr}. This procedure has not been used in any of the Illustrative Problems in this book.

TABLE 6-1

Time Days	Time Hours	Δf_{ss} ksi	Δf_{sr} ksi	Δf_{sc} ksi	Δf_s ksi	Δf_{cgs} psi	f_{cgs} psi
0	0	0	0	0	0	0	2740
15	360	0.54	8.59	14.90	24.03	−643	2097
30	720	0.38	1.01	4.00	5.39	−144	1953 / 1331
45	1080	0.29	0.59	1.48	2.36	−63	1268
60	1440	0.24	0.42	1.00	1.66	−44	1224
75	1800	0.20	0.33	0.75	1.28	−34	1190
90	2160	0.18	0.27	0.60	1.05	−28	1162
105	2520	0.16	0.23	0.49	0.88	−24	1138
120	2880	0.14	0.19	0.42	0.75	−20	1118
135	3240	0.13	0.17	0.36	0.66	−18	1100
150	3600	0.11	0.15	0.32	0.58	−16	1084
165	3960	0.11	0.14	0.28	0.53	−14	1070
180	4320	0.10	0.13	0.26	0.49	−13	1056
195	4680	0.09	0.12	0.23	0.44	−12	1044
210	5040	0.09	0.11	0.21	0.41	−11	1033
225	5400	0.08	0.10	0.20	0.38	−10	1023
240	5760	0.08	0.09	0.18	0.35	−9	1014
255	6120	0.07	0.09	0.17	0.33	−9	1005
270	6480	0.07	0.08	0.16	0.31	−8	997
285	6840	0.06	0.08	0.15	0.29	−8	989
300	7200	0.06	0.07	0.14	0.27	−7	982
315	7560	0.06	0.07	0.13	0.26	−7	975
330	7920	0.06	0.07	0.12	0.25	−7	968
345	8280	0.05	0.06	0.12	0.23	−6	962
360	8640	0.05	0.06	0.11	0.22	−6	956
375	9000	0.05	0.06	0.11	0.22	−6	950
390	9360	0.05	0.06	0.10	0.21	−6	944
405	9720	0.05	0.06	0.10	0.21	−6	938
420	10080	0.04	0.05	0.09	0.18	−5	933
435	10440	0.04	0.05	0.09	0.18	−5	928
450	10800	0.04	0.05	0.09	0.18	−5	923
465	11160	0.04	0.05	0.08	0.17	−5	918
480	11520	0.04	0.05	0.08	0.17	−5	913
495	11880	0.04	0.04	0.08	0.16	−4	909
510	12240	0.04	0.04	0.07	0.15	−4	905
525	12600	0.04	0.04	0.07	0.15	−4	901
540	12960	0.04	0.04	0.07	0.15	−4	897
555	13320	0.03	0.04	0.07	0.14	−4	893
570	13680	0.03	0.04	0.07	0.14	−4	889
585	14040	0.03	0.04	0.06	0.13	−3	886
600	14400	0.03	0.04	0.06	0.13	−3	883

ILLUSTRATIVE PROBLEM 6-2 Using the data of Illustrative Problem 6-1, compute the loss of stress if the prestressing steel has no relaxation and the creep coefficient is 1.5. All other details are the same.

SOLUTION:

The loss due to shrinkage of the concrete remains the same as before and the relaxation loss is taken to be zero. The creep loss at the level of the steel becomes

$$f_{sc} = (f_{cgs})\,1.5 \times 7 \times (0.157 \log_e t - 0.115)$$
$$= (f_{cgs})(1.65 \log_e t - 1.21)$$

and the incremental creep loss becomes

$$\Delta f_{sc} = 1.65 f_{cgs}(\log_e t - \log_e t - 1)$$

The results are summarized in Table 6-2. The summation of the loss in steel stress reveals that the total loss of prestress for these conditions is 21.63 ksi.

TABLE 6-2

ime ays	Δf_{ss} ksi	Δf_{sc} ksi	Δf_s ksi	Δf_{cgs} psi	f_{cgs} psi	Time Days	Δf_{ss} ksi	Δf_{sc} ksi	Δf_s ksi	Δf_{cgs} psi	f_{cgs} psi
0	0	0	0	0	2740	315	0.06	0.14	0.20	−5	1674
15	0.54	6.77	7.31	−195	2545	330	0.06	0.13	0.19	−5	1699
30	0.38	2.91	3.29	−88	2544 / 1922	345	0.05	0.12	0.17	−5	1664
45	0.29	1.29	1.58	−42	1880	360	0.05	0.12	0.17	−5	1659
60	0.24	0.89	1.13	−30	1850	375	0.05	0.11	0.16	−4	1655
75	0.20	0.68	0.88	−24	1826	390	0.05	0.11	0.16	−4	1655
90	0.18	0.55	0.73	−20	1806	405	0.05	0.10	0.15	−4	1647
05	0.16	0.46	0.62	−17	1789	420	0.04	0.10	0.14	−4	1643
20	0.14	0.39	0.53	−14	1775	435	0.04	0.10	0.14	−4	1639
35	0.13	0.34	0.47	−13	1762	450	0.04	0.09	0.13	−3	1636
50	0.11	0.31	0.42	−11	1751	465	0.04	0.09	0.13	−3	1633
65	0.11	0.28	0.39	−10	1741	480	0.04	0.09	0.13	−3	1630
80	0.10	0.25	0.35	−9	1732	495	0.04	0.08	0.12	−3	1627
5	0.09	0.22	0.31	−8	1724	510	0.04	0.08	0.12	−3	1624
0	0.09	0.21	0.30	−8	1716	525	0.04	0.08	0.12	−3	1621
5	0.08	0.20	0.28	−7	1709	540	0.04	0.08	0.12	−3	1618
0	0.08	0.18	0.26	−7	1702	555	0.03	0.07	0.10	−3	1615
5	0.07	0.17	0.24	−6	1696	570	0.03	0.07	0.10	−3	1612
0	0.07	0.16	0.23	−6	1690	585	0.03	0.07	0.10	−3	1609
5	0.06	0.15	0.21	−6	1684	600	0.03	0.07	0.10	−3	1606
0	0.06	0.14	0.20	−5	1679						

ILLUSTRATIVE PROBLEM 6-3 The loss of prestress computations from Illustrative Problems 6-1 and 6-2 can be approximated as follows:
At 14,400 hr

		I.P. 6-1	I.P. 6-2
$f_{sr} = \dfrac{f_{si} \log 14{,}400}{10}\left(\dfrac{f_{si}}{f_y} - 0.55\right)$			
$= 187\left(\dfrac{4.158}{10}\right)\left(\dfrac{187}{256} - 0.55\right)$	=	14.0	0.0
$f_{ss} = (\text{Shrinkage strain after stressing})\ E_s$			
$= 0.50 \times 300 \times 10^{-6} \times 28 \times 10^3$	=	4.2	4.2
$f_{sc} = f_{cgs} \times n \times C_t \left(\dfrac{fsi - \Delta fs}{fsi}\right)$	=		
$= \dfrac{2710}{1000} \times 7 \times 2.5 \left(\dfrac{187 - 52}{187}\right)$	=	34.6	
$= \dfrac{2740}{1000} \times 7 \times 1.5 \left(\dfrac{187 - 29}{187}\right)$	=		24.3
	Δfs	52.8	28.5

ILLUSTRATIVE PROBLEM 6-4 Compute the loss of prestress for the pre-tensioned beam of Fig. 6-4 assuming the span is 40 ft, the elastic modulii are 4×10^6 psi and 28×10^6 psi for the concrete and steel, respectively, the shrinkage of the concrete (under normal curing conditions) is 300×10^6 in./in., the creep ratio is 1.75 (normal cure), and the steel relaxation is as predicted by Eq. 2-1. The initial jacking force is 600,000 lb ($f_{si} = 187.5$ ksi). The member will be steam cured and stressed at the age of 24 hr. Assume $f_y = 256$ ksi. The shrinkage and creep coefficients for the concrete has been determined experimentally to be as given in Table 6-3.

SOLUTION:
The total shrinkage strain is assumed reduced 30% due to steam curing (*see* Sec. 3-9). Therefore, the total shrinkage strain becomes $0.70 \times 300 \times 10^6 = 210 \times 10^6$ in./in. and the loss of stress due to shrinkage becomes:

$$f_{ss} = (\text{shrinkage coefficient})\ (210 \times 10^6)(28 \times 10^3)$$
$$f_{ss} = 5.89\ (\text{shrinkage coefficient}) \text{ in ksi}$$

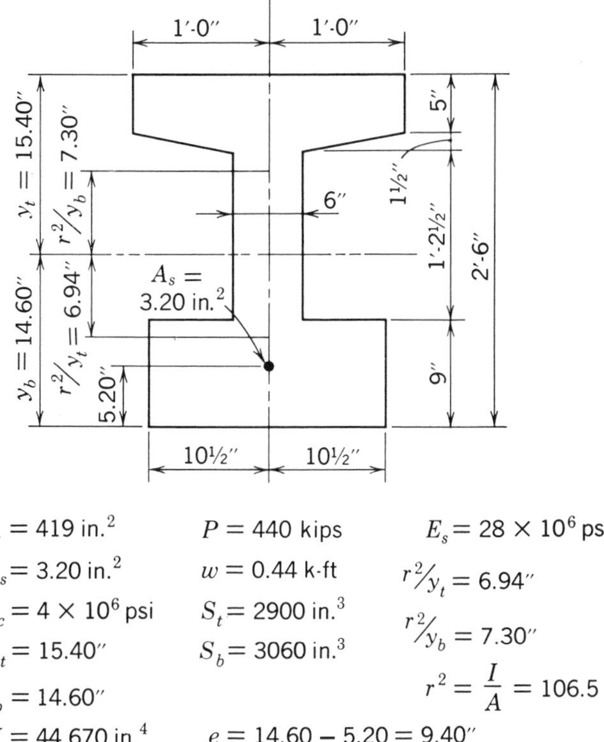

Fig. 6-4 Cross section and physical properties of the beam used in Prob. 6-4.

The creep strain is assumed reduced by 50% (*see* Sec. 3-11) due to the steam curing. Therefore, the creep ratio becomes $0.50 \times 1.75 = 0.875$ and the loss of stress due to creep is

$$f_{sc} = \frac{(f_{\text{cgs}})(0.875)\,(\text{incremental creep coefficient})}{E_c}\,E_s$$

The stress in the concrete at the level of the steel due to jacking force would be as follows:

$$\text{Due to } P_j = 600k\text{: } f_{\text{cgsi}} = \frac{600{,}000}{419}\left(1 + \frac{9.40}{106.5}\right) = 2650 \text{ psi}$$

TABLE 6-3

Time Days	*Total Shrinkage Coef.*	*Increm. Shrinkage Coef.*	Δf_{ss} *ksi*	f_{st} *ksi*	*Increm. Relax ksi*	*Total Creep Coef.*	*Increm. Creep Coef.*	Δf_{sc} *ksi*	Δf_s *ksi*	Δf_{cgs} *psi*	*f*
0	0	0		167.5	0	0	0	0	(20.0)	0	2
15	0.35	0.35	2.1	165.4	2.1	0.41	0.41	6.1	10.3	−146	2
30	0.50	0.15	0.9	164.9	0.5	0.52	0.11	1.5	2.9	−41	2
45	0.60	0.10	0.6	164.6	0.3	0.58	0.06	0.8	1.7	−24	2
60	0.66	0.06	0.4	164.4	0.2	0.62	0.04	0.5	1.1	−15	2
75	0.73	0.07	0.4	164.2	0.2	0.66	0.04	0.5	1.1	−16	2
90	0.78	0.05	0.3	164.1	0.1	0.69	0.03	0.4	0.8	−11	2
105	0.82	0.05	0.3	163.9	0.2	0.71	0.02	0.3	0.8	−11	2
120	0.86	0.04	0.2	163.8	0.1	0.73	0.02	0.3	0.6	−8	2
135	0.88	0.02	0.1	163.8	0.0	0.75	0.02	0.3	0.4	−6	2
150	0.91	0.03	0.2	163.7	0.1	0.76	0.01	0.1	0.4	−6	2
165	0.93	0.02	0.1	163.6	0.1	0.78	0.02	0.3	0.5	−7	2
180	0.94	0.01	0.1	163.5	0.1	0.79	0.01	0.1	0.3	−4	2
195	0.95	0.01	0.1	163.5	0.0	0.80	0.01	0.1	0.2	−3	2
210	0.96	0.01	0.1	163.4	0.1	0.81	0.01	0.1	0.3	−4	2
225	0.97	0.01	0.1	163.4	0.0	0.82	0.01	0.1	0.2	−3	2
240	0.98	0.01	0.1	163.3	0.1	0.83	0.01	0.1	0.3	−4	2
255	0.98	0.00	0.0	163.3	0.0	0.84	0.01	0.1	0.1	−1	2
270	0.99	0.01	0.0	163.2	0.1	0.85	0.01	0.1	0.2	−3	2
285	0.99	0.00	0.0	163.2	0.0	0.86	0.01	0.1	0.1	−1	2
300	1.00	0.01	0.0	163.1	0.1	0.87	0.01	0.1	0.2	−3	2

Due to dead load at midspan:

$$\text{Moment due to dead load is} = M_{DL} = \frac{0.44 \times 40^2}{8} = 88.0 \text{ ft-k}$$

$$f_{\text{cgs}}^D = -\frac{88.0 \times 12000 \times 9.40}{44{,}670} = -222 \text{ psi}$$

The relaxation loss of the steel in 24 hr is

$$\Delta f_{sr} = 187.5 \left(\frac{\log 24}{10}\right)\left(\frac{187.5}{256} - 0.55\right) = 4.75 \text{ ksi}$$

The loss of steel stress due to elastic shortening (Δf_{ses}) can be computed as follows:

$$n\left[\left(\frac{f_{si} - \Delta f_{sr} - \Delta f_{ses}}{f_{si}}\right) f_{\text{cgsi}}\right] - n f_{\text{cgs}}^D = \Delta f_{ses}$$

which can be rewritten

$$\Delta f_{ses} = \frac{nf_{cgsi}(f_{si} - f_{sr}) - nf^{D}_{cgs} f_{si}}{f_{si} + nf^{D}_{cgsi}}$$

In this example:

$$\Delta f_{ses} = \frac{7 \times 2.65(187.5 - 4.75) - 7 \times 0.222 \times 187.5}{187.5 + 7 \times 2.65}$$

$$= 15.0 \text{ ksi}$$

Hence, the stress in the steel immediately after release (f'_{si}) is approximately 167.5 ksi (187.5 – 20.0), and using Eq. 2-2, the stress in the steel at time n becomes

$$f_{st} = 167.5 \left[1 - \left(\frac{167.5}{256} - 0.55\right)\left(\frac{\log t'_n - \log 24}{10}\right)\right]$$

$$= 167.5 \left[1 - (0.105)\left(\frac{\log t'_n - 1.380}{10}\right)\right]$$

The incremental change in concrete stress at the level of the steel is

$$\Delta f_{cgs} = \frac{\Delta f_s}{167.5} \times 2367 \text{ psi} = -14.13\, \Delta f_s$$

Note: 2367 psi is the unit stress at the c.g.s. due to a steel stress of 167.5 ksi. The computations are summarized in Table 6-3.

It should be noted that the total loss of prestress for this example is 42,500 psi. If an additional load had been put on the member to reduce the concrete stress at the level of the steel, the total loss would be reduced.

6-3 Deflection and Camber

The computations of short term deflections in prestressed-concrete flexural members are made with the assumption that the concrete section acts as an elastic and homogeneous material. This assumption is only approximately correct, since the elastic modulus for concrete is not a constant value for all stress levels, and in addition, the elastic modulus varies with the age of the concrete and is influenced by other factors. As a result, deflection computations for prestressed concrete are approximate.

The deflections for dead and live loads are calculated in the same manner as they are for steel or reinforced concrete members. Normally the moment of inertia of the gross section is used in the computations. In members that have a large amount of reinforcing steel, the moment of inertia of the transformed section can be used. The deflection resulting from the prestressing can be readily calculated for prismatic members with known prestressing force and eccentricity by use of the area-moment principle. The results of such a calculation for a prismatic member with straight tendons (*see* Fig. 6-5) is

$$\delta = -\frac{PeL^2}{8\,EI}$$

where P is the prestressing force in pounds, e is the eccentricity in inches, L is the span in inches, E is the modulus of elasticity of the concrete in psi, and I is the moment of inertia of the gross section in inches to the fourth power. The negative sign indicates that the deflection of the beam due to prestressing alone is upward.

In Fig. 6-6, the moment diagram and corresponding deflection due to prestressing is shown for a member prestressed with a tendon that is on a parabolic curve having zero eccentricity at the ends. Finally, in Fig. 6-7, the moment diagram and corresponding deflection is indicated for a prismatic member prestressed with a tendon that has a trajectory composed of three straight lines that are symmetrical about midspan and has zero eccentricity at each end.

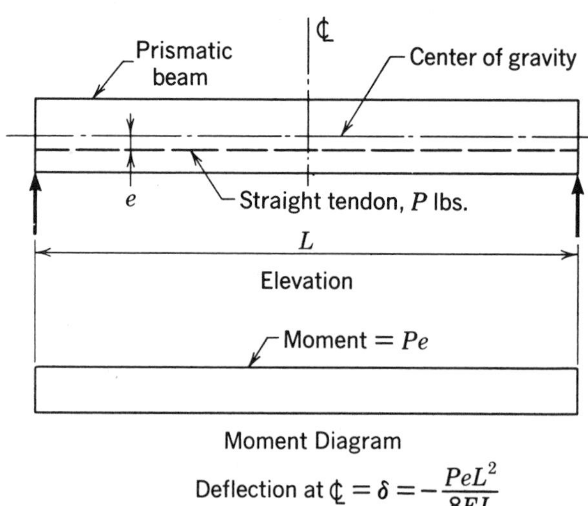

Fig. 6-5 Layout and prestressing-moment diagram for a beam and a straight tendon.

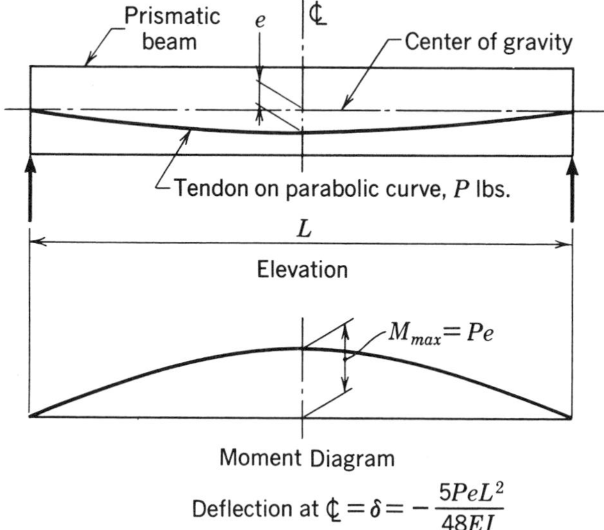

Deflection at ℄ $= \delta = -\dfrac{5PeL^2}{48EI}$

Fig. 6-6 Layout and prestressing-moment diagram for a beam with a parabolic tendon having no eccentricity at the supports.

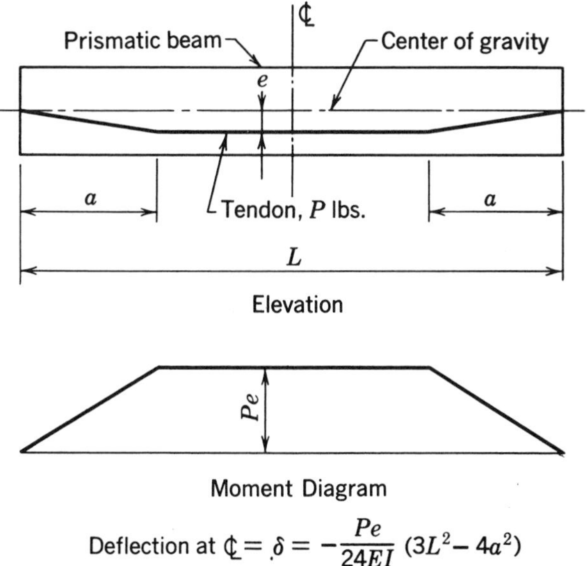

Deflection at ℄ $= \delta = -\dfrac{Pe}{24EI}(3L^2 - 4a^2)$

Fig. 6-7 Layout and prestressed-moment diagram for a beam with a deflected tendon having constant eccentricity in the center portoin and zero eccentricity at the ends.

It is assumed that the deflections due to the various loads can be superimposed in order to determine the net deflection or camber of the member. In this manner, the deflection of a beam under the effects of its own dead load and due to prestressing is computed as the algebraic sum of the deflections due to dead load of the beam and due to prestressing. In a similar manner, if a beam were prestressed with a tendon that was on a parabolic curve with an eccentricity at the ends, the deflection of the beam due to prestressing could be determined by computing the algebraic sum of the deflections indicated for a member prestressed with a straight tendon and for a member with a parabolic tendon as indicated in Figs. 6-5 and 6-6, respectively. In applying the principle of superposition as described, it is necessary to divide the moment due to prestressing into two portions that can be substituted into the appropriate relationships for the terms Pe. For unusual prestressing-moment diagrams, or if the designer questions the results obtained through the use of the superposition principle, the deflection can be easily calculated from the basic, area-moment principle. When members with variable moments of inertia are used, it is necessary to compute the deflections by use of basic principles.

The deflections at the ends of members that have overhanging ends are frequently large and often result in an undesired appearance. Because of this, many engineers and contractors avoid the use of cantilevers when ever possible. In the case of overhanging beams, the deflection at the end of the overhang is the algebraic sum of the deflection the cantilevered end would have if it were a fixed cantilever and the product of the length of the cantilever and the rotation at the support. It is frequently found that the effect of the rotation at the support is much greater than that of the cantilever deflection by itself. It is strongly recommended that deflection computations always be made for beams that have an overhanging end.

It is well known that in reinforced concrete the tendency is for the deflection of a member to increase with time as a result of creep of the compression flange. In prestressed concrete, the variation in deflection is a function of time as well as of the average distribution of stress in the member under the normal condition of loading. For example, if the effects of the prestressing and the external loads at the average section of a member were such that the distribution of stress was a uniform compression, the effect of creep would be to shorten the member without change in the deflection. If under the same conditions, the stress in the bottom flange were greater than the average compression, the tendency would be for the member to increase in camber, whereas, if the top-fiber compressive stress under the normal loading were higher than the average compression, the tendency would be for the deflection to increase as a result of the creep.

It is interesting to note that for the deflection due to prestressing alone, the effects of concrete shrinkage and steel relaxation are to reduce the deflection, since these two effects tend to reduce the prestressing force. The effect of creep is to alter the deflection for cases where the resultant force in the concrete is significantly eccentric, because the rotational changes due to creep are greater than is the creep shortening effect on reducing the prestress.

The computation of long term deflections in prestressed members is often facilitated by the use of the principle of superposition. In this way the effect of prestressing and transverse loads are considered separately and added algebraically.

It should be recognized that the rotation at any section of a beam is equal to

$$\phi = \frac{M}{EI} \tag{6-3}$$

This relationship is useful in estimating the time-dependent deflection changes that take place in a prestressed concrete member using the numerical integration method. The relationships for deflection due to prestressing of different types that are given in Figs. 6-5, 6-6 and 6-7 can be rewritten in the form

$$\delta = \frac{\phi L^2}{K} \tag{6-4}$$

in which K is a constant depending upon the trajectory of the tendon. From Fig. 6-8 it will be seen that the rotation at any section can be computed if

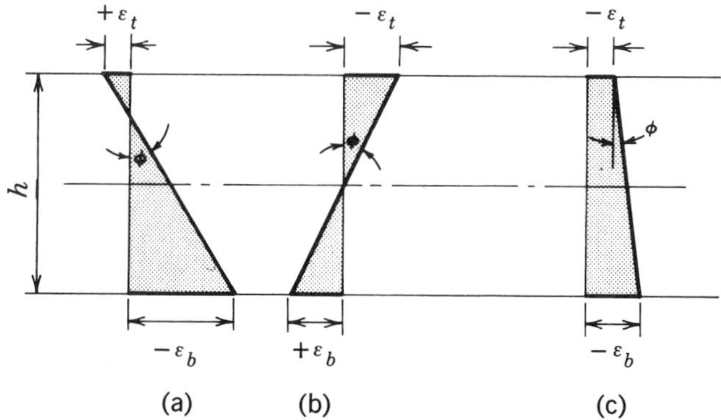

Fig. 6-8 Strain distributions due to: (a) prestressing only; (b) effects of a transverse load; and (c) the algebraic sum of (a) and (b).

the strain distribution is known. The rotation is

$$\phi = \frac{\epsilon_b - \epsilon_t}{h} \tag{6-5}$$

where compressive strains are negative and tensile strains and rotations that cause downward deflections are positive. This relationship applies equally to short and long terms rotations.

In computing long terms deflections, using the numerical integration method, the principles used in computing the loss of prestress in Illustrative Problems 6-1 and 6-2 can be used. All of the computations for the loss of prestress must be made in the deflection computations, because the variation in prestressing force must be taken into account. In addition, at each time increment, the strain in the top and bottom fibers must be determined so that the rotation can be computed. The computations must be made at several sections along the length of the beam and not just at the location of maximum moment, as in the computation for loss of prestress, because the conditions of stress and hence creep, loss of prestress, and rotation vary along the length of the beam. After the rotations have been determined at a number of locations along the beam length, the deflection can be determined by the classical methods.

The step by step procedure for computing the deflection of a precast beam with a composite cast-in-place slab (such as shown in Fig. 6-12) is as follows:

(1) Determine shrinkage and creep characteristics of the concrete to be used as a function of time. In addition, determine the relaxation characteristics of the prestressing steel as a function of time.

(2) Divide the beam into a number of incremental lengths. The computations that follow must be done for each increment. (Symmetry of beam and loading conditions will reduce the calculations required.) Determine the time increments to be used. Small increments are desirable from the standpoint of improving accuracy, but they also increase the number of computations.

(3) Compute the stresses in the concrete at the top and bottom fibers as well as at the center of gravity of the steel (c.g.s.). The stresses should include the effects of prestressing and all transverse loads. (The effect of transverse loads or restraints that are applied at later ages must be taken into account at the appropriate time, as is explained below in step 10.)

(4) Compute the initial strains in the top and bottom fibers and the rotation due to these strains.

(5) For the duration of the first time increment, compute the strains at the top and bottom fibers and at the c.g.s. due to creep, and shrinkage. In addition, compute the relaxation of the steel during the first time increment.

(6) Compute the total change in stress in the steel due to creep, shrinkage, and relaxation and, applying this force as a tensile force at the c.g.s., compute

the stresses in the top and bottom fibers as well as at the c.g.s. due to this tensile force.

(7) Add the stresses from step 6 to those of step 3 in order to find the stresses at the end of the time increment.

(8) Compute the strains in the top and bottom fibers at the end of the time interval as well as the rotation.

(9) Using the stresses from step 7, repeat the procedure (steps 5 through 8). This procedure is repeated until the total time being considered is covered.

(10) At the appropriate time, the effect of a superimposed load is taken into account by computing the changes in stress at the top and bottom fibers and at the c.g.s. due to the load. The stress at the c.g.s. should be multiplied by the modular ratio and the resulting steel stress applied as an incremental increase in the prestressing force. The effect of this increase in the stresses in the top and bottom fibers and at the c.g.s. due to the incremental increase in the prestressing force should be computed. From these computations, the resulting stresses in the concrete can be determined and the rotation computed.

(11) The procedure continues (steps 5 through 8) until the total time under consideration is covered.

(12) The effect of the differential strains in the cast-in-place concrete and the precast concrete are taken into account by first computing the difference in the unrestrained changes in strain in the cast-in-place concrete and the precast concrete at the interface between the two. Strain compatibility is then forced by applying equal and opposite forces to the cast-in-place concrete and the beam.

(13) The stresses from the forces computed in step 12 are to be computed in each subsequent time-interval computation and taken into account in the routine of steps 5 through 8.

(14) The deflection at the end of any time interval can be computed by the area-moment principle, since the rotation ϕ is equal to M/EI.

It should be apparent that a large amount of tedious computations must be made in applying this method. Because of this, the method can best be applied with the aid of programmable calculators or computers.

A close approximation of the deflection of a prestressed member can be made without using the numerical integration procedure by using the following relationship

$$\phi_{px} = \phi_{ix}\left[1 - \frac{\Delta f_s}{f_{si}} + C_t\left(1 - \frac{\Delta f_s}{2f_{si}}\right)\right] \tag{6-6}$$

Where ϕ_{px} and ϕ_{ix} are the ultimate and initial rotations at section x due to prestressing, respectively, Δf_s is a fictitious loss of stress, in the prestressing steel at section x, f_{si} is the initial stress in the prestressing steel, and C_t is the creep ratio. In using this relationship, the effects of prestressing and transverse loads must be considered separately and their effects superimposed. The accuracy of the results obtained with this method is improved by computing the rotations at a number of locations along the member and computing the deflection on the basis of the moment or rotation distribution obtained therefrom (Ref. 6-3).

ILLUSTRATIVE PROBLEM 6-5 Using the approximate method, compute the ultimate camber for the girder of Fig. 6-3 under prestress and dead load alone, assuming the tendon is on a parabolic trajectory with no eccentricity at the supports of the 180-ft span. For this calculation neglect variations in the initial prestressing stress due to friction, assume $E_c = 4 \times 10^6$ psi, $E_s = 28 \times 10^6$ psi, ultimate shrinkage strain after stressing is 200×10^{-6} the relaxation loss is 10% and $C_t = 2.00$.

SOLUTION:

Initial girder dead load defl. $\delta^{\mathrm{i}}_{\mathrm{dl}} = \dfrac{5wL^4}{384EI} = \dfrac{5 \times 1.82 \times 180^4 \times 1{,}728}{384 \times 4 \times 10^3 \times 1{,}908{,}000} = 5.63$ in.

Ultimate girder dead load defl. $\delta^{\mathrm{u}}_{\mathrm{dl}} = 5.63(1 + 2) = 16.9$ in.

The initial deflection due to prestressing from Fig. 6-6 is as follows:

$$\delta^{\mathrm{i}}_{\mathrm{p}} = -\frac{5PeL^2}{48EI} = -\frac{5 \times 2740\text{ k} \times 48.9'' \times 180^2 \times 144}{48 \times 4 \times 10^3 \times 1{,}908{,}000} = -8.53 \text{ in.}$$

The initial prestress at the midspan, quarter point and end of the member for the top and bottom fibers are as follows

midspan	*Top*	*Bottom*
$\dfrac{2740}{1743}\left(1 - \dfrac{48.9}{26.8}\right) =$	−1296 psi	
$\dfrac{2740}{1743}\left(1 + \dfrac{48.9}{20.2}\right) =$		+5377 psi
quarter point		
$\dfrac{2740}{1743}\left(1 - \dfrac{36.6}{26.8}\right)$	−575 psi	
$\dfrac{2740}{1743}\left(1 + \dfrac{36.6}{20.2}\right)$		+4420 psi
end (eccentricity = 0)		
$\dfrac{2740}{1743}$	+1572	+1572

The instantaneous deformations at midspan are (tensile strains are positive):

$$\text{top} = +\frac{1296}{4 \times 10^6} = +0.000324 \text{ in./in.}$$

$$\text{bottom} = -\frac{5377}{4 \times 10^6} = -0.001344 \text{ in./in.}$$

and the rotation is

$$\phi = \frac{-0.001344 - 0.000324}{95''} = -0.0000176$$

at the quarter point, the strains are

$$\text{top} = \frac{+575}{4 \times 10^6} = +0.000144$$

$$\text{bottom} = \frac{-4420}{4 \times 10^6} = -0.001105$$

and the rotation is

$$\phi = \frac{-0.001249}{95} = -0.0000131$$

At the end, the rotation is zero.

The fictitious losses of prestress must be determined in order to compute the time dependent deflection. The losses of prestress will be evaluated at the center, quarter point, and end of the span for the purpose of this computation. More points would give better accuracy. The concrete stress at the center of gravity of the steel is computed as follows:

at midspan: $f_{\text{cgs}} = \dfrac{P}{A}\left(1 + \dfrac{e^2}{r^2}\right) = \dfrac{2740}{1743}\left(1 + \dfrac{48.9^2}{1095}\right) = 5000 \text{ psi}$

at the quarter point $e = 36.6$ in.

$$f_{\text{cgs}} = \frac{P}{A}\left(1 + \frac{e^2}{r^2}\right) = \frac{2740}{1743}\left(1 + \frac{36.6^2}{1095}\right) = 3680 \text{ psi}$$

at the end $e = 0$

$$f_{\text{cgs}} = \frac{P}{A} = 1570 \text{ psi}$$

The losses of prestress are computed, using the creep loss, as follows:

$$\Delta f_{sc} = C_t n f_{\text{cgs}}\left(1 - \frac{C_t n f_{\text{cgs}}}{2f_i'}\right)$$

which for this example becomes

$$\Delta f_{sc} = 2 \times 7 f_{cgs}\left(1 - \frac{2 \times 7 f_{cgs}}{2 \times 187}\right)$$

$$= 14 f_{cgs}\left(1 - \frac{14 f_{cgs}}{187{,}000}\right)$$

at midspan

$$\Delta f_{sc} = 14 \times 5000\left(1 - \frac{14 \times 5000}{187{,}000}\right) = 43{,}700 \text{ psi}$$

at the quarter point

$$\Delta f_{sc} = 14 \times 3680\left(1 - \frac{14 \times 3680}{187{,}000}\right) = 37{,}300 \text{ psi}$$

at the end

$$\Delta f_{sc} = 14 \times 1570\left(1 - \frac{14 \times 1570}{187{,}000}\right) = 19400 \text{ psi}$$

The losses of prestress due to relaxation and shrinkage are $0.10 \times 187{,}000 + 200 \times 10^{-6} \times 28 \times 10^{6} = 18{,}700 + 5{,}600 = 24{,}300$ psi
and the total losses of prestress are

at midspan	$43.7 + 24.3 = 68.0$ ksi
at the quarter point	$37.3 + 24.3 = 61.6$ ksi
at the end	$19.4 + 24.3 = 43.7$ ksi

The ratio of the stress losses to the initial stress is as follows:

$$\text{at midspan} = \frac{68.0}{187} = 0.364$$

$$\text{at the quarter point} = \frac{61.6}{187} = 0.329$$

$$\text{at the end} = \frac{43.7}{187} = 0.234$$

The rotation due to the time dependent deformations and the instantaneous deformations from Eq. 6-6 are as follows:
at midspan

$$\phi_{px} = -0.0000176\left[1 - 0.364 + 2\left(1 - \frac{0.364}{2}\right)\right] = -0.0000400$$

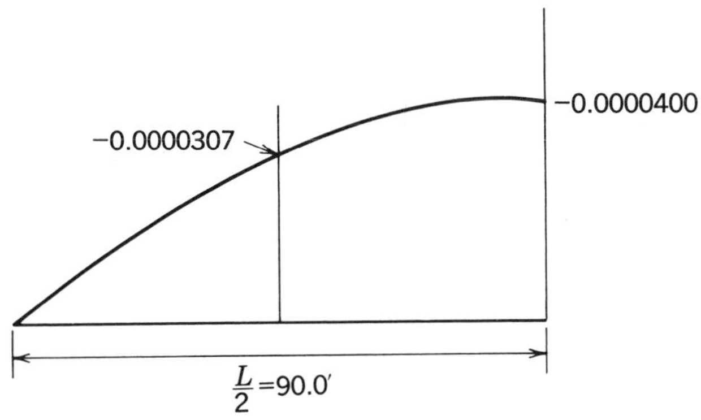

Fig. 6-9 *M/I* diagram for Prob. 6-5.

at the quarter point

$$\phi_{px} = -0.0000131 \left[1 - 0.329 + 2\left(1 - \frac{0.329}{2}\right)\right] = -0.0000307$$

The ratios are plotted in Fig. 6-9 and it will be seen that the curve is a parabola for all practical purposes. Since $\phi = M/EI$, the deflection at midspan can be computed as follows:

$$\delta_C = \frac{M_{\max}}{EI} \times 90.0 \times 12 \times \tfrac{2}{3} \times \tfrac{5}{8} \times 90 \times 12 =$$

$$-0.0000400 \times 90 \times 12 \times \tfrac{2}{3} \times \tfrac{5}{8} \times 90 \times 12 = -19.44 \text{ in.}$$

The net initial deflection is: $\delta i = +5.63 - 8.53 = -2.90$ in. (upwards) and the net long term deflection is

$$\delta = +16.9 \text{ in.} - 19.4 = -2.50 \text{ in. (upwards).}$$

ILLUSTRATIVE PROBLEM 6-6 Compute the midspan deflection for the beam of Fig. 6-10 immediately after stressing and after creep has taken place, assuming the initial prestress is 520 kips, $C_t = 1.75$, the shrinkage strain is 300×10^{-6} in./in., $E_c = 4 \times 10^6$ psi, and $E_s = 28 \times 10^6$ psi. The design span is 40 ft and the beam weight is 440 plf. The tendon is straight. Assume a relaxation loss of 12%.

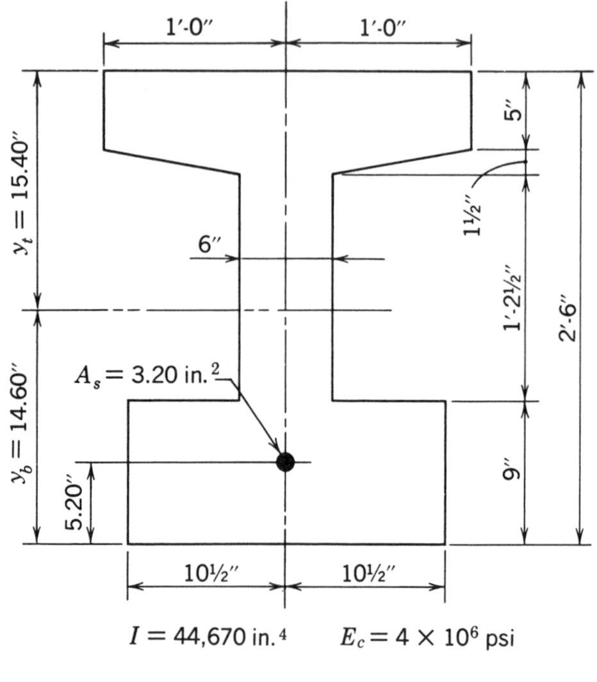

$A = 418.5$ in.^{2} $r^2 = 106.7$

Fig. 6-10 Cross section of beam used in Prob. 6-2.

SOLUTION:

Instantaneous Deflections

$$\text{Prestress Deflection} = \frac{-PeL^2}{8EI} = \frac{520 \times 9.40 \times 40^2 \times 144}{8 \times 4 \times 10^3 \times 44{,}670} = -0.788$$

$$\text{DL Deflection} = \frac{5wL^4}{384EI} = \frac{5 \times 0.44 \times 40^4 \times 1728}{384 \times 4 \times 10^3 \times 44{,}670} = +0.142$$

Net Initial Deflection $= -0.646$ in. (upwards).

The fictitious loss of prestress is as follows:

$$f_{si} = \frac{520}{3.20} = 162.5 \text{ ksi}$$

$$f_{cgs} = \frac{520}{418.5}\left(1 + \frac{9.40^2}{106.7}\right) = +2271 \text{ psi}$$

$$\Delta f_{cs} = 1.75 \times 7 \times 2271 \left(1 - \frac{1.75 \times 7 \times 2271}{2 \times 162{,}500}\right) = 25{,}438 \text{ psi}$$

The total loss $= 0.12 \times 162.5 + 300 \times 10^{-6} \times 28 \times 10^6 + 25{,}438 = 53{,}338$ psi. The ratio of the steel stress loss to the initial stress is

$$\frac{53{,}338}{162{,}500} = 0.328$$

And the long term deflection due to prestressing becomes

$$\delta_p = -0.788 \left[1 - 0.328 + 1.75\left(1 - \frac{0.328}{2}\right)\right] = 1.682 \text{ in.}$$

The long term deflection due to dead load is

$$\delta_{DL} = +0.142(1 + 1.75) = +0.391$$

The net long term deflection is equal to

$$\delta = +0.391 - 1.682 = -1.292 \text{ in. (upwards).}$$

ILLUSTRATIVE PROBLEM 6-7 Determine the deflection of the beam shown in Fig. 6-12 using the numerical integration method. The following data is given.

For the precast section:

$A_c = 573$ in.^{2}
$A_s^* = 3.90$ in.^{2}
$C_t = 2.00$
$E_c = 4 \times 10^3$ ksi
$E_s = 28 \times 10^3$ ksi
$S_t = 4450$ in.^{3}
$S_b = 5950$ in.^{3}
$S_{cgs} = 8240$ in.^{3}
$f_{si} = 187$ ksi
$M_{sdl} = 246$ Ft/Kips

$y_t = 24.0$ in.
$y_b = 18.0$ in.
$e = 13.0$ in.
$\epsilon_u = 300 \times 10^{-6}$
$M_{gdl} = 422$ Ft/Kips
$r^2/y_t = 7.80$ in.
$r^2/y_b = 10.3$ in.
$r^2/e = 14.4$ in.
$f_y = 256$ ksi

Steel stress loss relationships at time t (days), $t = 0$ at time of stressing:

Shrinkage $\quad f_{ss} = \epsilon_u E_s [0.157 \log_e t - 0.115]$

Creep $\quad f_{sc} = nC_t f_{cgs} [0.157 \log_e t - 0.115]$

*Relaxation $\quad f_{sr} = \dfrac{f_{si} \log_e 24t}{10 \log_e 10} \left[\dfrac{f_{si}}{f_y} - 0.55\right]$

Incremental steel stress loss relationships for the period to time (t) starting from time ($t - \Delta t$) are

Shrinkage $\quad \Delta f_{ss} = 0.157\epsilon_u E_s [\log_e (t + t_s) - \log_e (t + t_s - \Delta t)]$

Creep $\quad \Delta f_{sc} = 0.157nC_t f_{cgs}^{t-\Delta t} [\log_e (t) - \log_e (t - \Delta t)]$

Relaxation $\quad \Delta f_{sr} = \dfrac{f_{si}}{10 \log_e 10} \left[\dfrac{f_{si}}{f_y} - 0.55\right] \log_e 24\,(t) - \log_e 24(t - \Delta t)$

Time t_s is the age of the concrete at the time of stressing.
The precast section is 15 days old at the time of stressing and 45 days old at the time the slab is cast. The slab does not shrink for the first 15 days after placing, due to water curing during that period.

For the cast-in-place slab:

$A_{cs} = 336 \text{ in.}^2 \qquad h = 6.0 \text{ in.} \qquad E_{cs} = 3.2 \times 10^3 \text{ ksi}$

$\epsilon_{us} = 300 \times 10^{-6}$ Shrinkage relationships same as for the precast beam section.

Initial concrete stresses due to prestressing are

$$\text{Top fiber } f_{ti} = \frac{f_{si} A_s^*}{A_c}\left(1 - \frac{ey_t}{r^2}\right) = 1.2728(-0.6667) = -0.8485 \text{ ksi}$$

$$\text{Bottom fiber } f_{bi} = \frac{f_{si} A_s^*}{A_c}\left(1 + \frac{ey_b}{r^2}\right) = 1.2728(2.2621) = +2.8792 \text{ ksi}$$

$$\text{At the cgs } f_{cgsi} = \frac{f_{si} A_s^*}{A_c}\left(1 + \frac{e^2}{r^2}\right) = 1.2728(1.9028) = +2.4218 \text{ ksi}$$

Stresses in the precast section due to girder dead load are

Top fiber = + 1.1380 ksi

Bottom fiber = −0.8511 ksi

At the cgs = −0.6146 ksi

* Note: This is Eq. 2-1 converted to use the natural logarithms and the time in days. This has been done in order to avoid confusion in the computations.

Net initial stresses in the precast section are

$$\text{Top fiber} = +0.289 \text{ ksi}$$

$$\text{Bottom fiber} = +2.028 \text{ ksi}$$

$$\text{At the cgs} = +1.807 \text{ ksi}$$

Initial strains in the top and bottom fibers are

$$\epsilon_{ti} = \frac{f_t}{E_c} = \frac{0.289}{4000} = 0.00007225$$

$$\epsilon_{bi} = \frac{f_b}{E_c} = \frac{2.028}{4000} = 0.000507$$

The initial rotation is

$$\phi = \frac{\epsilon_{bi} - \epsilon_{ti}}{h} = 0.00001035$$

The stresses due to slab dead load are

$$\text{Top fiber} = +0.663 \text{ ksi}$$

$$\text{Bottom fiber} = -0.496 \text{ ksi}$$

$$\text{At the cgs} = -0.358 \text{ ksi}$$

The increase in the prestressing force due to the slab load is

$$\Delta P = A_s^* n f_{\text{cgs}} = 3.90 \times 7 \times 0.358 = 9.77 \text{ kips}$$

and the change in the concrete stresses due to ΔP are

$$\text{Top fiber} = \frac{9.77}{573}(-0.6667) = -0.011 \text{ ksi}$$

$$\text{Bottom fiber} = \frac{9.77}{573}(+2.2621) = +0.039 \text{ ksi}$$

$$\text{At the cgs} = \frac{9.77}{573}(1.9028) = +0.032 \text{ ksi}$$

Therefore, the net stress change in the precast section due to the application of the slab dead load is

$$\text{Top fiber} = +0.652 \text{ ksi}$$

$$\text{Bottom fiber} = -0.457 \text{ ksi}$$

$$\text{At the cgs} = -0.326 \text{ ksi}$$

The losses in steel stress due to shrinkage, creep and relaxation for the first time increment are

($t = 15$ days, $\Delta t = 15$ days and $t_s = 15$ days):

$$\Delta f_{ss1} = 0.157\epsilon_u E_s [\log_e (t + t_s) - \log_e (t + t_s - \Delta t)]$$

$$= 1.3188 [\log_e (30) - \log_e (15)] = +0.914123 \text{ ksi}$$

$$\Delta f_{sc1} = nC_t f_{\text{cgs}}^{(t-t\Delta)} [0.157 \log_e t - 0.115]$$

$$= 7 \times 2 \times 1.807[0.157 \times \log_e 15 - 0.115] = +7.84652 \text{ ksi}$$

$$\Delta f_{sr1} = \frac{f_{si}}{10 \log_e 10} \left[\frac{f_{si}}{f_y} - 0.55\right] \log_e 24t$$

$$= 8.1213[0.1805] \log_e 360 = 8.62692 \text{ ksi}$$

The summation of losses for the first time increment is

$$\Delta f_{s1} = 17.3875 \text{ ksi}$$

The change in the concrete stresses due to Δf_{s1} are:

$$\Delta f_{t1} = -\frac{\Delta f_{s1}}{f_{si}} (f_{ti}) = +0.07890 \text{ ksi}$$

$$\Delta f_{b1} = -\frac{\Delta f_{s1}}{f_{si}} (f_{bi}) = -0.26759 \text{ ksi}$$

$$\Delta f_{\text{cgs}1} = -\frac{\Delta f_{s1}}{f_{si}} (f_{\text{cgs}i}) = -0.22518 \text{ ksi}$$

The concrete stresses at the end of the first time interval are

Top fiber $= +0.3679$ ksi

Bottom fiber $= +1.7604$ ksi

At the cgs $= +1.58182$ ksi

The strain in the top and bottom fibers at the end of the first time interval are equal to the initial strain plus the effects of shrinkage, creep, and elastic rebound. They are

$$\epsilon_{t1} = \epsilon_{ti} + \frac{\Delta f_{ss1}}{E_s} + f_t^{t=0} \left(\frac{\Delta f_{sc1}}{f_{\text{cgs}}^{t=0} \times E_s}\right) + \frac{\Delta f_{t1}}{E_c} = 0.00016945$$

$$\epsilon_{b1} = \epsilon_{bi} + \frac{\Delta f_{ss1}}{E_s} + f_b^{t=0} \left(\frac{\Delta f_{sc1}}{f_{\text{cgs}}^{t=0} \times E_s}\right) + \frac{\Delta f_{b1}}{E_c} = 0.00078725$$

and the rotation at the end of the first time interval becomes:

$$\phi_1 = \frac{\epsilon_{b1} - \epsilon_{t1}}{h} = 0.00001471$$

For the second time interval, $t = 30$, the stress losses are

$$\Delta f_{ss2} = 1.3188[\log_e 45 - \log_e 30] = 0.534727 \text{ ksi}$$

$$\Delta f_{sc2} = 0.157nC_t f_{\text{cgs}}^{t-\Delta t}[\log_e (t) - \log_e (t - \Delta t)]$$

$$= 0.157 \times 7 \times 2f_{\text{cgs}}^{t=15}[\log_e (30) - \log_e (15)]$$

$$= 2.198 \times 1.58182[\log_e (30) - \log_e (15)] = 2.40996 \text{ ksi}$$

$$\Delta f_{sr2} = \frac{f_{si}}{10 \log_e 10}\left[\frac{f_{si}}{f_y} - 0.55\right] \log_e (24t) - \log_e 24(t - \Delta t)$$

$$= 8.1213 \times 0.1805 \times [\log_e (720) - \log_e (360)]$$

$$= 1.46589[\log_e 720 - \log_e 360]$$

$$= 1.01591 \text{ ksi}$$

and the stress loss in the second interval is

$$\Delta f_{s2} = 3.96060 \text{ ksi}$$

The change in concrete stresses and the stresses at the end of the second time interval are summarized in Table 6-4. The strains at the end of the second time interval are:

$$\epsilon_{t2} = \epsilon_{t1} + \frac{\Delta f_{ss2}}{E_s} + f_t^{t=15}\left(\frac{\Delta f_{sc2}}{f_{\text{cgs}}^{t=15} \times E_s}\right) + \frac{\Delta f_{t2}}{E_c} = 0.00021306$$

$$\epsilon_{b2} = \epsilon_{b1} + \frac{\Delta f_{ss2}}{E_s} + f_b^{t=15}\left(\frac{\Delta f_{sc2}}{f_{\text{cgs}}^{t=15} \times E_s}\right) + \frac{\Delta f_{b2}}{E_c} = 0.00088689$$

and the rotation becomes

$$\phi_2 = 0.00001604$$

The computations for the third time interval are made as for the second, and the results are summarized in Table 6-4. At the end on the third time interval, the slab dead load is applied, the effects of which are shown in Table 6-4.

The effects of subsequent intervals are determined in the same way as for the previous intervals, except the additional shrinkage effect from the slab should now be included. This effect is taken into account by computing the difference in unrestrained strains for the slab concrete and the top of the beam

TABLE 6-4 Summary of Calculations for Illustrative Problem 6-7

	Incremental Losses of Steel Stress				Net Stress at Time t Without Strain Compatibility			Incremental Changes in Stress due to Steel Loss (and S.D.L.)		
Time t	Incremental Shrinkage ksi	Incremental Relaxation ksi	Incremental Creep ksi	Total ksi	f_{cgs} ksi	f_t ksi	f_b ksi	Δf_{cgs} ksi	Δf_t ksi	Δf_b
0					+1.80724	+0.28946	+2.0126			
	0.914123	8.62692	7.84652	17.3875				−0.22518	−0.07890	−0.26759
15					+1.58182	+0.36790	+1.7604			
	0.534727	1.01591	2.40996	3.96060				−0.05129	+0.01797	−0.06098
30					+1.53053	+0.38587	+1.6994			
	0.379395	0.594266	1.36403	2.33769				−0.03028	+0.01061	−0.03560
45					+1.50025	+0.39648	+1.6638			
								(−0.326)	(+0.652)	(−0.457)
45′					+1.17425	+1.04848	+1.2068			
	0.294282	0.421638	0.742508	1.45843				−0.01889	+0.00662	−0.02244
60										
	0.240446	0.327048	0.570396	1.137890				−0.014735	+0.0051626	−0.017512
75										
	0.203294	0.267218	0.45459	0.925104				−0.011980	+0.0041972	−0.014237
90										
	0.176101	0.225929	0.37968	0.78507				−0.01017	+0.003561	−0.012082
105										

TABLE 6-4 (Continued)

	Top Fiber Incremental Strain Changes					Top Fiber	Bottom Fiber Incremental Strain Changes				
Time *t*	*Shrinkage* $\times 10^6$	*Creep* $\times 10^6$	*Elastic Rebound* $\times 10^6$	*Total Free Strain* $\times 10^6$	*Strain Compat. Corr.* $\times 10^6$	*Net Strain* $\times 10^6$	*Shrinkage* $\times 10^6$	*Creep* $\times 10^6$	*Elastic Rebound* $\times 10^6$	*Total Free Strain* $\times 10^6$	*Strain Compat. Corr.* $\times 10^6$
0						+72.25					
	+32.65	+44.82	+19.73	+97.20	0		+32.65	+314.50	−66.90	+280.25	0
15						+169.45					
	+19.10	+20.02	+4.49	+43.61	0		+19.10	+95.79	−15.25	+99.64	0
30						+213.06					
	+13.35	+12.28	+2.65	+28.28	0		+13.35	+54.09	−8.90	+58.54	0
45						+241.34					
	—	—	+163.0	+163.0	0		—	—	−114.25	−114.25	0
45′						+404.34					
	+10.51	+23.68	+1.66	+35.85	−11.59		+10.51	+27.25	−5.61	+32.15	+3.78
60						+428.60					
	+8.587	+17.67	+1.29	+27.547	+21.19		+8.587	+21.01	−4.37	+25.227	−6.93
75						+477.34					
	+7.261	+15.724	+1.049	+24.034	+2.787		+7.267	+16.52	−3.56	+20.221	−0.909
90						+504.16					
	+6.289	+13.347	+0.89	+20.526	−0.46		+6.289	+13.42	−3.02	+16.689	+0.1505
105						+524.23					

TABLE 6-4 (Continued)

	Bottom Fiber	Free Shrinkage	$\Delta \varepsilon_F$ Differential	Stress changes due to Strain Compatibility			Net Stresses at Time t With Strain Compatibility			Rotation $\times 10^6$		
Time t	Net Strain $\times 10^6$	Slab $\times 10^6$	Free Strain $\times 10^6$	Δf_{cgs} ksi	$\Delta f_t'$ ksi	$\Delta f_b'$ ksi	f_{cgs} ksi	f_t ksi	f_b ksi	W/o Compat.	With Compat.	P_s kips
0	507									10.35		
		0								14.71		
15	787.25											
		0								16.04		
30	886.89											
		0								16.76		
45	945.43											
		0								10.16		
45′	831.18											+6.518
		0	−35.85	+0.0076	−0.0464	+0.0151	+1.16296	+1.00870	+1.19946	10.07	10.44	
60	867.11											−11.911
		+93.05	+65.503	−0.01385	+0.08476	−0.02771	+1.134375	+1.09862	+1.15423		9.71	
75	885.40											−1.57
		+32.65	+8.616	−0.00182	+0.00013	−0.00363	+1.120575	+1.102947	+1.109176		9.53	
90	+904.71											+0.26
		+19.10	−1.426	+0.0003	−0.00184	+0.00060					9.46	
105	+921.55											

during the time interval and forcing strain compatibility between the two elements. In the case of this example, there would be no shrinkage strain in the slab concrete between 45 and 60 days, since the slab would be undergoing water curing, and hence, would not shrink. Therefore, for the fourth time interval, slab shrinkage will be considered to be zero. The changes in stresses and strains in the beam are computed as before and are shown in the Table 6-4.

Since the top slab is assumed to have no shrinkage, temperature or creep strain during the time interval, the differential strain is equal to the free strain change for the top of the precast section. The force compatibility relationships for the slab and beam are computed from the strain and force conditions shown in Fig. 6-11.

For the slab

$$\Delta f_{sb} = \frac{P_s}{A_{cs}}\left(1 + \frac{h/2}{h/6}\right) = \frac{4P_s}{A_{cs}}$$

$$\varepsilon_{sb} = \frac{4P_s}{E_{cs}A_{cs}}$$

For the beam

$$\Delta f'_{bt} = -\frac{P_b}{A_c}\left(1 + \frac{y_t^2}{r^2}\right)$$

$$\Delta f'_{bb} = -\frac{P_b}{A_b}\left(1 - \frac{y_b y_t}{r^2}\right)$$

$$\varepsilon'_{bt} = -\frac{P_b}{E_c A_c}\left(1 + \frac{y_t^2}{r^2}\right)$$

$$\varepsilon'_{bb} = \frac{\Delta f'_{bb}}{E_c}$$

Since force $P_s = -P_b$

$$\frac{\varepsilon_{sb} E_{cs} A_{cs}}{4} = -\frac{\varepsilon'_{bt} E_c A_c}{1 + \dfrac{y_t^2}{r^2}}$$

which for this example will give

$$\varepsilon_{sb} = -2.0915\varepsilon'_{bt}$$

The difference between the free shrinkage of the slab and beam top fiber ($\Delta\varepsilon_F$) is equal to:

$$\Delta\varepsilon_F = \varepsilon_{bt} - \varepsilon_{sb}$$

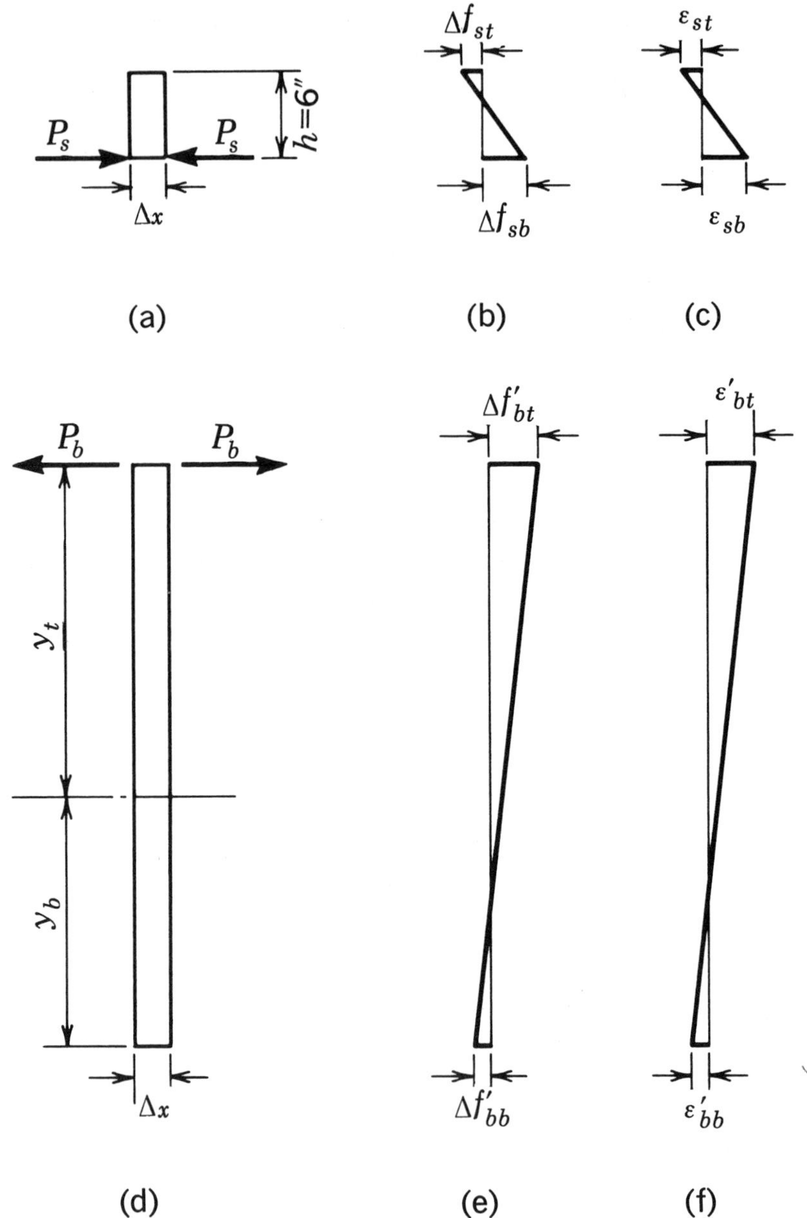

Fig. 6-11 (a) Free-body for a short length of slab. (b) Slab stress distribution due to force P_s. (c) Strain distribution for the slab. (d) Free-body for a short length of beam. (e) Beam stress distribution due to force P_b. (f) Strain distribution due to force P_b.

From the above two expressions, the force resulting from strain compatibility as well as the strains can be determined. The results of these computations for $t = 60$ and $t = 75$ are shown in Table 6-4. The computations can be continued for as many time increments as necessary. A review of the computations shown in Table 6-4 will convince the reader of the value of utilizing calculators and computers when this computation is required.

The deflection computation is completed by continuing this computation for a time interval of from 600 to 1500 days, depending upon the precision desired. The computations must then be repeated at several more locations along the length of the beam, after which, the deflection can be computed from the rotations thus obtained.

6-4 Composite Beams

Flexural members that are formed of precast and cast-in-place elements are frequently employed in bridge and building construction. Beams of this type are referred to as composite beams. An illustration of a typical, composite, bridge beam is given in Fig. 6-12.

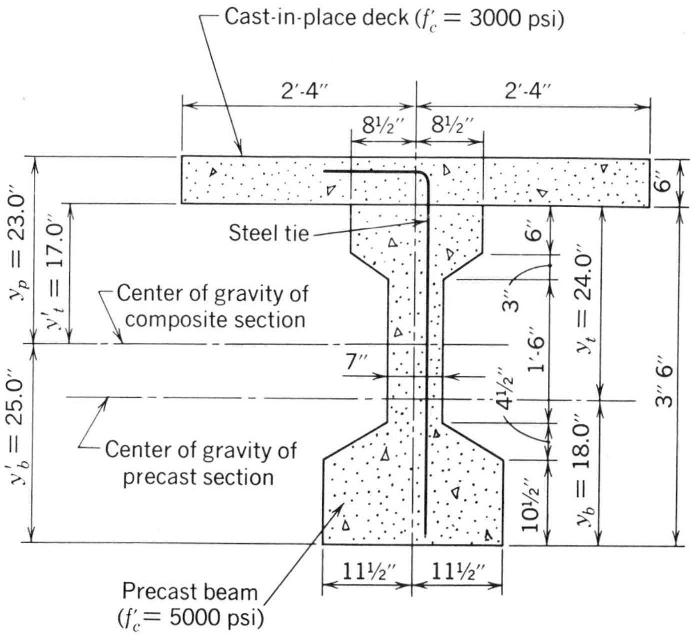

Fig. 6-12 Typical cross section of a composite beam.

Composite beams are used to permit the precasting of the elements that are difficult to form and that have the bulk of the reinforcing. Falsework is not normally needed with composite construction, since the precast elements are usually designed to carry the dead load of the precast beams, the diaphragms as well as the composite slab, without the composite top flange. Dead load and live load, which are applied after the cast-in-place deck has hardened, are carried by the composite beam.

The use of large, composite top flanges results in high, ultimate resisting moments and greater flexural strength in the elastic range, but does not improve the shear strength of a prestressed beam to a significant degree. For this reason, there is little if any advantage to be gained by using composite construction for short-span members in which shear stresses are more significant than flexural stresses.

In designing composite beams, it is necessary to compute the section properties of the precast and composite sections. The flexural stresses due to the various causes at the critical fibers are then computed using the appropriate section properties. In addition, the designer must provide adequate means of transferring the shear stress from the precast to the cast-in-place elements. Nominal shear stresses can be transferred by bond alone, if the concrete surface receiving the composite topping is clean and rough. Higher shear stresses can be transferred if, in addition to the above, steel dowels are extended from the precast section into the cast-in-place section. Shear keys are not required to transfer shear in composite members.

The current ACI 318 permits the full transfer of horizontal shear forces to be assumed without calculations if: (1) the contact surfaces are clean and intentionally roughened; (2) the minimum amount of reinforcing connecting the components conform to Eqs. 5-48 or 5-49, with a spacing not exceeding four times the least dimension of the supported element nor 24 in. (b' used in Eqs. 5-48 or 5-49 should be the width of the cross section being investigated for horizontal shear); (3) web members are designed to resist the entire vertical shear; and (4) all stirrups are fully anchored into all intersecting components. If this is not done, the horizontal shear stresses must be fully investigated.

The horizontal shear stress may be computed from

$$v_h = \frac{V_u}{\phi b' d} \tag{6-7}$$

in which V_u is the total design (ultimate) shear force at the section, ϕ is the strength reduction factor and equals 0.85, b' is the width of the cross section being investigated for horizontal shear, and d is the depth of the entire composite section. In lieu of using Eq. 6-7, the designer may compute the actual force that must be transferred within any incremental length and make the necessary provisions for transferring the force by horizontal shear to the supporting element.

Permissible design (ultimate) shear stresses are:

(1) When ties are not provided, but the contact surfaces are clean and intentionally roughened — 80 psi

(2) When vertical bars or extended stirrups, proportioned to be equal to or exceed the requirements of Eqs. 5-48 or 5-49, are provided at spacings that do not exceed four times the least dimension of the supported element nor 24 in., the contact surfaces are clean but not intentionally roughened — 80 psi

(3) When the conditions of (2) are met, but the contact surfaces are intentionally roughened — 350 psi

(4) If an area of reinforcing equal to

$$A_{vf} = \frac{V_u}{\phi f_y \mu} \tag{6-8}$$

is provided ($\phi = 0.85$, $\mu = 1.0$ and $f_y = 60{,}000$ psi max.) — $0.2f_c'$ (800 psi max.)

Of course, all ties must be fully anchored into the components.

It is recommended that the reader review the complete requirements of ACI 318-71 regarding composite concrete flexural member, since only the more important points have been discussed here.

Differential-shrinkage stresses in composite construction can result in tensile stresses being developed in the cast-in-place concrete and a reduction in the precompression of the tensile flange of the precast element. The differential shrinkage has no effect on the ultimate flexural strength of the composite beam, but it does slightly reduce the load required to crack the tensile flange of the precast element. This effect should be considered in structures in which the cracking load is significant. The effect is normally ignored.

When computing the properties of the transformed section, the difference in the elastic properties of the cast-in-place concrete and the concrete in the precast element should be taken into account by adjusting the width of the composite flange in proportion to the modular ratio of the two concretes. This is exemplified in Prob. 6-8.

ILLUSTRATIVE PROBLEM 6-8 Compute the flexural stresses in the precast and cast-in-place concrete for the composite section shown in Fig. 6-12, if the moment due to the dead load of the beam, slab, and diaphragms is 673 k-ft and the moment due to the future wearing surface, live load, and impact is 830 k-ft. The section properties for the precast and composite sections are as follows:

Precast section

$$y_t = 24.0 \text{ in.} \qquad S_t = 4450 \text{ in.}^3 \qquad y_b = 18.0 \text{ in.} \qquad S_b = 5950 \text{ in.}^3$$

Composite section properties are based upon the transformed cast-in-place flange width of 0.6 × 56 in. = 33.6 in. It is assumed that the ratio of the elastic modulus of the slab and girder concrete is 0.60.

$$y_p = 23.0 \text{ in.} \qquad S_p = 9400 \text{ in.}^3 \qquad I = 216{,}000 \text{ in.}^4$$
$$y_t = 17.0 \text{ in.} \qquad S_t = 12{,}700 \text{ in.}^3$$
$$y_b = 25.0 \text{ in.} \qquad S_b = 8650 \text{ in.}^3$$

SOLUTION:

		Top Fiber Stress in Cast-in-Place Slab	Stresses in Precast Section	
			Top	Bottom
Dead load:	$\dfrac{673 \times 12{,}000}{4450/5950}$		+ 1810 psi	− 1360 psi
$L + I$:	$\dfrac{830 \times 12{,}000}{9400/12{,}700/8650}$	+ 1060 psi	+ 785 psi	− 1150 psi
Totals :		+ 1060 psi	+ 2595 psi	− 2510 psi

6-5 Beams with Variable Moments of Inertia

The magnitude of the prestressing moment in a simple beam can be made to vary along the length of the beam by varying the depth of the member. Members of variable depth may be prestressed with either a straight or curved tendon. This is illustrated by the sloped beam of Fig. 6-13. This type of beam is obviously adaptable to roof construction where the pitch of the roof required for drainage will be provided by the sloping top flange. Although beams of this shape have been produced in this country by a few

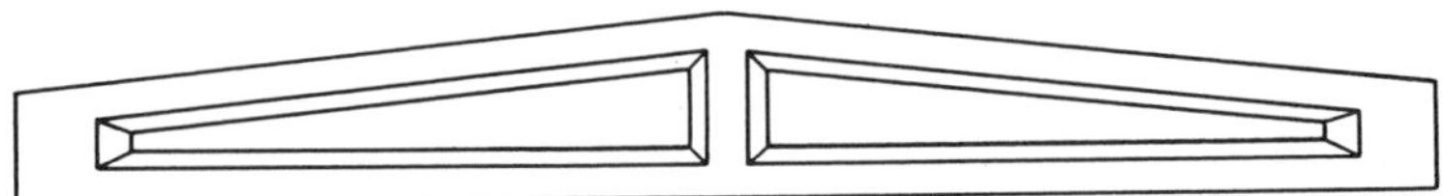

Fig. 6-13 Elevation of beam with sloping top flange.

prestressing plants and on a few special jobs, they are not used to a great extent. The following disadvantages, characteristic of this type of beam, account for the limited use of this mode of framing:

(1) The design of such beams must be done with care, since the maximum moment and maximum flexural stresses may not (probably do not) occur at the same section. Therefore, in order to be certain that the most severe conditions are investigated, the stresses must be determined at several points along the span and plotted to determine the most severe cases. This is a refinement not normally required in simple beam design and is a not always recognized characteristic of this type of beam.

(2) The sloping top flanges intersect at the center of the beam, which is the point of maximum moment. The large inclined compressive forces that are carried by these flanges intersect at an angle as shown in the vector diagram of Fig. 6-14. Provision must be made for the vertical component of these forces; otherwise, there is danger the top flange may buckle upward under load. The danger of this occurring is even greater if a Vierendeel truss is used, rather than a beam, or if an opening is made in the web of the beam at the center line for the passage of utilities.

(3) Forms for members with sloping flanges are expensive and are not easily converted for manufacturing the many different span lengths encountered in modern commercial and industrial building construction.

Another type of beam with a variable moment of inertia and depth that can be used to advantage in roof as well as bridge construction is illustrated in Fig. 6-15. This beam can be stressed with a straight tendon, and since the depth of the section is greater at the ends, the relative eccentricity at the ends will be less at the ends than at the center. As a result, the stresses due to prestressing will not be as high at the ends. In this manner, an effect similar to curving the tendons in a prismatic beam is obtained.

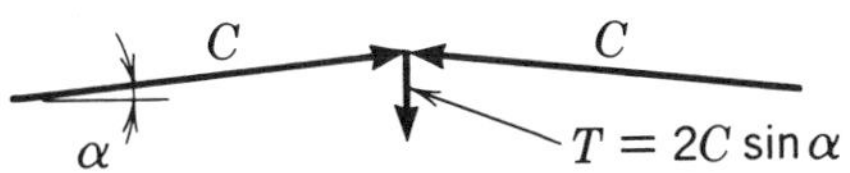

Fig. 6-14 Vector diagram of forces at ridge of beam with sloping top flange.

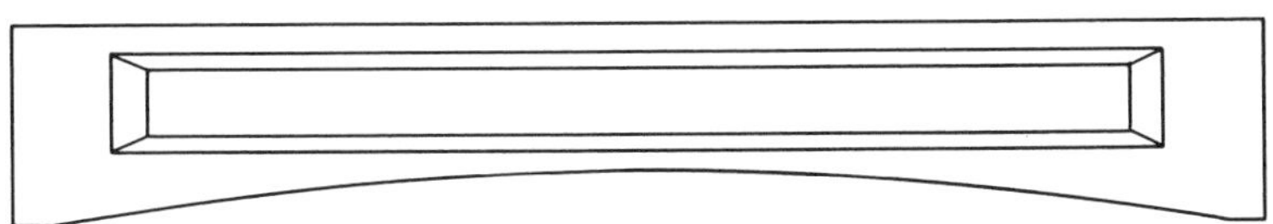

Fig. 6-15 Elevation of beam with variable bottom-flange thickness.

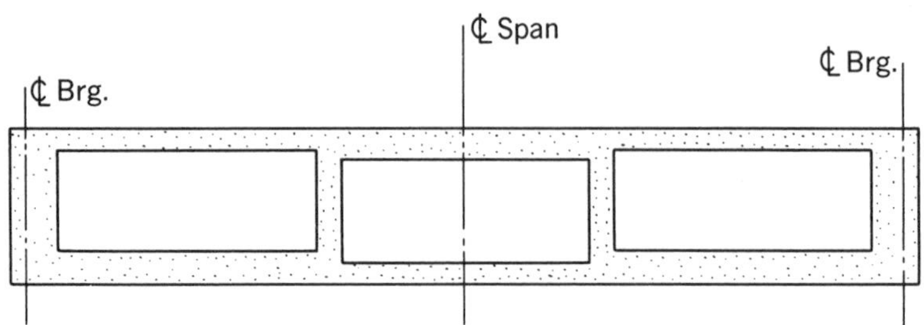

Fig. 6-16 Longitudinal cross section of a hollow box beam showing a method of varying the moment of inertia.

Another economical method of forming a beam with a variable moment of inertia is to use a box section, as illustrated in Fig. 6-16. In such a beam, the hollow core is frequently made with inexpensive plywood or paper forms that can be placed lower at the center than at the ends. In this manner, the larger concrete flanges are placed where needed to resist the large compressive stresses due to pre-tensioning at the ends and to the flexural stresses at the center.

Beams with variable moments of inertia and depth are frequently used in continuous prestressed-concrete structures, for the same reasons variable depths are employed in continuous members made of other materials. Continuity in prestressed concrete construction is discussed in detail in Chapter 8.

6-6 Segmental Beams

Post-tensioned beams consisting of two or more elements or components held together by prestressing are used occasionally to facilitate fabrication transportation, erection, or for economic considerations. An example of a multi-element beam formed of three precast units is shown in Fig. 6-17.

In beams on which this method is employed, the prestressing force is generally very large in comparison with the shear force that must be developed between the elements. This is true because this method is most often used on large, long beams in which shear forces are not as important as in short beams. As a result, the friction that can be developed between the elements due to the prestressing is normally sufficiently large to provide very high factors of safety against slipping. Keys are frequently provided in order to facilitate assembly of the units.

A precast segment of the Downstream Bridge at Autevil in Paris during erection is shown in Fig. 6-18.

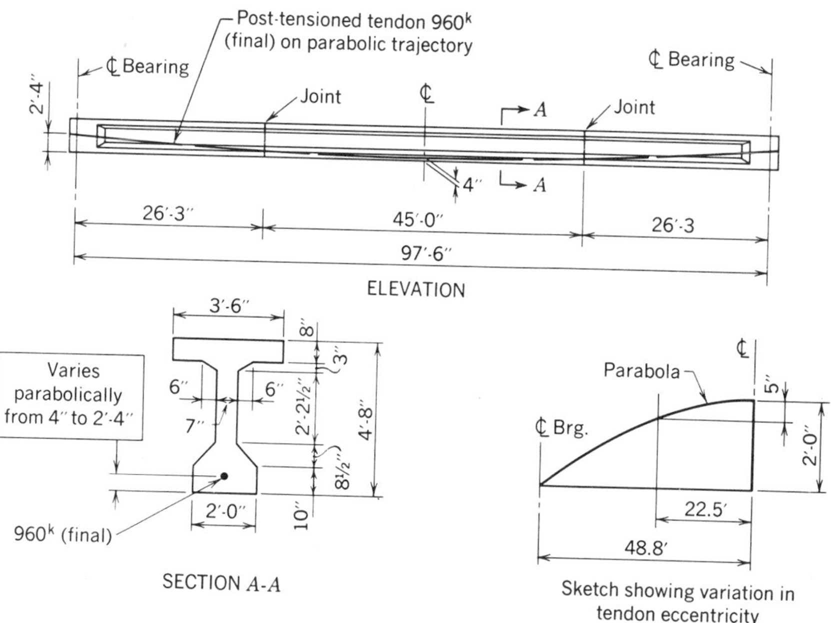

Fig. 6-17 Segmental post-tensioned beam. Adapted from bridge over Naugatuck River, Route 68 in Connecticut.

Fig. 6-18 A precast segment of the Downstream Bridge at Autevil, Paris. (*Courtesy of the Freyssinet Co., New York.*)

ILLUSTRATIVE PROBLEM 6-9 For the beam shown in Fig. 6-17, compute the factor of safety against slipping at the joint if the maximum shear load at the joint is 70 k. Include the effect of the inclination of the tendons that are on a parabolic trajectory. Assume that the coefficient of friction between the concrete units is 1.0.

SOLUTION: The vertical displacement through which the tendon moves between the center line and the joint is

$$\left(\frac{22.5}{48.8}\right)^2 \times 24 \text{ in.} = 5 \text{ in. } (\textit{see} \text{ sketch})$$

At the joint, if α is the inclination of the tendons

$$\tan \alpha = \frac{2 \times 5 \text{ in.}}{12 \times 22.5} = 0.037$$

$$P \sin \alpha = 0.037 \times 960 = 35.5 \text{ k}$$

$$V_{conc} = V - P \sin = 34.5 \text{ k}$$

$$\mu P = 1.0 \times 960 = 960 \text{ k}$$

$$F/S \text{ slipping} = \frac{960}{34.5} \cong 27.8$$

6-7 Partial Prestressing

When tensile stresses are permitted in the tensile flange of a prestressed flexural member under the condition of design live load, a member is said to be partially prestressed. Partial prestressing was first suggested as a means of permitting the use of higher stresses in normal reinforcing bars to which supplementary pre-tensioned tendons were added as a means of reducing the cracking of the concrete. Later, Abeles suggested the use of high-tensile steel for the entire tensile reinforcement, but with only a portion of the steel being prestressed (Ref. 5). In this manner, economy would result from the reduction in labor required to stress and grout the tendons. In addition, the use of high-tensile steel for the entire tensile reinforcement is the most economical, since the cost per pound of ultimate tensile strength is less for high-tensile steel than for lower quality steels.

Currently there are three motivations for the use of partial prestressing. The first is economy of labor and steel costs, as was explained above. Partial prestressed beams made in this manner will have somewhat lower ultimate bending moments than fully-prestressed members, since the average effective stress in the tendon will be lower than in fully-prestressed members. Therefore, the ultimate must be computed from basic strain computations. The fundamental principles of ultimate moment analysis developed in Sec. 5-2 are applicable to this condition of reinforcing.

If partial prestressing is used in a member because the concrete cross section to be used is inefficient, but all of the steel is stressed to the normal values, the ultimate moment will not be affected as a result of the use of partial prestressing. This can be better understood when the basic reason for using I- and T- shaped members (Secs. 4-8 and 4-9) is analyzed and compared to a rectangular section. The use of I and T shapes is based upon elastic-design considerations and not upon ultimate load conditions. If a rectangular section, which is easier to manufacture and more resistant to large shear stresses, can be found that will work satisfactorily at service loads with moderate tensile stresses in the tensile flange, the ultimate load may very well be as high as would be found for an I or T beam designed for the same loads, but without tensile stresses in the bottom flange under full load. This is particularly true for short-span members in which the dead weight of the member itself is not important in comparison to the total moment. The motivations for using partial prestressing of this second type is the reduction of form costs and labor that can be derived through the use of simple rectangular or tapered sections.

Camber due to prestressing and differentials in camber between members have been a significant problem in the manufacture of prestressed-concrete members. The total camber due to prestressing as well as the variation of camber between members, which is assumed to be a function of the total camber, can be reduced by not fully prestressing the members. Assuming the ultimate moment capacity is still adequate, this procedure will result in satisfactory construction for many types of applications. Hence, the third motivation, which is the principal one, is the desire to achieve better performance at service loads without sacrificing the minimum safety requirements of the codes.

In most roof and floor members, partial prestressing can be used without risk, since the loads can generally be predicted with reasonable accuracy, and fatigue or frequent overloading is not normally a possibility. In bridge construction, due to the uncertainty of impact loads, vibration and fatigue, or repetition of loads, partial prestressing has not been allowed to the same extent it has been in building construction. Nominal tensile stresses are currently permitted in the design of prestressed-concrete highway bridges. There is reason to believe that more use will be made of partial prestressing in bridge construction in the future.

6-8 End Blocks

In post-tensioned beams, it is customary and often necessary to curve the tendons up at the ends of the beams as a means of reducing the eccentricity of the prestressing force. In order to have sufficient concrete section in which to slope the tendons to the end anchorages and in which to embed the anchorages, a short section at the end of the beam is often enlarged and made rectangular in cross section. This rectangular section, which is called an end block, is illustrated in Fig. 6-19.

Cracking has occurred along the trajectory of the tendons or between tendons near the ends of post-tensioned beams, as illustrated in Fig. 6-20. This cracking is due to secondary tensile stresses that result from the distribution of the highly concentrated compressive stresses at the end of the member. This can be visualized by considering the schematic diagram of stress trajectories that result from a large concentrated load acting upon a short prism over a small bearing area that extends across the width of the beam, as is shown in Fig. 6-21. From this diagram, it will be seen that the stress trajectories are closely spaced near the ends and spread to the distribution that results from the normal elastic analysis at a distance of approximately one times the depth of the beam from the end of the beam.

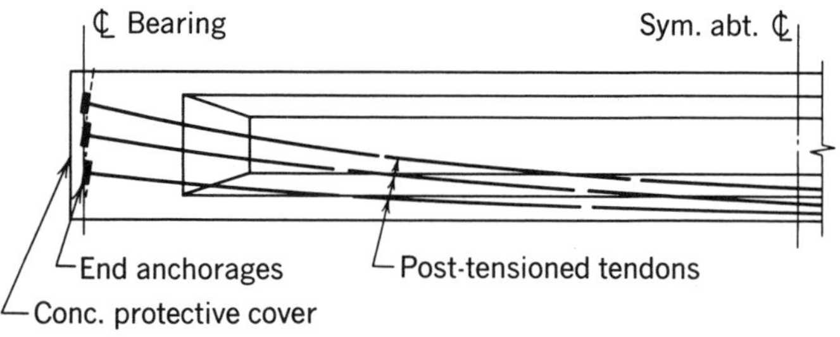

Fig. 6-19 Half-elevation of a post-tensioned beam.

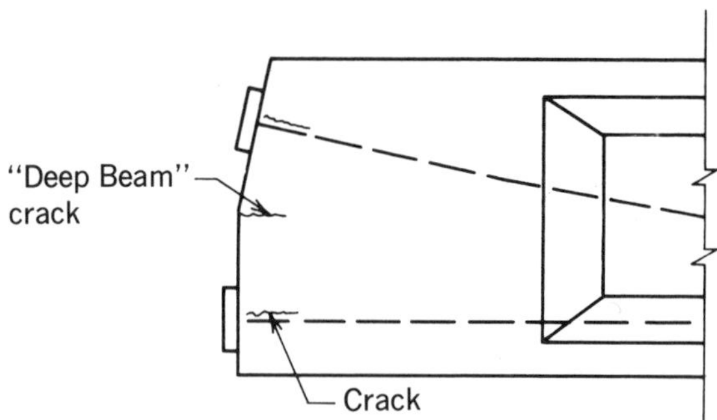

Fig. 6-20 Elevation of an end block showing cracks due to secondary tensile stresses.

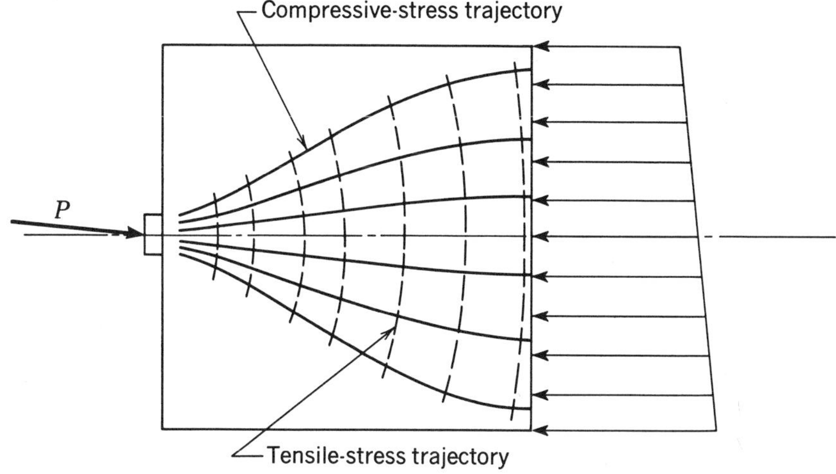

Fig. 6-21 Idealized stress trajectories in end block with a single load.

If the same total prestressing force is applied to the beam by a number of smaller tendons distributed over the end of the beam rather than by one large tendon, the condition of stress can be approximated by the diagram of Fig. 6-22, in which the load is represented as several small forces acting on a common bearing plate. Under this condition of loading, the stress trajectories are seen to be further apart, and the same distribution of stress is obtained at one times the depth of the beam from the end of the beam.

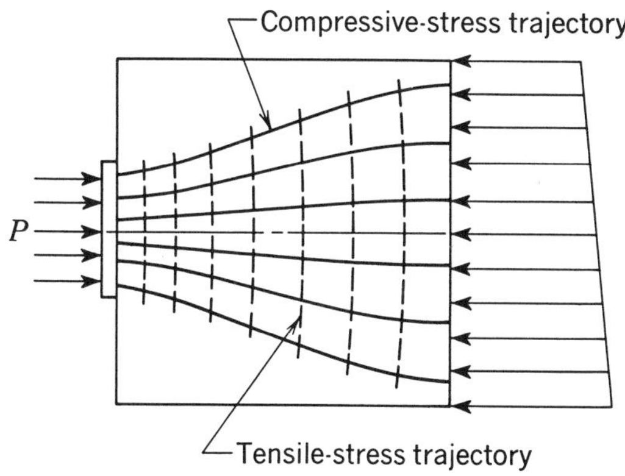

Fig. 6-22 Idealized stress trajectories in end block with several small loads.

The above examples are an oversimplification of the problem since: (1) the bearing plates or end anchorages do not normally extend across the entire width of the beam, but are of small width; and (2) highly stressed concrete acts as a plastic material rather than an elastic material, with the result that the more highly stressed areas exhibit large plastic deformations. These two factors result in the actual stresses that occur in the ends of post-tensioned beams. being indeterminate by elastic methods of analysis. Hence, the design of end blocks and end-block reinforcing must be done by empirical methods.

Experience has shown that relatively small post-tensioned tendons (up to 120 kips ± initial force) can be anchored in concrete that has a cylinder strength of the order of 4000 psi, without end-block cracks occurring, if the average bearing stress under the anchorages does not exceed the cylinder strength of the concrete, if the bearing area of the anchorages does not exceed one-third of the total end surface, and if grids, composed of small reinforcing bars placed at right angles to each other, are placed in the concrete under the anchorages or bearing plates (Fig. 6-23). Experience has shown that grids of small-diameter bars allow the concrete to withstand large plastic deformations without cracking. Little if any end-block reinforcing is actually needed in addition to the grids if the end anchorages can be distributed uniformly over the ends of the beam.

When large tendons are used, the stress conditions at the ends are more severe, and it is customary to provide vertical and horizontal reinforcing in the end blocks to restrict cracking. The amount of reinforcing can be proportioned on the basis of recommendations resulting from three-dimensional tests. (Refs. 7 and 8.) These tests revealed that for an end block subjected to a single concentrated force, the distribution of the tensile stresses parallel to the tendon under the anchorage are primarily dependent on the ratio of the loaded area to the area of the cross section of the end block. End blocks loaded by more than one tendon can be analyzed by treating the individual tendons as if they existed on individual prisms, as shown in Fig. 6-24. A

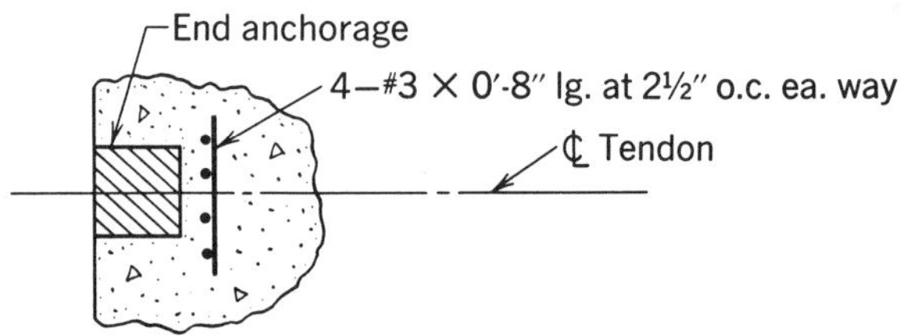

Fig. 6-23 Section at post-tensioning anchorage showing grid reinforcing.

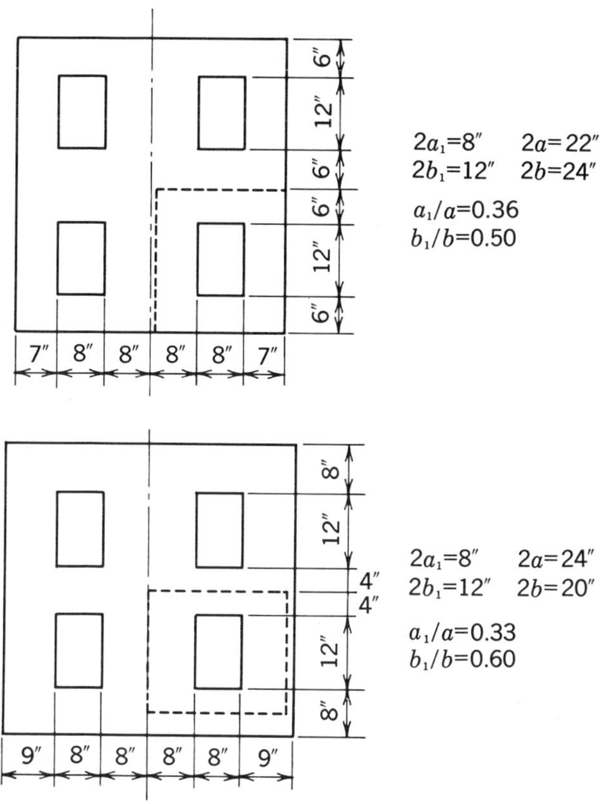

Fig. 6-24 End view of end anchorages showing prism dimensions.

rectangular anchor having dimensions of 2a′ by 2b′ is assumed to act on a prism that is 2a by 2b in dimension. The dimensions 2a and 2b are determined by inspection and are normally assumed to be the distance between the center lines of adjacent anchors or twice the distance between the center line of the anchor and the edge of the concrete, which ever is less. The ratios of a′/a and b′/b are used for computing the tensile stresses in each direction. For square anchor, only one dimension of a′/a would be used. Round anchorages are treated as square anchorages having an area equal to that of the round anchorage.

The distribution of the stress is not significantly affected by whether the anchorage is internal or external, nor if it is made of steel or concrete. The distribution of the tensile stresses shown in Fig. 6-25 is typical for all, except

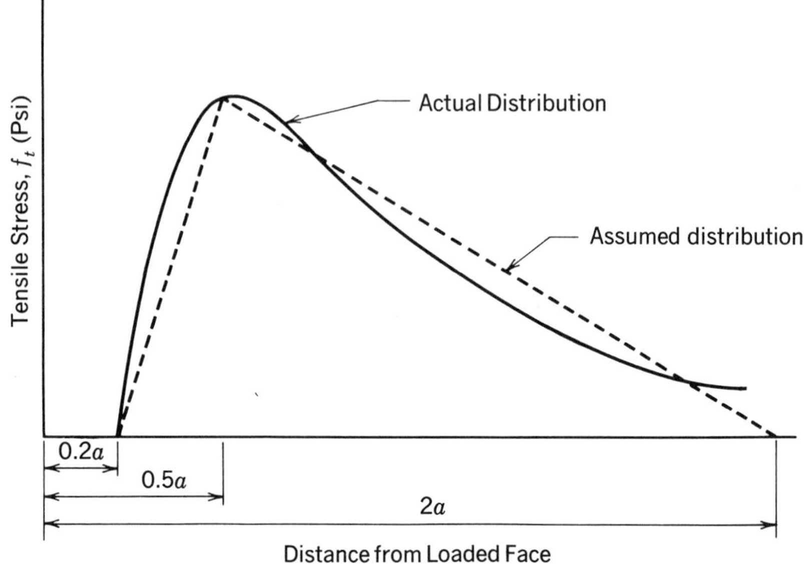

Fig. 6-25 Distribution of tensile stresses at end anchorage.

internal anchorages which have external spiral fins. The maximum stress for the latter type occurs at 0.88a from the loaded face rather than at 0.50a. Wedge type internal anchorages result in unit tensile stresses about 12% higher than those from external or internal non-wedge type, but the total tensile force is about the same for all types of anchors.

The maximum tensile stress and the total tensile force, as determined experimentally, shown as a function of the ratio of the side of the loaded area to the side of the prism are shown in Fig. 6-26.

The maximum tensile stress can be expressed mathematically as follows:

$$\frac{f_t}{P/A_n} = 0.46\beta^2 - 1.30\beta + 1.10 \tag{6-9}$$

in which f_t is the tensile stress indicated in the Fig. 6-25, P is the total prestressing force applied to the anchorage, A_n is the net area of the prism after deducting the area of the tendon duct, and $\beta = a'/a$ or b'/b. For purposes of determining the required reinforcing, the tensile stress distribution can be assumed to be as shown in Fig. 6-27, in which f_{ta} is the allowable tensile stress on the unreinforced concrete. f_{ta} is taken as 0.8 Kr in which K is the coefficient from Fig. 6-28 and r is the splitting tensile stress (*see* below). This

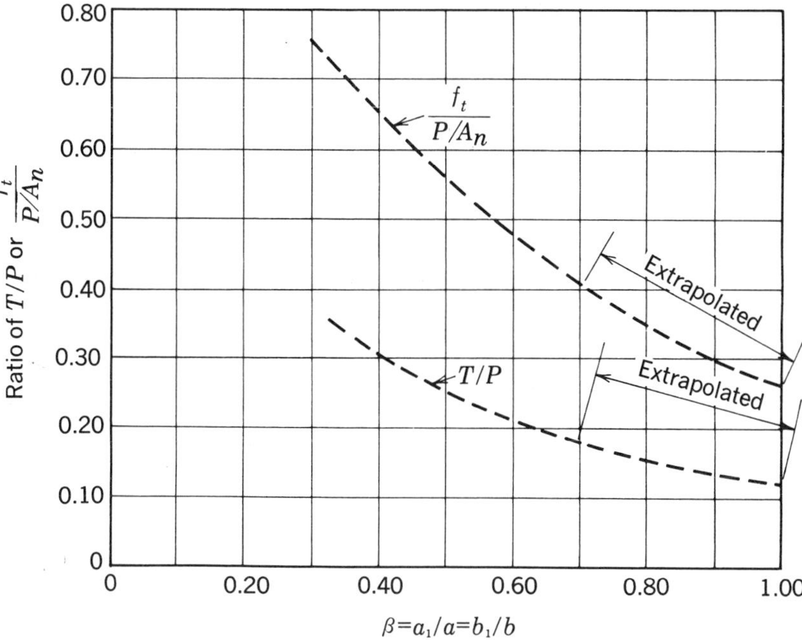

Fig. 6-26 Ratio of the total tensile force, T, to the prestressing force, P, and maximum tensile stress, f_t, to the average bearing stress, P/A_n, vs the ratio of the side of the loaded area to the side of the prism.

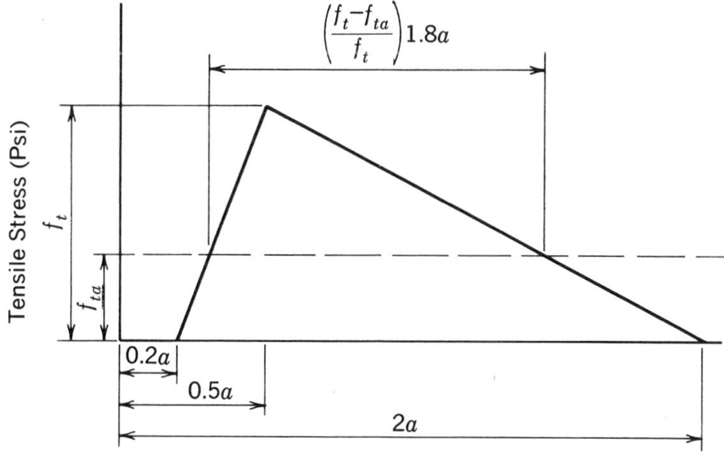

Fig. 6-27 Distribution of tensile stress, f_t, vs distance from the face of anchor (expressed in terms of the prism dimension).

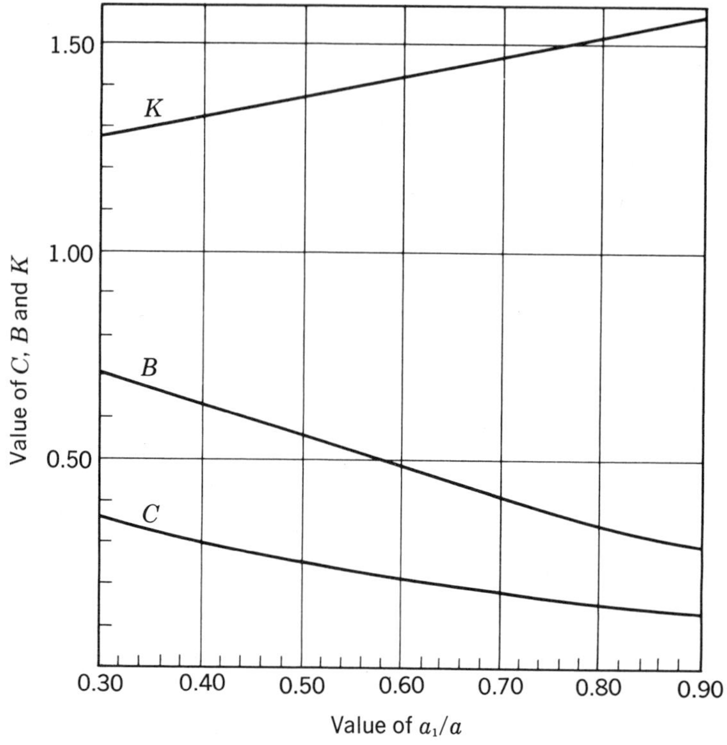

Fig. 6-28 Coefficients *K*, *B*, and *C* for computation of end block reinforcing.

value should give a load factor against cracking of about 1.25. The tensile force that must be resisted by the steel can be computed by

$$T_s = T\left[1 - \left(\frac{f_{ta}}{f_t}\right)^2\right] \tag{6-10}$$

Theoretically, the reinforcing should be distributed in proportion to the stress distribution shown in Fig. 6-27, but in practice, this is difficult if not impossible to do since the steel must be spaced in such a manner that concrete can be placed.

The required area of steel (A_s) can be computed using the factors, C, B, and K from Fig. 6-28 with the relationship

$$A_s = \frac{CP}{f_s}\left[1 - \left(\frac{0.8\ KrA_n}{BP}\right)^2\right] \tag{6-11}$$

P is the prestressing force, f_s the allowable steel stress, r is the concrete tensile splitting strength and A_n is the net area of the prism less the area of the duct. The splitting tensile strength r for normal concrete can be taken as 400 psi for $f_c' = 4000$ psi and 600 psi for $f_c' = 7200$ psi. Intermediate values can be interpolated.

The area of reinforcing provided should be checked to see that the stress in the reinforcing is less than the yield stress when the tendon is stressed to ultimate. This can be done with the following relationship:

$$f_{su} = \frac{CP_u}{A_s} \leqq f_y \tag{6-12}$$

Reinforcement of rectangular stirrups or weld mesh is preferred and should be placed in such a manner that all prisms are connected. The reinforcing should start as near the face of the anchorage as possible. Care must be taken to space the reinforcing in such a manner that the concrete near the anchorage can be placed and be well compacted. It is considered better to have an under-reinforced (but well compacted) end block than to produce an over-reinforced section that cannot be completely compacted.

Some reinforcing should be provided in the loaded face when groups of anchorages are used, since tensile stresses are produced between the anchorages near the loaded face. These tensile stresses tend to produce "deep beam" cracks, as are illustrated in Fig. 6-20. The end block tends to span between the anchorage groups as a deep beam. The reinforcing required to resist the deep beam stresses can be computed using the normal deep beam methods of analysis (Ref. 10).

The total tensile force due to deep beam action can be shown to be

$$F = 0.2S\frac{P_i}{A}b \tag{6-13}$$

in which S is the clear spacing between concentrated tendon groups, P_i is the total initial prestressing force, A is the area of the concrete section at a distance of S from the loaded face, and b is the width or thickness of the end block or web at distance S from the loaded face. The area of the steel required in the loaded face is then

$$A_s = \frac{0.2SP_i b}{f_s A} \tag{6-14}$$

In which f_s is the allowable steel stress (Ref. 11).

ILLUSTRATIVE PROBLEM 6-10 Compute the anchorage reinforcing required for an anchorage 10.5 in. × 10.5 in. having 12-1/2 in. diameter 270 k strand and stressed to 300 kips. $P_u = 496$ kips, the duct is 2.75 in. in diameter, the prism dimensions are 18 in. × 18 in., $r = 400$ psi, $f_s = 20{,}000$ psi, and $f_y =$ 40,000 psi.

$$A_n = [10.5 \times 10.5] - 0.784 \times 2.75^{-2} = 104 \text{ in.}^2$$

$$\frac{2a'}{2a} = \frac{10.5}{18.0} = 0.583 \qquad f_t = 0.48 \times \frac{300{,}000}{104} = 1384 \text{ psi}$$

$$C = 0.21 \qquad K = 1.42 \qquad B = 0.50$$

$$f_{ta} = 0.8Kr = 0.8 \times 1.42 \times 400 = 454 \text{ psi}$$

$$A_s = \frac{0.21 \times 300}{20}\left[1 - \left(\frac{0.8 \times 1.42 \times 0.40 \times 104}{0.50 \times 300}\right)^2\right]$$

$$= 2.84 \text{ in.}^2$$

or 7 – No. 4 ties

$$f_{su} = \frac{0.21 \times 496}{2.84} = 36.7 \text{ ksi} < 40.0 \text{ ksi}$$

The distance over which the reinforcing should be provided (from Fig 6-27) is

$$\text{distance} = \left(\frac{f_t - f_{ta}}{f_t}\right) 1.8a = \left(\frac{1384 - 454}{1384}\right) 1.8 \times 9.0 = 10.88 \text{ in.}$$

Use No. 4 ties at 2 in. o.c.

6-9 Spacing of Pre-tensioning Tendons

The bond between the tendons and the concrete section at the ends of pre tensioned members is relied upon to transfer the pre-tensioning force fron the tendons to the concrete section. Adequate flexural bond strength i necessary in order to develop maximum resistance to cracking and maximun ultimate moment, as well as to ensure good crack distribution if the crackin load is exceeded. In addition, the secondary tensile stresses in the concret at the end of the member should be controlled so that the tensile strength o the concrete is not exceeded and cracking does not occur at the ends of th member. In order to achieve these properties, it is essential that the concret be placed and compacted properly around the tendons. Hence, the designe must exercise care in selecting the tendon spacing specified in order to facili tate the placing and vibrating of the concrete, as well as to reduce the possi bility of developing tensile cracks at the ends of members due to the con centration of stress.

In the manufacture of pre-tensioned concrete in this country, interna vibration is relied upon to a high degree to ensure that the concrete is ade quately compacted. For this reason, particularly in deep beams, it is importan that the pre-tensioning tendons be spaced in such a manner that the head of the internal vibrator can penetrate to the extreme bottom of the forms. In addition, the tendons should not be placed in such a manner that the flow and compaction of the plastic concrete is unduly restricted in areas not directly accessible to the vibrator head.

In order to save labor in handling and stressing the strands in the manufacture of pre-tensioned concrete, the trend has been toward the use of a few of the larger seven-wire strands in lieu of many lower strength, solid wires commonly used abroad. As progress has been made in prestressed-concrete manufacturing techniques and as more has been learned about the action of transfer bond and fatigue on flexural bond stresses, the size of tendons commonly used has been increased from 1/4 to 1/2 in., with strands as large as 0.60 in. in diameter being used occasionally. It should be recognized that the head of the internal vibrator used in compacting the concrete in beams with 0.50 in. tendons at 2 in. on centers, which is a common spacing with tendons of this size, is restricted to 1 1/2 in. in diameter. Because large, internal vibrators are generally much more effective than smaller vibrators, placing of the concrete is often materially facilitated if at least one, wide, vertical opening (through which a large internal vibrator can be inserted) is left down the center of deep members. This is illustrated in Fig. 6-29, in which

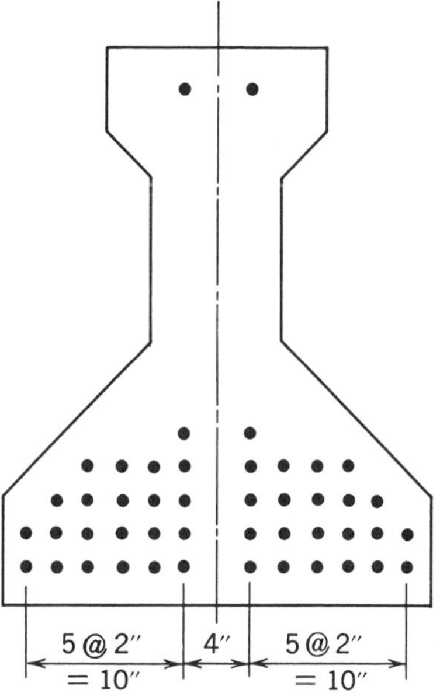

Fig. 6-29 Section through a pre-tensioned beam showing tendon spacing to facilitate placing and vibrating concrete.

it will be seen that omission of the strands in the center row allows the use of a 3-in. vibrator head.

Heavy concentrations of two groups of tendons at the ends of members, when deflected pre-tensioned tendons are used, can result in cracking at the ends of the members, just as such concentrations of load may cause cracking in the ends of post-tensioned members.

The area of reinforcing steel required to control cracking at the ends of pre-tensioned members that do not have end blocks can be computed by the relationship

$$A_s = \frac{0.021\ P_i h}{f_s L_t} \tag{6-15}$$

In which A_s is the area of steel required, P_i is the initial prestressing force, h is the depth of the member, f_s is the allowable steel stress, and L_t is the transfer length. The steel should be placed as near the ends of the beams as possible. This relationship was developed experimentally and was justified for values of h/L_t of up to about 2. For higher values of h/L_t, Eq. 6-15 is considered to be conservative with the degree of conservatism increasing with the ratio of h/L_t. Furthermore, the relationship was developed from tests of girders in which the tendons were divided into two groups located in the top and bottom fibers, respectively. It is believed Eq. 6-15 will give conservative results for members in which the tendon groups are close together or in which the tendons are uniformly distributed over the ends of the members.

The distance from the edge or center line of the tendons to the surface of the concrete in the area in which transfer bond is developed must also be chosen with care. Placing the tendons too close to the surface can cause cracking or splitting along the tendon. In general, the smaller tendons can be placed as close as 1 in. to the surface without ill effects, while the larger tendons should not be closer than 2 in. to the surface. The center-to-center spacing of seven-wire strands is frequently restricted to four times the nominal strand diameter, for example, 1 1/2 in. for a 3/8-in. strand.

6-10 Pre-tensioning Stresses at Ends of Beams

In simple prismatic beams, pre-tensioned with straight tendons, the eccentric pre-tensioning force results in a high compressive stress in the bottom flange and a tendency for tensile stresses in the top flange. Furthermore, in well proportioned beams, the compressive stresses in the top flange at midspan under full load are not normally a problem. The amount of prestressing required, as a rule, is controlled by the flexural, tensile stress in the bottom flange, which the prestressing force must fully or partially nullify. The usual specifications permit some tensile stresses resulting from prestressing in the top flange without reinforcing, and higher-tensile stresses if unstressed reinforcing steel is provided to carry the entire tensile force. As a means of illustrating the effect of top-flange tensile stresses on the quantities required for a given design, consider the beam shown in Fig. 6-30. Assume this beam is to be used on a span of 70 ft, no tensile stress is to be allowed in the bottom fibers under full load, and the flexural stresses due to dead load of the girder and due to the superimposed load are as follows:

	Top Fiber	*Bottom Fiber*
Stress due to dead load of girder only	+ 524 psi	− 780 psi
Stress due to superimposed load only	+ 826 psi	−1220 psi
Total Stress	+1350 psi	−2000 psi

Under these conditions, the minimum prestressing force required to produce the required 2000 psi compression in the bottom fibers with various amounts of tensile stress in the top fiber is as follows:

Allowable top-fiber tensile stress	zero	160 psi	320 psi
Minimum prestressing force required	644 kip	570 kip	491 kip
Number of tendons, 11 kip each, required	59	52	45

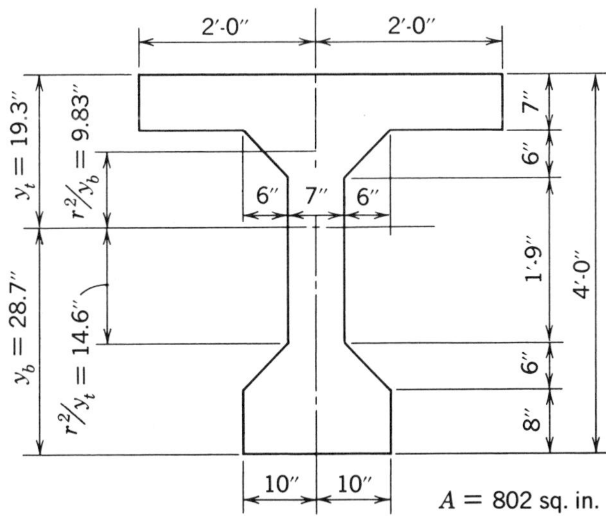

Fig. 6-30 Beam cross-section used in illustrating the effect of top-fiber tensile stresses on the prestressing force.

From this table, it can be seen that a reduction of 12% can be made in the amount of prestressing steel required if a tensile stress of 160 psi is allowed and 23.7% if 320 psi is allowed.

Another factor to be considered is that, as was explained in Sec. 5-7, the stress in a pre-tensioned tendon is zero at the end of the tendon and the stress increases to a maximum value at a distance, from the end of the beam, that is referred to as the transmission length. The transmission length is primarily a function of the type and size of the tendon. Values of 50 and 100 diameters are recommended for use in estimating the transmission length for strand or wire tendons, respectively. The initial stress in the tendon varies along the transmission length approximately as shown in Fig. 6-31.

An additional consideration is that a force applied to an elastic body causes stresses in the body which flow out along smooth curves or stress trajectories. A large force applied to the end of a prism, such as shown in Fig. 6-32, results in principal compressive stresses that follow a pattern similar to the solid lines and principal tensile stresses that follow along lines similar to the dashed lines. At a distance of about one times the depth of the block from the end, the stresses are approximately equal to the values that would be computed from the usual combined stress relationship used in prestressed-concrete design. In other words, the effect of the concentration of the load is virtually reduced to zero at a distance of one times the depth of the prism from the end, under normal conditions.

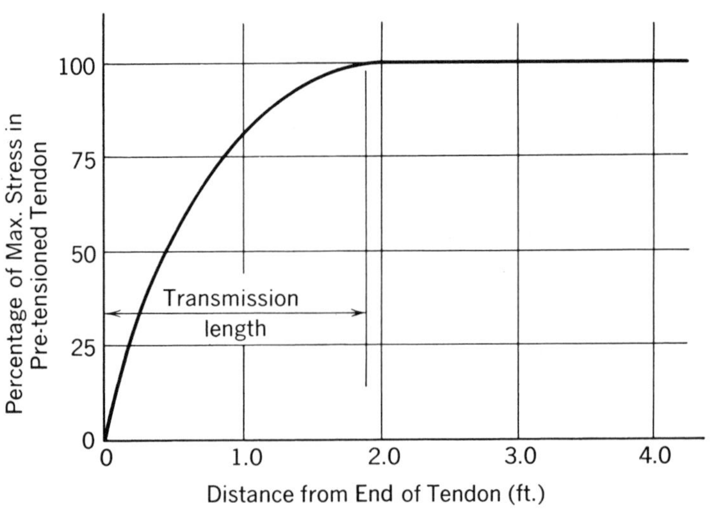

Fig. 6-31 Variation in initial stress in a pre-tensioned tendon near the end of a beam.

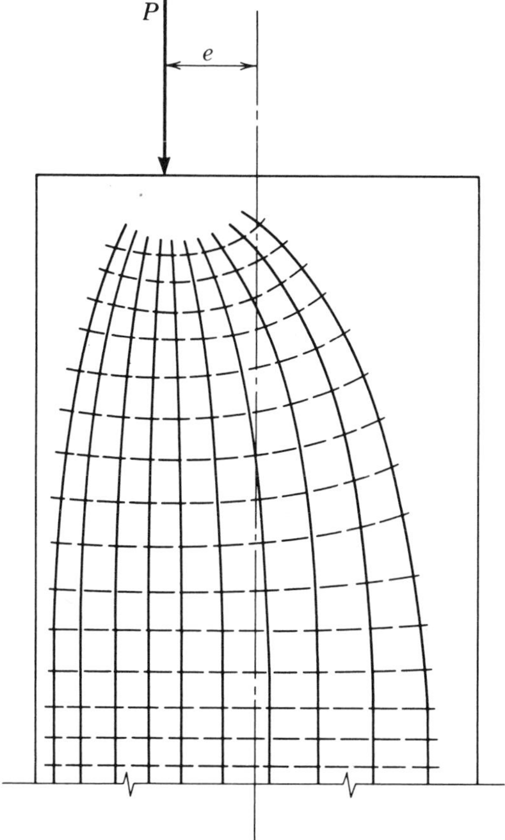

Fig. 6-32 Approximate trajectories of principal tensile and compressive stresses in a prism that is eccentrically loaded.

As a result of the combined effects of the transmission length required to develop full bond and the distance required for the concentrated prestressing force to distribute fully, the maximum tensile stress in the top fibers does not occur at the immediate end of the beam. This is a significant fact that can affect the economy of a design.

Returning to the example used above, if the effects of transmission length and distribution of the concentrated force are taken into account on the beam acting on a span of 70 ft, the 320-psi tension resulting from the 45 tendons would not be acting at the immediate end of the beam, but at a distance of from 4 to 8 ft from the end of the beam. The actual tensile stress in the top fibers of the beam would be less than 320 psi, due to the effect of the dead load of the beam. The effect of the stress in the top fiber of the beam resulting from the dead weight of the beam on the net, final tensile stress in the top fibers as well as the amount of unstressed reinforcing that would be required is summarized as follows:

Distance from the end to the point under consideration (ft)	zero	4	6	8
Theoretical top-fiber tensile stress (psi)	−320	−320	−320	−320
Top-fiber stress due to girder dead load (psi)	0	+113	+164	+213
Net tension under dead load of girder plus prestress (psi)	−320	−207	−156	−107
Area of unstressed reinforcing steel required (sq in.)	3.00	1.42	0	0

From this study it is quite apparent that, in taking all of these factors into consideration, the designer may be able to reduce the pre-tensioning steel required in a specific design as much as 25% without adding any mild steel in the top flange.

6-11 Bond Prevention in Pre-tensioned Construction

In Sec. 4-6, it was shown that the stresses at the ends of a member prestressed with straight tendons may limit the capacity of the member and that, by varying the eccentricity of the prestress, these stresses at the ends can be reduced and the capacity of the member thereby increased. The prestressing moment and stresses at the ends can also be reduced in a pre-tensioned member by varying the prestressing force. This can be done by preventing a portion of the tendons from bonding to the concrete at the immediate ends of the member and, in so doing, preventing these tendons from stressing the concrete at the ends.

This principle can best be explained by considering an example such as the beam shown in Fig. 6-33. The pre-tensioning tendons, as located in the figure, result in an initial, top-fiber tensile stress due to prestressing of 384 psi. It can be shown that by preventing bond on five tendons in the bottom row and four tendons in the second row, as indicated in the figure, the initial tensile stress in the top fibers at the ends can be reduced to 270 psi.

The length over which the bond must be prevented is a function of the beam dead-load stresses. In most cases, the beam dead-load stresses reduce the initial prestressing stresses to permissible values at only a few feet from the end of the beam. The transmission length required for the tendons to develop full tension, as well as the distance required for the pretensioning force to distribute, which are discussed in Sec. 6-10, should also be taken

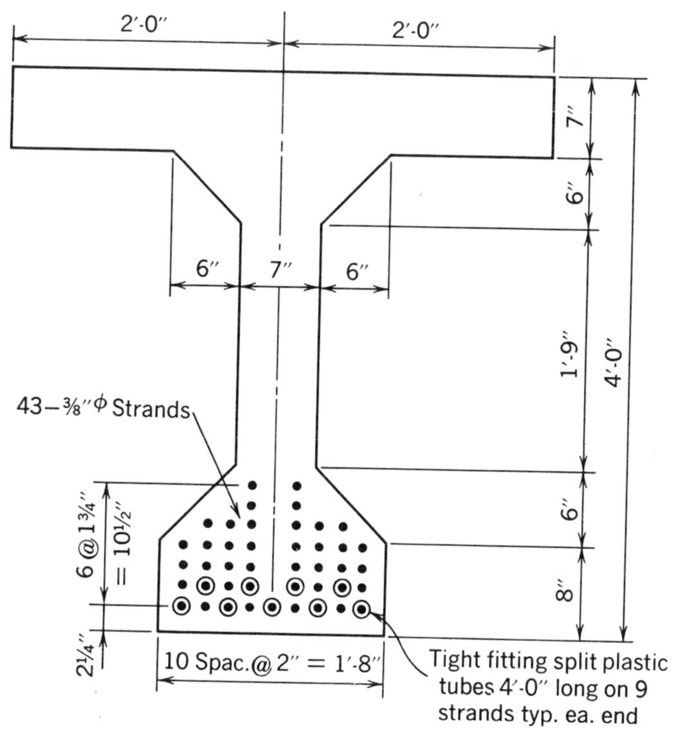

Fig. 6-33 Beam section indicating method of preventing bond on pre-tensioned tendons.

into account when calculating the maximum tensile stresses at the ends of members.

It is believed that bond prevention can be used to advantage with complete safety if a split, plastic tube, or a heavy paper or cloth tape with a waterproof adhesive is used as the bond-preventing media. Grease and chemicals that retard the concrete set have been used in lieu of plastic tubes or tape as a means of preventing bond. Because of the obvious danger of the workmen inadvertently or carelessly applying the grease or retarder to the incorrect tendons or number of tendons, this procedure should only be permitted when adequate supervision and inspection will be provided to prevent errors.

ILLUSTRATIVE PROBLEM 6-7 Compute the stresses due to a prestressing force of 11 k per tendon for the AASHO-PCI bridge stringer, type III, pre-tensioned as shown in Fig. 6-34, sections AA and BB. The plastic tubes indicated are

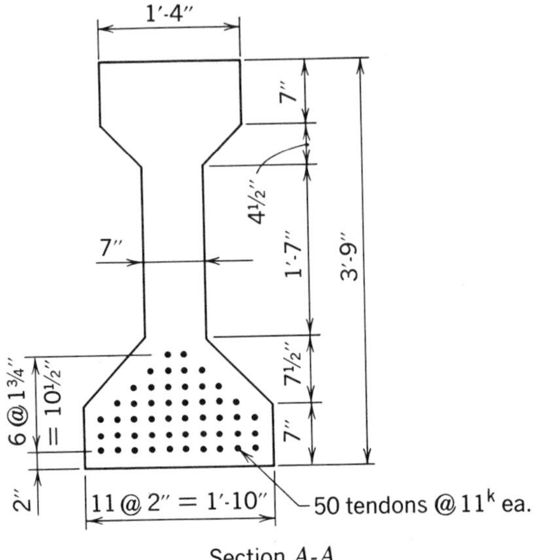

Section A-A

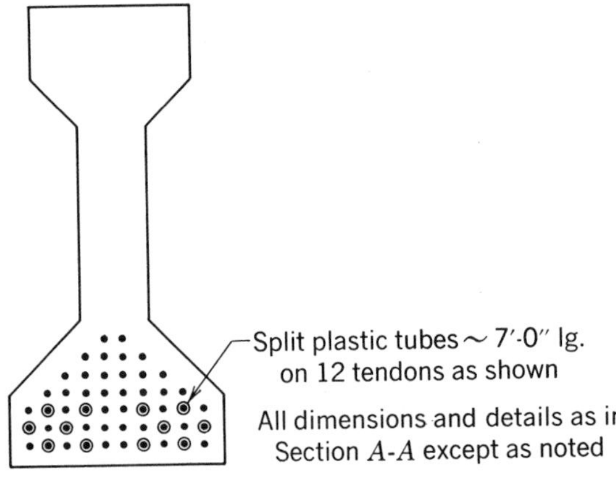

Section B-B

Fig. 6-34 AASHO-PCI type III bridge stringer pre-tensioned with 50 tendons. Section A-A shows details in typical section. Section B-B shows details near end where plastic tubes are used to prevent 12 tendons from bonding.

used to reduce the stresses due to prestressing. The section properties of the concrete section are:

$$A = 560 \text{ in.}^2 \qquad y_t = 24.73 \text{ in.} \qquad r^2/y_t = 9.06 \text{ in.}$$

$$I = 125{,}400 \text{ in.}^4 \qquad y_b = 20.27 \text{ in.} \qquad r^2/y_b = 11.03 \text{ in.}$$

Compute the center of gravity of the tendon for section AA by taking moments about the bottom

No. tendons	× Distance	= N.D.
10	× 2.00	= 20.00
10	× 3.75	= 37.50
10	× 5.50	= 55.00
8	× 7.25	= 58.00
6	× 9.00	= 54.00
4	× 10.75	= 43.00
2	× 12.50	= 25.00
50		292.50

$$e = 20.27 - \frac{292.5}{50} = 14.43 \text{ in.}$$

$$f_t = \frac{50 \times 11{,}000}{560}\left(1 - \frac{14.43}{9.06}\right) = -580 \text{ psi}$$

$$f_b = \frac{50 \times 11{,}000}{560}\left(1 + \frac{14.43}{11.03}\right) = +2270 \text{ psi}$$

At section B-B:

50	292.50
− 4 × 2.00 =	− 8.00
− 4 × 3.75 =	− 15.00
− 4 × 5.50 =	− 22.00
38	247.50

$$e = 20.27 - \frac{247.5}{38} = 13.75 \text{ in.}$$

$$f_t = \frac{38 \times 11{,}000}{560}\left(1 - \frac{13.75}{9.06}\right) = -388 \text{ psi}$$

$$f_b = \frac{38 \times 11{,}000}{560}\left(1 + \frac{13.75}{11.03}\right) = +1680 \text{ psi}$$

6-12 Deflected Pre-tensioned Tendons

For the reasons explained in Sec. 4-6, it is frequently desirable to have pre-tensioned tendons follow a trajectory more eccentric at the center line than at the ends of a beam. This method is also used as a means of reducing the camber due to prestressing in light roof slabs.

When applied to roof slabs, it has been customary to deflect the tendon at one or two points so that the tendon spacing at the ends and center line of the member is similar to that shown in Fig. 6-35, in which it will be seen that the tendons are allowed to touch each other at the center and are spaced out at the ends. In this manner, the tendons are spaced out where they must develop the all important transfer bond and are bundled at the center where only flexural bond stresses must be developed. This construction practice has been used a great deal with satisfactory results. The flexural bond strength at the center of such members is considered as good or better than that which is achieved in grouted, post-tensioned construction.

In recent years, deflected tendons have been used extensively in the construction of bridge beams. The theoretical principles involved in the use of deflected tendons in bridge construction is the same as in roof slabs. It can be shown that the same flexural strength that is obtained with spaced, deflected tendons can be achieved by using bond prevention, although some unstressed reinforcing may be required at the ends of the members, and by so doing, the large capital investment required for deflecting equipment, as well as the labor required in the deflecting operation, can be avoided. This is illustrated in Fig. 6-36, in which two AASHO-PCI, type III, bridge stringers of equivalent flexural strength, but detailed for deflected tendons and bond prevention, are shown.

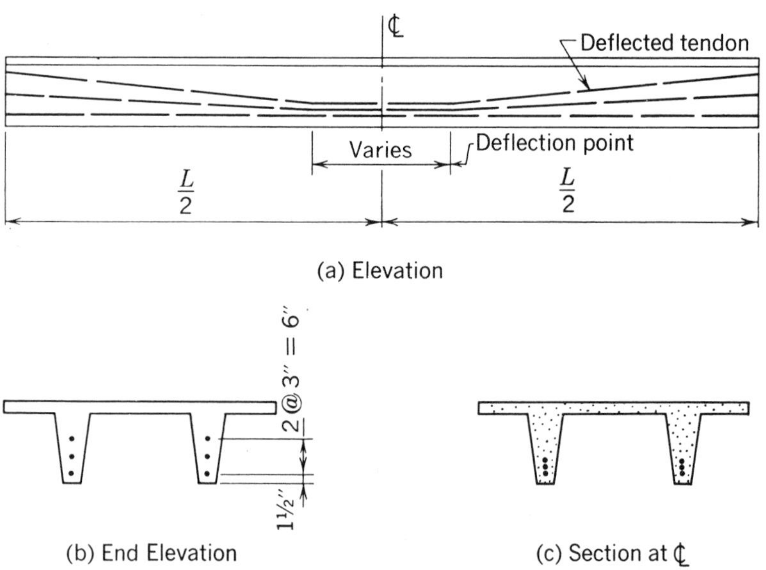

Fig. 6-35 Pre-tensioned double-T roof slab with deflected tendons.

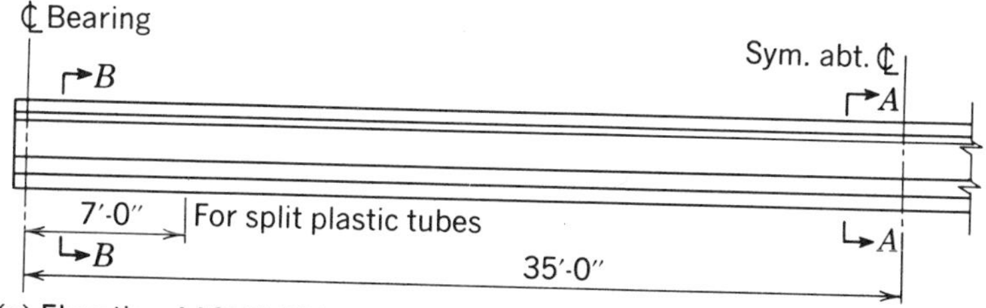

(a) Elevation AASHO-PCI, Type III, Bridge Stringer with Bond Prevention

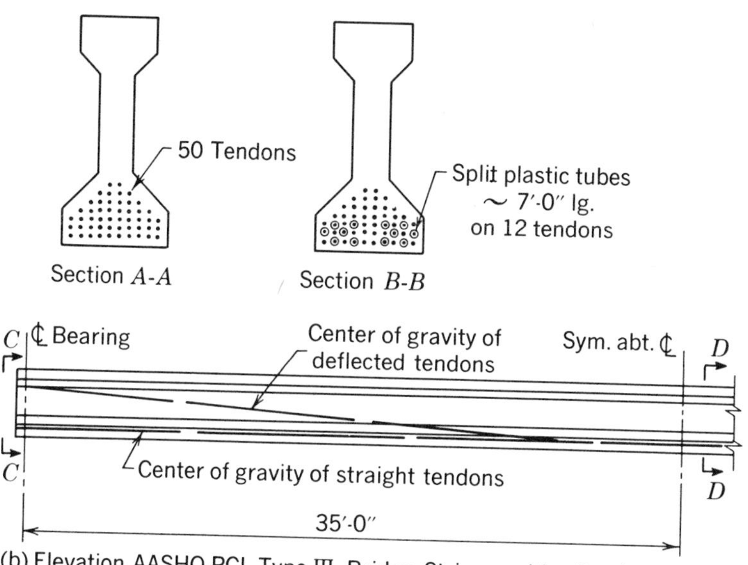

(b) Elevation AASHO-PCI, Type III, Bridge Stringer with a Portion of the Tendons Deflected.

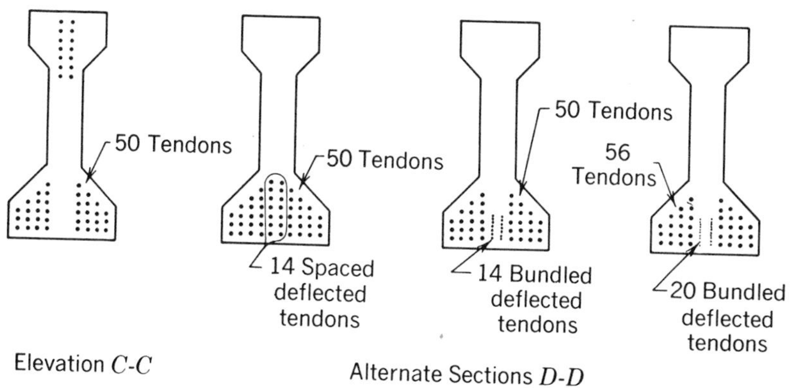

Fig. 6-36 Elevations and sections of AASHO-PCI type III bridge stringer shown (a) pre-stressed with pre-tensioned tendons utilizing bond prevention, and (b) pre-tensioned with a portion of the tendons deflected.

If the deflected, pre-tensioned tendons in the AASHO-PCI, type III, stringer were bundled at the center rather than being spaced out, the stress in the bottom flange at the center line due to prestressing alone, and therefore the capacity of the stringer, could be increased 4% without additional materials or labor being required. If the number of bundled, deflected tendons were increased to 20 and each tendon had an initial prestressing force of 13,000 lb, the initial net bottom-fiber compressive stress (prestress-dead load) at the center of the girder would be of the order of 2425 psi if the girder were to have a span of 70 ft. This latter tendon layout would develop the maximum practical capacity of this concrete section for the span of 70 ft and would not be possible with spaced tendons of the same size.

The mechanics of tendon deflection are discussed in Chapter 14, but it should be pointed out here that there are several methods currently being used for deflecting tendons in bridge girders. Most of these methods include deflection of the tendon over a rivet, roller, or spacer of small diameter. As mentioned in Sec. 9-6, the bending of tendons and wire over small spacers in post-tensioned beams has resulted in the wires breaking at these points when the beams are subjected to fatigue tests. It is not known if a similar effect results for the small-diameter spacer used in deflecting tendons in pre-tensioned construction, since no tests have been made to investigate this possibility, however, it is considered to be a possible point of low-fatigue resistance. Large-radii deflecting saddles, such as shown in Fig. 6-37, can be

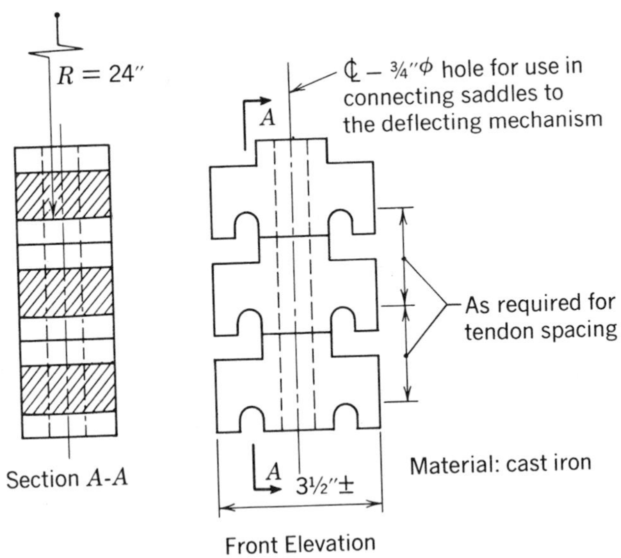

Fig. 6-37 Large-radius tendon saddle composed of small cast-iron elements that can be stacked as required to obtain desired number of tendon positions.

used to reduce the secondary stresses resulting from the curvature, in order to minimize the possibility of fatigue failure in the tendon at the deflecting points.

6-13 Combined Pre-tensioned and Post-tensioned Tendons

The structural advantages of draped tendons can be obtained without materially reducing the economy of pre-tensioned construction with straight tendons by using a combination of pre-tensioned and post-tensioned tendons. This is illustrated in Fig. 6-38, in which the details of the AASHO-PCI, type III, bridge stringer are shown with two combinations of pre-tensioned and post-tensioned tendons.

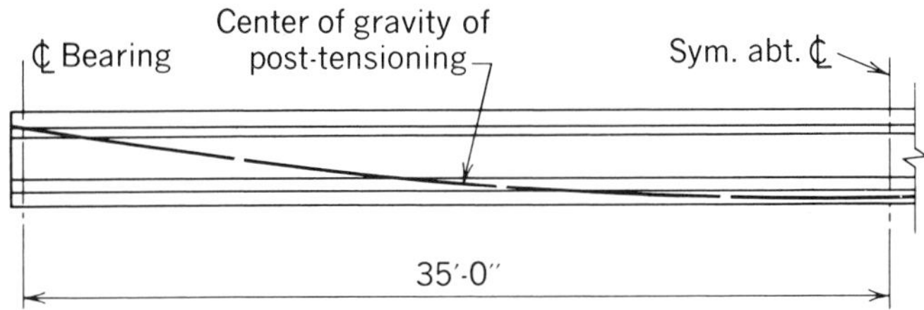

(a) Half Elevation AASHO-PCI, Type III, Bridge Stringer with Combined Pre- and Post-tensioned tendons

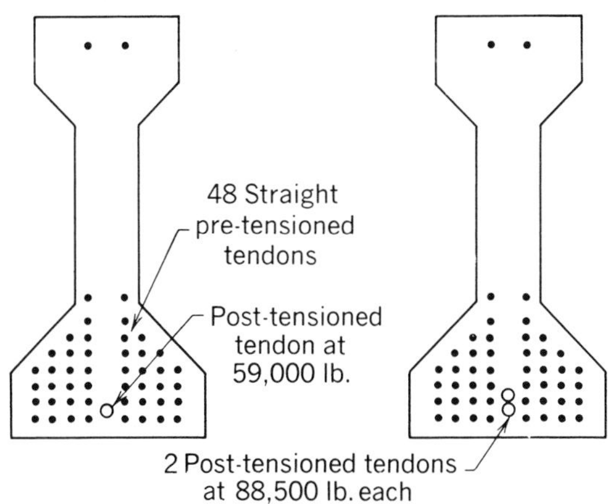

(b) Alternate Sections at Center Line

Fig. 6-38 Half-elevation and section of beam prestressed with pre-tensioned and post-tensioned tendons.

The use of the 48 pre-tensioned tendons and one small post-tensioned tendon (59 k initial force) results in a distribution of prestressing stresses that is equivalent to that which is obtained with 36 straight and 14 deflected tendons (50 tendons total) and with bond prevention in combination with 50 tendons, as shown in Fig. 6-36. The number of pre-tensioned tendons in this solution could be reduced to 42 tendons if the post-tensioned tendons were stressed before the pre-tensioning force were completely released on the concrete section. In such a procedure, if steam curing were used, it would be necessary to release the pre-tensioned tendons partially and to allow the girders to cool somewhat before post-tensioning and the complete release of the pre-tensioning force; this is done to eliminate the possibility of vertical cracks forming in the girder as a result of the strain changes that take place in the concrete and in the pre-tensioning tendons during curing and cooling. (This latter effect is discussed further in Sec. 10-2.) With the 48 tendons as shown, however, the pre-tensioning force could be released when the concrete attains a strength of 4000 psi and the girders could be removed from the casting bed immediately and post-tensioned when it is convenient to do so.

If the two larger post-tensioned tendons were used rather than the one small tendon (Fig. 6-38), the stresses due to prestressing would be nearly equivalent to those in the same beam section that would result from 56 tendons of which 20 are deflected, as shown in Fig. 6-36, and which, as was explained previously, would be the maximum stresses that could normally be imposed on this section when it is to be used on a 70-ft span.

When using combined pre-tensioning and post-tensioning, it is often not necessary to use end blocks if the post-tensioned tendons are terminated at the top of the member rather than at the end. Small parallel wire and strand post-tensioning tendons are readily adaptable to this detail.

6-14 Buckling Due to Prestressing

The danger of buckling of columns or other long, slim compression members is known to all structural engineers. The question of the possibility of a prestressed member buckling as a result of the prestressing force is frequently raised. Obviously, when prestressing is done by the application of external load such as jacking against abutments and where no tendons are used through the member, a possibility exists that the member may buckle. In such a case, it is essential that buckling be investigated in the conventional manner. In addition, if tendons are used to prestress the member and the tendons are placed externally in such a manner that they are in contact with the member at the ends only, there would be a possibility of buckling.

When the tendons are placed internally in a way that they are in contact with the member at points between the ends of the member, the tendency to buckle is reduced a significant degree. When the tendons are in intimate contact with the member throughout the length of the member, as is the normal case, in post-tensioning and in pre-tensioning, there is no possibility of buckling. This fact has been demonstrated experimentally and mathematically and can be understood by considering the difference between the action of prestressing and column action.

Column action is characterized by an increase in eccentricity of the load as the load is increased over a critical value. This is illustrated in Fig. 6-39, in which it is seen that the column load has an eccentricity of e at load P, and if the load is increased to ΔP, the member deflects an additional amount, Δe. This action continues until the critical value of $P + \Delta P$ is reached, and the column buckles. If there were no eccentricity of the load, the column would fail by crushing, as is indeed the case for short columns.

Prestressing action results in a specific distribution of stresses in a member. The eccentricity of the prestressing force remains constant, even if the member is deflected laterally, providing, as was mentioned above, the tendons and concrete are in intimate contact with each other. If the concrete section were

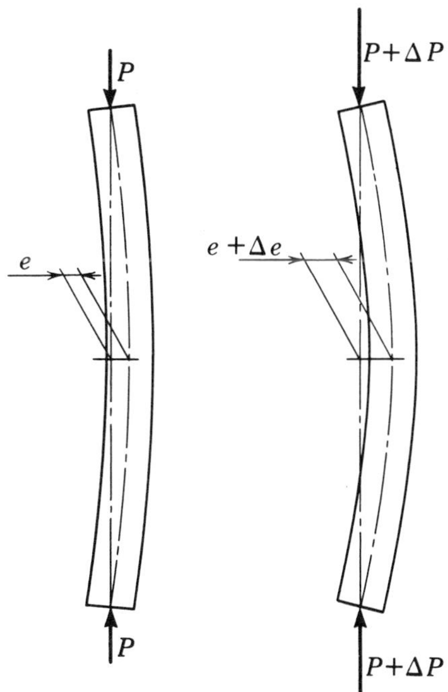

Fig. 6-39 Illustration of column action.

cast slightly curved or crooked, as is often the case, the effect of the prestressing alone would be to straighten the concrete member (opposite to column action), since the taut tendon would attempt to assume a straight path.

Prestressed columns and piles, which are pre-tensioned or post-tensioned with the tendons in ducts through the members in the normal manner, can of course buckle, due to externally applied loads, and these members must be designed with care. Prestressed columns and prestressed piles are treated in Sec. 9-3.

The top flanges of flexural members that do not have adequate lateral support can also fail as a result of buckling. For this reason, the designer should give attention to the conditions of support and loading when selecting the dimensions of the concrete section. This is discussed in Sec. 10-5.

6-15 Secondary Stresses Due to Tendon Curvature

In considering a short segment of a curved post-tensioned tendon, such as is shown in Fig. 6-40, neglecting friction between the tendon and the concrete, it will be seen that the forces acting upon the tendon are the tension P, which acts at each end of the tendon, and the unit stress c, which is between the tendon and the concrete and which holds the tendon in the curved trajectory. If the segment under consideration is infinitesimal, the length of the segment is ds, the angular change in length ds is $d\alpha$, and the radius of curvature of the tendon is ρ. Since a very small angle is equal to the tangent of the angle, one can write

$$\tan \alpha = d\alpha = \frac{ds}{\rho}$$

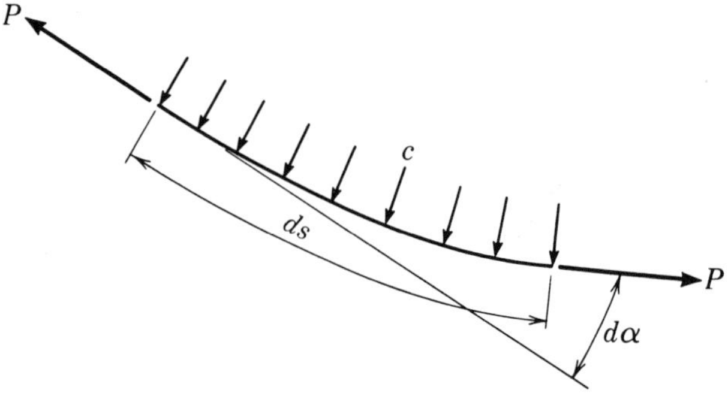

Fig. 6-40 Free-body diagram of an infinitesimal length of a curved tendon.

and

$$\rho = \frac{ds}{d\alpha}$$

It is evident from the vector diagram, Fig. 6-41, that the unit stress exerted by the steel on the concrete is

$$cds = P\,d\alpha$$

which can be rewritten

$$c = \frac{P\,d\alpha}{ds}$$

and since

$$\rho = \frac{ds}{d\alpha}$$

the expression becomes

$$c = \frac{P}{\rho}$$

This expression is useful in determining the secondary stresses that result when a tendon is placed on a curve in thin webs or on a horizontal curve in an end block. Only on rare occasions are the unit stresses between the concrete and the tendon of such magnitude to cause difficulty. The curvatures must be high in order to produce high stresses, a condition which is normally avoided in order to minimize the friction between the tendon and concrete during stressing.

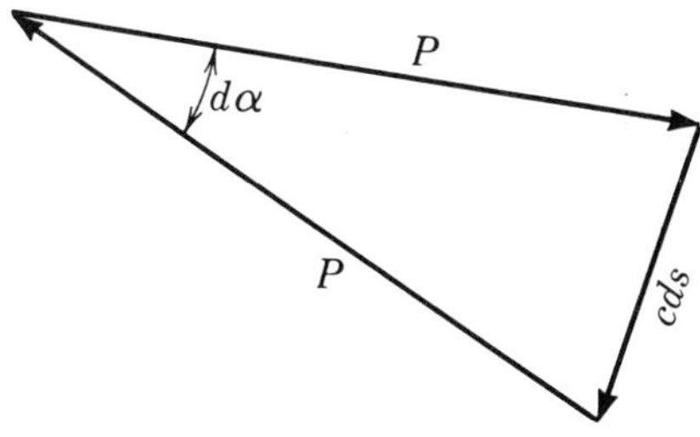

Fig. 6-41 Vector diagram of forces acting on curved tendon.

ILLUSTRATIVE PROBLEM 6-8 Compute the secondary stress between a curved tendon and the duct if the radius of curvature is 25 ft and the force in the tendon is 500 k.

SOLUTION:

$$c = \frac{P}{\rho} = \frac{500\ \text{k}}{25\ \text{ft}} = 20\ \text{k/ft}$$

6-16 Variation in Tendon Stress

The question of the effect of the variation in the stress in the individual wires of a parallel-wire, post-tensioning tendon or in the individual tendons in a pre-tensioned member is often raised. It can be stated that the normal variations in stress encountered in practice do not exceed the normal tolerances expected in structural design.

Consider the case of a parallel-wire, post-tensioned tendon composed of n wires stressed to a total force, P. The force P is measured during construction by determining the elongation of the tendon during stressing as well as by observing the hydraulic pressure required to stress the tendon. The value of P can normally be controlled within the required tolerance without difficulty.

The average stress in the individual wires is P/n. There will be a variation in the unit stresses between the individual wires for the following reasons:

(1) The wires are not connected to the jack in such a manner that the length of each wire is precisely equal, and yet all wires are elongated the same amount by the jack. This effect is small in almost all instances, since the wires are very stiff and are confined in a relatively small duct or sheath which renders it physically impossible for one wire to have significantly more curvature than the average wire, as would be required if the wire were to be materially longer than the average length.

(2) The difference in the length between individual wires is only important with respect to the magnitude of the elongation of the tendon that is obtained during stressing. If, for example, the elongation of a tendon that is 100 ft long is 7 in., a difference in length of 1/4 in. between an individual wire and the average wire length will only result in a stress variation of $\pm 3\ 1/2\%$ from the average stress in the tendon. The total force in the tendon will not be affected, since the average elongation is not affected.

(3) Variation in the modulus of elasticity of prestressing wire, along the length of one wire and from coil to coil, as much as $\pm 4\%$ is not uncommon.

(4) Although the relaxation loss of a wire that is stressed higher than the

average will be higher than the average relaxation loss, this will be offset by the wires stressed less than the average.

(5) The estimate of losses of stress in the tendons is generally not as precise as the initial prestressing stresses in the tendon.

In general, the same factors that affect parallel-wire, post-tensioned tendons affect pre-tensioned tendons. The exception is that the pre-tensioned tendons are not confined in a sheath or duct, and therefore, the variation in length could be significant if the tendons were not laid out approximately parallel, prior to stressing. Although the wires are usually sufficiently parallel before stressing, a small force is normally applied to each tendon before the tendons are stressed to their final value. This force straightens the tendons and equalizes the lengths. This procedure is not necessary when the tendons are stressed individually, as is discussed in Chapter 14.

6-17 Standard vs Custom Prestressed Members

In recent years, there has been a trend in the prestressing industry toward the adoption of standard beams and slab cross sections. There are several reasons why this is taking place, but the primary reasons are that the manufacturers of prestressed concrete prefer to use the same forms many times in order to reduce the amount of form cost that must be charged to each unit. In addition, the labor required to produce the prestressed concrete can be reduced to a minimum if the workmen perform the same duties each day and are not confronted with variable duties and operations. Furthermore, when standard products are used, load tables and advertising literature can be prepared for distribution to architects and engineers, and the manufacturers can often operate with a smaller sales-engineering force than would be required if custom products were used exclusively.

There are many instances where standard prestressed-concrete members have been used on small structures on which the use of custom-made members would have been out of the question, due to the high cost of the special forms required. Since all structural methods and framing schemes have their limitations, and because many large structures have peculiar framing or loading requirements, the designer should carefully consider the economy that could result from the use of custom-made members on large projects.

6-18 Precision of Elastic Design Computations

Prestressed-concrete flexural members are normally designed with the assumption that the concrete is an elastic material under the service loads and the stresses under such conditions of loading are made to conform to a standard or criteria. In addition, as has been pointed out, it is essential that the ultimate moment be computed, in order to be sure that the elastic design has resulted in adequate safety factors.

It is well known that concrete is not an elastic material and that stresses computed on the basis of elastic assumptions can only be considered as approximations. Furthermore, in order to facilitate the design of prestressed members, most engineers base their computations upon the gross concrete section rather than using the net and transformed sections. Errors in the elastic computations are introduced as a result of this simplification, as is apparent from the discussion of Sec. 4-10. These considerations can only lead to the conclusion that normal elastic-design computations can only be considered approximate and nothing is gained by using more than three significant figures in such computations.

It is significant that ultimate-moment computations of bonded construction can be made with good precision if the characteristics of the steel are known. Ultimate-moment computations are virtually independent of the elastic properties of the concrete and are not materially influenced by variations in the effective prestress. For these reasons, the ultimate-moment computations are usually more important and precise than the elastic design computations.

6-19 Load Balancing

Consider a tendon that is placed on a parabolic trajectory in a simple beam in such a manner that the sag of the tendon, as measured vertically from a straight line which connects the ends of the tendon, is equal to e as shown in Fig. 6-42. If the total uniform load on the beam is equal to w, the load

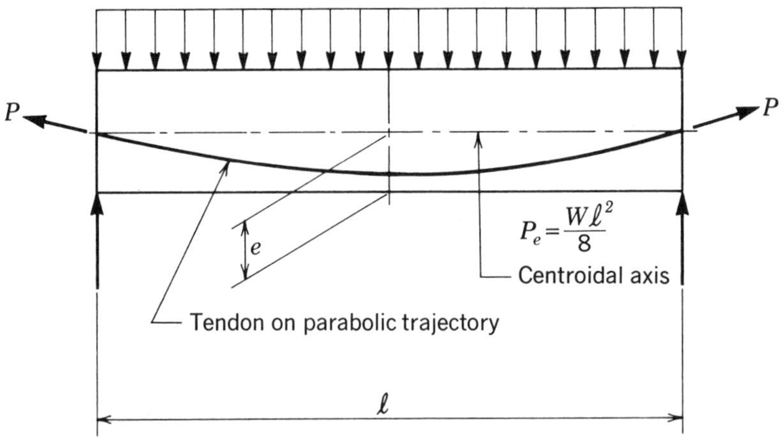

Fig. 6-42 The principle of load balancing.

will be exactly balanced if

$$Pe = \frac{wl^2}{8}$$

because the parabolic tendon results in a uniform upward force (neglecting friction) as has been explained. In such circumstances, the pressure line acts along the centroidal axis of the members if the end eccentricities are equal to zero. This is the principle of load balancing and is a useful design aid in certain circumstances.

REFERENCES

1. Bureau of Public Roads, "Criteria for Prestressed Concrete Bridges," U.S. Department of Commerce, Superintendent of Documents, Washington, D.C., 1954.
2. Curley, W. G., Sozen, M. A. and Siess, C. P. "Time Dependent Deflections of Prestressed Concrete Beams," p. 937, Highway Research Board, Bulletin 3-7, National Research Concrete Publication, 1961.
3. Subcommittee 5, ACI Committee 435, "Deflection of Prestressed Concrete Members," *Journal of the American Concrete Institute*, **60**, No. 12 (Dec. 1963).
4. Hanson, J. A., "Restress Loss as Affected by Type of Curing," *Journal of the American Concrete Institute*, **9**, No. 2, 69–93 (April 1964).
5. Ables, P. W. "Principles and Practice of Prestressed Concrete," p. 46, Crosby Lockwood and Son, Ltd., London, 1949.
6. Highway Research Board. "The AASHO Road Test Report 4 Bridge Research," Pub. 953, National Academy of Sciences, National Research Council, Washington, D.C., 1962.
7. Zielinski, J. and Rowe, R. E. "An Investigation of the Stress Distribution in the Anchorage Zones of Post-Tensioned Concrete Members," Research Report No. 9, Cement and Concrete Association, London, 1960.
8. Zielinski, J. and Rowe, R. E. "The Stress Distribution Associated with Groups of Anchorages in Post-Tensioned Concrete Members," Research Report No. 13, Cement and Concrete Association, London, 1962.
9. Rhodes, B. and Turner, F. H. "Design of End Block for Post-Tensioned Cables." *Journal of the Concrete Society*, **1**, No. 12 (Dec. 1967).
10. Chow, L., Conway, H. D. and Winter, G. "Stress in Deep Beams," *Transaction of the A.S.C.E.*, **118**, p. 686 (1953).
11. Fountain, Richard S., "A Field Inspection of Prestressed Concrete Bridges," Portland Cement Association, Chicago, Illinois, 1963.
12. Marshall, W. T. and Mattock, A. H. "Control of Horizontal Cracking in the Ends of Pre-tensioned Prestressed Concrete Girders." *Journal of the Prestressed Concrete Institute*, **7**, No. 5, 56–74 (Oct. 1962).

13. Furr, Howard L. and Sinno Raouf. "Creep in Prestressed Lightweight Concrete," Research Report Number 69–2, Texas Transportation Institute, Texas A & M University, College Station, Texas (Oct. 1967).

14. Yetteram, A. L. and Robbins, K. "Anchorage Zone Stresses in Axially Post-Tensioned Members of Uniform Rectangular Section," *Magazine of Concrete Research*, **21**, No. 67 103–111 (June 1969).

15. Kaar, Paul H. and Magura, Donald D. "Effect of Strand Blanketing on Performance of Pre-tensioned Girders," *Journal of the Prestressed Concrete Institute*, **10**, No. 6 20–33 (Dec. 1965).

16. ACI-ASCE Joint Committee 323, "Tentative Recommendations for Prestressed Concrete, "*Journal of the American Concrete Institute*, **29**, No. 7, 545–578 (Jan. 1958).

7 Design Expedients and Computation Methods

-1 Introduction

ne of the major deterrents to the use of prestressed concrete in all forms of onstruction has been the relatively great amount of time required to design restressed structures in comparison to the time required to design rein- rced-concrete or structural-steel structures. This has been due to the fact at the average designer has not been familiar with the basic design principles f prestressed concrete, nor has the average designer had sufficient experience prestressed design to develop short cuts and design expedients.

This chapter is intended to bridge the gap between theoretical considera- ons and practical design methods. The theorems and methods explained ere can be applied in many different ways and can be modified by the indi- idual designer for any special conditions. Proper use of these design methods nd expedients will greatly reduce the time and labor required to prepare conomical prestressed designs.

These design methods were developed to facilitate design calculations eing made with a slide rule. The use of modern calculators and computers ill render some of these methods unnecessary.

7-2 Computation of Section Properties

In order to determine the elastic stresses that result in the concrete sectio from the prestressing force and from the external loads, the physical prope ties of the concrete section must be known. The basic dimensions and prope ties to be determined are the location of the center of gravity, the area, an the moment of inertia of the section. The other properties used to facilita the computation of stresses are determined from these basic properties.

The computation of the basic properties of a section can be done by sever methods, all of which produce the same results. These methods differ only i the organization of the computations, the datum for taking the static moment and the procedure used in computing the moment of inertia. One convenie method is to use the top of the section as the datum for taking moments i computing the location of the center of gravity of the section. The moment c inertia of the section can then be computed about the center of gravity c the gross section. Whichever datum is used in any design office or by an individual, it should be used consistently in order to facilitate the checkin and reviewing of the computations.

All moment-of-inertia computations are made using one or more varia tions of the basic relationship

$$I_{xx} = I_o + Ay^2 \quad (7\text{-}1)$$

which can be expressed verbally as: The moment of inertia of any section c shape about the axis, $x - x$, is equal to the moment of inertia of that shap or section about an axis parallel to axis $x - x$ and passing through the cente of gravity of the section (I_o), plus the product obtained from multiplying th

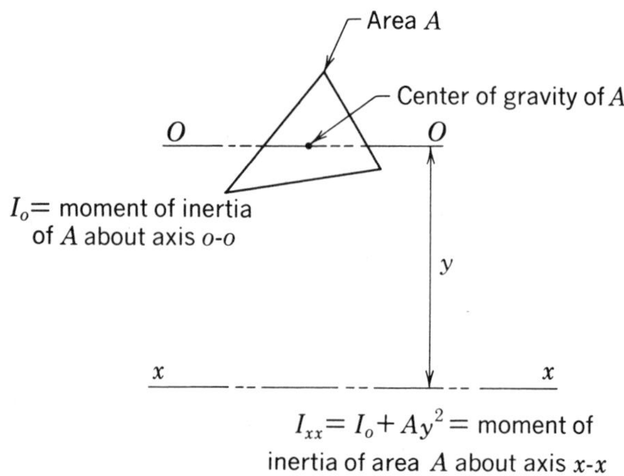

Fig. 7-1 Notation for moment-of-inertia computations.

area of the shape or section by the square of the distance between the axis $x - x$ and the center of gravity of the section. This is illustrated in Fig. 7-1.

The location of the center of gravity and the moment of inertia, about the center of gravity, of various shapes frequently encountered in prestressed concrete design are given in Fig. 7-2. The location of the center of gravity and moments of inertia of other, less-common sections that may be encountered can be found in standard engineering references or calculated by fundamental mathematical relationships. It should be noted that the moments of inertia given in Fig. 7-2 are expressed in terms of the dimensions of the section as well as a function of the area of the section and the height or diameter of the section. The expression that gives the moment of inertia in terms of the area of the section is a device used to facilitate the moment-of-inertia computation for complex shapes when done in tabular form. This is illustrated in the following discussion.

This method of computation can best be explained with an example, and the location of the center of gravity, area, moment of inertia, and other section properties for the AASHO-PCI standard bridge beam, type IV, is computed as an illustration of the recommended procedure. Referring to Fig. 7-3 and Table 7-1, the procedure is as follows:

(1) Divide the cross section into shapes of known area, centers of gravity, and moments of inertia, such as the rectangles and triangles numbered 1 through 5 in Fig. 7-3. Note that to facilitate the computations and reduce the number of component areas that must be extended in the table, the

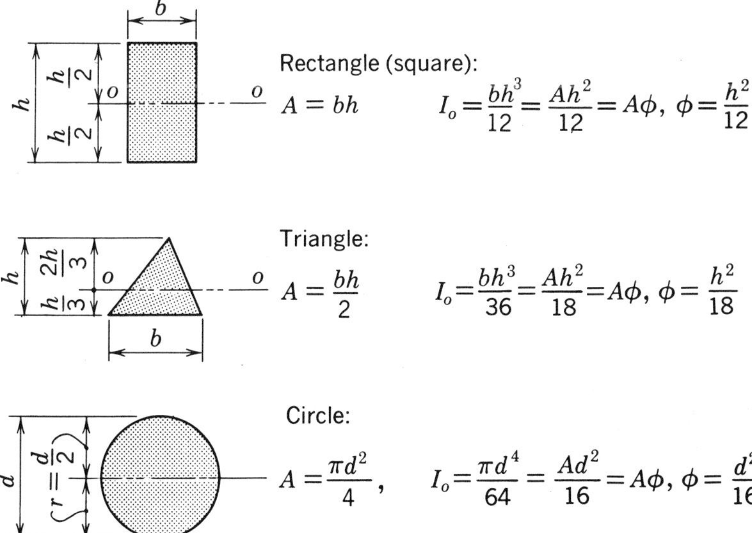

Rectangle (square):
$A = bh$
$I_o = \frac{bh^3}{12} = \frac{Ah^2}{12} = A\phi,\ \phi = \frac{h^2}{12}$
Triangle:
$A = \frac{bh}{2}$
$I_o = \frac{bh^3}{36} = \frac{Ah^2}{18} = A\phi,\ \phi = \frac{h^2}{18}$
Circle:
$A = \frac{\pi d^2}{4}$,
$I_o = \frac{\pi d^4}{64} = \frac{Ad^2}{16} = A\phi,\ \phi = \frac{d^2}{16}$
b
h
h/2
o
2h/3
h/3
d
d/2
(r = d/2)

Fig. 7-2 Area and moment of inertia for common geometric shapes.

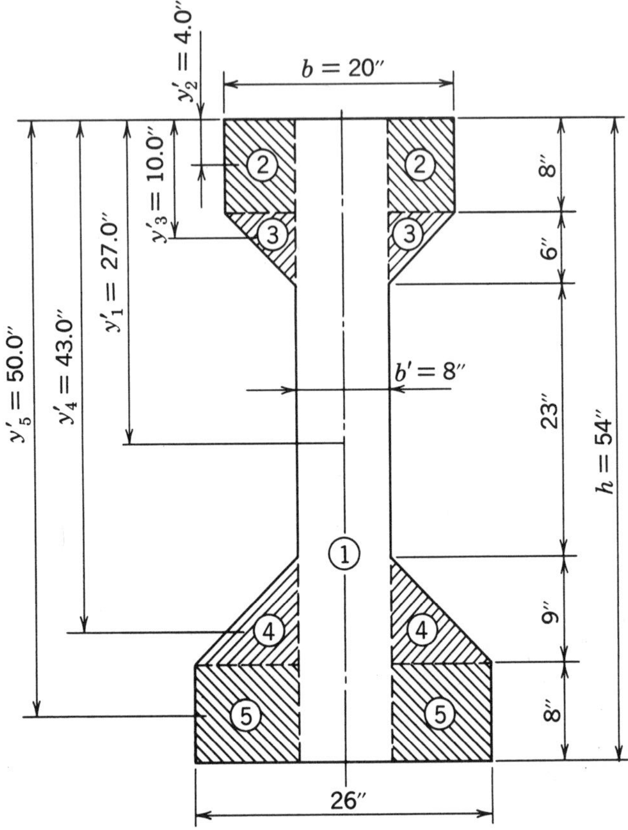

Fig. 7-3 AASHO-PCI bridge beam, type IV, divided into rectangles and triangles to facilitate the computation of section properties.

rectangular areas listed as parts 2 and 5 in Table 7-1, as well as the triangular areas listed as parts 3 and 4, consist of two pieces.

(2) Prepare a table, such as Table 7-1, and compute the areas of the component parts as well as the moments of these areas about the top of the section Ay'.

(3) Divide the sum of the static moments of the various component parts ($\Sigma Ay'$) by the total area of the section (ΣA), to obtain the distance from the top fibers to the center of gravity of the section (y_t).

(4) Compute and tabulate the distance from the center of gravity of the section to the center of gravity of each of the component areas (y) by subtracting $y_t - y'$ or $y' - y_t$.

(5) Square and tabulate the distances obtained (y^2).

TABLE 7-1 Computation of Section Properties of the AASHO Type IV Bridge Beam, in Tabular Form.

Part	Area (A) Computation (in. × in. = in.²)	y' (in.)	Ay' (in.³)	y (in.)	y^2 (in.²)	ϕ (in.²)	$y^2+\phi$ (in.²)	$A(y^2+\phi)$ (in.²)
1	8 × 54 = 432	27.0	11,700	2.30	5.3	243	248	107,000
2	12 × 8 = 96	4.0	384	25.3	640	5.33	645	62,000
3	12 × 6/2 = 36	10.0	360	19.3	373	2.0	375	13,500
4	18 × 9/2 = 81	43.0	3,480	13.7	188	4.5	193	15,600
5	18 × 8 = 144	50.0	7,200	20.7	428	5.33	433	62,300

$\Sigma A = 789$ $\quad\Sigma Ay' = 23{,}124$ $\quad I = 260{,}400$ in.^{4}

$y_t = 23{,}124/789 = 29.3$ in. $\quad S_t = 260{,}400/29.3 = 8880$ in.^{3}

$$r^2/y_t = \frac{260{,}400}{789 \times 29.3} = 11.3 \text{ in.}$$

$y_b = 54.0 - 29.3 = 24.7$ in. $\quad S_b = 260{,}400/24.7 = 10{,}500$ in.^{3}

$$r^2/y_b = \frac{260{,}400}{789 \times 24.7} = 13.4 \text{ in.}$$

(6) Compute and tabulate the factors by which the areas of the component parts can be multiplied in order to obtain their respective moments of inertia about their centers of gravity (ϕ) (*see* Fig. 7-2).

(7) Tabulate the sums of $(y^2 + \phi)$.

(8) Multiply and tabulate the terms $A(y^2 + \phi)$ for each of the component areas. The summation of the $A(y^2 + \phi)$ is the moment of inertia of the section.

(9) The remaining properties used in the design computations are computed as follows:

$$y_b = h - y_t, \qquad S_t = I/y_t, \qquad S_b = I/y_b, \qquad r^2/y_t = I/Ay_t, \qquad r^2/y_b = I/Ay_b$$

Bridge stringers, such as the AASHO-PCI standard, prestressed-concrete beams for highway bridges (*see* Chapter 12), are frequently used with a cast-in-place deck slab that acts compositely with the stringers, as a result of shear stresses that develop between the slab and the top of the stringers (*see* Sec. 6-4). The computation of the composite section properties for the AASHO-PCI bridge beam, type IV, with a 6 × 36 in. cast-in-place top flange, as is illustrated in Fig. 7-4, can be done using the same fundamental procedure described above. This is illustrated in Table 7-2.

The concrete in the deck slab does not have the same elastic properties as the prestressed stringers under normal conditions, since the quality of the concrete normally employed in cast-in-place bridge decks is not as high as

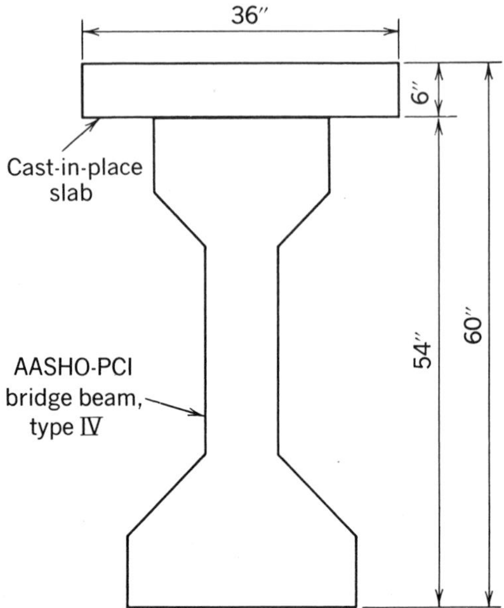

Fig. 7-4 Composite section composed of AASHO-PCI bridge beam, type IV, and 6 × 36 in. cast-in-place slab.

TABLE 7-2 Computation of the Section Properties of the Composite Beam Illustrated in Fig. 7-4.

Part	*Area Computation (in. × in. = in.²)*	*y′ (in.)*	*Ay′ (in.³)*	*y (in.)*	*y² (in.²)*	*ϕ (in.²)*	*y² + ϕ (in.²)*	*A(y + ϕ²) (in.⁴)*
Slab	36 × 6 in. = 216	3.0	648	25.3	640	3.0	643	139,000
Beam	= 789	35.3	27,800	7.0	49	*	49	38,600
								260,400*

$\Sigma A = 1005$ in.² $\Sigma Ay' = 28{,}448$

$I = 438{,}000$ in.⁴

$y_p = 28.3$ in $\quad S_p = 15{,}500$ in.³

$y_t = 22.3$ in. $\quad S_t = 19{,}600$ in.³

$y_b = 31.7$ in. $\quad S_b = 13{,}800$ in.³

* The value of I_0 is known to equal 260,400 in.⁴ for the precast section. Hence, the value of ϕ is not computed for this portion of the composite section and the value of I_0 for the precast section is simply added in the $A(y^2 + \phi)$ column.

that used in prestressed stringers. This effect is taken into consideration by using a transformed cross section when computing the properties of the composite section. The transformed cross-section to be used consists of the gross section of the prestressed beam and a slab section having a depth equal to that of the actual slab (less any allowance for wearing surface) and the effective slab width, equal to the actual width normally assumed to be effective, multiplied by the ratio of the elastic modulus of the slab concrete to the elastic modulus of the concrete in the beam. If the slab and beam concretes are assumed to have moduli of 3.5×10^6 psi and 5×10^6 psi, respectively, the width of the slab that should be used in the composite section would be 3.5/5.0 or 0.70 of the actual effective width.

It should be noted that y_p and y_t are used to denote the distances from the center of gravity of the composite section to the top fibers of the cast-in-place deck and the precast stringer, respectively. This procedure is recommended in order to avoid confusion, because the subscript t is used to denote the distance to the top fibers of the precast section during the computation of the section properties of the precast section.

Finally, in Table 7-2, it will be noted that there is no entry for the second part in the ϕ column. The moment of inertia of the precast section about its center of gravity is known, and hence, is simply added to the last column.

After the designer becomes familiar with this tabular form, the computation of section properties becomes rapid and routine. The effects of minor adjustments in the concrete sections are often determined by subtracting or adding areas to the section which is being modified, rather than by entirely recomputing the section properties of the modified section. In addition, the column marked $(y^2 + \phi)$ is often eliminated in actual calculations and the terms y^2 and ϕ are added mentally as their sum is multiplied by the area.

7-3 Allowable Concrete Stresses to be Used in Design Computations

Most prestressed-concrete design criteria specify maximum allowable initial, or temporary, compressive and tensile stresses, as well as maximum allowable final, or permanent, compressive and tensile stresses. This procedure is generally considered necessary or justified for the following reasons:

(1) In order to obtain an economical and realistic production schedule under many conditions, it is essential that the prestress be applied to the member before the concrete attains the minimum required 28-day cylinder strength. Hence, it is normal practice to apply the prestress to the concrete when the strength of the concrete is of the order of 4000 psi, although the required minimum cylinder strength of the concrete at the age of 28 days is generally of the order of 5000 psi. Therefore, the temporary, or initial, allowable stresses are reckoned from a cylinder strength lower than that used in determining the final, or permanent, allowable concrete stresses.

(2) The initial prestressing force is the maximum prestressing force that will ever be imposed on the member. This force is subject to a reduction in the amount of 10% to 30%. The reduction or relaxation of the prestressing force starts to take place immediately after stressing and requires 3 years or more to reach its maximum value.

(3) The stresses imposed on the member due to prestressing are of opposite direction to those imposed by the service loads, i.e., the prestressing normally causes small tensile stresses in the top fibers and large compressive stresses in the bottom fibers of simple beams, while the superimposed loads that are to be carried by the beams cause tensile stresses in the bottom fibers and compressive stresses in the top fibers.

(4) The stresses resulting from prestressing can be controlled by the fabricator to a relatively high precision, but usually neither the designer nor the fabricator can control or predict with high precision the loads that will be imposed on the structure while it is in service. For this reason, and in view of the reasons listed under 2 and 3, the safety factor required to guard against failure of the concrete during stressing does not need to be as high as that required for the design loads.

In a beam pre-tensioned with straight tendons, the highest initial stresses occur near the ends where there is no moment, due to dead load, to counteract the effects of the prestressing (*see* Sec. 4-6). Therefore, the restrictions of the temporary allowable stresses for a pre-tensioned member can be expressed mathematically as follows:

$$\frac{f_{si} A_s^*}{A}\left(1 - \frac{e y_t}{r^2}\right) \leqq -3\sqrt{f'_{ci}} \tag{7-2}$$

and

$$\frac{f_{si} A_s^*}{A}\left(1 + \frac{e y_b}{r^2}\right) \leqq 0.60 f'_{ci} \tag{7-3}$$

in which the minus sign denotes a tensile stress, f_{si} is the initial stress in the prestressing steel, A_s^* is the area of the tendons, A is the area of the concrete, e is the eccentricity of the tendon, r is the radius of gyration, f'_{ci} is the concrete cylinder strength at the time of stressing, and y_t and y_b are the distances from the centroidal axis to the top and bottom fibers, respectively.* If f_{se} is the effective stress in the tendons, f'_c is the cylinder strength at 28 days, f_{tt} and f_{tb} designate the total stresses due to dead and live loads in the top and bottom fibers at the section of maximum moment,

* In Eqs. 7-2 through 7-5 the terms on the right side are stresses allowed by the AASHO Standard Specifications given in Sec. 3-16.

respectively, the restrictions of the final allowable stresses can be expressed mathematically as follows:

$$\frac{f_{se} A_s^*}{A}\left(1 - \frac{ey_t}{r^2}\right) + f_{tt} \leqq 0.40 f_c' \tag{7-4}$$

and

$$\frac{f_{se} A_s^*}{A}\left(1 + \frac{ey_b}{r^2}\right) + f_{tb} \geqq -3\sqrt{f_c'} \tag{7-5}$$

It should be noted that f_{tb} is a negative stress and the negative sign is included in this symbol.

For an assumed concrete section and an assumed ratio between the effective steel stress and the initial steel stress $m = f_{se}/f_{si}$, the values of f_{tt} and f_{tb} can be computed and substituted in Eqs. 7-2 through 7-5, in which case, all of the terms that appear in the equations will be known or assumed, except the values of $f_{se} A_s^*$ and e. Since a number of combinations of e and $f_{se} A_s^*$ will normally satisfy each of the four equations, the combinations that will satisfy all of the equations can be determined by plotting each of the four relationships, as shown in Fig. 7-5. The shaded area of Fig. 7-5 indicates the combinations of $f_{se} A_s^*$ and e that satisfy the conditions of the allowable stresses for the assumed section.

Although the procedure of plotting a figure similar to that shown in Fig. 7-5, first suggested by Magnel (Ref. 1), will yield accurate results, it is obviously too cumbersome and time consuming to be used as a general design

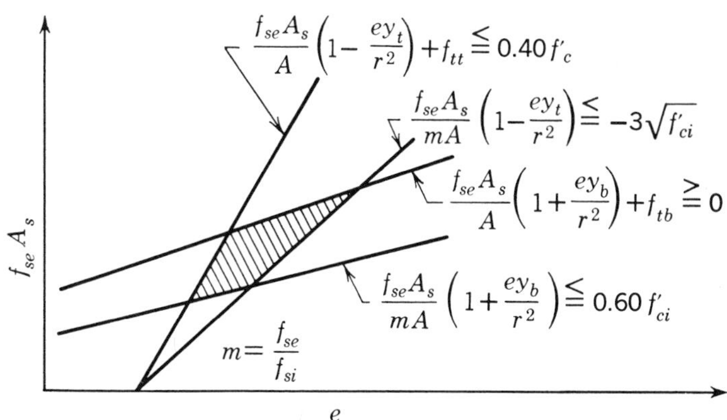

Fig. 7-5 Graphical solution of four equations in solving for the prestressing force and eccentricity.

procedure. It does, however, illustrate the fact that there is frequently a combination of prestressing forces and eccentricities that will yield a satisfactory solution with a specific beam section.

The most frequently used allowable stresses for prestressing steel and concrete in U.S. practice are given in Secs. 2-11 and 3-16, respectively.

7-4 Limitations of Sections Prestressed with Straight Tendons

It should be apparent that fully bonded, straight pretensioning tendons can only be used in prismatic beams in which the maximum flexural stress in the bottom fibers, due to the total load, does not exceed the arithmetic sum of either the allowable tensile stress and the final bottom fiber stress due to prestressing or the sum of the allowable tensile stress and the allowable compressive stress. If, for example the maximum stress in the bottom fiber due to the total external load were 2300 psi and the allowable tensile stress and the final compressive stress due to prestressing (assuming the final stress due to prestressing is less than the allowable final compressive stress) were -400 psi and $+2000$ psi, respectively, the design would not be restricted by the bottom-fiber stress. The maximum load that could be applied without exceeding the allowable stresses would be one that results in a maximum bottom-fiber stress of -2400 psi.

In a similar manner the top-fiber stress may limit the capacity of a prismatic beam section if the maximum flexural stress in the top fiber, due to the total load, is greater than the arithmetic sum of the allowable compressive stress in the completed member and the top fiber tensile stress due to final prestressing.

As a result of these limitations, the designer can normally determine if a specific concrete section can be used with straight tendons without calculating the magnitude and eccentricity of the prestressing force. It is only necessary to determine the stresses in the section due to the total load and compare these values with the sum of the appropriate, allowable stresses.

ILLUSTRATIVE PROBLEM 7-1 Determine the maximum moment the section in Fig. 7-6 can withstand if pretensioned with straight tendons having an initial and final stress of 180 and 154 ksi, respectively, if the AASHO requirements given in Sec. 3-16 are applicable and f'_{ci} an f'_c are 4000 and 5000 psi, respectively.

The allowable stresses are:

Top fiber:

$$\text{Initial tension} = -3\sqrt{4000} = -190 \text{ psi}$$

$$\text{Final compression} = 0.40 \times 5000 = 2000 \text{ psi}$$

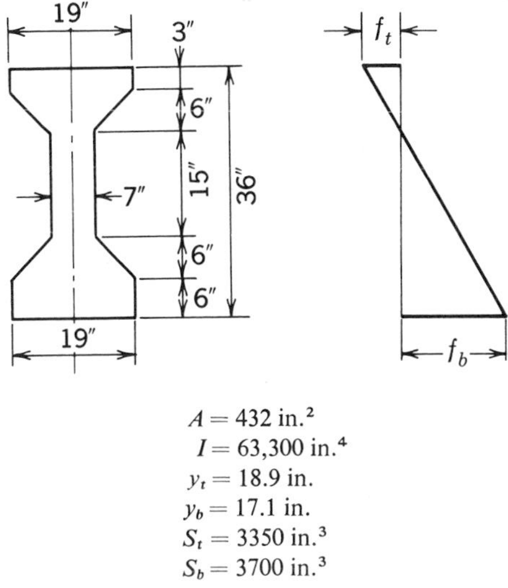

$$A = 432 \text{ in.}^2$$
$$I = 63{,}300 \text{ in.}^4$$
$$y_t = 18.9 \text{ in.}$$
$$y_b = 17.1 \text{ in.}$$
$$S_t = 3350 \text{ in.}^3$$
$$S_b = 3700 \text{ in.}^3$$

Fig. 7-6 Beam section and stress distribution for Prob. 7-1.

Bottom fiber:

$$\text{Initial compression} = 0.60 \times 4000 = 2400 \text{ psi}$$

$$\text{Final tension} = -3\sqrt{5000} = -212 \text{ psi}$$

$$\text{Final compression} = 0.40 \times 5000 = 2000 \text{ psi}$$

The stresses at the time of stressing will reduce to 154/180 = 0.856 of their initial value as a result of the prestressing losses. Therefore, the maximum stress due to all loads will be limited as follows:

$$\text{Top fiber} = 2000 + 0.856 \times 190 = 2163 \text{ psi}$$

$$\text{Bottom fiber} = 212 + 0.856 \times 2400 = 2266 \text{ psi}$$

$$\text{Bottom fiber} = 212 + 0.40 \times 5000 = 2120 \text{ psi}$$

The allowable moments as controlled by the top and bottom fibers are

$$M_T = \frac{2163 \times 63{,}300}{18.9 \times 12000} = 604 \text{ 'k}$$

$$M_B = \frac{2120 \times 63{,}300}{17.1 \times 1200} = 654 \text{ 'k}$$

The stresses in the top fiber control the capacity of this section with these design conditions.

7-5 Limitations of Sections Prestressed with Curved Tendons

In considering the stress in the bottom fiber at any specific section of a simple beam prestressed with a curved tendon, it should be apparent that the maximum stress due to external loads must not exceed the arithmetic sum of the stress due to the effective prestressing force and the allowable tensile stress in the completed structure. In addition, the algebraic sum of the stress due to the initial prestressing force and the stress due to the minimum loading condition must not exceed the allowable, inital compressive stress.

ILLUSTRATIVE PROBLEM 7-2 Determine the maximum possible allowable total moment that could be permitted on the beam section of Fig. 7-6 if a curved tendon is used and the initial and final prestressing steel stresses are 187 and 162 ksi, respectively. The design span is 50.0 ft. The allowable stresses are those permitted by AASHO and $f'_{ci} = 4000$ psi, $f'_c = 5000$ psi.

SOLUTION:

$$w_d = \frac{432 \times 0.15}{144} = 0.450 \text{ ft/kips} \qquad M_d = \frac{0.450 \times 50^2}{8} = 141 \text{ ft/k}$$

Dead load flexural stresses:

$$f_t = \frac{141 \times 12}{3350} = +0.505 \text{ ksi}$$

$$f_b = \frac{141 \times 12}{3700} = -0.457 \text{ ksi}$$

Maximum allowable top fiber stress and moment:

$$f_t = 0.40 \times 5000 + 505 + \frac{162}{187}(3\sqrt{4000}) = 2669 \text{ psi}$$

$$M_T = \frac{2669 \times 3350}{1200} = 745 \text{ 'k}$$

Maximum allowable bottom fiber stress and moments:

$$(0.60 \times 5000)\frac{162}{187} = 2599 > 0.40 \times 5000 = 2000 \text{ psi}$$

$$f_b = 2000 + 457 + 3\sqrt{5000} = 2669 \text{ psi}$$

$$M_T = \frac{2669 \times 3700}{1200} = 823 \text{ 'k}$$

It should be recognized that the condition of prestress assumed by this type of analysis may not be attainable due to practical considerations. (*See* Sec. 7-7.)

The initial tensile stresses in the top fibers of beams prestressed with curved tendons are not normally critical at the section of maximum moment in beams of good proportions. If the top-fiber stresses limit the design of beams with curved tendons, it is usually due to excessive, compressive stress under maximum loading conditions. Top fiber stresses are much more apt to be a problem in a beam with a narrow top flange than in a beam with a wide top flange.

7-6 Determination of Minimum Prestressing Force for Straight Tendons

In the trial and error procedure used in designing prestressed flexural members, a beam with known or calculated cross sectional properties is tentatively adopted and reviewed to determine the stresses due to the external loads. If the external loads result in stresses within practical limits (*see* Sec. 7-4), the magnitude and eccentricity of the prestressing force required to develop the desired net concrete stresses must then be determined. When straight tendons are used in prismatic simple beams subjected to usual loading conditions, the maximum stresses under minimum loading conditions (dead load of the beam alone, usually) occur at the ends of the beam where there is no moment due to external loads. The maximum stresses under the maximum loading conditions occur near midspan. The procedures recommended for determining the magnitude of the minimum prestressing force and the required eccentricity for different specific conditions are illustrated with explanation in Problems 7-3 through 7-5.

ILLUSTRATIVE PROBLEM 7-3 Determine the prestressing force and eccentricity required to prestress a slab, 4 ft wide and 8 in. deep, with straight tendons. The slab is to be used, simply supported, on a span of 30 ft and is to be composed of normal concrete (150 pcf) with $f'_{ci} = 4000$ psi and $f'_c = 5000$ psi. The superimposed load is 45 psf and the member will be exposed to a corrosive atmosphere in service.

SOLUTION:

Loads and Moments:

$$\text{Slab dead load} = 4 \times 100 = 400 \text{ plf}$$

$$\text{Superimposed load} = 4 \times 45 = 180 \text{ plf}$$

$$\text{Total load} = 580 \text{ plf}$$

$$\text{Total bending moment} = 580 \times \frac{30^2}{8} = 65{,}400 \text{ ft/lb}$$

Section modulus $S = bd^2/6 = 48 \times 8^2/6 = 512$ in.^{3}

Top and bottom fiber stresses due to the total bending moment

$$= \frac{65{,}400 \times 12}{512} = \pm 1530 \text{ psi}$$

Assume non-prestressed reinforcement is not to be used in the top flange to resist tensile stresses in the concrete. The final prestressing stress in the bottom fiber must be equal to +1530 psi. (Since the slabs are to be exposed to a corrosive atmosphere, the net bottom-fiber stress should not be tensile under full load.) If non-prestressed reinforcement is not to be provided at the ends, the top-fiber tensile stress at the ends, due to initial prestress, should be equal to or less than $3\sqrt{f'_{ci}} = 190$ psi. Assuming the ratio of $f_{se}/f_{si} = 0.85$, the tensile stress in the top fiber resulting from the effective prestress must be less than $0.85 \times -190 = -160$ psi. Therefore, the stress distribution due to the final prestressing force should be as shown in Fig. 7-7.

The prestressing stress at the center of gravity of a section is equal to P/A, since the familiar equation for stress due to prestressing

$$f = \frac{P}{A}\left(1 + \frac{ey}{r^2}\right)$$

becomes

$$f = \frac{P}{A}$$

for the fibers located at the center of gravity of the section, where $y = 0$. This is a fundamental principle that holds true for sections that are symmetrical or asymmetrical about an axis which passes through the center of gravity of the section. The average stress, which is at the center of gravity of the section (in this case, mid-depth of the section), can be rapidly computed by

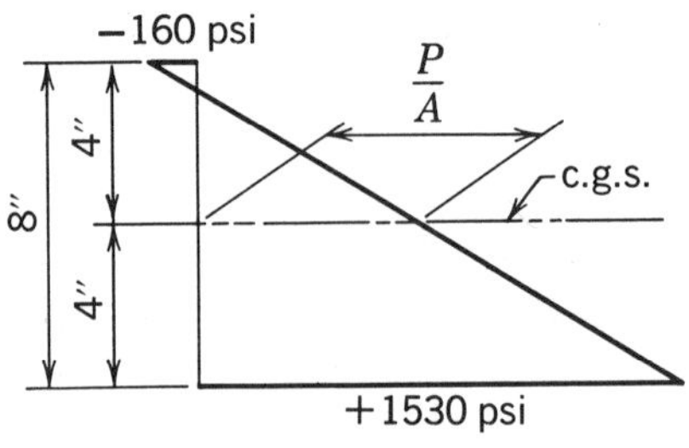

Fig. 7-7

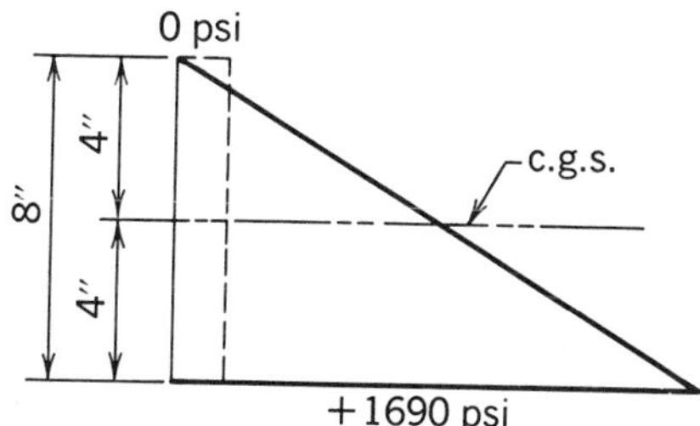

Fig. 7-8

use of the proportional triangles indicated in Fig. 7-8, and from which it will be seen that

$$\frac{P}{A} = (f_t + f_b)\frac{y_t}{d} - f_t = 1690\left(\frac{4.0}{8.0}\right) - 160 = 685 \text{ psi}$$

Therefore, the final prestressing force P can be computed by

$$P = 685 \times 48 \times 8 = 263{,}000 \text{ lb}$$

Since this force must develop +1530 psi in the bottom fiber, the familiar relationship for the bottom-fiber stress due to prestressing

$$f_b = \frac{P}{A}\left(1 + \frac{ey_b}{r^2}\right)$$

can be rewritten

$$e = \left(\frac{f_b}{P/A} - 1\right)\frac{r^2}{y_b} = \left(\frac{1530}{685} - 1\right)\frac{8}{6} = 1.65 \text{ in.}$$

It should be pointed out that $r^2/y = d/6$ for a rectangular section.

If it is decided to use non-prestressed reinforcement in the top fibers to resist the tensile stresses in the concrete due to prestressing, the initial, top-fiber tensile stress could be as high as $6\sqrt{f'_{ci}} = -380$ psi and the top-fiber tensile stress due to the effective prestressing force could be as high as $0.85 \times -380 = -320$ psi. Assuming it is desired to limit the top-fiber stress to -300 psi, the required distribution of prestress would be as shown in Fig. 7-9, and the computation of P and e becomes

$$\frac{P}{A} = 1830 \times \frac{4.0}{8.0} - 300 = 615 \text{ psi}$$

$$P = 236{,}000 \text{ lb}$$

$$e = \left(\frac{1530}{615} - 1\right)\frac{8}{6} = 1.98 \text{ in.}$$

Compare the simplicity of this computation to that required using the classical relationship demonstrated in Prob. 4-8.

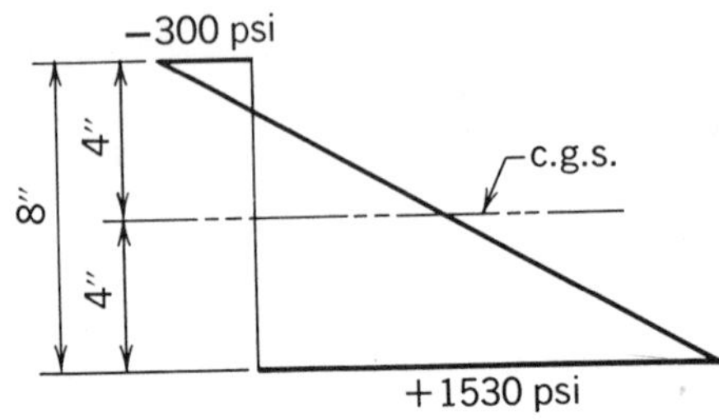

Fig. 7-9

ILLUSTRATIVE PROBLEM 7-4 For the slab of Prob. 7-3 assume the superimposed load is to be 100 psf. Compute the required prestressing force and eccentricity assuming: (1) The final top-fiber stress due to prestressing must not exceed −300 psi and no tension is to be allowed in the bottom fibers, and (2) the final top-fiber stress due to prestressing must not exceed −300 psi and the net stress in the bottom fiber under full load must not exceed −400 psi.

SOLUTION:

Loads and moments:

$$\text{Slab dead load} = 4 \times 100 = 400 \text{ plf}$$

$$\text{Superimposed load} = 4 \times 100 = 400 \text{ plf}$$

$$\text{Total load} = 800 \text{ plf}$$

$$\text{Total moment} = 800 \times 30^2/8 = 90{,}000 \text{ ft lb}$$

$$\text{Top and bottom fiber stresses} = \frac{90{,}000 \times 12}{512} = \pm 2110 \text{ psi.}$$

Part (1): The bottom fiber stress of +2110 is too high for $f'_{ci} = 4000$ psi, since $0.85 \times 0.60 \times 4000$ psi $= 2040$ psi. If this solution is to be used and if the design is to conform to the allowable stresses specified above, the value of f'_{ci} must be $2110/(0.85 \times 0.60)$, which is equal to 4150 psi. It should be pointed out that the net compression in the top fiber will be $2110 - 300$ psi $=$ 1810 psi. If the final net compressive stress is to be $0.40\, f'_c$ or less, the minimum value of f'_c is 4500 psi. Assuming these values of f'_{ci} and f'_c are to be used, the required values of P and e are (*see* Fig. 7-10)

$$\frac{P}{A} = 2410 \times \frac{4.0}{8.0} - 300 = 905 \text{ psi}$$

$$P = 348{,}000 \text{ lb}$$

$$e = \left(\frac{2110}{905} - 1\right)\frac{8}{6} = 1.77 \text{ in.}$$

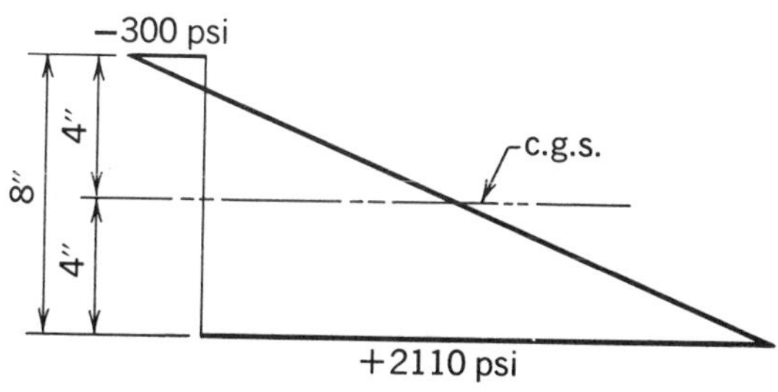

Fig. 7-10

Part (2): The desired distribution of stresses due to prestressing are as shown in Fig. 7-11, and the values of P and e become

$$\frac{P}{A} = 2010 \times \frac{4.0}{8.0} - 300 = 705 \text{ psi}$$

$$P = 271{,}000 \text{ lb}$$

$$e = \left(\frac{1710}{705} - 1\right)\frac{8}{6} = 1.89 \text{ in.}$$

It should be noted that this example illustrates a procedure that can be adopted if the initial concrete stresses (or final stresses) are nominally higher than would be allowable for the quality of concrete which was at first assumed; thus, the designer can increase the value of f'_{ci} and f'_c (within reasonable limits) in order to confine the stresses within the allowable limits. Also illustrated is the procedure used in the calculation of the prestressing required for members in which tensile stresses are permitted in the bottom fibers of the members under full load.

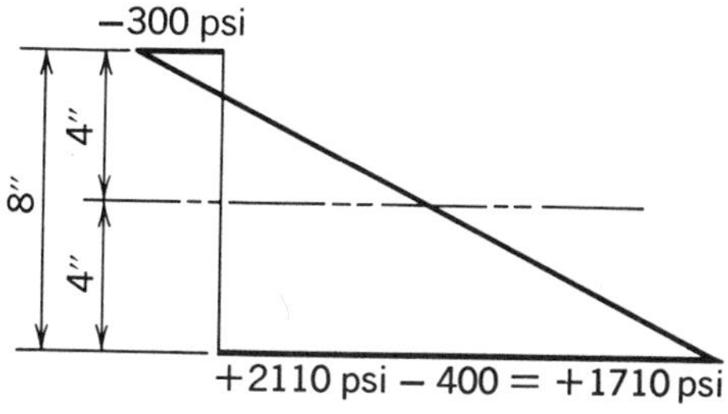

Fig. 7-11

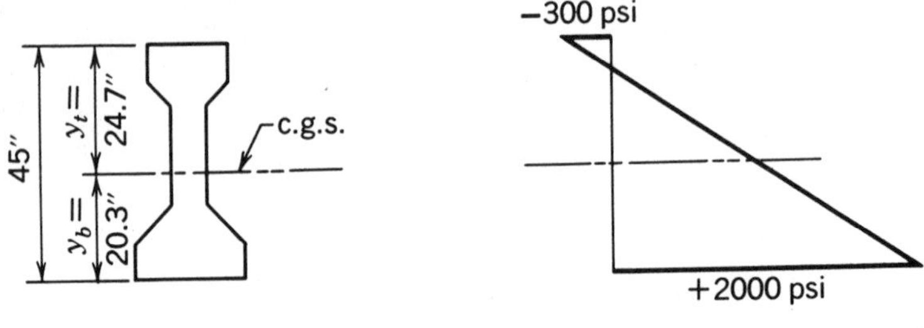

Fig. 7-12 (a) AASHO-PCI type III bridge beam. (b) Required effective prestress distribution.

ILLUSTRATIVE PROBLEM 7-5 Compute the prestressing force and eccentricity required to produce a final stress of −300 psi in the top fibers and +2000 psi in the bottom fibers of the AASHO-PCI type III bridge beam as shown in Fig. 7-12. The required section properties are: $A = 560$ in.^{2}, $y_t = 24.7$ in., and $r^2/y_b = 11.03$ in.

SOLUTION: The desired distribution of stresses is illustrated in Fig. 7-12 and P and e are computed as follows:

$$\frac{P}{A} = 2300 \times \frac{24.7}{45.0} - 300 = 960 \text{ psi}$$

$$P = 538{,}000 \text{ lb}$$

$$e = \left(\frac{2000}{960} - 1\right) 11.03 = 11.9 \text{ in.}$$

7-7 Determination of Minimum Prestressing Force for Curved Tendons

Since the dead weight of a simple prismatic beam is normally acting at the time of stressing, the eccentricity of prestressing can be greater near the center of the beam than at the ends, without the net concrete stresses exceeding the allowable values, as was explained in Sec. 4-6. This is the principle reason for draping or curving the tendons, and because this method results in a variable prestressing moment along the length of the beam, the concrete stresses resulting from curved prestressing tendons must be considered at several locations. For prismatic members, the magnitude of the prestressing force required with curved tendons is determined from the condition of stress at the position of maximum moment. For members of variable depth, the magnitude of the prestressing force required may be controlled by conditions at a section which is not the section of maximum moment.

In detailing an actual design, the prestressing force must be developed by a specific number of tendons. For reasons of economy, it is desirable to use tendons stressed near their maximum allowable values.

ILLUSTRATIVE PROBLEM 7-6 For the AASHO-PCI bridge beam, type III, which is to be used on a span of 70 ft, compute the minimum prestressing force and eccentricity that can be used if the member must withstand a superimposed moment of 800 k-ft at midspan. The superimposed moment varies parabolically from a maximum value at midspan to zero at the support. Assume $f'_{ci} = 4000$ psi and $f'_c = 5000$ psi. The section properties of the AASHO-PCI bridge beam, type III, are

$$\text{Area} = 560 \text{ in.}^2 \qquad \text{Moment of Inertia} = 125{,}400 \text{ in.}^4$$

$$y_t = 24.7 \text{ in.} \qquad S_t = 5080 \text{ in.}^3 \qquad r^2/y_t = 9.06 \text{ in.}$$

$$y_b = 20.3 \text{ in.} \qquad S_b = 6180 \text{ in.}^3 \qquad r^2/y_b = 11.03 \text{ in.}$$

$$\text{Dead weight of the beam} = 0.585 \text{ k/ft}$$

SOLUTION:

$$\text{Moment due to dead load of beam} = 0.585 \times 70^2/8 = 358 \text{ k-ft}$$

$$\text{Moment due to the superimposed load} = 800 \text{ k-ft}$$

$$\text{Total moment at midspan} = 1158 \text{ k-ft}$$

	Top Fiber	*Bottom Fiber*
Stresses due to total moment	+2740 psi	−2250 psi
Stresses due to dead load only	+ 845 psi	− 695 psi

The distribution of concrete stresses due to the effective prestress at midspan must be as shown in Fig. 7-13(a), if the net top-fiber stress due to total load plus effective prestress is to be held to $0.40 f'_c = 0.40 \times 5000 = 2000$ psi and if the tensile stresses in the bottom fiber due to the total load are to be exactly nullified by the effective prestress. Assuming non-prestressed reinforcement is not to be used, the net, top-fiber concrete stress due to initial prestressing plus dead load of the beam would be at the allowable value of

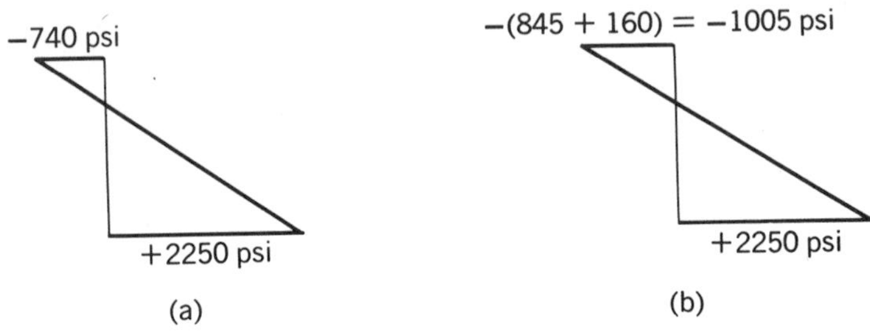

Fig. 7-13.

-190 psi (assuming $f_{se}/f_{si} = 0.85$) and the bottom-fiber stress due to the total load would be exactly nullified if the effective prestress resulted in a stress distribution as shown in Fig. 7-13(b). The bottom-fiber stress, due to the effective prestress, could be as high as $0.40f_c' + 695 = 2695$ psi. The most economical design will result from a prestressing force that can develop the required minimum effective prestress in the bottom fibers ($+2250$ psi), without exceeding the allowable, initial tensile stress in the top fibers, such as is shown in Fig. 7-13(b), if such a stress distribution can be obtained with a practical eccentricity.

For the distribution of stress shown in Fig. 7-13(b) the values of P and e are

$$\frac{P}{A} = 3255 \times \frac{24.7}{45.0} - 1005 = 780 \text{ psi}$$

$$P = 436 \text{ k}$$

$$e = \left(\frac{2250}{780} - 1\right) 11.03 = 20.80 \text{ in.}$$

It is apparent that this is not a practical solution for this case, because the required eccentricity is greater than the distance from the center of gravity of the section to the bottom fibers (y_b); hence, if this solution were used, the center of gravity of the tendons would be below the bottom of the beam. Therefore, the distribution of stress due to the effective prestress must be revised in such a manner that the eccentricity is reduced.

Using the distribution of stress indicated in Fig. 7-13(a), the values of P and e are

$$\frac{P}{A} = 2990 \times \frac{24.7}{45.0} - 740 = 900 \text{ psi}$$

$$P = 504 \text{ k}$$

$$e = \left(\frac{2250}{900} - 1\right) 11.03 = 16.5 \text{ in.}$$

This solution is reasonable for the conditions at midspan and should be adopted, because the stresses are allowable and the eccentricity of 16.5 in. results in the center of gravity of the tendons being 3.8 in. from the bottom of the beam, allowing adequate concrete cover.

Because the dead-load moment and the moment due to the superimposed loads vary parabolically (from maximum at the center of the span to zero at the supports), it can be specified that the eccentricity of the prestressing is zero at the supports. Nominal eccentricities above or below the center of gravity of the section could be allowed at the supports without exceeding the allowable stresses.

At the quarter point, the stresses due to the dead and superimposed loads are only 75% of the stresses due to these loads at midspan, or +2060 and −1690 psi in the top and bottom fibers, respectively. Assuming the prestressing force is 504 k, the eccentricity is 16.5 in. at midspan, and, assuming the eccentricity varies parabolically to zero at the supports, the eccentricity would be 75% of 16.5 in. at the quarter point and the stresses due to the effective prestress would be

$$f_t = \frac{504{,}000}{560}\left(1 - \frac{12.4}{9.06}\right) = -332 \text{ psi}$$

$$f_b = \frac{504{,}000}{560}\left(1 + \frac{12.4}{11.03}\right) = +1910 \text{ psi}$$

It can be shown that these stresses will result in net concrete stresses under minimum and maximum loading conditions within the allowables, as follows:

	Top Fiber	*Bottom Fiber*
Stress due to dead load of beam	+634	−521
Effective prestress	−332	+1910
Net, minimum loading	+302	+1389
Stress due to superimposed load	+1420	−1165
Net, maximum loading	+1722	+224

ILLUSTRATIVE PROBLEM 7-7 For the beam and the conditions specified in Prob. 7-6, determine the number of 1 1/2-in. diameter, high-tensile, steel rods that could be used, if the rods were to be used with an effective prestress of 82 k each and $f_{se}/f_{si} = 0.85$.

SOLUTION:

Assume six rods.

$$P = 6 \times 82 = 492 \text{ k}$$

$$M_T = P(e + r^2/y_b) = 1158 \text{ k-ft.} = 492\ (e + 11.03)$$

$$e = 17.3 \text{ in.}$$

$$f_t = \frac{492{,}000}{560}\left(1 - \frac{17.3}{9.06}\right) = -800 \text{ psi}$$

$$f_b = \frac{492{,}000}{560}\left(1 + \frac{17.3}{11.03}\right) = +2250 \text{ psi}$$

This solution is satisfactory, since the stress distribution is between those of Fig. 7-13(a) and (b) and the distance from the bottom of the beam to the center of gravity of the tendon is 3.0 in., which will allow a clear concrete cover of 2 in. for the 1 1/2 in. sheaths if placed as shown in Fig. 7-14.

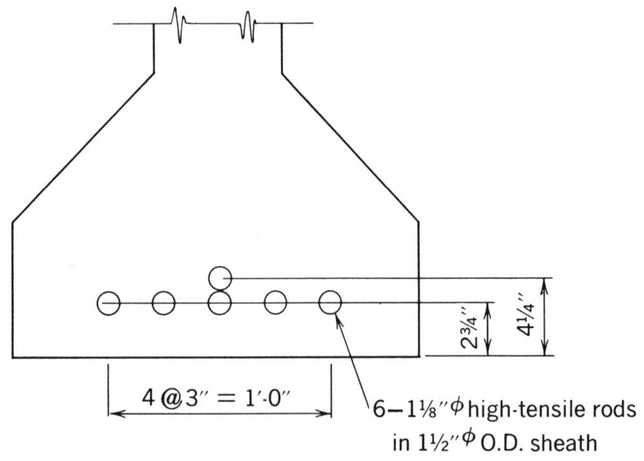

Fig. 7-14

ILLUSTRATIVE PROBLEM 7-8 Assume the beam for Prob. 7-6 is to be stressed with a combination of pre-tensioning and post-tensioning. Determine the amount and eccentricity of the prestressing required for each of these methods, if it is assumed the maximum, initial, tensile and compressive concrete stresses are -350 psi and 2400 psi, respectively, and $f_{se}/f_{si} = 0.85$. The magnitude of the post-tensioning force is to be kept as small as possible, in the interest of economy. Assume the post-tensioned tendon is not to be stressed until the beam is pre-tensioned and removed from the casting bed.

SOLUTION:

The maximum distribution of stress that can be allowed by the effective prestress in the straight pretensioned tendons is as shown in Fig. 7-15, for which the values of P and e are

$$\frac{P}{A} = 2340 \times \frac{24.7}{45.0} - 300 = 985 \text{ psi}, \qquad P = 551 \text{ k}$$

$$e = \left(\frac{2040}{985} - 1\right) 11.03 = 11.8 \text{ in.}$$

Assuming the supplementary post-tensioning is developed by one tendon at an eccentricity of 17.5 in. (2.8 in. from the bottom of the bottom of the beam to the center of gravity of the tendon), the magnitude of the prestressing force required to increase the bottom-fiber stress, 210 psi to a total value of 2250 psi, is calculated as follows:

$$\frac{P}{A} = \frac{210}{1 + \left(\frac{17.5}{11.03}\right)} = 81.3 \text{ psi} \qquad f_t = \frac{45{,}500}{560}\left(1 - \frac{17.5}{9.06}\right) = -76 \text{ psi}$$

$$P = 45.5 \qquad f_b = \frac{45{,}500}{560}\left(1 + \frac{17.5}{11.03}\right) = +210 \text{ psi}$$

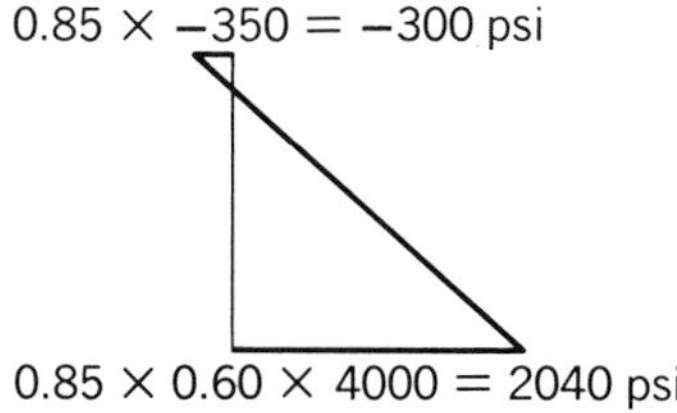

Fig. 7-15

Summarizing the net concrete stresses for this solution reveals the following:

	Top Fiber	*Bottom Fiber*
Stress due to total moment	+2740 psi	−2250 psi
Effective prestress: pre-tension	−300 psi	+2040 psi
Effective prestress: post-tension	−76 psi	+210 psi
Net concrete stresses	+2364 psi	0 psi

If this combination of prestressing is allowed, the value of f_c' required for conformance to the recommendations is 2364/0.40 = 5900 psi. This value is too high for general use and the member must be redesigned with a larger top flange, in order to resist the compressive stresses or the tensile stresses in the top flange due to prestressing must be increased.

Since the top-fiber tensile stress resulting from the effective pre-tensioning is confined to −300 psi, in order to satisfactorily revise the prestressing so that the value of f_c' can be held to 5000 psi, the post-tensioning must develop −440 psi. In this manner, the top-fiber compressive stress can be confined to 2000 psi when the member is subjected to maximum loading conditions. Assuming the center of gravity of the post-tensioning tendons is 4.40 in. above the bottom of the beam (e = 15.9 in.), the force required to develop the required tension of −440 psi in the top fibers and the corresponding compressive stress in the bottom fibers are

$$f_t = -440 = \frac{P}{560}\left(1 - \frac{15.9}{9.06}\right)$$

$$P = 328 \text{ k}$$

$$f_b = \frac{328{,}000}{560}\left(1 + \frac{15.9}{11.03}\right) = +1430 \text{ psi}$$

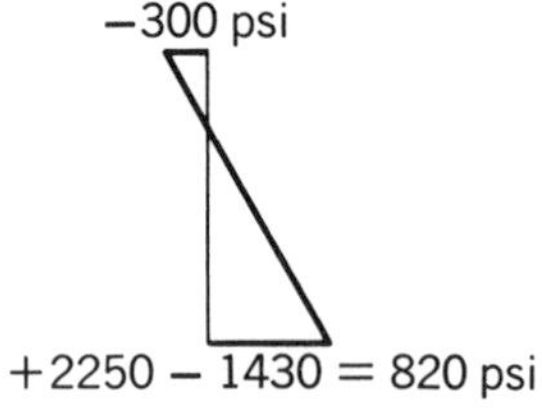

Fig. 7-16

Hence, the required stress distribution due to the supplementary effective pre-tensioning is as shown in Fig. 7-16 and the P and e required are as follows:

$$\frac{P}{A} = 1120 \times \frac{24.7}{45.0} - 300 = 315 \text{ psi}, \qquad P = 117 \text{ k}$$

$$e = \left(\frac{820}{315} - 1\right) 11.03 = 17.6 \text{ in.}$$

The net concrete stresses resulting from this combination of prestress are summarized as follows:

	Top Fiber	*Bottom Fiber*
Stresses due to M_T	+2740 psi	−2250 psi
Stresses due to post-tensioning	−440 psi	+1430 psi
Stresses due to pre-tensioning	−297 psi	+822 psi
Net concrete stresses	+2003 psi	+2 psi

This problem illustrates the fact that the capacity and method of prestressing used in a beam may be limited by compressive stresses in the top fiber of the beam. This condition is particularly acute in beams of long span with narrow top flanges.

7-8 Estimating Prestressing Force and Cross-Sectional Characteristics

In Sec. 4-4, it was shown that for a simple beam that has zero bottom-fiber stress in the loaded state, the total moment that the beam can withstand is expressed by

$$M_T = M_D + M_{SL} = P(e + r^2/y_b) \tag{7-6}$$

In a similar manner, if the stress in the top fiber due to prestressing alone is to be zero, it can be shown that the eccentricity of the prestressing force must not be greater than r^2/y_t below the center of gravity of the section. For this limitation of the concrete stress due to the effective prestress, the relationship of Eq. 7-6 becomes

$$M_T = M_D + M_{SL} = P(r^2/y_t + r^2/y_b) \tag{7-7}$$

These relationships are useful in making preliminary designs since, by assuming values for M_D and P, the required value of the quantity $(e + r^2/y_b)$ can be computed. In employing these relationships, the designer must bear in mind the following fundamental factors:

(1) Most economical designs of simple beams result in values of P/A (the average compressive stress) between 800 and 1200 psi. This gives a means of

making a rough check on the estimated dead weight of the beam without assuming a specific cross section.

(2) The dead weight of the beam itself is a small portion of the total load for short-span beams, whereas for long-span beams, the dead load of the beam itself is of great importance.

(3) When straight tendons are used in short-span members, the value of the quantity $(e + r^2/y_b)$ approaches the lower limit given in Eq. 7-7 as $(r^2/y_t + r^2/y_b)$.

(4) When curved or draped tendons are used in beams of moderate to long spans, the value of e is frequently limited by the dimensions of the concrete section, rather than by top- or bottom-fiber stresses, and Eq. 7-6 approaches

$$M_T = P(y_b + r^2/y_b) \tag{7-8}$$

(5) When tensile stresses are allowed in the bottom fibers in the fully loaded state, the pressure line goes higher than r^2/y_b above the center of gravity of the section and the relationship given by Eq. 7-6 can be rewritten

$$M_T = \psi P(e + r^2/y_b) \tag{7-9}$$

The value of ψ to be assumed in the above relationship must be estimated by considering the magnitude of the allowable bottom-fiber tensile stress with respect to the bottom-fiber stress resulting from prestressing alone. For example, if the bottom fiber stress due to the effective prestress must be confined to 2000 psi and the allowable tensile stress in the bottom fiber is 400 psi, the value of $(e + r^2/y_b)$ must be increased by the ratio $\psi = 2400/2000 = 1.20$ to give an accurate estimate of the movement of the pressure line that will take place when the beam is loaded from the condition of zero bottom-fiber stress (due to external loads) to the point where the stress in the bottom fiber is -400 psi. If the bottom-fiber stress due to the effective prestress alone is as high as 3000 psi, as it frequently is in long-span, post-tensioned members, and allowable tensile stress of 400 psi in the bottom fiber would result in a ratio for ψ of only 3400/3000, which is equal to 1.13.

(6) The value of the term $(e + r^2/y_b)$ varies from 33 to 80% of the depth of the beam, depending upon the efficiency of the cross section, the allowable stresses, and the dead-load moment. Average values of this factor for use in estimating the preliminary design of roof and bridge girders are between 60 and 75% of the depth of the member, with the larger values being applicable to the longer spans and to members with relatively large flanges.

(7) The average value for depth-to-span ratio for most simple beams can be assumed to be 1/20. This ratio does vary between relatively wide limits, but for simple beams it is rarely greater than 1/16 or less than 1/24, except for solid and cored slabs.

ILLUSTRATIVE PROBLEM 7-9 Estimate the depth of a beam required to carry a superimposed moment of 800 k/ft on a span of 70 ft, using curved tendons.

SOLUTION

Assume the weight of the girder will be 400 plf.

$$\text{Total moment} = 0.40 \times \frac{70^2}{8} + 800 = 1045 \text{ k/ft}$$

If $(e + r^2/y_b) = 0.70\, d$ and $P = 400$ k,

$$d = \frac{1045}{0.70 \times 400} = 3.73 \text{ ft}$$

This amounts to a depth-to-span ratio of 1 to 18.8. If $P/A = 1000$ psi, $P = 400$ k, and $A = 400$ in.^{2}, then the weight of the girder will be 415 plf. The estimated dead weight of the beam is reasonably close to the assumed value, but the depth is somewhat greater than normal for this span. Therefore, try $P = 450$ k, $A = 450$ in.^{2} and $w_d = 470$ plf.

$$\text{Total moment} = 0.47 \times \frac{70^2}{8} + 800 = 1088 \text{ k/ft}$$

$$d = \frac{1088}{0.70 \times 450} = 3.46 \text{ ft}$$

The depth-to-span ratio for this prestressing force is 1 to 20.2, which is reasonable and slightly more slender than average. It is apparent that a preliminary estimate can be made with the data developed here, since the magnitude of the prestressing force and the concrete quantity are known approximately.

ILLUSTRATIVE PROBLEM 7-10 For the conditions stated for Problem 7-9, assume that the depth of the beam cannot exceed 3.5 ft, due to headroom restrictions. Estimate the prestressing force required as well as determine a preliminary cross-sectional shape. Assume that the member is to be post-tensioned, $f_c' = 5000$ psi, and that no tensile stresses are to be allowed in the bottom fibers. Check the estimate.

SOLUTION:

Assume

$$w_d = 0.50 \text{ and } (e + r^2/y_b) = 0.70\, d$$

$$M_T = 0.50 \times \frac{70^2}{8} + 800 = 1106 \text{ k/ft}$$

$$P = \frac{1106}{0.70 \times 3.5} = 452 \text{ k}$$

For a beam 42 in. deep, the web thickness is normally about 7 in. The average stresses in the flanges can be estimated at 2000 psi for the purpose of selecting dimensions of the trial section. For the top flange, the total force to be resisted is 452 k, since the pressure line will be quite high when the beam is fully loaded. Furthermore, the width of the top flange of a member that is 70 ft long should be about 70/35 or 2 ft wide. Therefore, the thickness of the top flange can be computed by

$$t = \frac{452{,}000}{2000 \times 24} \cong 9.40 \text{ in.}$$

The bottom flange must resist a smaller force than the top flange, since the dead load of the beam is acting at the time of stressing. The force the bottom flange must resist can be approximated by multiplying the estimated prestressing force by the ratio of the moment due to the superimposed load and the moment due to the total load, or

$$452 \text{ k} \times \frac{800}{1106} \cong 330 \text{ k}$$

Assuming the width of the bottom flange is to be 18 in., the thickness of the bottom flange can be computed by

$$t = \frac{330{,}000}{2000 \times 18} \cong 9.20 \text{ in.}$$

The assumed trial section is shown in Fig. 7-17, where the estimated values of the thicknesses for the top and bottom flanges calculated above are shown superimposed.

To check the estimated prestressing force and concrete area, the section properties of the trial section are computed to be as follows:

$$A = 576.5 \text{ in.}^2 \qquad I = 115{,}680 \text{ in.}^4$$

$$y_t = 19.7 \text{ in.} \qquad S_t = 5860 \text{ in.}^3 \qquad r^2/y_t = 10.2 \text{ in.}$$

$$y_b = 22.3 \text{ in.} \qquad S_b = 5180 \text{ in.}^3 \qquad r^2/y_b = 9.00 \text{ in.}$$

$$M_D = 0.60 \times \frac{(70)^2}{8} = 368 \text{ k/ft}$$

$$M_{SL} = 800 \text{ k/ft}$$

$$M_T = 1168 \text{ k/ft}$$

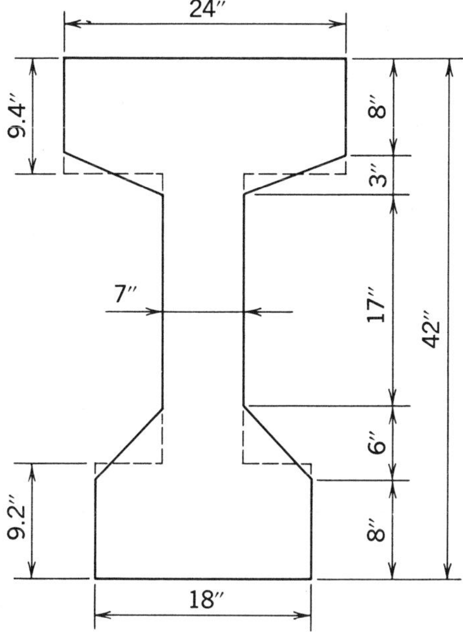

Fig. 7-17

The section is reviewed as follows:

Assume $e = 22.3$ in. $-$ 4.5 in. $= 17.8$ in.

$$P = \frac{1168 \text{ k/ft} \times 12}{17.8 + 9.0} = 523 \text{ k}$$

	Top Fiber	*Bottom Fiber*
Stress due to M_D	+753 psi	−853 psi
Stress due to M_{SL}	+1640 psi	−1850 psi
Total	+2393 psi	−2703 psi
Stress due to P	−680 psi	−2700 psi
Net Stress	+1713 psi	−3 psi

Examination of these stresses will reveal that the net compressive stress in the top fiber is 1713 psi when the beam is under full load. This value is substantially below the value of 2000 psi, which is allowable for the assumed concrete strength of 5000 psi. In addition, if f'_{ci} is to be 4000 psi, since the dead load of the beam is acting at the time of stressing, the 1850 psi compression in the bottom fibers is below the allowable initial stress. Therefore, the area of the flanges can be reduced.

In addition to considerations of stress, the following factors must be considered in selecting the final shape of the section.

(1) Flanges must not be so thin that they might be broken during handling and transportation of the members.

(2) The top flange should have sufficient width to protect against undue lateral flexibility during transportation and erection, as well as to ensure that the flange will not buckle under load if it is to be used in the completed structure without supplementary lateral support.

(3) The bottom flange must be of such shape that the prestressing tendons can be positioned with adequate cover and spacing to protect them against corrosion and to facilitate placing of the concrete, and, when post-tensioning is used, the shape of the bottom flange must allow curving of the tendons up into the web while the minimum cover is maintained.

(4) For reasons of economy, the shape should be as simple as possible to facilitate the fabrication.

(5) The slopes provided as transitions between the flanges and the webs should be of such size and shape that danger of honeycomb in the bottom

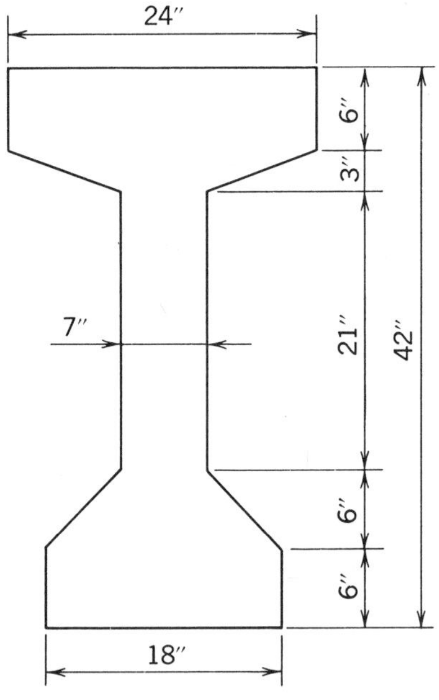

Fig. 7-18

flange is minimized. In addition, stripping of the form is facilitated by large slopes on the flanges.

In this example, it is assumed that the flanges can be reduced to 6 in. in thickness at their extremities, as shown in Fig. 7-18, and the revised section properties and stresses are computed as follows:

$$A = 502.5 \text{ in.}^2 \qquad I = 108{,}090 \text{ in.}^4$$

$$y_t = 19.9 \text{ in.} \qquad S_t = 5430 \text{ in.}^3 \qquad r^2/y_t = 10.4 \text{ in.}$$

$$y_b = 22.1 \text{ in.} \qquad S_b = 4900 \text{ in.}^3 \qquad r^2/y_b = 9.39 \text{ in.}$$

$$M_D = 0.542 \times \frac{(70)^2}{8} = 332 \text{ k/ft}$$

$$M_{SL} = 800 \text{ k/ft}$$

$$M_T = 1132 \text{ k/ft}$$

The section is evaluated as follows:

Assume $e - 22.1 \text{ in.} - 4.5 \text{ in.} = 17.6 \text{ in.}$

$$P = \frac{1132 \times 12}{9.39 + 17.6} = 503 \text{ k}$$

	Top Fiber	*Bottom Fiber*
Stress due to M_D	+735 psi	−814 psi
Stresses due to M_{SL}	+1770 psi	−1960 psi
	+2505 psi	−2774 psi
Stresses due to P	−665 psi	+2770 psi
Net stresses	+1840 psi	−4 psi

It should be noted that the prestressing force required in the final design is about 11% higher than the preliminary estimate and the concrete quantity is about 4% higher in the final design. These errors are the result of an error in the assumed value of $(e + r^2/y_b)$, which was assumed to be $0.70d$ and which is only $0.644d$ in the final design.

The initial stress in the bottom fiber should be checked. Assuming $f_{se}/f_{si} = 0.85$, the initial bottom-fiber stress is approximately +2445 psi, in which case, the value of f'_{ci} should be 4450 psi, if the initial compression is confined to $0.55\,f'_{ci}$, and 4100 psi and if the maximum allowable initial compression is $0.60\,f'_{ci}$.

7-9 Reduction in Shear Force Due to Curvature of Parabolic Tendons

In Sec. 4-6, it was shown that the curvature of prestressing tendons results in a reduction in the shear force the concrete must withstand. Furthermore, it was shown that this reduction is equal to the vertical component of the prestressing force at the point under consideration. The vertical component of the prestressing force is equal to $P \sin \alpha$, in which α is the angle of inclination of the tangent to the prestressing tendon at the point under consideration.

Because the angle α is small in almost all instances, the sine and tangent are practically equal. Hence, the tangent can be used in computing the vertical component of the prestressing force without introducing significant error.

The computation of the tangent of the angle of inclination for tendons placed on a series of chords is basic and requires no explanation. For tendons on parabolic curves, the computation of the tangent of the angle of inclination is equally simple, when the properties of a parabola are understood.

A parabola is shown in Fig. 7-19 with the dimensions and tangents that are most important in the analysis of prestressing shear forces. It will be seen from the figure that the tangent to the parabola at the end is at an angle of α to the horizontal and the tangent of the angle α is equal to

$$\tan \alpha = \frac{2E}{L/2} = \frac{4E}{L} \tag{7-10}$$

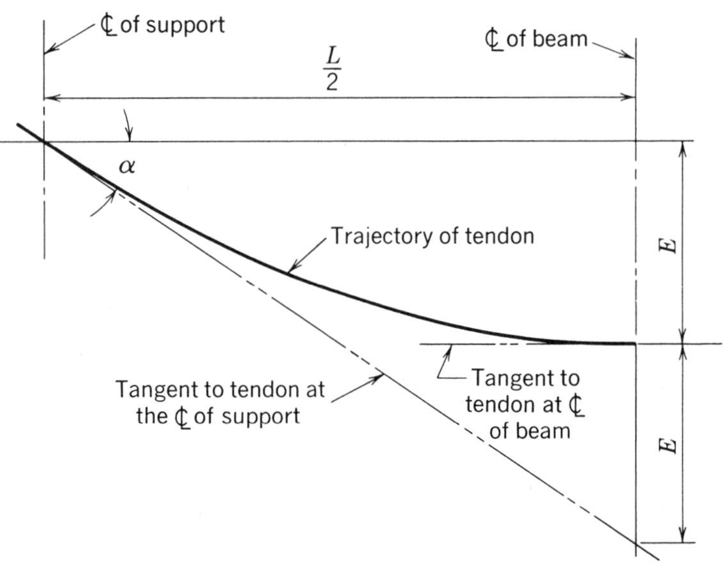

Fig. 7-19 Diagram illustrating fundamental properties of a parabola.

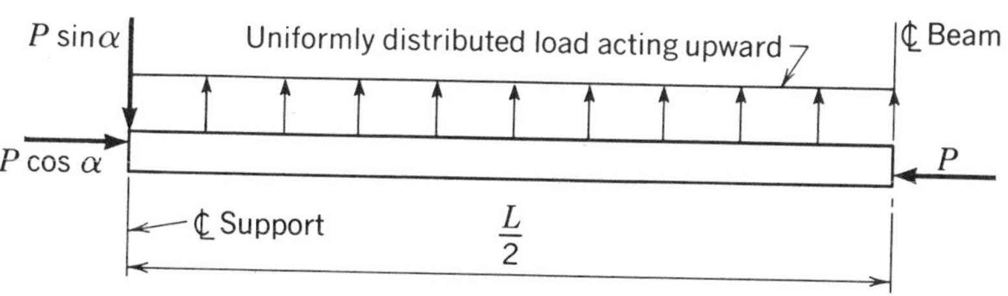

Fig. 7-20 Free-body diagram of the forces on a beam prestressed by a parabolic tendon.

The dimension E is the total displacement of the prestressing force and is equal to the normal eccentricity of the force only when the eccentricity of the prestressing force is zero at the ends. The units of E and L must be the same.

It is apparent from the free-body diagram of Fig. 7-20, in which the forces that act on the concrete as a result of stressing with a parabolic tendon are shown, that the shear stresses resulting from the tendon vary uniformly from a maximum value at the end to zero at midspan, just as the shear stresses in a simple beam subjected to a uniform load vary from zero at midspan to a maximum value at the support. Hence, the vertical component of the shear stress carried by the tendon at points between the end and the midspan of the beam can be determined by the following relationship:

$$P \sin \alpha_x = P \sin \alpha \times \frac{(L/2 - x)}{L/2} \tag{7-11}$$

in which x is the distance from the support to the point under consideration.

Although the relationships presented here are derived for tendons on parabolic trajectories, they can normally be applied to tendons on other curved trajectories encountered in prestressed concrete, without introducing significant error. If the displacement of a tendon is very large in comparison to the span, as is sometimes the case in post-tensioned folded plates or shells, it is advisable to compute the reduction in shear using the sine of the actual angle as determined from the tendon layout.

An example of the computation of the shear component carried by a tendon on a parabolic curve is given in Prob. 4-10.

7-10 Computing the Location of Pre-tensioning Tendons

The selection of the location or pattern of the pre-tensioning tendons must be made after the cross-sectional shape, the prestressing force, and the eccentricity have been determined. This is done by trial and error and can generally be accomplished quickly if the trial and error computations are made according to a specific procedure. This procedure consists of first determining the number of tendons required by dividing the required effective prestressing force by the maximum allowable effective prestressing force that one tendon can withstand. Secondly, locate the position of the required number of tendons, by computing moments of the tendons at assumed locations, that are adjusted by trial and error, until the center of gravity of the tendons is at the desired location. The procedure can be illustrated by the sketch of Fig. 7-21 which represents the cross section of a beam that requires N tendons placed with their center of gravity at a distance d' from the bottom of the beam. If n_1, n_2, $n_3 \ldots n_n$ represent the number of tendons in the first, second, third, and the nth rows from the bottom of the beam and y_1, y_2, y_3, $\ldots y_n$ represent the distances from these rows to the bottom of the beam, respectively, it is apparent that in order to obtain the desired location of the center of gravity, the following relationship must be satisfied

$$Nd' = \sum (y_1 n_1 + y_2 n_2 + y_3 n_3 + \cdots y_n n_n) \tag{7-12}$$

Since the values of N and d' are known, the majority of the tendons can be

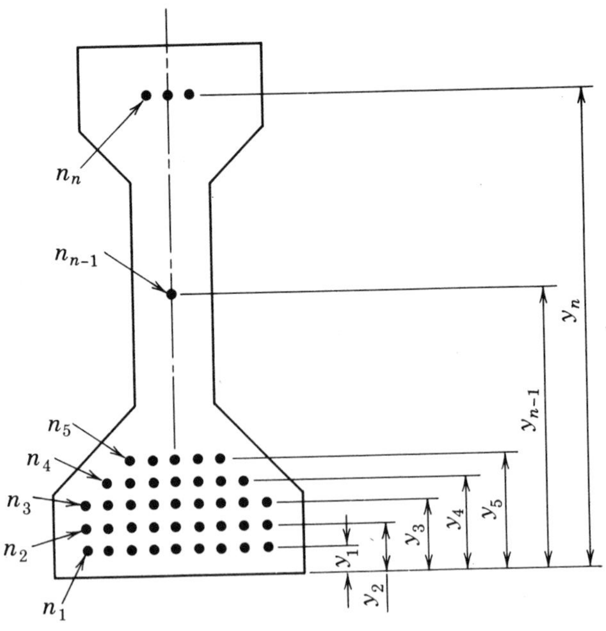

Fig. 7-21

located and the value of $\sum (y_1 n_1 + y_2 n_2 + y_3 n_3, \ldots y_n n_n)$ adjusted to the desired value with the remaining tendons.

It is apparent that the majority of the tendons will be near the bottom of the member in order to achieve the required eccentricity. It is desirable that some tendons be supplied near the top of most members for the purpose of supporting web reinforcing, inserts, and other accessories. This can frequently be done with the required number of tendons and without supplying additional tendons specifically for this purpose.

In tendon patterns which have some tendons high in the section, the upper tendons should be disregarded when computing the ultimate moment capacity of the section. The distance to the center of gravity of the lower group of tendons, which would be very highly stressed at ultimate load, must be determined for use in calculating the ultimate moment capacity.

ILLUSTRATIVE PROBLEM 7-11 Compute the location of the tendons required to produce a prestressing force of 538 k at an eccentricity of 11.9 in. in the AASHO-PCI bridge beam, type III, if the tendons to be used have an effective force of 11 k each and $y_b = 20.3$ in.

SOLUTION: The number of tendons required is: 538/11 = 49 each. The distance from the bottom of the beam to the center of gravity of the steel is: $d' = 20.3 - 11.9 = 8.4$ in. $Nd' = 49 \times 8.4 = 412$. The summation of the moments of the tendons in their final location should equal 412. Forty-five of the tendons are tentatively positioned as shown in Fig. 7-22 and the moment of the tendons computed about the bottom of the section is equal to 278. Therefore, the remaining four tendons must have an average distance from the bottom of the beam equal to:

$$\text{Average } y = \frac{412 - 278}{4} = 33.5 \text{ in.}$$

This average distance is of the order of 75% of the depth of the beam and it appears that the tendon layout should be adjusted in order to include one more tendon in the bottom group. Therefore, the pattern is revised by increasing the number of tendons in the sixth row, from the bottom of the section, to 5 and reducing the number of tendons in the seventh row, to 2. The moment of the 45 tendons in the revised tendon pattern is 287.25. The average distance required for the 3 remaining tendons is computed as follows:

$$\frac{412 - 287.25}{3} = 41.5$$

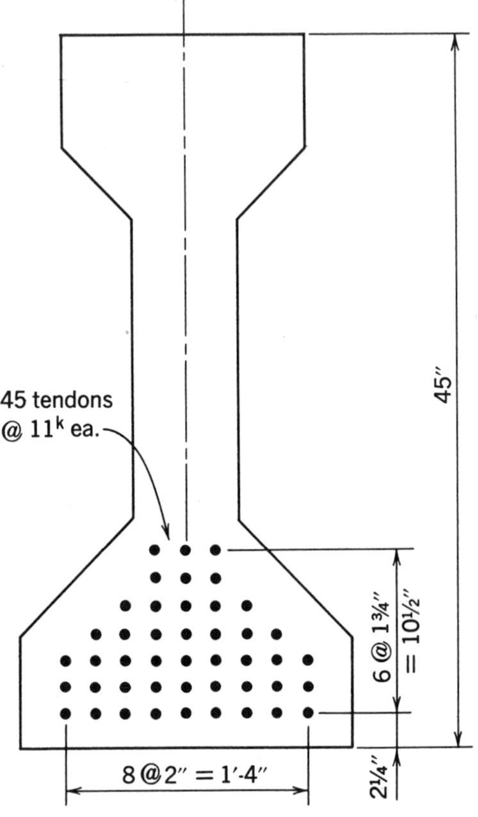

Fig. 7-22

The final tendon layout is illustrated in Fig. 7-23. The values of d' and d'_u are computed as follows:

$$d' = \frac{411.75}{49} = 8.40 \text{ in.}$$

$$d'_u = \frac{287.25}{46} = 6.25 \text{ in.}$$

The ultimate moment should be computed on the basis of 46 tendons, 6.25 in. from the bottom of the beam.

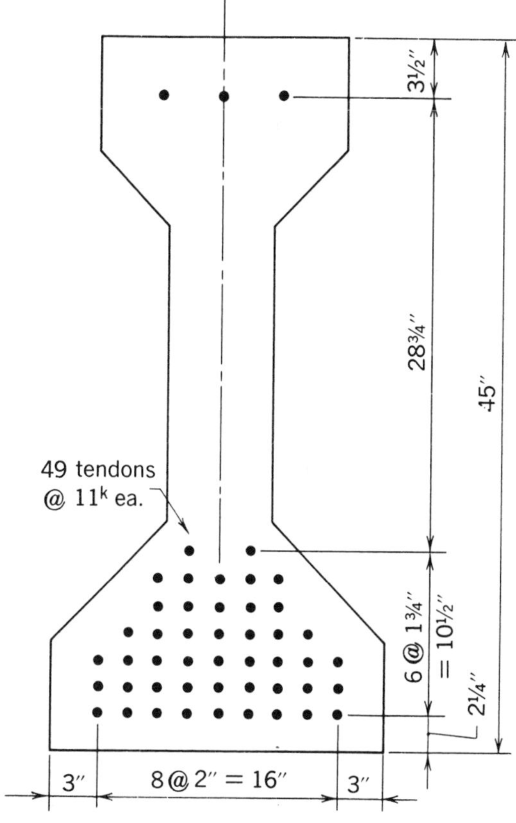

Fig. 7-23

7-11 Fiber Stresses at Ends of Prismatic Beams

In employing bond prevention or in using non-prestressed reinforcing as a means of controlling the stresses resulting from the initial prestress at the end of a beam, it is necessary to analyze the flexural stresses resulting from the dead weight of the beam near the end of the beam, in order to determine the limits over which the bond must be prevented or over which the special end reinforcement should be provided. This can be done using the fundamental principles of strength of materials, through the use of factors selected from unit parabolic curves, or by computing the location of the required dead load stresses through the use of the known properties of parabolas. Each of these methods should yield identical results. The use of the latter method is shown in the following problem.

ILLUSTRATIVE PROBLEM 7-12 Compute the length over which non-prestressed reinforcing is required at the ends of a simple prismatic beam in which the top-fiber stresses due to initial prestressing are equal to −360 psi and the maximum, allowable tensile stresses without non-prestressed reinforcing is −190 psi. The stress in the top fibers at midspan of the beam due to dead load alone is +730 psi. The beam is 70 ft long.

SOLUTION: The stress due to dead load in the top fiber varies parabolically as shown in Fig. 7-24. It is required to determine the distance from the end of the beam, where the top-fiber stress is 360 − 190 = 170 psi, since at this location, the net concrete stress will be −190 psi, which can be allowed without non-prestressed reinforcement. The ordinates of the parabola vary according to the relationship:

$$f_{\mathrm{¢}} - f_x = \left(\frac{L/2 - x}{L/2}\right)^2 f_{\mathrm{¢}}$$

or

$$x = \frac{L}{2}\left[1 - \left(\frac{f_L - f_x}{f_{\mathrm{¢}}}\right)^{1/2}\right]$$

Using the values given in the example

$$x = 35 \text{ ft}\left[1 - \left(\frac{730 - 170}{730}\right)^{1/2}\right] = 35 \text{ ft} \times 0.125 = 4.37 \text{ ft}$$

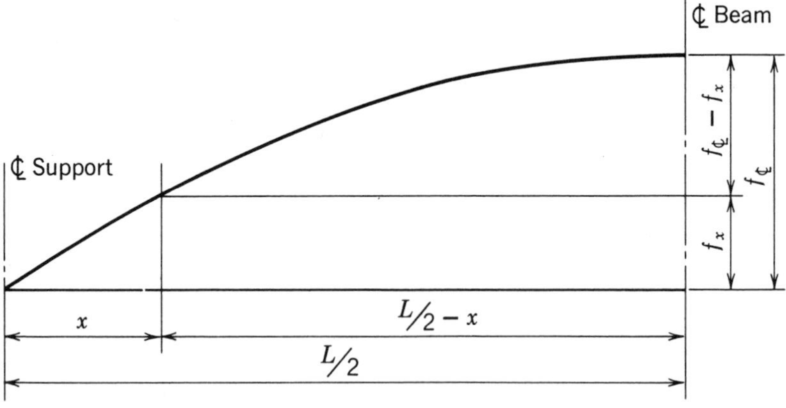

Fig. 7-24 Diagram used to compute stresses at different locations along the length of a beam having a parabolic moment diagram.

7-12 Computating the Effects of Bond Prevention

When the initial prestressing stresses at the ends of a simple pre-tensioned beam exceed the allowable stresses, the effect of the prestressing can be reduced by preventing bond on a specific number of tendons over a specific length, as is explained in detail in Sec. 6-11. The length over which the bond must be prevented can be computed according to the methods suggested in Sec. 7-11. The number and location of tendons that should be prevented from bonding to the concrete can be determined by computing the effect of one tendon in each of the lower rows of the tendon pattern and then, by trial and error, determining the number of tendons in each row that should be prevented from bonding to the concrete.

7-13 Organization and Abbreviation of Computations

Each designer develops his own methods for organizing his computations and in taking short cuts to facilitate and expedite the computations. In addition to the methods already enumerated, the author has found it convenient and expeditious to abbreviate the computation of stresses due to prestressing as follows:

$$f_t/f_b = \frac{P}{A}\begin{pmatrix} 1 - \dfrac{ey_t}{r^2} \\ \\ 1 + \dfrac{ey_b}{r^2} \end{pmatrix}$$

This expression is used in lieu of writing

$$f_t = \frac{P}{A}\left(1 - \frac{ey_t}{r^2}\right)$$

and

$$f_b = \frac{P}{A}\left(1 + \frac{ey_b}{r^2}\right)$$

Because the values of r^2/y_t and r^2/y_b are computed at the time the section properties are determined, and since e and P/A are the same values in each equation, the computation of the quantities $[1 - (ey_t/r^2)]$ and $[1 + (ey_b/r^2)]$ can be rapidly performed on the slide rule by dividing the value of e first by r^2/y_t and adding 1(algebraically) to the result and then dividing e by r^2/y_b and adding 1 to the result.

Another method which the author uses in computing and summarizing the stresses in the top and bottom fibers of member that are not symmetrical is to write

$$f_t/f_b = \frac{M}{S_t/S_b}$$

in lieu of

$$f_t = \frac{M}{S_t}$$

and

$$f_b = \frac{M}{S_b}$$

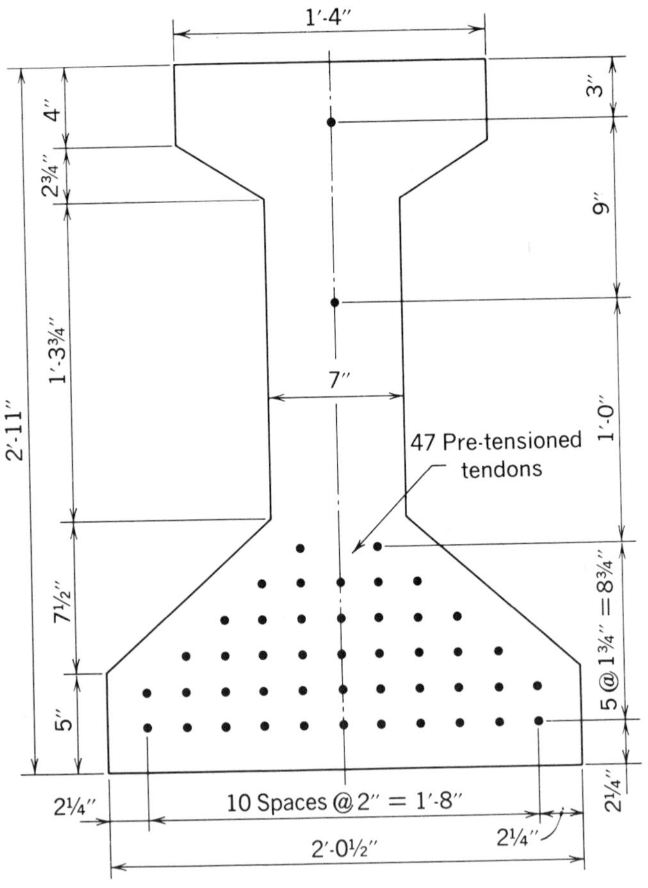

Fig. 7-25

This procedure abbreviates the computations and allows the result to be summarized in tabular form as it is obtained. These computation methods are illustrated in the following problem.

ILLUSTRATIVE PROBLEM 7-13 Compute the stresses due to prestressing and due to an external moment of 600 k-ft for the beam section shown in Fig. 7-25, if the member is prestressed with 47 tendons stressed at 13 k each at $e = 8.40$ in. For the cross section shown in Fig. 7-25, the following section properties are given

$$\frac{r^2}{y_t} = 6.50 \text{ in.} \qquad S_t = 2880 \text{ in.}^3 \qquad A = 445 \text{ in.}^2$$

$$\frac{r^2}{y_b} = 8.99 \text{ in.} \qquad S_b = 3970 \text{ in.}^3$$

SOLUTION:

$$\text{Prestress:} \quad f_t/f_b = \frac{47 \times 13{,}000}{445}\begin{pmatrix} -0.29 \\ +1.935 \end{pmatrix} = \begin{matrix} -398 \text{ psi} \\ +2660 \text{ psi} \end{matrix}$$

$$\text{Moment:} \quad f_t/f_b = \frac{600 \times 12{,}000}{2880/3970} = \begin{matrix} +2500 \text{ psi} \\ -1815 \text{ psi} \end{matrix}$$

REFERENCES

1. Magnel, G. "Prestressed Concrete," pp. 20–25, Concrete Publications, Ltd., London, 1948.
2. ACI Committee 318. "Proposed Revision of ACI 318–63: Building Code Requirements for Reinforced Concrete," *Journal of the American Concrete Institute*, **67**, No. 2, 77–186 (Feb. 1970).
3. The American Association of State Highway Officials. "Standard Specifications for Highway Bridges," Published by the Association General Offices, Washington, D.C., 1969.

8 Continuity in Prestressed Concrete Flexural Members

8-1 Introduction

The theoretical reduction in moments, stresses and hence cost of materials, achieved through the use of continuity in prestressed concrete, is comparable to that which can theoretically be made with other structural materials. The actual economy of materials and cost of construction resulting from the use of continuity is greatly influenced by the design criteria used, the magnitude of the spans involved, the type of structure under consideration, the type of loading to be carried by the structure in question, and the methods of prestressing available.

The economy associated with the majority of prestressed structural elements would be non-existent if the elements were not precast. Although precast elements can be field-connected in order to form fully continuous prestressed members, this procedure has proved economical only under special conditions.

One of the important uses of cast-in-place, continuous prestressed concrete in the United States has been in the construction of roof and floor slabs. The slabs may be flat plates which are continuous in both directions or may be "one-way" slabs supported by beams or walls. Another important use of cast-in-place continuous prestressed concrete members has been in the construction of highway and railroad bridges. Continuity has been used extensively in cast-in-place long span bridge construction in the Western United States.

Precast, prestressed elements, which are rendered continuous by cast-in-place concrete and normal reinforcing steel, have been widely used. In this type of construction, the precast, prestressed elements act as simple beams resiting a portion of the dead load, but the live load, as well as the dead load which is applied after the hardening of the cast-in-place concrete, is carried by the continuous beam. This type of construction has proved to be economical in the American market, and it is expected that the use of this method will continue. The method results in a structure with the fundamental advantages of precast, prestressed concrete, as well as those resulting from the use of continuity. The basic structural analysis of this type of construction does not involve principles unfamiliar to structural engineers.

Continuous prestressed spans frequently have depth-to-span ratios of the order of 1 to 30 for prismatic members and as little as 1 to 80 at the section of minimum depth in members which have variable depths. The greater rigidity of continuous prestressed members also results in less vibration from moving or alternating loads. A significant advantage gained through the use of continuity in prestressed concrete construction, as is the case with other materials, is the reduction of deflections. Over-all structural stability and resistance to longitudinal and lateral loads is normally improved through the introduction of continuity, in any structural system.

8-2 Disadvantages of Continuity

The construction of cast-in-place, continuous, prestressed-concrete box girder bridges of conventional design do not present any unusual or serious construction problems. The more sophisticated methods of constructing continuous prestressed bridges, such as cantilevered construction (*see* Sec. 8-13), require a great deal of technical skill on the part of the contractor. Additionally, these methods frequently require more construction labor than is required with conventional construction methods. These factors have deterred the use of almost all of the more sophisticated methods of constructing continuous prestressed concrete members in the United States. Except for very special cases, it is expected this situation will continue as the industry seeks methods of reducing the amount of labor, as well as the degree of skill required in construction.

Many engineers are under the impression that continuous prestressed structures are difficult to design and analyze, because of the secondary moments that result from the deformation of the structure during prestressing. As will be seen in the following discussion, the analysis of continuous prestressed structures is not particularly complex and involves only the familiar principles used in the analysis of ordinary statically indeterminate structures.

8-3 Methods of Framing Continuous Beams

There are a number of different methods that have been used or suggested for framing continuous beams. The possible variations are infinite and are limited only by the imagination of the designer. In general, these methods can be divided into three categories, cast-in-place monolithic structures, structures formed of precast elements, which may or may not have cast-in-place joints, and supplementary elements and members constructed of precast or cast-in-place segments utilizing the cantilever method of erection.

The cast-in-place monolithic, structure is generally used on the longer spans where the use of falsework is feasible and where precast beams cannot be used with ease, because of the dead weight of the beam and resulting difficulties in transportation and erection. Such construction is generally confined to one of the five types illustrated in Fig. 8-1. The construction can be described briefly by the following characteristics.

(1) Prismatic beam with curved tendon (Fig. 8-1(a)). This type of beam is simple to analyze, since the moments can be computed without the mathematical complications that result from a variable moment of inertia. This type of beam requires relatively simple formwork and has a relatively high friction loss. In this mode of framing, the concrete is not used as efficiently as it might be, but the over-all economy may remain good in spite of this, due to the simplicity of the member. Designs of this type normally have good appearance.

(2) Beams with variable depth and straight tendons (Fig. 8-1(b)). The structural analysis of such a beam is complicated by the variable moment of inertia and difficulty is frequently experienced in determining the beam proportions that will result in the desired balance of moments, due to dead and live loads, in combination with the moments, due to prestressing. Although this shape of beam results in minimum friction loss in the tendons for a specific length of structure, the formwork required to construct beams of this type is somewhat more complicated than that of prismatic beams. In addition, it is frequently difficult to achieve the required eccentricity of prestress at the several critical sections in this type of framing without resorting to impractical amounts of haunching.

(3) Beams with variable depth and curved tendons (Fig. 8-1(c)). This type of beam, which is among the more complex to analyze, frequently results in the best solution, since the curvature (and hence friction loss) required is generally not as severe as in the case of the prismatic beam, and the proportions of the beam and tendon trajectory can be adjusted during the design, in order to obtain an efficient use of materials.

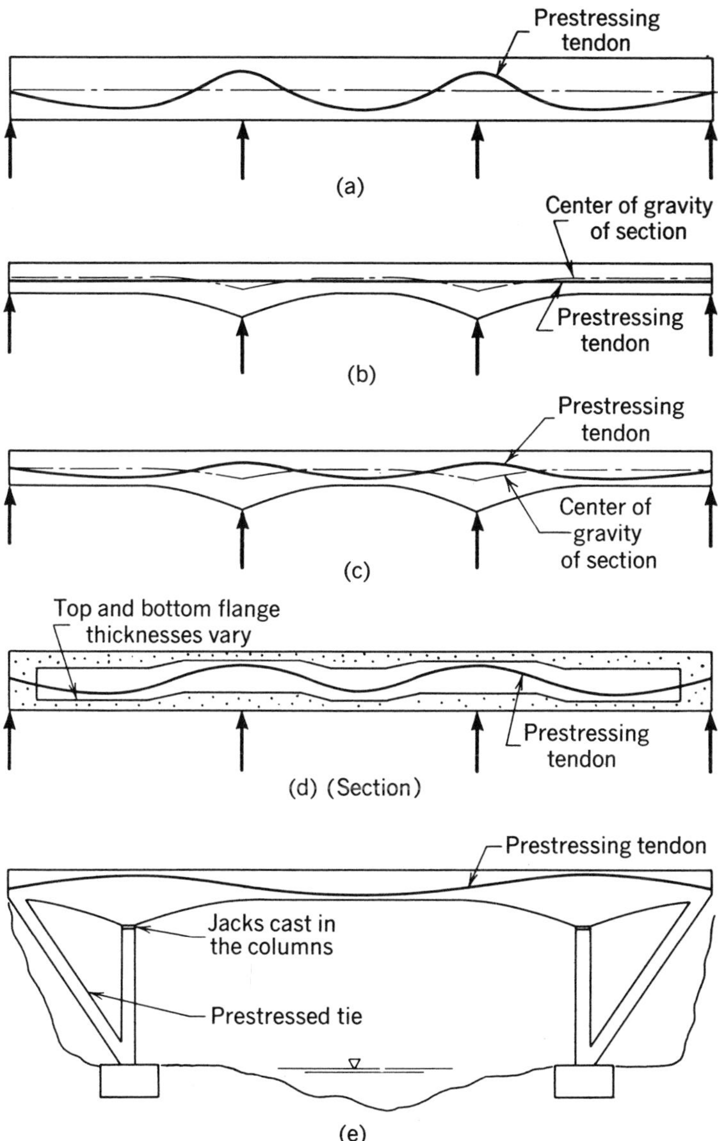

Fig. 8-1 Typical monolithic framing.

(4) Beams of uniform depth and variable moment of inertia (Fig. 8-1(d)). The structural efficiency and advantages obtained from a variable moment of inertia can be achieved without loss of the appearance of a prismatic beam by employing a beam that has uniform depth and variable flange and web thickness. Such members are easily formed from a hollow box section with the variable dimensions on the inside. The amount of curvature of the prestressing tendons is not normally excessive in such designs and the prestressing force can be made variable quite easily and without excessively complicating the construction procedure.

(5) Beams with special substructures used to induce controlled end moments (Fig. 8-1(e)). The design and construction procedure, as well as the maintenance of this type of structure, is influenced by the plastic properties of the concrete to a more significant degree than in structures prestressed by tendons alone. The time-dependent deformation of the concrete in the substructure has a large effect on the end moments and end reactions. The design of structures of this type requires a good knowledge of the physical properties of the concrete to be used in the construction. The analysis is similar in nature to the type used in determining the effects of differential settlement.

Precast elements can be used to construct continuous structures in a number of ways. The use of precast elements is generally more feasible in applications that have short to medium spans, since there is little difficulty in transporting and erecting beams or elements of moderate length. Precast beams as long as 250 ft have been shown to be practical and economical under certain conditions on large multi-span bridges. Various schemes utilizing precast elements are shown in Fig. 8-2. The characteristics of the various schemes can be listed as follows:

(1) Overlapping beams in areas of high moment (Fig. 8-2(a)). A method that has been used in bridges abroad is characterized by overlapping precast beams in the areas of high negative moment. An advantage to this method is that no high-quality concrete must be produced on the job site, since the job site concrete is confined to diaphragms, shear keys, and deck. A major disadvantage to the method is that, due to the staggered beams, a fascia beam must be provided for architectural purposes.

(2) Precast beams made continuous by prestressing (Fig. 8-2(b)). This method of developing continuity has been used abroad to a limited degree and, although the scheme appears efficient from the theoretical design viewpoint, the short cap cables are difficult to place and stress. A relatively large quantity of anchorage devices are required for the cap cables and a delay period is required for the cast-in-place concrete to gain sufficient strength for stressing.

(3) Continuity developed by ordinary reinforcing steel (Fig. 8-2(c)). This method is equally applicable to prismatic beams and beams with variable depths, and it has been used extensively in the United States in bridge

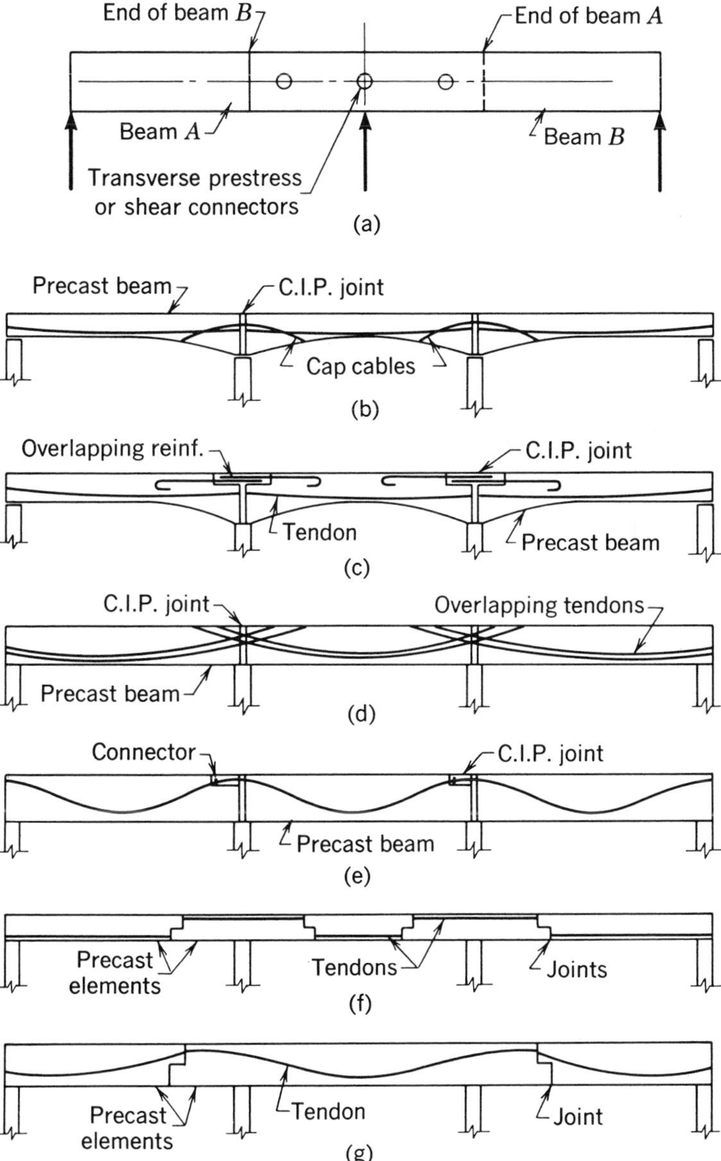

Fig. 8-2 Continuous beams formed of precast elements.

construction. The cast-in-place joint is required to assure uniform stress transfer at the section of negative moment. In order to attain economy of materials with this method, when used with prismatic precast beams, it is necessary to use curved tendons or bond prevention as a means of reducing the bottom-fiber stresses, due to prestressing at the ends of the precast member. No job site post-tensioning is required after erection of the precast elements with this method of framing.

(4) Continuity due to overlapping tendons (Fig. 8-2(d)). The method of developing continuity with tendons that overlap from one beam to the next has been used abroad, but it is not generally considered practical here because of the large amount of job site labor required to erect, thread tendons, stress, and grout. In addition, the continuity is developed for the superimposed load only, and some prestressing or other reinforcing must be provided to resist the dead load of the precast elements during handling and erection.

(5) Continuity developed by coupling tendons (Fig. 8-2(e)). This method has been used to a limited extent in slabs and beams in building construction, but its use is limited to the post-tensioning systems that are efficiently and economically coupled and that have flexible tendons that can withstand relatively sharp curvatures. This basic scheme can also be used without the tendon couplers, in which case, the "continuity" tendon extends from one end of the structure to the other.

(6) Precast elements connected at the inflection points of a continuous structure (Fig. 8-2(f)). Members that are not subject to large reversals of moment and, hence, have relatively stable inflection points, can be composed of simple elements that are stressed with straight tendons and that are joined together at the inflection points with cast-in-place concrete. The number of cast-in-place joints required can be reduced by using curved tendons to form overhanging beams, as illustrated in Fig. 8-2(g).

Precasting can also be applied to all of the schemes and methods indicated in Fig. 8-1, which were previously explained to be practicable for cast-in-place construction, by casting the beams in short elements, assembling them on falsework at the job site, inserting post-tensioning tendons through ducts preformed in the elements, and stressing and grouting the tendons. The joints between the individual elements may have to be packed with mortar during assembly of the precast elements in order to ensure uniform bearing between the elements, unless the elements are made by precasting one against the other.

The cantilever method, in which the segments can either be cast-in-place or precast, is illustrated in Fig. 8-3. In this method, which has been used widely on long span bridge construction in Europe, the members are constructed from the piers towards the center of the spans in increments that are prestressed to the previously completed sections. The members are simple

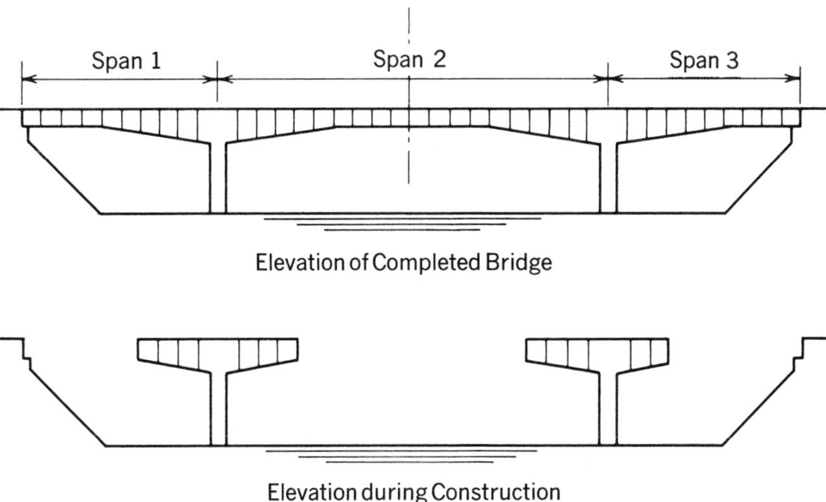

Fig. 8-3 Cantilever bridge construction.

cantilevers until they are joined at the center of the spans, after which, they may be rendered continuous by additional tendons or simply be hinge-connected (Ref. 4).

As was mentioned above, the number of schemes and methods that can be used to develop continuity in prestressed construction is limited only by the imagination of the designer. This is the primary advantage of concrete construction, since the designer need not confine his thinking to standard shapes. A complete discussion of all the schemes that have been used in developing continuity is beyond the scope of this book. The interested reader will find considerable detailed descriptions of various other methods in the literature pertaining to prestressed concrete.

8-4 Continuous Prestressed Slabs

The methods of fabricating continuous prestressed slabs can also be divided into cast-in-place monolithic slabs and precast slabs, in order to simplify consideration of these structural elements. The cast-in-place monolithic slabs, which may be cast-in-place in the conventional manner or which may utilize the lift-slab technique, are most frequently one of the following:

(1) Solid slabs of uniform thickness prestressed in one or in each direction. Slabs of this type are normally from 6 to 10 in. thick and are post-tensioned with curved tendons.

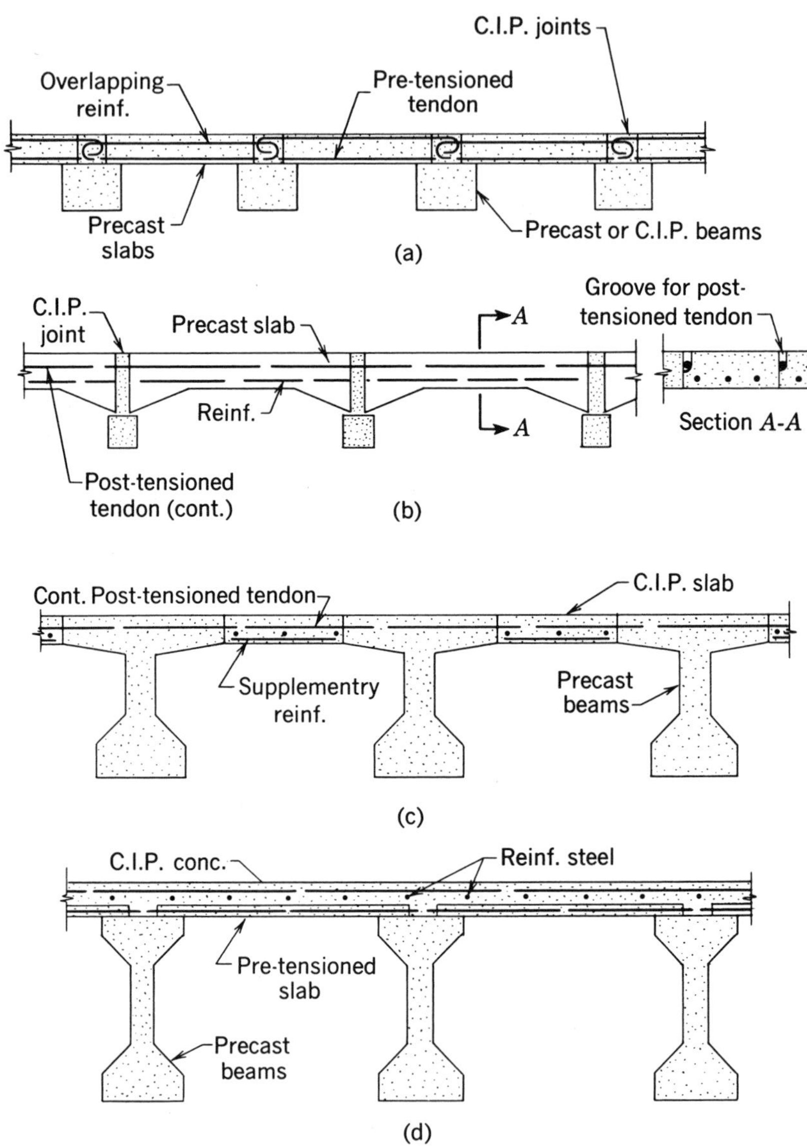

Fig. 8-4 Various types of continuous prestressed slabs.

(2) Monolithic joist and slab construction.
(3) Two-way, monolithic joist and slab construction (waffle slabs).
(4) Hollow slabs of variable or uniform dimensions.

Monolithic slabs are generally prestressed with the tendons placed in the same general shape that is used in monolithic beams and, from the designer's viewpoint, can be considered a special case of a continuous beam.

Precast continuous slabs have been used extensively. The slabs frequently contain prestressed and ordinary reinforcing and can be divided into the following general types:

(1) Solid and hollow slabs, reinforced for positive moments with ordinary reinforcing or with pre-tensioning in which the negative moments are carried by ordinary reinforcing or, in some instances when limit design is used, in which the ends of the members are partially restrained by post-tensioned tendons, which pass through ducts or grooves in the members. These are illustrated in Fig. 8.4(a), (b), and (c).

(2) Slabs, such as bridge decks, which are formed of precast units that carry the positive moments, and cast-in-place elements, which transfer shear and carry negative moment. This is illustrated in Fig. 8-4(d).

8-5 Elastic Analysis of Beams with Straight Tendons

The moments due to dead and live loads in an indeterminate prestressed-concrete structure are calculated using the same classical methods employed in analyzing indeterminate structures composed of other materials. The one significant difference in a prestressed structure is that secondary moments may or may not result from the prestressing. The secondary moments are the result of the deformation of the structure and are also calculated by the usual methods of indeterminate analysis. In most areas of structural design, the term secondary moments denotes undesirable moments that are to be avoided if possible. In prestressed concrete design, the secondary moments are not always undesirable and can be quite helpful. It is essential that the designer be aware that such moments do exist and that they must be included in the design of indeterminate prestressed structures.

In the design and analysis of continuous prestressed beams, the following assumptions are generally made:

(1) The concrete acts as an elastic material within the range of stresses permitted in the design.

(2) Plane sections remain plane.

(3) The effects of each cause of moments can be calculated independently and superimposed to attain the result of the combined effect of the several causes (the principle of superposition).

(4) The effect of friction on the prestressing force is small and can be neglected.

(5) The eccentricity of the prestressing force is small in comparison to the span and, hence, the horizontal component of the prestressing force can be considered uniform throughout the length of the member.

(6) Axial deformation of the member is assumed to take place without restraint.

Research into the performance of continuous prestressed-concrete beams has revealed that these assumptions do not introduce significant errors in normal applications. If the cracking load of a beam is exceeded, and in cases where the effect of friction during stressing is significant, special attention should be given to the effects of these conditions. The axial deformation that results from prestressing can have a significant effect on the moments and stresses when such deformations are restrained, as in rigid frames; hence, special investigation into the effects of this phenomenon may be required. Some of these effects will be considered subsequently, but for the general discussion which follows, the above assumptions will be assumed to be valid.

The magnitude and nature of secondary moments can be illustrated by considering a two-span, continuous, prismatic beam that is not restrained by its supports, but which must remain in contact with them, as is illustrated in Fig. 8-5(a). This beam is prestressed with a straight tendon that has a force of P and eccentricity of e. This would tend to make the beam deflect away from the center support by the amount

$$\delta = \frac{Pe(2l)^2}{8EI} = \frac{Pel^2}{2EI} \tag{8-1}$$

Because the beam must remain in contact with the center support, a downward reaction must exist at the location of the center support to cause an equal but opposite deflection. The deflection at the center of a beam which has a span of $2l$, due to a concentrated load (R_b) applied at the center, is equal to

$$\delta = \frac{R_b l^3}{6EI} \tag{8-2}$$

Since the deflections must be equal in magnitude, by equating Eq. (8-1) and (8-2), the value of R_b is found to be

$$R_b = \frac{3Pe}{l} \tag{8-3}$$

Therefore, by applying the rules of statics, it can be shown that the forces that are acting upon the beam must be as shown in Fig. 8-5(d) in order to

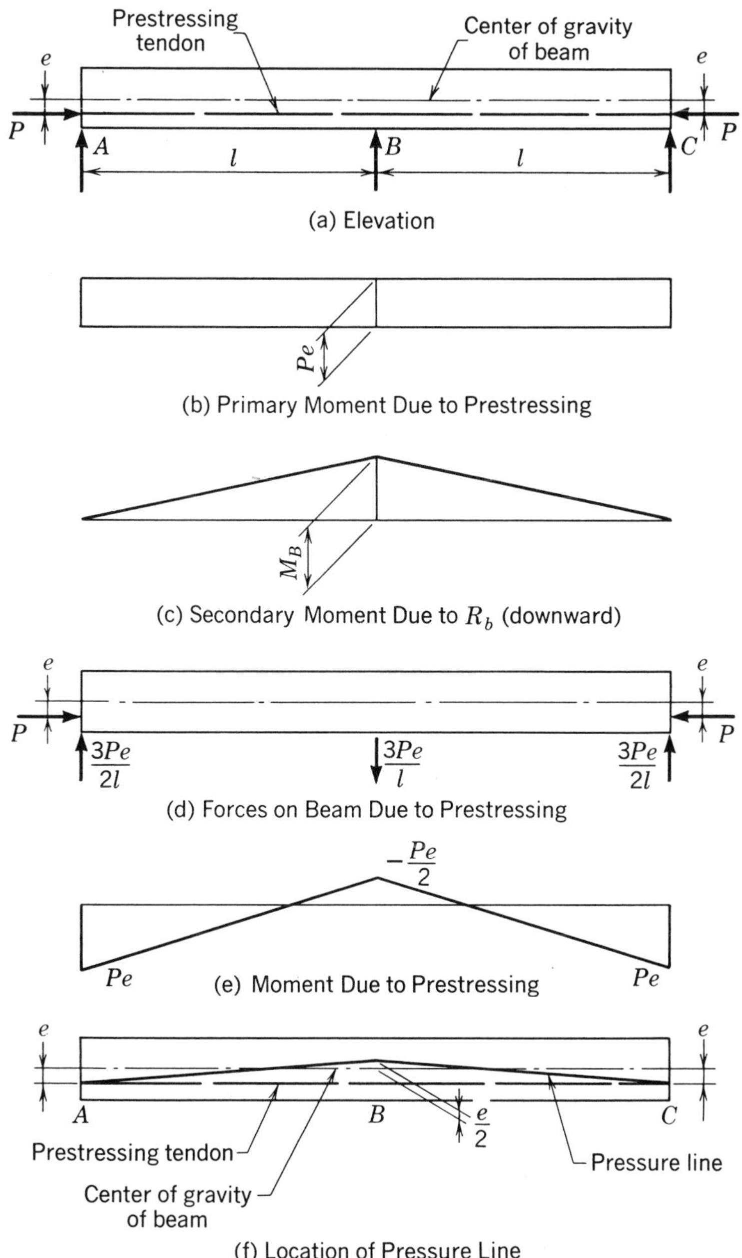
Prestressing tendon
Center of gravity of beam
e
P
A
B
C
l
(a) Elevation
Pe
(b) Primary Moment Due to Prestressing
M_B
(c) Secondary Moment Due to R_b (downward)
$\frac{3Pe}{2l}$
$\frac{3Pe}{l}$
$\frac{3Pe}{2l}$
(d) Forces on Beam Due to Prestressing
$-\frac{Pe}{2}$
Pe
(e) Moment Due to Prestressing
$\frac{e}{2}$
Prestressing tendon
Center of gravity of beam
Pressure line
(f) Location of Pressure Line

Fig. 8-5

maintain equilibrium, and the moment diagram due to prestressing alone is as shown in Fig. 8-5(e) (Ref. 5).

By dividing the moment due to prestressing at each section by the prestressing force P, the eccentricity of the pressure line, e, can be found and plotted (Fig. 8-5(f)). At the ends, as would be expected, the pressure line is seen to be coincident with the location of the prestressing force, while at the center support, the pressure line is at $-e/2$ above the center of gravity of the section. The effect of the secondary reaction, R_b, at the center support has been to move the pressure line from an eccentricity of e below the center of gravity of the section to an eccentricity of $-e/2$ above the center of gravity. Since there are no additional loads between the end support and the center support, the pressure line is a straight line, as shown in Fig. 8-5(f). It will be seen from this example that the secondary moment is secondary in nature, but not in magnitude.

As the above example illustrates, one of the effects of the prestressing may be the creation of secondary reactions, and these reactions cause a linear moment diagram, as shown in Fig. 8-5(c), which, when combined with the primary, prestressing-moment diagram, Fig. 8-5(b), results in the actual moment due to prestressing, Fig. 8-5(e). Because the secondary reactions can only cause a moment which varies linearly, the effect of the secondary moment is to displace the pressure line linearly from the center of gravity of the steel, in direct proportion to the distance from the support. It is also apparent from this example that, if the reactions which result from the prestressing are known, the location of the pressure line can be determined and the stresses due to prestressing at any point can be determined thereby.

The stresses due to prestressing in continuous beams are calculated from the basic relationship

$$f = \frac{P}{A}\left(1 \pm \frac{ey}{r^2}\right) \tag{8-4}$$

in which e is the eccentricity of the pressure line and not the eccentricity of the tendon (although it may be both, as will be seen). An important axiom that is illustrated by the above is that the pressure line due to prestressing is not necessarily coincident with the center of gravity of the steel in indeterminate prestressed structures.

If the position of the tendon in the above example is revised so that it is coincident with the location of the pressure line, which was computed above, the resulting tendon location and moment diagram due to prestressing alone would be as shown in Fig. 8-6(a) and (b), respectively. Removing the reaction at B, in order to render the structure statically determinate, and using the principle of elastic weights, the reactions and forces acting on the beam are as

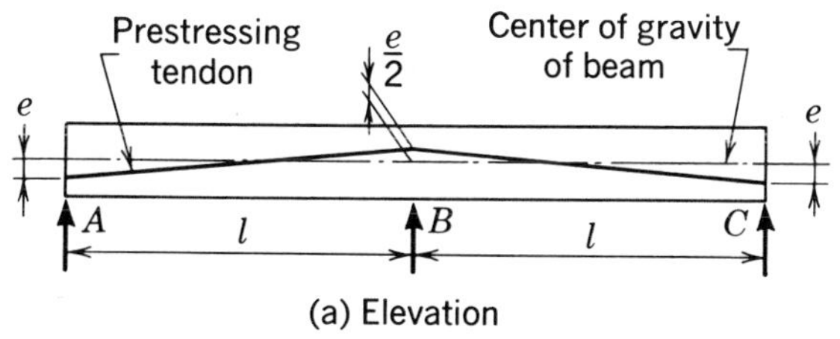

(a) Elevation

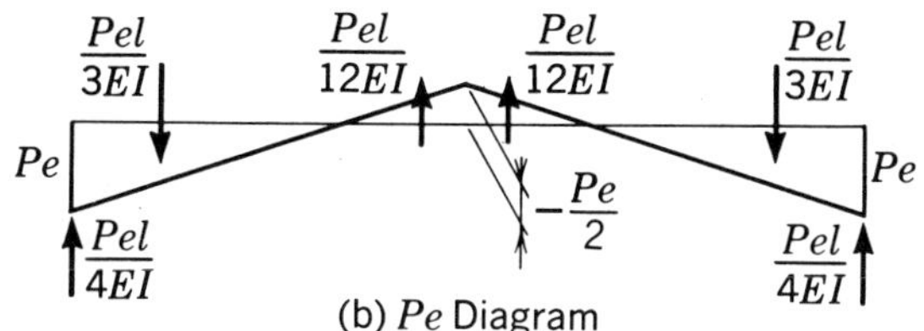

(b) Pe Diagram

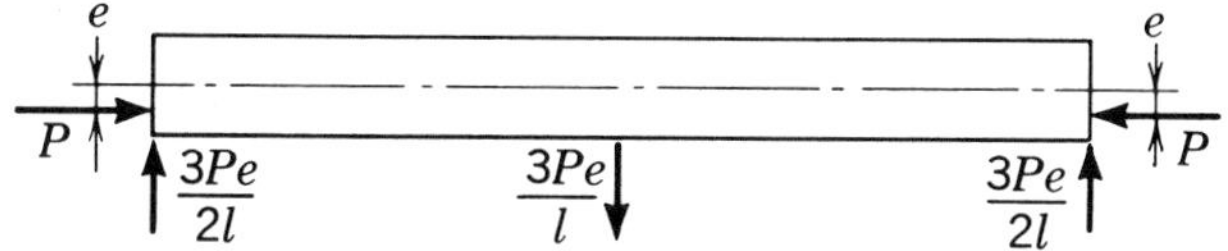

(c) Forces on Beam Due to Prestressing

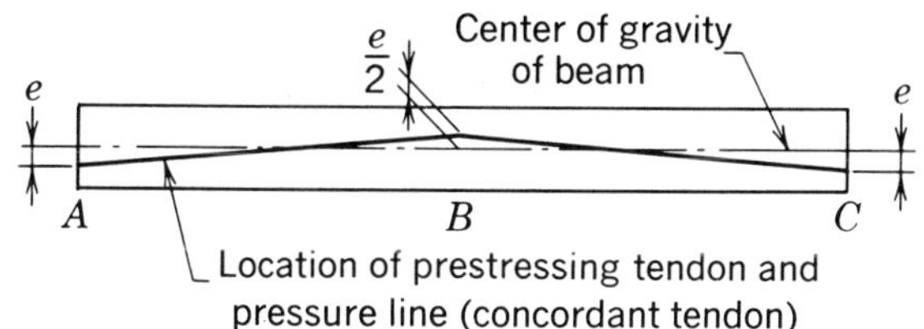

(d) Location of Pressure Line

Fig. 8-6

shown in Fig. 8-6(c). The deflection of the beam at B due to the prestressing is equal to the moment at B resulting from the elastic weights or

$$\delta = \frac{Pel}{4EI} \times l - \frac{Pel}{3EI} \times \frac{7l}{9} + \frac{Pel}{12EI} \times \frac{l}{9}$$

$$= \frac{27Pel^2}{108EI} - \frac{28Pel^2}{108EI} + \frac{Pel^2}{108EI} = 0$$

Since the deflection due to prestressing at B is equal to zero, no secondary reaction is required at B to keep the center support in contact with the beam, and there are no secondary moments. In this example, the pressure line and the center of gravity of the steel are coincident and the tendon is said to be "concordant." In the first example, the pressure line and the center of gravity of the prestressing were not coincident and the tendon is said to be "non-concordant."

If the tendon is placed in the trajectory shown in Fig. 8-7(a), the elastic weights are as shown in Fig. 8-7(b) and the reactions and moments due to prestressing are found to be as shown in Fig. 8-7(c) and (d). It will be seen that in this case the moment diagram due to prestressing, and hence the location of the pressure line, is identical to that in the above two examples. This tendon is also "non-concordant."

This series of three examples illustrates several principles that are extremely useful in the elastic design and analysis of statically indeterminate prestressed structures. In the three examples, the only variable is the eccentricity of the prestressing tendon at the center support. The force in the tendon, P, as well as the eccentricity of the tendon, e, at the end supports was held constant. The inspection of the three solutions reveals that the moments on the concrete section which resulted from each trajectory of the tendon are identical, although the secondary reactions and secondary moments are not equal. It is apparent, from inspection of the moment diagrams used in the three examples, that if the eccentricity at the end of the member were changed to another value, e', the moment due to prestressing and, therefore, the location of the pressure line, would be changed. Free-body diagrams for each of the three examples (Fig. 8-8) reveal that the net forces that act on the members (combination of components of the prestressing force and the secondary reactions) are identical for the three conditions of prestressing, as of course would be expected if the moments are equal.

The above examples also illustrate the very important principle of linear transformation, which can be defined as follows: The trajectory of the prestressing force in any continuous prestressed beam is said to be linearly transformed when the location of the trajectory at the interior supports is altered without altering the position of the trajectory at the end supports and without changing the basic shape (straight, curved, or series of chords) of the trajectory between any supports. Linear transformation of any tendon can be made without altering the location of the pressure line.

The only difference between the three tendon trajectories in the above three examples is that they are displaced from each other, at every section, by an amount which is in direct (linear) proportion to the distance of the section from the end of the member. The eccentricity of the tendon at the end supports and the shape of the tendon between the supports were not changed. It

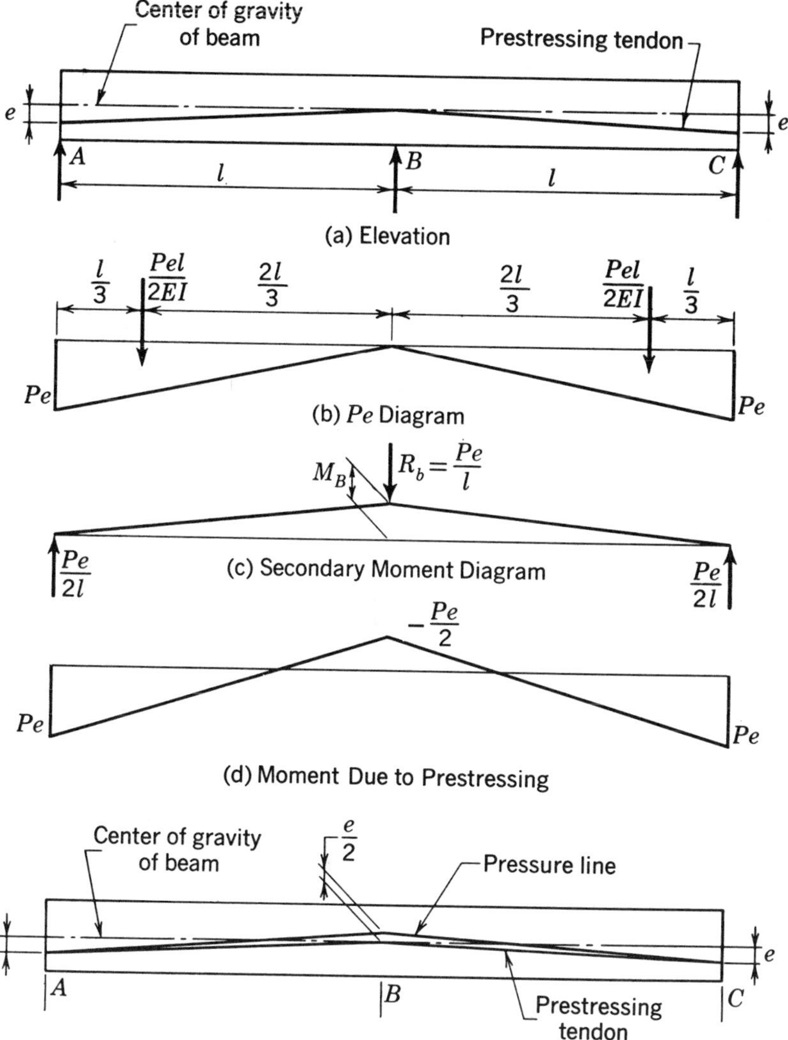
Center of gravity
of beam
Prestressing tendon
e
e
A
B
C
l
l
(a) Elevation
l/3
Pel/2EI
2l/3
2l/3
Pel/2EI
l/3
Pe
Pe
(b) Pe Diagram
M_B
R_b = Pe/l
Pe/2l
(c) Secondary Moment Diagram
Pe/2l
−Pe/2
Pe
Pe
(d) Moment Due to Prestressing
Center of gravity
of beam
e/2
Pressure line
e
e
A
B
Prestressing
tendon
C
(e) Location of Pressure Line

Fig. 8-7

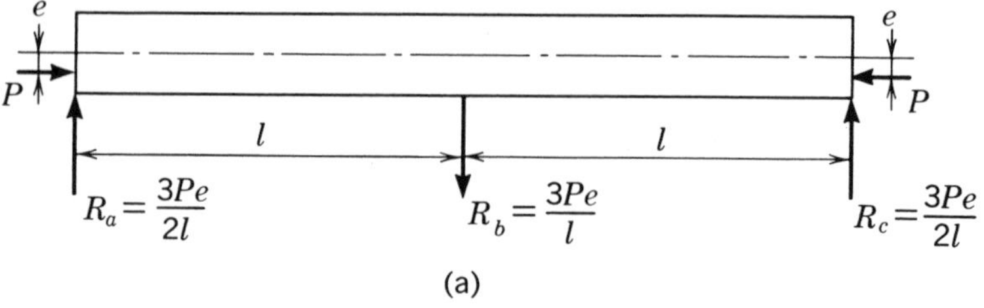

(a)

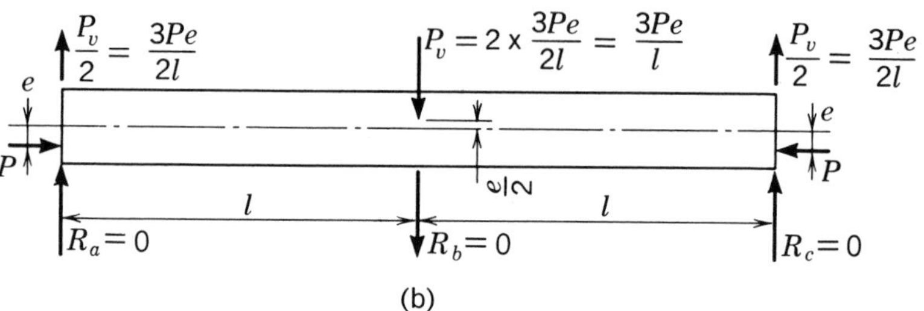

(b)

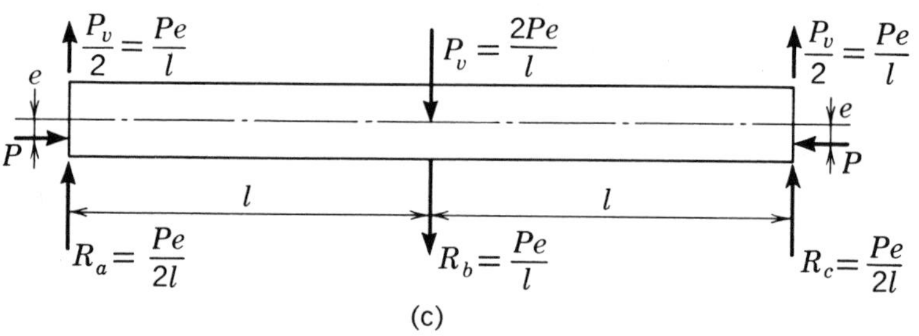

(c)

Fig. 8-8 Free-body diagrams with (a) straight tendon of eccentricity e, (b) tendon sloping from eccentricity of e at each end to $-e/2$ at the center support, and (c) tendon sloping from eccentricity of e at the ends to 0 at the center support.

will be subsequently shown that this principle of linear transformation is equally applicable to tendons that are curved and, from the above examples, it is apparent that the principle applies to concordant tendons as well as to non-concordant tendons.

The principle of linear transformation is particularly useful in designing continuous beams when it may be desirable to adjust the location of the tendon in order to provide more protective cover over the prestressing tendons, without altering the location of the pressure line.

Another significant principle, apparent from these examples, is that the location of the pressure line in beams stressed by tendons alone is not a function of the elastic properties of the concrete. The elastic modulus of the concrete did not appear in the values of the secondary reactions. Therefore, the only effect of changes in the elastic properties of the concrete (i.e., creep) is a reduction in the magnitude of the prestressing force, just as it is in statically determinate structures. This has been proven by tests (Ref. 6).

On the other hand, if the location of the pressure line were to be altered by the application of an additional reaction to a beam, such as by moving the beam up or down at one or more supports with jacks, the reactions and moments induced thereby are a function of the elastic properties of the concrete. Therefore, if such methods are used, the effect of creep must be included in the design of the structures by employing an analysis similar to the type used in studying the effects of differential settlement.

Additional understanding of the action of secondary moments can be gained by considering a prismatic beam fixed at each end in such a manner that rotation of the ends of the beam cannot take place. A beam of this type, prestressed with a straight tendon which is stressed to a force P and at eccentricity e, is illustrated in Fig. 8-9. Neglecting the dead weight of the beam itself, if the ends were released and allowed to rotate, the beam would camber or deflect upward as a normal, simple prestressed-concrete beam would

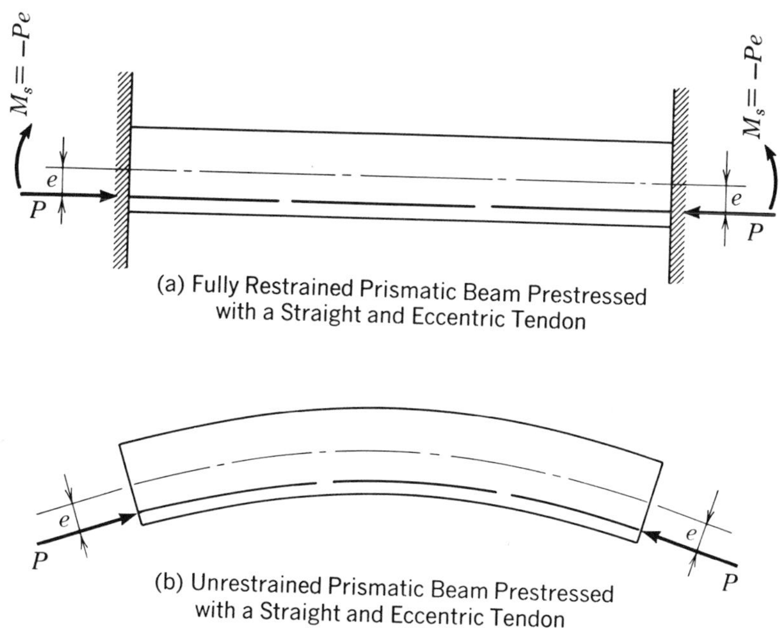

Fig. 8-9 Effect of end restraint on prismatic prestressed beams. (a) Fully restrained prismatic beam with a straight eccentric tendon. (b) Unrestrained prismatic beam with a straight eccentric tendon.

do, as illustrated in Fig. 8-9(b). Since the ends of the beam are fixed and cannot rotate, it is apparent that another moment must be present at the ends to nullify the end rotation that would be caused by the prestressing moment *Pe*. This secondary moment must cause a rotation at each end which is equal in magnitude, but opposite in direction, from that which results from the prestressing moment. It can therefore be concluded that the secondary moment is equal to $-Pe$. It is of significance to note that in this particular case (prismatic fixed beam stressed by a straight tendon), the secondary moment has the effect of nullifying the eccentricity of the prestressing force and results in a uniform compressive stress due to prestressing at every cross section. Again in this example, as has been noted previously, the secondary moment is secondary in nature, but not in magnitude.

If a haunched beam of rectangular cross section, as shown in Fig. 8-10, is prestressed with a straight tendon located at the elastic center of the member, the prestressing will result in stresses of the proper direction for resisting negative moments at the ends of the beam and positive moments at the center of the beam, and there will be no secondary moments. The elastic center is defined as the location through which a force may be applied without causing rotations at the ends of the beam. Applying the same principles used in the above discussion of the prismatic beam, it can be shown that if the straight, prestressing tendon is not located at the elastic center, secondary moments will result in combined stresses that are equal to the stresses that would occur if the tendon were at the elastic center.

The location of the elastic center of the haunched beams of Fig. 8-11 can be determined by solving the relationship

$$\int_0^{L/2} \frac{e\,dx}{I} = 0 \tag{8-5}$$

The value of y, which satisfies Eq. 8-5, is the location of the elastic center.

The example given above for straight tendons are useful in all design problems, but are of particular use in short-span continuous slabs, such as those which occur in bridge decks, viaducts, and waterfront structures.

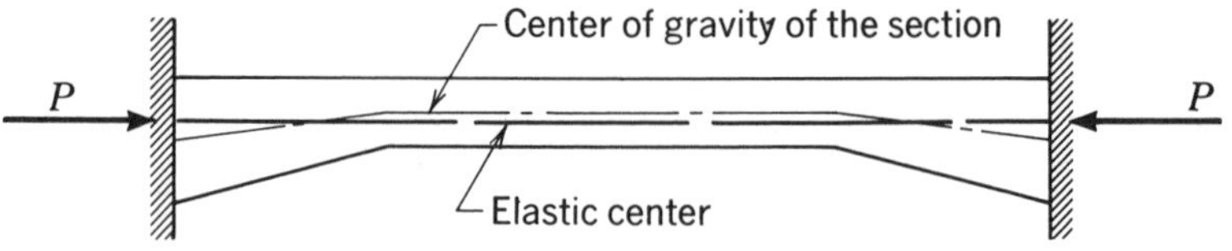

Fig. 8-10 Haunched beam of rectangular cross section prestressed with a straight tendon.

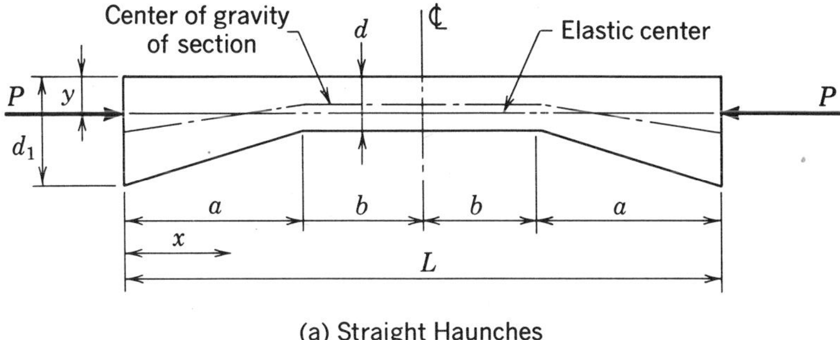

(a) Straight Haunches

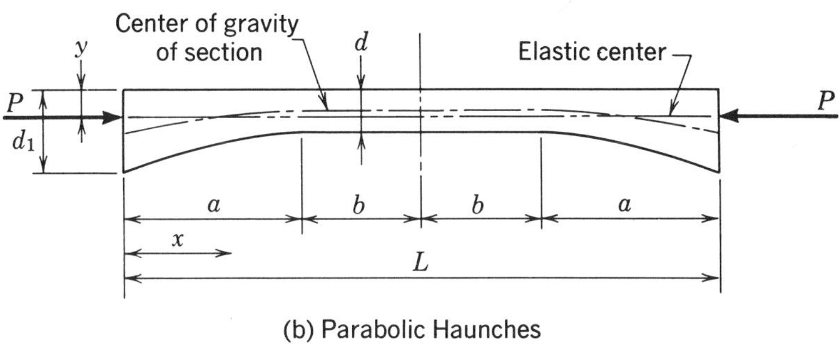

(b) Parabolic Haunches

Fig. 8-11 Beams with (a) straight haunches and (b) parabolic haunches.

8-6 Elastic Analysis of Beams with Curved Tendons

The introduction of curvature to the tendon does not affect the basic methods or principles of analysis in any way; the calculation of the secondary reactions or moments can be made using the principle of elastic weights, the theorem of three moments, or other classical methods, if desired. The familiar moment distribution method is considered among the easier methods for analyzing prismatic beams that have simple tendon trajectories (*see* Ref. 7). For beams with variable moments of inertia, the theorem of three moments, which may be simplified by using a semigraphical method of computing the static moments of the M/I diagram, is rapid and easily understood. Each of these methods is used subsequently in problems, and a brief explanation of the procedures that are followed in applying the methods is given.

The method used in the analysis of a continuous prestressed-concrete member does not affect the results obtained, and the selection of the methods to be used in actual design should be determined by the designer on the basis of the ease of application of the various methods for the particular conditions at hand.

In using the moment distribution method, the end eccentricities, curvature and abrupt changes in slope of the prestressing tendons are converted into specific, equivalent end moments, uniformly applied loads, and concentrated loads, respectively, and the fixed end moments that result from the equivalent loads are distributed in the usual manner. The conversion of end eccentricity and curvature of the tendon into equivalent loading is illustrated by considering the beam shown in Fig. 8-12(a). This two-span continuous beam is prestressed by a tendon having an eccentricity of e_1 at each support. The tendon deflects downward parabolically between supports with a total vertical deflection of e_t.

As was shown in Sec. 7-9, the tangent to the parabolic tendon at the support is equal to

$$\tan \theta = \frac{4e_t}{l}$$

Since the curvatures are small (*see* design assumptions in Sec. 8-5), $\tan \theta = \sin \theta = \theta$ and the vertical component of the tendon at each end of each span is:

$$V_p = P \sin \theta = \frac{4Pe_t}{l}$$

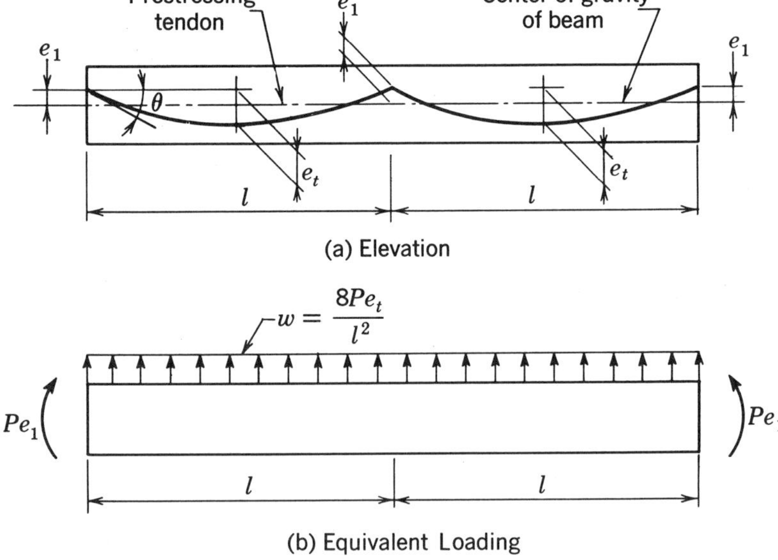

Fig. 8-12 (a) Beam continuous over two spans—prestressed with tendon on parabolic trajectory. (b) Equivalent loading for beam in (a).

Therefore, the total vertical component of the prestressing force that acts on each span of the beam is equal to twice the force that acts at each end, and the equivalent uniformly applied load is found to be

$$wl = \frac{2 \times 4Pe_t}{l}$$

$$w = \frac{8Pe_t}{l^2} \qquad (8\text{-}6)$$

The vertical components of the prestressing force that occur at the supports do not cause moments in the beam, but pass directly through to the supports, and, for this reason, these forces are disregarded in the equivalent loading. The horizontal component of the prestressing tendon is eccentric by an amount equal to e_1 at each end of the beam and the equivalent loading must, therefore, include end moments in the amount of Pe_1. The end moment Pe_1 is treated just as the moment due to a cantilevered end would be treated in the analysis. The equivalent loading is shown in Fig. 8-12(b).

When the tendon curves parabolically over the length of the span but with the ends of the curve at different elevations, the value of e_t and l to be used in Eq. 8-6 are as shown in Fig. 8-13. If the tendon trajectory is formed of compounded parabolas, as shown in Fig. 8-14, the equivalent loads due to tendon curvature is computed by

$$w_n = \frac{2Pe_n}{{x_n}^2} \qquad (8\text{-}7)$$

It is usually sufficiently accurate to assume all curves are parabolic even though they may be circular or of other shape. Since the eccentricity is normally small in comparison to the span, the error introduced by this assumption is small.

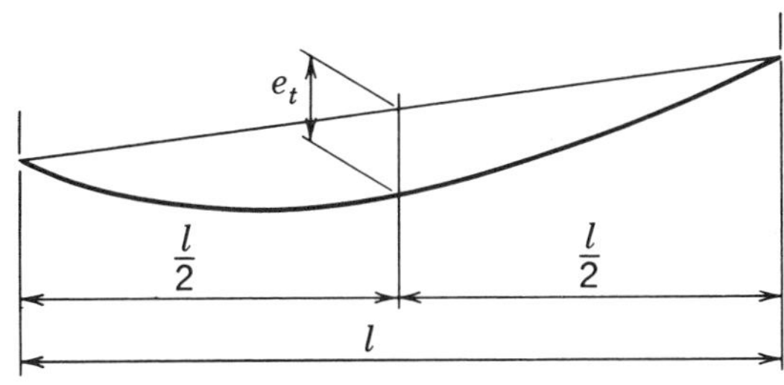

Fig. 8-13 Value of e_t to be used for parabolic tendon trajectory terminated at different elevations.

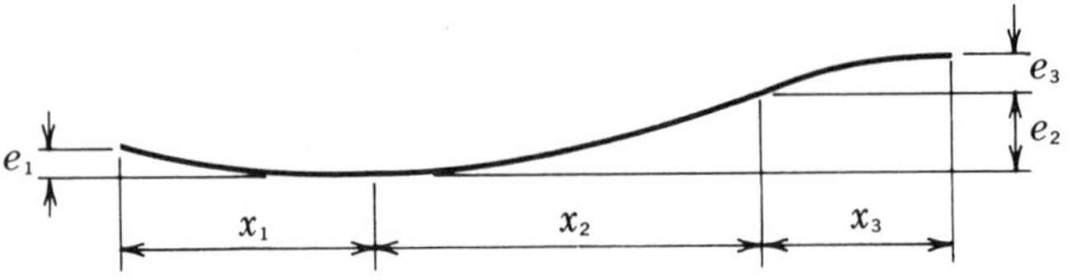

Fig. 8-14 Tendon trajectory composed of compounded parabolas.

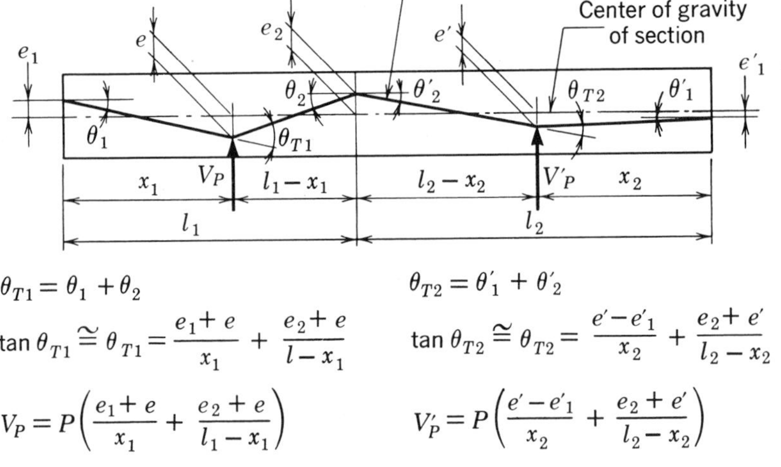

$$\theta_{T1} = \theta_1 + \theta_2$$

$$\tan \theta_{T1} \cong \theta_{T1} = \frac{e_1 + e}{x_1} + \frac{e_2 + e}{l - x_1}$$

$$V_P = P\left(\frac{e_1 + e}{x_1} + \frac{e_2 + e}{l_1 - x_1}\right)$$

$$\theta_{T2} = \theta'_1 + \theta'_2$$

$$\tan \theta_{T2} \cong \theta_{T2} = \frac{e' - e'_1}{x_2} + \frac{e_2 + e'}{l_2 - x_2}$$

$$V'_P = P\left(\frac{e' - e'_1}{x_2} + \frac{e_2 + e'}{l_2 - x_2}\right)$$

Fig. 8-15 Equivalent loads for tendons placed on a series of straight slopes.

The vertical load resulting from an abrupt change in slope of the tendons is computed as follows:

$$V_p = P(\sin \theta_t) \cong P(\tan \theta_t)$$

The value of $\tan \theta$ is determined by the dimensions of the tendon trajectory, as is illustrated in Fig. 8-15.

ILLUSTRATIVE PROBLEM 8-1 Compute the moments due to prestressing and draw the pressure line for the prismatic beam shown in Fig. 8-16. Use moment distribution and the theorem of three moments.

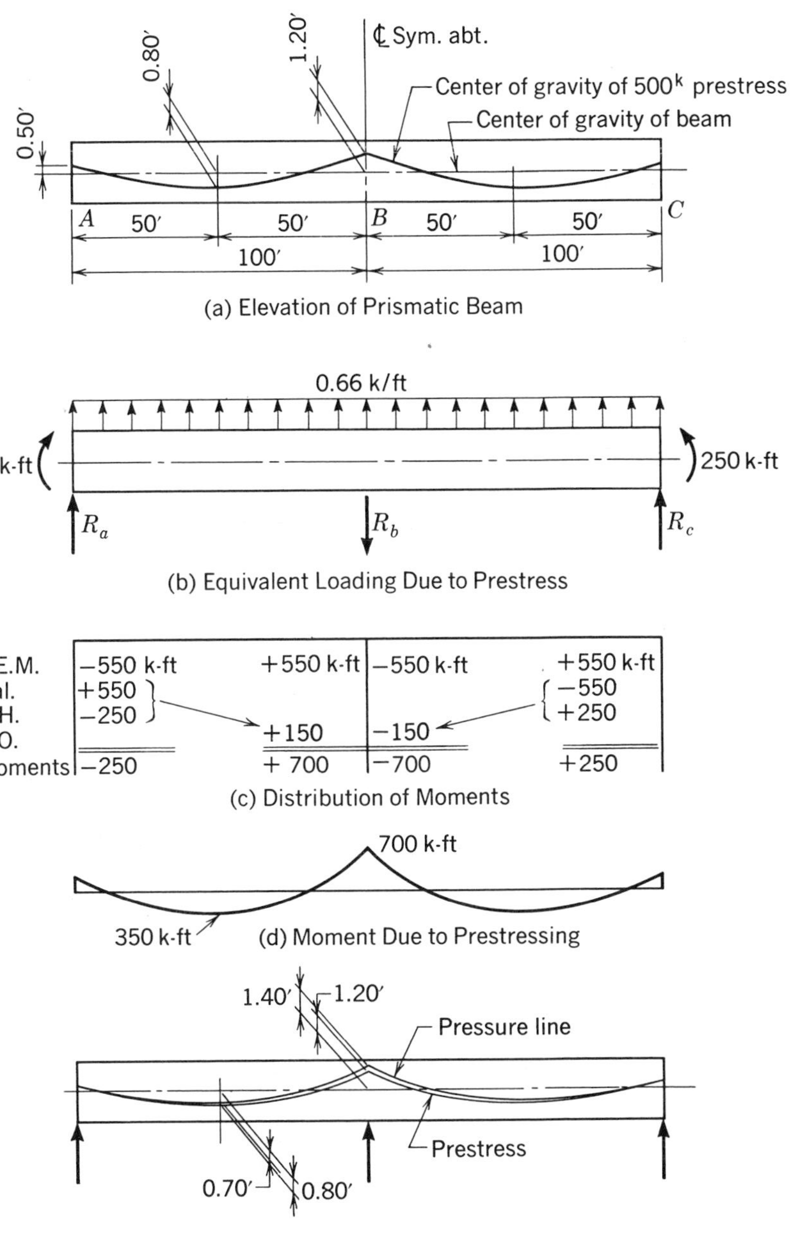

Fig. 8-16 Analysis of two-span continuous beam with moment distribution of eqiva-lent loading.

SOLUTION:

Equivalent uniform load $= w = \dfrac{8Pe}{l^2}$

$$e = 0.80 + \frac{0.50 + 1.20}{2} = 1.65 \text{ ft} \qquad w = \frac{8 \times 500 \times 1.65}{(100)^2} = 0.66 \text{ k/ft}$$

Fixed end moments:

$$M_{AB}{}^F = -\frac{0.66 \times (100)^2}{12} = -550 \text{ k-ft} = M_{BC}{}^F$$

$$M_{BA}{}^F = +\frac{0.66 \times (100)^2}{12} = +550 \text{ k-ft} = M_{CB}{}^F$$

Overhang moments:

$$M_A = -0.5 \text{ ft} \times 500 = -250 \text{ k-ft}$$

$$M_C = +0.5 \text{ ft} \times 500 = +250 \text{ k-ft}$$

The distribution of moments is performed in Fig. 8-16 and the momen diagram due to prestressing is plotted in Fig. 8-16(d). The computed momen at B is 700 k-ft and the eccentricity of the pressure line is computed as follows

$$e = \frac{700}{500} = 1.40 \text{ ft}$$

The pressure line is then 1.40 ft − 1.20 ft = 0.20 ft above the tendon at B, sinc the pressure line is linearly transformed from the tendon trajectory. At th center of the span, the pressure line is 0.10 ft higher than the tendon trajector since (50 × 0.20)/100 = 0.10 ft. This is shown in Fig. 8-16(e). Due to th symmetry of the beam and the loading, the three-moment equation for thi example is reduced to

$$M_A + 4M_B + M_C = -\frac{wL^2}{2}$$

$$250 + 4M_B + 250 = -\frac{0.66 \times (100)^2}{2}$$

$$M_B = -700 \text{ k-ft}$$

ILLUSTRATIVE PROBLEM 8-2 Compute the moments due to prestressing fo the prismatic beam and condition of loading illustrated in Fig. 8-17. Us moment distribution and the theorem of three moments.

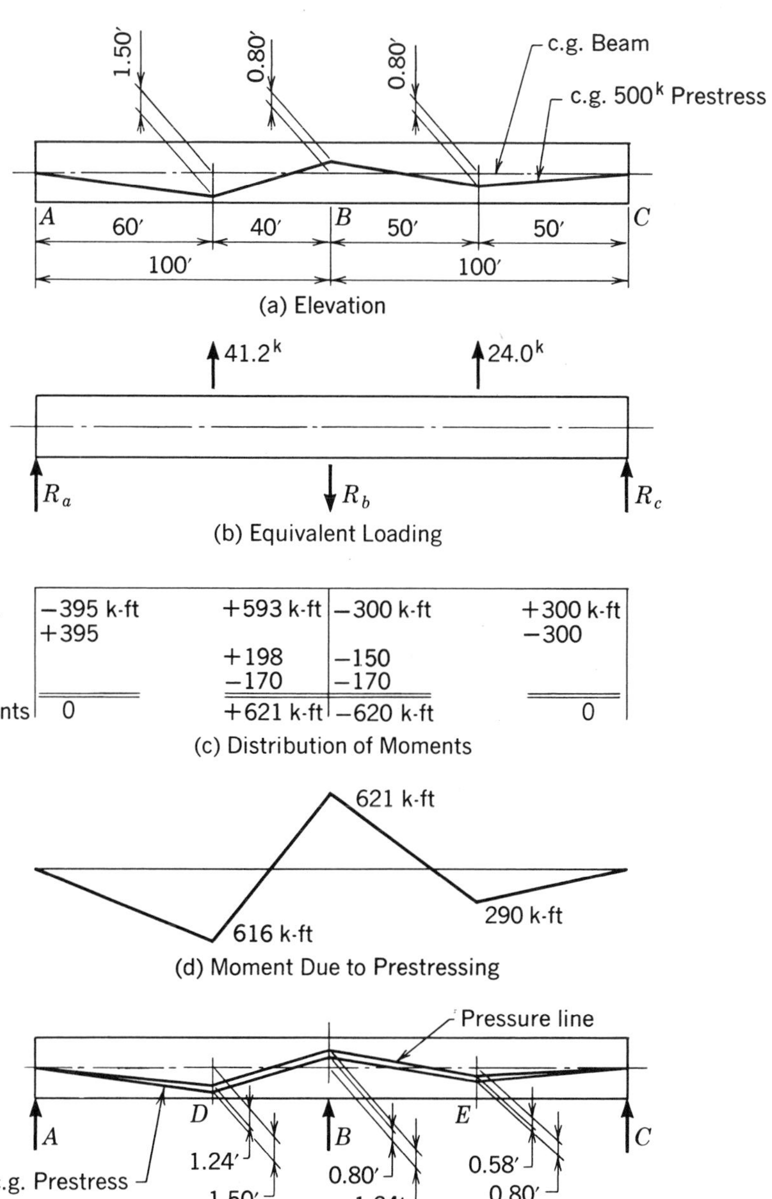

Fig. 8-17 Analysis of two-span continuous beam with moment distribution of equivalent loading.

SOLUTION:

$$\text{Concentrated load in span } AB = 500\left(\frac{1.50}{60} + \frac{2.30}{40}\right) = 41.2 \text{ k}$$

$$\text{Concentrated load in span } BC = 500\left(\frac{0.80 + 1.60}{50}\right) = 24.0 \text{ k}$$

Fixed end moments:

$$M_{AB}{}^F = \frac{41.2 \times 60 \times (40)^2}{(100)^2} = -395 \text{ k-ft}$$

$$M_{BA}{}^F = \frac{41.2 \times (60)^2 \times 40}{(100)^2} = +593 \text{ k-ft}$$

$$M_{BC}{}^F = \frac{24 \times 50 \times (50)^2}{(100)^2} = -300 \text{ k-ft} \qquad M_{CB}{}^F = +300 \text{ k-ft}$$

The moments are distributed in Fig. 8-17, and the moment at B is found to b 621 k-ft. The eccentricity of the pressure line at B is then

$$e = \frac{621}{500} = 1.24 \text{ ft}$$

Therefore, the pressure line is 1.24 ft – 0.80 ft = 0.44 ft above the trajector of the steel at B. The distance from the pressure line to the tendon trajector at points D and E are found from the principle of linear transformation a

$$\text{At point } D \qquad \frac{0.44 \text{ ft} \times 60}{100} = 0.264 \text{ ft}$$

$$\text{At point } E \qquad \frac{0.44 \text{ ft} \times 50}{100} = 0.220 \text{ ft}$$

Using the theorem of three moments, the computation of the moment at for the above example is as follows:

$$M_A L_1 + 2M_B(L_1 + L_2) + M_C L_2 = -\frac{41.2 \times 60}{100}[(100)^2 - (60)^2]$$

$$-\frac{24 \times 50}{100}[(100)^2 - (50)^2]$$

$$400\, M_B = -158{,}000 - 90{,}000$$

$$M_B = -620 \text{ k-ft}$$

ILLUSTRATIVE PROBLEM 8-3 Compute the moments due to prestressing fc the beam illustrated in Fig. 8-18. Note that the relative moment of inertia

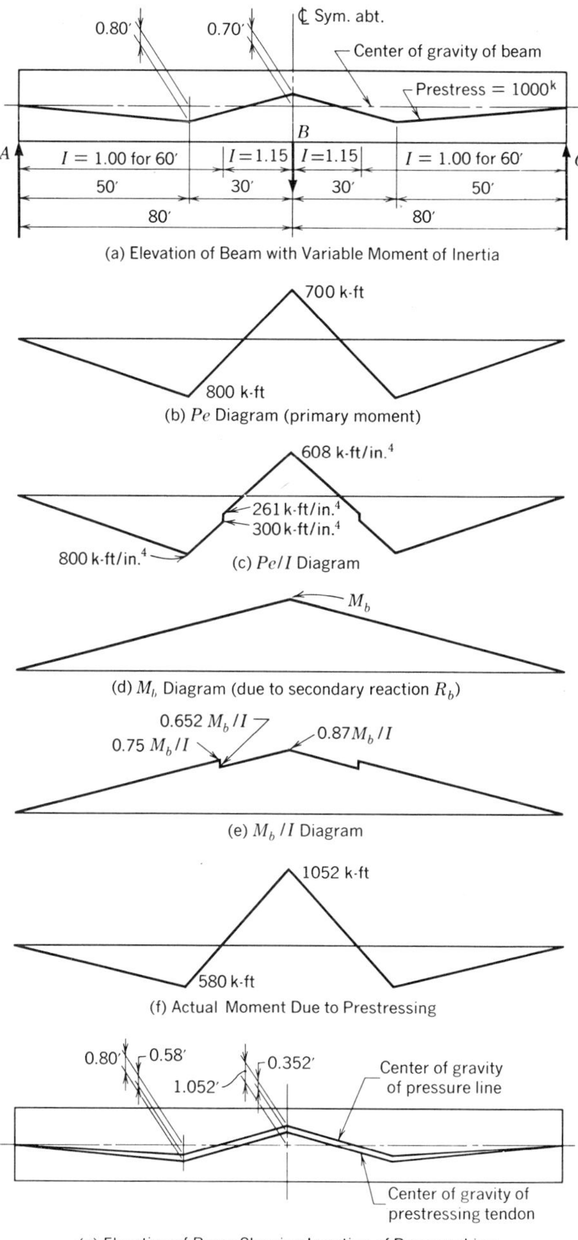

Fig. 8-18 Analysis of two-span continuous beam with a variable moment of inertia, using the principle of elastic weights.

1.00 for the outermost 60 ft of each span and 1.15 for the center 40 ft of the beam. The center of gravity of the section is a straight line (the variable moment of inertia is the result of an abrupt change in web thickness or abrupt, symmetrical change in top and bottom flange thicknesses, or both).

SOLUTION: In Fig. 8-18(b), (c), (d), and (e) are plotted the *Pe* diagram, the *Pe/I* diagram, the assumed M_b diagram, and the M_b/I diagram, respectively. The magnitude of M_b can be determined rapidly by employing the principle of elastic weights. The former is used here by computing and equating the moments (deflections) at the center of the span *AC* of the beam loaded with the *M/I* diagrams for *Pe* and M_b.

The upward deflection due to *Pe/I* diagram (δpe) is

$$
\begin{array}{rcrcr}
-800 \times 50/2 & = & -20{,}000 \times 46.70 & = & -\ \ 933{,}000 \\
-300 \times 10 & = & -\ \ 3{,}000 \times 25.00 & = & -\ \ 75{,}000 \\
-500 \times 10/2 & = & -\ \ 2{,}500 \times 26.70 & = & -\ \ 66{,}700 \\
-261 \times 6/2 & = & -\ \ 783 \times 18.00 & = & -\ \ 14{,}100 \\
+608 \times 14/2 & = & +\ \ 4{,}256 \times 4.67 & = & +\ \ 19{,}800 \\
\hline
 & & -22{,}027 & & -1{,}069{,}000
\end{array}
$$

$$\delta_{pe} = +22{,}027 \times 80 \text{ ft} - 1{,}069{,}000 = 691{,}000 \text{ k-ft}^3/\text{in.}^4$$

The downward deflection due to M_b/I diagram (δM_b) is

$$
\begin{array}{rcrcr}
+0.75\ M_b \times 60/2 & = & 22.5M_b \times 40.00 & = & +\ \ 900.0M_b \\
+0.652M_b \times 20 & = & 13.0M_b \times 10.00 & = & +\ \ 130.0M_b \\
+0.218M_b \times 20/2 & = & 2.2M_b \times 6.67 & = & +\ \ 14.6M_b \\
\hline
 & & +37.7M_b & & +1044.6M_b
\end{array}
$$

$$\delta_{M_b} = -37.7M_b \times 80 + 1044.6M_b = -1965M_b$$

Equating the two deflections

$$\delta_{pe} = \delta_{M_b}$$

$$M_b = \frac{691{,}000}{-1965} = -352 \text{ k-ft}$$

The combined or actual moment diagram due to prestressing and the location of the pressure line are as shown in Fig. 8-18(f) and (g) respectively.

It should be noted that the total moment due to prestressing and not the secondary moment was computed in Illustrative Problems 8-1 and 8-2. When equivalent loads are used for the analysis of moment due to prestressing, the total moment due to prestressing is computed. In Illustrative Problem 8-3, the secondary moment and not the total moment due to prestressing was computed. The secondary moment due to prestressing is computed when the effects of prestressing are analyzed, using the basic principles of indeterminate structural analysis with the primary moment due to prestressing being considered as the initial loading condition.

8-7 Additional Elastic-Design Considerations

From the previous discussions, it should be recognized that in applying the moment distribution method in the analysis of the moments due to prestressing, the effect of the prestressing is analyzed by determining an equivalent loading that is a function of the eccentricity of the tendon at the end supports and of the intrinsic shape of the tendon between supports, but is independent of the eccentricity of the tendon at the interior supports. There is a family of curves or trajectories that will result in this equivalent loading but, for this loading, there will be only one pressure line. It should also be apparent that the pressure line that results from the distribution of the moments, due to the equivalent loading, is the location of the concordant tendon for the particular condition of end eccentricities and intrinsic shape of the tendon. Finally, it should be recognized that the location of the pressure line is determined by dividing the moment diagram, resulting from the equivalent loading, by the effective prestressing force. From these considerations, the following corollary should be apparent: A moment diagram for a particular beam, due to any condition of loading on the beam, defines a location for a concordant tendon and, if a tendon is placed on a trajectory that is proportional to any moment diagram for the beam, the tendon will be concordant. This principle is useful in selecting a trial trajectory for a tendon for which it is not necessary to compute the effects of secondary moments.

From the discussion of Sec. 8-5, it should be apparent that revision of the eccentricities of the tendon at the end support, without revision of the shape of the tendon between supports, results in a linear change in the pressure line. This is evident when the pressure lines of the example problems are studied. From this, it is seen that if the end eccentricity is changed from e to e' the location of the pressure line at the center support is changed from $-e/2$ to $-e'/2$; moreover, the location of the pressure line is independent of the eccentricity of the tendon at the interior support. This principle is useful in the design of continuous beams, when it is desired to modify the location of the tendon and the pressure line, since the revision of the end eccentricity can be treated as a linear adjustment in the pressure line.

When continuity is used in bridges or other structures where the loading conditions are quite variable, the complication of designing for specific conditions of maximum and minimum moment, which do not occur simultaneously at any one section and which vary along the span, is introduced.

The effect of maximum and minimum moments on the relationships for the minimum prestressing force and the corresponding eccentricity that are adequate at any particular section, if tensile stresses are not to be allowed in the section, can be developed by considering the conditions shown in Fig. 8-19, which illustrates the movements of the pressure line that take place as a result of the applied moments.

If the stress in the top fiber of the beam under the action of the minimum moment is to be zero, the relationship shown in Fig. 8-19(a) must exist. If it is assumed that the minimum moment is negative, the condition can be expressed mathematically as follows:

$$r^2/y_t = e - \frac{M_{min}}{P} \tag{8-7}$$

In the above equation, the minimum moment is assumed to be negative and the negative sign is included in the symbol M_{min}. This relationship is equally applicable to minimum moments that are positive, as illustrated in Fig. 8-17(c).

In a similar manner, if the stress in the bottom fiber is zero when the maximum positive moment is applied to the member, the condition of Fig. 8-18(b) must exist. The mathematical relationship for P and e becomes

$$\frac{M_{max}}{P} = e + \frac{r^2}{y_b} \tag{8-8}$$

Equating (8-7) and (8-8) and solving for P, the relationship for the minimum prestressing force that will satisfy the requirements set forth above is as follows:

$$P = \frac{M_{max} - M_{min}}{r^2/y_t + r^2/y_b} \tag{8-9}$$

and the corresponding relationship for e becomes

$$e = \frac{M_{min} r^2/y_b + M_{max} r^2/y_t}{M_{max} - M_{min}} \tag{8-10}$$

In the above relationships, the values of maximum and minimum moments include the effects of the dead load of the beam. The maximum and minimum moments applicable at any section are the algebraic maximum and minimum moments, respectively, and are not necessarily maximum or minimum numerically. It should be kept in mind that tensile stresses are frequently permitted and, under such conditions, Eqs. 8-9 and 8-10 do not apply. They can, however, be useful in determining values of P and e to be used in the initial computations and for locating the most critical sections. Furthermore,

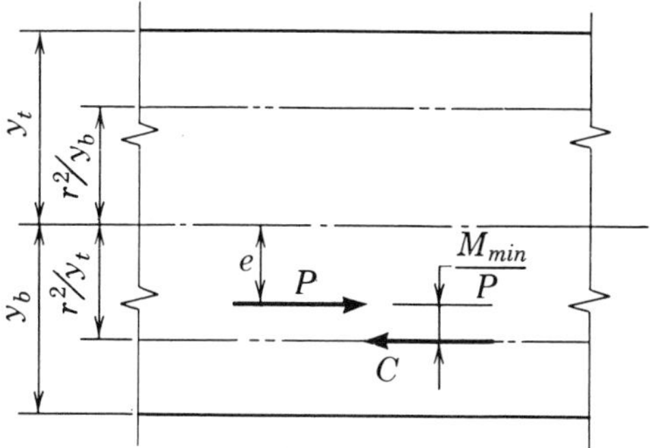

(a) Section with Negative Minimum Moment

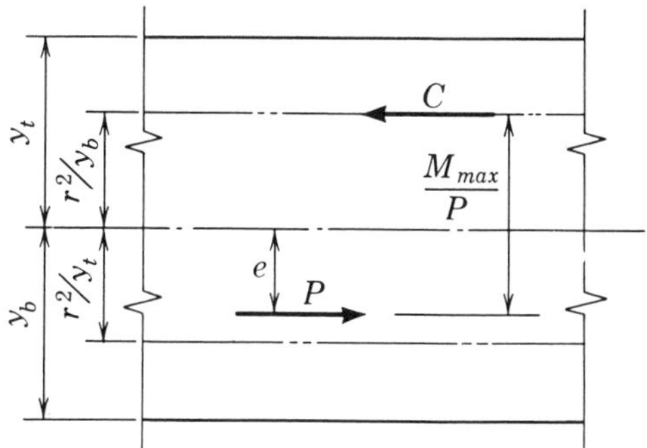

(b) Section with Positive Maximum Moment

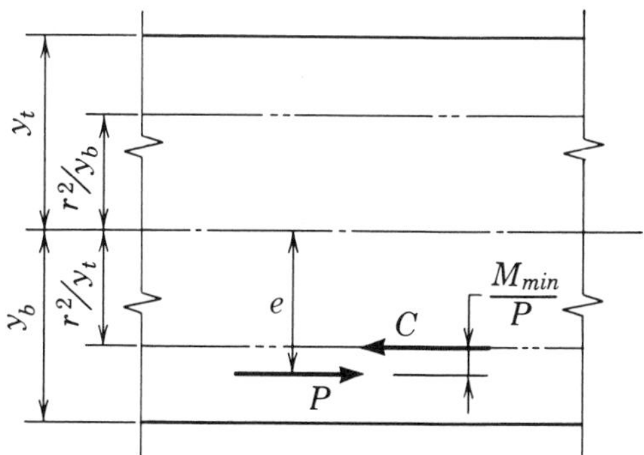

(c) Section with Positive Minimum Moment

Fig. 8-19 Vector diagrams illustrating the possible conditions in beams subject to reversals of moment.

since the value of e required for the minimum prestressing force as given in Eq. 8-10 may not be obtainable for a particular beam section, the prestressing force required may be controlled by the familiar relationship

$$P = \frac{M_{max}}{e + r^2/y_b} \quad \text{or} \quad \frac{M_{min}}{e + r^2/y_t} \tag{8-11}$$

Therefore, the normal design procedure is to determine the minimum prestressing force by Eq. 8-9 and check the result with Eq. 8-11, using a reasonable value for e. The larger force must be used.

It must be emphasized again that because these relationships apply to continuous beams in which the pressure line is not always coincident with the trajectory of the prestressing force, as well as to beams with concordant tendons, the designer must exercise care to ensure that the eccentricity of the pressure line is used in the above relationships.

From the discussions of linear transformation in Sec. 8-5, from the principles of elastic design that are the basis for the analysis of statically indeterminate structures, and from the discussion at the beginning of this article, the following axioms and corollaries applicable to most continuous beams can be stated (*see* Sec. 8-13 for an exception):

(1) For any particular beam, a specific shape or trajectory of the prestressing tendon between the supports will result in a specific pressure-line location for each condition of eccentricity at the end supports.

(2) Alteration of the eccentricity at one or both end supports will result in a shift in the location of the pressure line for any particular continuous beam.

(3) Alteration of the eccentricity of the tendon trajectory at the interior supports will not affect the location of the pressure line, if the eccentricities at the ends of the beam are not changed and if the intrinsic shape of the tendon trajectory is maintained.

(4) The conditions of continuity require that the deflection of a beam at the supports is equal to zero and that the end rotations of each span that adjoin at a common interior support are equal. These conditions apply equally to the effect of the prestressing force (pressure line) and to the effect of external loads.

(5) The moments and the location of the pressure line due to prestressing can be determined by resolving the primary prestressing moment into equivalent loads and end moments, then calculating the resulting moment in the continuous structure.

(6) The location of a pressure line is directly proportional to the moment in the continuous beam which results from a specific condition of (equivalent) loading.

(7) Since the pressure line defines the location of a concordant tendon, which

has specific end eccentricities and trajectory between supports, a tendon trajectory can be made to be concordant, if it is located along a trajectory proportional to the moment diagram that results from any specific loading for the beam under consideration.

(8) The location of the pressure line determined for any non-concordant tendon trajectory is the pressure-line location that will result from all non-concordant tendons in the particular beam, if the eccentricity of the tendons are equal at the end supports and the basic shape of the tendon is not changed between supports.

(9) The location of the pressure line determined for any non-concordant tendon trajectory is the only location for a concordant tendon for the particular beam, basic shape of the tendon trajectory between supports, and eccentricities at the end supports.

(10) The combination of two concordant tendon trajectories will result in a concordant tendon, since the principle of superposition applies and each concordant tendon satisfies the requirements of continuous structures, i.e., the deflection at the supports is zero and the end rotations of adjacent spans are equal at common supports.

(11) Superposition of a non-concordant tendon with a concordant tendon will result in a non-concordant tendon that has a pressure-line location which can be determined by superimposing the trajectory of the concordant tendon on the pressure line resulting from the non-concordant tendon.

It should be pointed out that concordant tendon locations are not more desirable than non-concordant tendon locations. It is a fact that non-concordant tendon locations often result in a more efficient design, since the protective cover over the prestressing steel may be greater for non-concordant tendon locations than for equivalent concordant tendon locations. It is often desirable to start a design with an assumed tendon trajectory that is concordant, in order to avoid the necessity of determining secondary moments. The tendon location can then be linearly transformed, or otherwise altered, into a non-concordant tendon that may result in a better design.

8-8 Elastic Design Procedure

It is difficult to generalize on the best procedure to be followed in the complete design of a structure, since there are so many considerations that must be taken into account. These include the theoretical elastic-design considerations, as well as practical construction and economic considerations, such as the methods of construction that are feasible on the site and clearance requirements. Therefore, the procedure outlined here must be considered for the general case of "design," by reviewing a concrete section which has been selected after due consideration of all governing factors.

The computation of the maximum and minimum live-load moments that act at various sections along a continuous prestressed beam is no different than it is for other types of construction. The unique procedures in the design of continuous prestressed beams are the determination of the minimum prestressing force that will perform satisfactorily with the assumed concrete section and the determination of a satisfactory trajectory for the tendon. The normal design procedure is as follows:

(1) Using an assumed dead load, compute and plot the maximum and minimum moments that result from the various possible combinations of live load and dead load.

(2) Using the computed moments as a guide, adopt a trial section and check the assumed dead weight. If necessary, revise the dead-load moment computations and adopt a revised trial section.

(3) Compute the section properties of the assumed section. If the trial section has a variable moment of inertia, the section properties of the assumed section must be determined at several locations. The stresses in the concrete should be checked at several locations in order to be sure that the concrete stresses are not excessive under the conditions of maximum or minimum moment.

(4) Using the relationship developed above for the minimum prestressing force in members subjected to maximum and minimum moments (Eqs. 8-9 and 8-11), determine the minimum prestressing force that will satisfy the condition of loading for the assumed section.

(5) At several sections along the length of the member, compute the minimum and maximum eccentricities that can be allowed with the minimum prestressing force, without tensile stresses developing under either condition. This can be done with Eqs. 8-7 and 8-8, developed in Sec. 8-7.

Negative moments have negative signs and positive eccentricities are below the center of gravity of the section. It is recommended that these values be calculated, since they are useful in determining precisely whether or not a particular pressure line lies within the required limits. The values should also be plotted using an exaggerated vertical scale, since such a plot reveals the general shape that a satisfactory pressure line must have. The plot can be made by plotting the values of $e_{\min}$ from the upper kern line (r^2/y_b), and $e_{\max}$ from the lower kern line (r^2/y_t), in which negative values are plotted above and positive values below the respective kern line.

(6) The plotted limits of the eccentricity should resemble the sketch shown in Fig. 8-20, in which the shaded area indicates the area in which the pressure line must lie. Using the shaded area as a guide, a tendon trajectory is assumed, and the location of the pressure line is computed. If the shaded area resembles the moment diagram for a known condition of loading, the tendon can be placed on a trajectory proportional to that moment diagram and, since it will

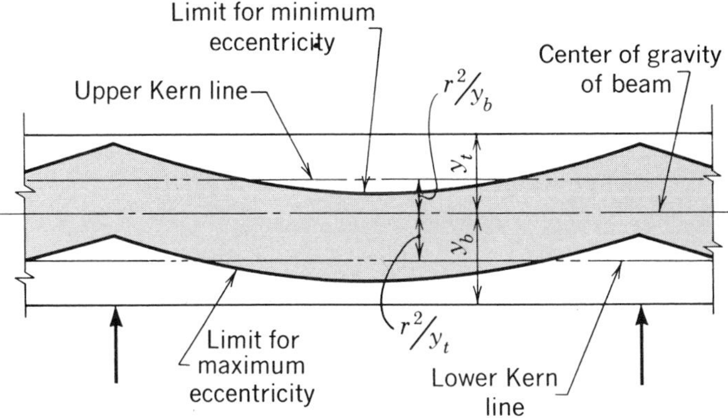

Fig. 8-20 Graphical representation of the solution of Eqs. 8-7 and 8-8. The shaded area indicates the area in which the pressure line must be confined.

be a concordant tendon, it will not be necessary to compute the location of the pressure line (or secondary moments).

(7) If the pressure line does not fall within the shaded area, adjustments in the location of the pressure line (tendon trajectory) must be made. The tendon location can be revised by changing the shape of the trajectory or by revising the eccentricity at the end supports. It was pointed out that a tendon trajectory proportional to a moment diagram for the structure under the action of any condition of external load is the location of a concordant tendon. Therefore, the effects of revising the end eccentricities can be obtained by determining the distribution of moments along the member for a specific end moment alone and adding the effect of the end moment to the pressure line that is obtained for the tendon that had a different end eccentricity.

(8) If a satisfactory location or trajectory cannot be found for the magnitude of the prestressing force selected in step 4, a larger prestressing force may have to be used. The adoption of a larger force will have the tendency of spreading the limits in which the pressure line must be confined and which were computed and plotted in step 5. It also may become apparent that the design may be improved by revising the concrete section in order to obtain more desirable limits for the pressure line. This procedure is repeated until a satisfactory solution is found.

(9) After the prestressing tendons have been selected and located, the shear stresses at critical sections must be investigated.

(10) The final step, which is not part of the elastic design but which must not be overlooked, is to investigate the ultimate load capacity of the member in order to ensure the design is adequate. This is discussed in Secs. 8-9 and 8-10.

ILLUSTRATIVE PROBLEM 8-4 Determine a satisfactory means of prestressing the beam shown in Fig. 8-21. The beam is a hollow box and varies in depth from 24 to 49 in. The superimposed live load is 600 plf. Each span may be loaded independently of the other span.

SOLUTION: The section properties of the beam are computed at 10 ft intervals along the beam and the moment of inertia at the different locations is summarized in Table 8-1 (p. 312). The relative moment of inertia is computed listed in line 2 of the table, and plotted in Fig. 8-21. The relative moment of inertia is used in determining the redundant reactions and moments, rather than the absolute values, as a means of facilitating the computations. The moments at the different points along the beam are then computed for the following conditions of loading:

(1) A uniformly distributed load ($w = 0.08$ k/ft) which results in a maximum simple span moment (at point 5) of 100 k-ft in span *AB*. By using this unit moment diagram, the moments due to uniformly distributed loads of other magnitudes are easily and rapidly determined.

(2) The triangular loading due to the dead load resulting from the haunch.

(3) The moments due to M_b.

The ordinates of the M/I diagrams for the various moments are then determined and plotted as shown in Fig. 8-21(d), (e), and (f). The M/I diagrams are used in determining the static moments required in solving for the moment M_b, using the theorem of three moments.

The distribution of moments in the continuous beam are determined through the use of the theorem of three moments for the following conditions of loading:

(1) Beam dead load alone.

(2) Dead load plus live load on one span only.

(3) Dead load plus live load on each span simultaneously. The maximum and minimum moments that may occur at the various points along the beam are then determined. The values of the maximum and minimum moments are listed in Table 8-1, lines 4 and 5.

The value of $M_{max} - M_{min}$ is determined (line 6) and it is seen to reach maximum values at points 5 and 10. The more critical section is point 5, since the value of $r^2/y_t + r^2/y_b$ at point 10 is greater than twice that of the smaller section at point 5. Therefore, the minimum prestressing force which will satisfy the conditions of loading is controlled by the moments at point 5 and is found to be

$$P = \frac{749 \text{ k-ft}}{0.475 + 0.475} = 790 \text{ k}$$

The values of e_{max} and e_{min} are determined for the value of $P = 790$ k, and

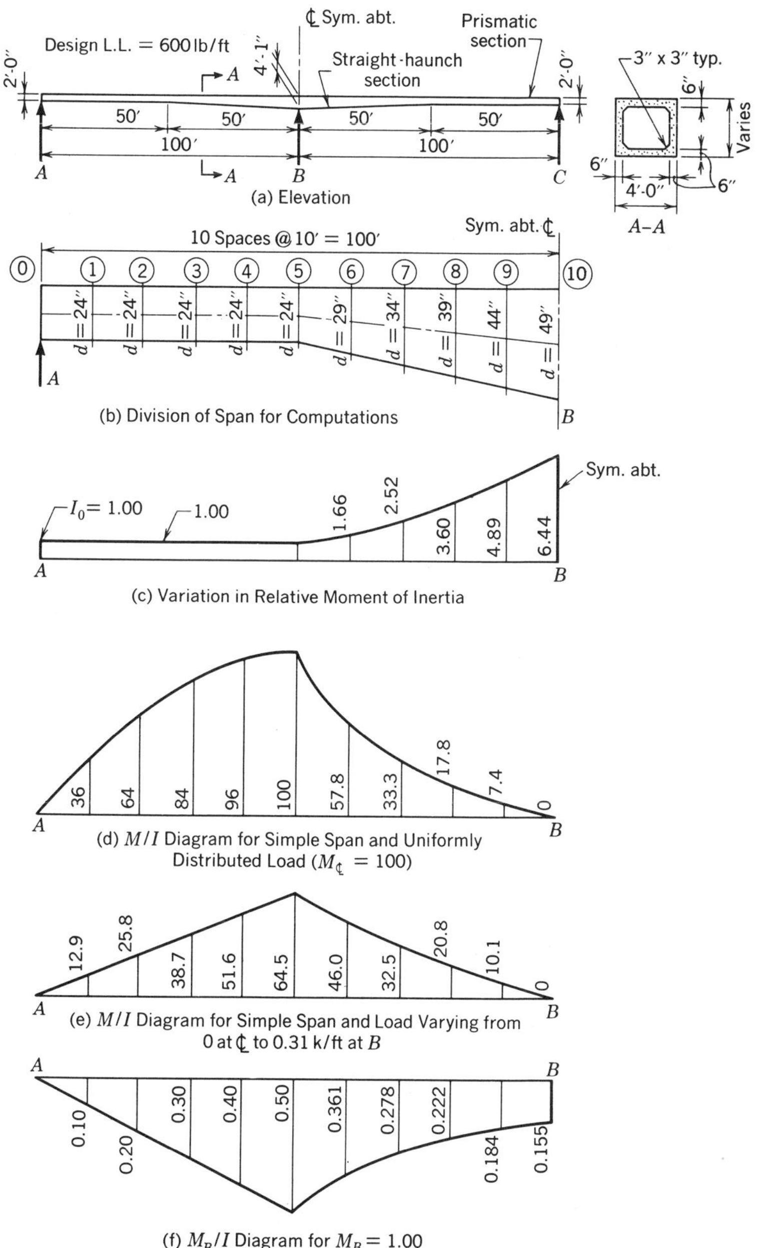

Fig. 8-21 Semigraphical solution of a two-span continuous beam having a variable moment of inertia.

TABLE 8-1 Summary of Computations Made in the De

		0	1	2	3
1	Depth (in.)	24 ←			
2	Moment of Inertia, I	50,460 ←			
3	I_n/I_0	1			
4	M_{max}	0	+416	+696	+838
5	M_{min}	0	+146	+217	+209
6	$M_{max} - M_{min}$	0	270	479	629
7	Moment, due to $w = 0.08$, over entire beam	0	+20.9	+33.8	+38.7
8	Relative Moment, due to $w = 0.08$ on both spans	0	+0.853	+1.380	+1.58
9	Moment due to $M_a = M_c = 100$	+100	+80.8	+61.7	+42.7
	FOR $P = 790^k$				
10	e_{min} (ft)	−0.475	+0.052	+0.407	+0.58
11	e_{max} (ft)	+0.475	+0.660	+0.749	+0.74
12	*P.L.*, Unif. load	0.000	+0.365	+0.591	+0.67
13	*L.T.* ordinates $B = 1.0$	0.000	+0.100	+0.200	+0.30
14	(12) + (13)	0.000	+0.465	+0.791	+0.97
15	$e_0 = -0.300$	−0.300	−0.242	−0.185	−0.12
16	(12) + (15)	−0.300	+0.123	+0.406	+0.54
	FOR $P = 1000^k$				
17	e_{min} (ft)	−0.475	−0.059	+0.221	+0.36
18	e_{max} (ft)	+0.475	+0.621	+0.692	+0.68
19	*P.L.* Unif. load	+0.000	+0.281	+0.455	+0.52
20	*L.T.* ord. $B = 0.50$	+0.000	+0.050	+0.100	+0.15
21	Tendon Loc. (19) + (20)	0.000	+0.341	+0.555	+0.67
22	$e_0 = 0.35$	−0.350	−0.283	−0.216	−0.14
23	Concordant Ten. (19) + (22)	−0.350	+0.008	+0.239	+0.37

L Symmetrical about ℄

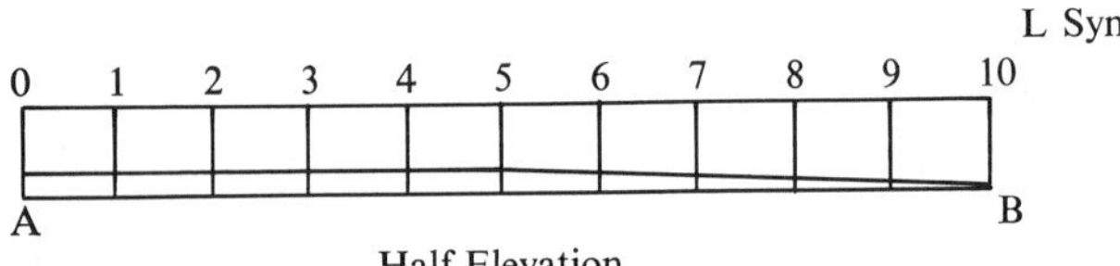

Half Elevation

f the Two-Span Continuous Beam Shown in Fig. 8-20

Point						
4	5	6	7	8	9	10
	→ 24	29	34	39	44	49
	→ 50,460	83,620	127,210	181,320	246,060	324,000
	1	1.66	2.52	3.60	4.89	6.44
+845	+714	+445	+32	−528	−1001	−1557
+125	−35	−275	−598	−1008	−1750	−2687
720	749	720	630	480	749	1130
+35.6	+24.5	+5.4	−21.7	−56.8	−99.9	−151.0
+1.450	+1.000	+0.220	−0.885	−2.32	−4.07	−6.16
+23.5	+4.3	−14.7	−33.8	−53.0	−72.0	−91.0
+0.595 +0.633 +0.622 +0.400 +1.022	+0.428 +0.431 +0.428 +0.500 +0.928	−0.040 +0.255 +0.094 +0.600 +0.694	−0.687 −0.029 −0.379 +0.700 +0.321	−1.509 −0.439 −0.995 +0.800 −0.195	−2.240 −1.260 −1.740 +0.900 −0.84	−3.040 −2.330 −2.64 +1.00 −1.64
−0.075 +0.547	−0.013 +0.415	+0.044 +0.138	+0.102 −0.277	+0.159 −0.836	+0.216 −1.524	+0.273 −2.367
+0.370 +0.600 +0.478 +0.200 +0.678	+0.239 +0.440 +0.330 +0.250 +0.580	−0.158 +0.328 +0.073 +0.300 +0.373	−0.695 +0.129 −0.292 +0.350 +0.058	−1.369 −0.167 −0.766 +0.400 −0.366	−1.961 −0.790 −1.340 +0.450 −0.890	−2.627 −1.617 −2.030 +0.500 −1.530
−0.082 +0.396	−0.015 +0.315	+0.051 +0.124	+0.118 −0.174	+0.185 −0.581	+0.252 −1.088	+0.318 −1.712

: + Moment causes tension on bottom fibers.
− Moment causes tension on top fibers.
+ *e* is measured below center of gravity.
− *e* is measured above center of gravity.

the results are tabulated in Table 8-1, lines 10 and 11, and are plotted to an exaggerated vertical scale in Fig. 8-22(a).

Because the pressure-line limits plotted in Fig. 8-22(a) indicate a trajectory similar to the curve which would be expected to result from a moment diagram for a uniformly applied load, the trial trajectory will be a tendon positioned in proportion to the ordinates (which are listed in line 7 of the table) of the unit-moment diagram for a uniformly applied load. Since the pressure line must pass through a very confined space at point 5, the ordinates of line 7 are expressed in terms of the ordinate at point 5, to facilitate computation of the eccentricities along the beam. Adopting a trial eccentricity of +0.428 ft at point 5, the ordinates of the tendon trajectory are computed and entered on line 12 in the table and the results are plotted in Fig. 8-22(a). Inspection of the plot, as well as comparison of the tendon ordinates with the limits of the eccentricity in lines 10 and 11 of Table 8-1, reveals that the tendon is located within the required limits. The tendon is concordant, since its ordinates were selected in proportion to a moment diagram for the beam, and hence, the pressure line is coincident with the tendon. Secondary moments need not be computed.

Inspection of the tendon trajectory shown in Fig. 8-22(a) will also reveal that the tendon falls outside of the dimensions of the beam at point *B*. This obviously cannot be allowed, and the trajectory of the tendon must be revised so that it will fall within the dimensions of the beam by an amount that will allow adequate concrete cover of the tendons to protect them from corrosion.

It is decided to transform the tendon linearly 1 ft at point 10, which will give the tendon 2.04 ft − 1.64 ft = 0.4 ft protective cover, which should be adequate. The ordinates of the linear transformation are given in line 13 and the transformed trajectory is given in line 14 of Table 8-1, from which it will be seen that the tendon does not fall within the limits of the beam at point 4. It is apparent that this trajectory cannot be linearly transformed in such a manner that the result will be satisfactory.

Another alternative, making the tendon eccentric at the ends, is then studied by computing the ordinates of the moment diagram that result in the beam when a unit moment is applied at each end (*A* and *C*) of the beam. The M/I diagram and the resulting moment diagram for this condition are shown in Fig. 8-22(b) and (c), and the ordinates at different points are summarized in Table 8-1, on line 9.

Because the ordinates which were determined for the unit moment at *A* and *C* are the ordinates of a moment diagram for the beam under study, they represent the location of a concordant tendon. By combining the ordinates of two concordant tendons, another concordant tendon location is obtained. Therefore, the pressure line of line 12 (proportional to the moment diagram

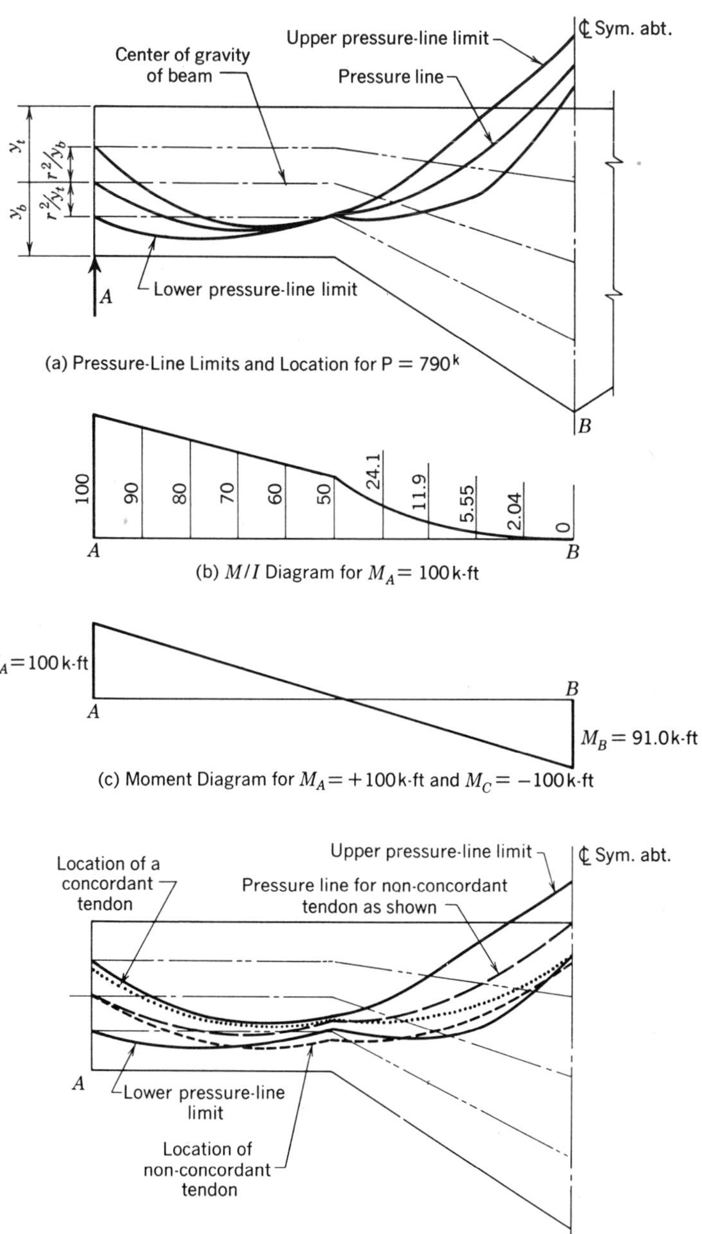

Fig. 8-22 Diagrams used in the semigraphical solution of the two-span continuous beam of Fig. 8-21.

of a uniformly applied load) is combined with the ordinates due to an eccentricity of the prestressing tendon of $-.30$ ft at point A, and the location of the resulting concordant tendon is given in line 16 of Table 8-1. Comparison of the results of this adjustment of the trajectory of the pressure line with the required limits of the pressure line (lines 10 and 11) reveals that the new trajectory is too high at points 3, 4, and 5.

The requirements of the pressure line for a force of 790 k will lead one to conclude that the shape of the trajectory must be changed from one that is proportional to the moment diagram, due to a uniformly applied load. Due to the narrowness of the limits for the pressure line near points 5 and 10, difficulty will be experienced in selecting another curve which will fall within these limits. Another consideration is that the required eccentricities at points 5 and 10 are so large, in comparison with the dimensions of the beam (for $P = 790$ k), that the usefulness of linear transformation is greatly restricted. Although a satisfactory curve can probably be found for a force of 790 k, to illustrate the devices available to the designer, a larger force will be used. Another, and perhaps better, alternative would be to alter the configuration of the beam.

Adopting a prestressing force of 1000 k, the limits of the eccentricity and the pressure line that is proportional to the uniformly applied load moment diagram (indicated as the pressure line for the non-concordant tendon in Fig. 8-22(d)) are computed and listed in lines 17, 18 and 19 of Table 8-1 and, in addition, are plotted in Fig. 8-22(d). It should be noted that the use of a larger force has greatly increased the area in which the pressure line can be placed with satisfactory results.

The pressure line that is proportional to the uniformly applied load was selected on the basis of having an eccentricity of $+0.330$ at point 5. The pressure line falls within the required limits. In order to provide adequate cover for the tendons, it is necessary to transform the tendon linearly 0.50 ft at point 10, which is done in line 20 of the table. The ordinates of the resulting tendon trajectory are indicated in line 21 of the table and are plotted in Fig. 8-20(d). This solution is satisfactory, since the concrete cover is adequate at all sections and the pressure line of the non-concordant tendon is within the required limits.

Another satisfactory, concordant tendon trajectory is computed in Table 8-1, lines 22 and 23. This trajectory is the combination of the pressure line which results from the non-concordant tendon and the pressure line from an eccentricity of the tendon at the ends of -0.35 ft. The tendon is concordant, since it is the result of adding the trajectory of a concordant tendon to the trajectory of a pressure line.

8-9 Limitations of Elastic Action

As has been stated previously, prestressed-concrete continuous structures perform substantially as elastic structures under loads which do not result in stresses that exceed the normal working stresses permitted by recognized prestressed-concrete design criteria. Adequate experimental data are available to substantiate this.

Under normal conditions, when the design load is exceeded, the concrete stresses remain reasonably elastic up to the load which causes visible cracking in the structure. The first cracks, which are not visible to the unaided eye, do not materially affect the performance of the structure. As a matter of fact, when the load at which the first crack observed with the unaided eye during a test of a beam is plotted on a load-deflection curve, it is often below the point at which pronounced deviation of the tangent (or plasticity) to the elastic deflection curve takes place.

In most continuous structures, the cracking load is not reached simultaneously in all highly stressed sections, since the magnitude of the moments vary at different sections along the member and, when members are designed for moving live loads, the largest moments that may occur at each section under the assumed design loads do not occur under the same condition of loading. Furthermore, once the cracking load has been significantly exceeded at a particular section, there is a reduction in the effective moment of inertia, and hence, stiffness of the member in the cracked area. At loads which result in one or more areas of a beam being stressed substantially above cracking, the effective modulus of elasticity of the concrete in the cracked areas may be considerably lower than in areas that are still stressed in the elastic range—this further contributes to the reduction in the stiffness of the member in the cracked areas.

Because of the localized changes in stiffness of a continuous member subjected to significant overload, the distribution of moments are no longer proportional to the distribution of moments in the elastic range. This is explained by the fact that, after cracking has reached a significant degree at one or more areas in a beam, the moment that results from the application of additional loads is carried in greater proportion by the portion of the member that remains uncracked. The areas first to attain a highly cracked and highly stressed condition yield more upon the application of additional load than areas that remain uncracked, and hence, the cracked areas resist less of the additional loads than would be indicated by purely elastic analysis. The phenomenon is called "redistribution of the moments."

Redistribution of the moments is the phenomenon that results in continuous beams which are designed on a purely elastic basis, frequently, but not

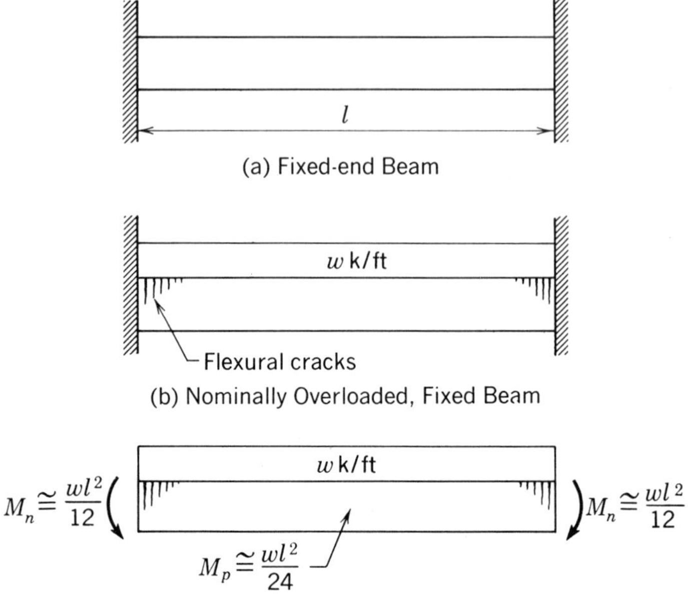

(a) Fixed-end Beam

(b) Nominally Overloaded, Fixed Beam

(c) Moments in Nominally Overloaded, Fixed Beam

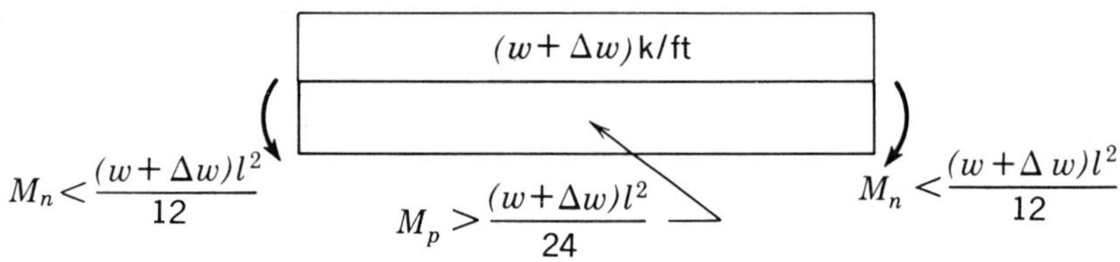

(d) Moments in Significantly Overloaded, Fixed Beam

Fig. 8-23 Variation in moments in a fixed beam subjected to various conditions of overload.

always, having very high factors of safety. Redistribution of the moments can be illustrated by considering the fixed beam of Fig. 8-23, which, when subjected to loads slightly above the cracking, has a distribution of moments that is virtually identical to the distribution that would result from an elastic analysis. When an additional increment of load is applied, the cracked sections are not as stiff in proportion to the uncracked sections as they were previously, and hence, the distribution of moments deviates from the elastic distribution. The cracked section is called a "plastic hinge" when the concrete and steel stresses at that section reach the plastic range and yield (rotate) with virtually no increase in stresses as additional increments of moment are applied.

The redistribution of moments takes place to various degrees. Redistribution is said to be complete when the various critical sections of a beam all attain a high degree of plasticity and attain the ultimate moments that would be indicated by the ultimate moment analysis developed in Sec. 5-2. Some beams, if loaded to destruction, would fail before redistribution is complete. It is not completely understood why this occurs. One theory that has been proposed is that the premature failures (Ref. 4) are the result of the requirement that the rotations of each span of a continuous beam that adjoins at a common support must be equal. This requirement may restrict the redistribution of moment and prevent it from becoming complete.

Additional research is needed into the phenomena that control the redistribution of moments, but at the present time, it is believed the following characteristics improve the redistribution of moments that take place in a beam, loaded to destruction.

(1) Low values of the steel index (percentages of reinforcement).

(2) Good bond between the prestressing reinforcement and the concrete.

(3) The use of untensioned reinforcement, in areas of high moment in beams with very low values of steel index, and in beams that have unbonded tendons, in order to improve the cracking patterns at high loads.

(4) Prevention of large inclined shear cracks in areas of high moment, since such cracks reduce the ultimate moment and rotation capacities of the section. The use of untensioned shear reinforcement in the areas of high shear and high moment is considered necessary.

(5) Redistribution appears to be more complete in members having large differences in ultimate moment capacity at the various sections than it is in members in which the moment capacity is nearly equal at all critical sections (Ref. 9).

In most applications, when the design is based upon the elastic theory using normal stresses and economical percentages of steel, the redistribution of moments could be assumed to be complete at ultimate. The load which will cause collapse can be determined according to the procedure outlined in the following article. The Building Code Requirement for Reinforced Concrete (ACI 318) permits limited redistribution of the moments in continuous beams. These provisions are as follows:

Continuous beams and other statically indeterminate structures shall be designed for adequate strength and satisfactory behavior. Behavior shall be determined by elastic analysis, taking into account the reactions, moments, shear, and axial forces produced by prestressing, the effects of temperature, creep, shrinkage, axial deformation, restraint of attached structural elements, and foundation settlement.

The negative moments due to design dead and live loads calculated by elastic theory for any assumed loading arrangement, at the supports of continuous prestressed concrete beams with sufficient bonded steel to assure control of cracking, may be increased or decreased by not more than $20[1 - (q + q^* - q')/0.30]$ percent, provided these modified negative moments are also used for final calculations of the moments at other sections in the span corresponding to the same loading condition. Such an adjustment shall only be made when the section at which the moment is reduced is so designed that q^*, $(q + q^* - q')$, or $(q_w + q_w^* - q_w')$, whichever is applicable, is equal to or less than 0.20. The effect of moments due to prestressing shall be neglected when calculating the design moments.

The following definitions are applicable to the above:

A_s = area of nonprestressed tension reinforcement.

A_s^* = area of prestressed reinforcement.

A_s' = area of compression reinforcement, sq in.

$$q = \frac{A_s f_y}{b\,df_c'}$$

$$q^* = \frac{A_s^* f_{su}}{b\,df_c'}$$

$$q' = \frac{A_s' f_y}{b\,df_c'}$$

q_w, q_w^*, q_w' = reinforcement indices for flanged sections computed as for q, q^*, and q', except that b shall be the web width and the steel area shall be that required to develop the compressive strength of the web only.

From these provisions it will be seen the redistribution of moments would be a maximum value of 20% for a reinforcement index of 0 and a minimum value of 6.67% when the applicable steel index is 0.20.

For designs that have very high or very low steel indices or that have unbonded tendons, special study and perhaps tests are justified. With the present understanding of the phenomena that control redistribution of moments, it is not recommended that designs be based upon a limit analysis with applicable load factors. The recommended procedure is to base the design upon an elastic analysis and review the safety of the structure using the limit analysis. Again, in continuous structures, as in statically determinate structures, it is recommended that the ultimate moments and safety factors always be investigated, regardless of the elastic-design methods and criteria employed.

8-10 Analysis of Ultimate Loads

The computation of the ultimate load which continuous beams can withstand is based upon the plastic-hinge (limit analysis) theory. This theory, very simply, is that a continuous member loaded near ultimate will acquire sections that are more highly stressed than other sections, due to the variation in the magnitude of the moments in the member. When the most highly stressed sections become stressed in the plastic range, application of additional load results in these sections yielding rather than resisting additional moment. A continuous prestressed beam subjected to a steadily increasing load is stable and will continue to develop plastic hinges until the number of hinges is equal to the degree of indeterminacy in any one span or in the beam as a whole. The appearance of one plastic hinge, more than the degree of indeterminacy, will result in instability in the structure, and it will collapse.

This is illustrated by the fixed beam shown in Fig. 8-24 which, when subjected to a very high uniform load, develops plastic hinges at each end and, for the purposes of analyzing the distribution of moments at higher loads, the beam can be assumed to be a simple beam with negative end moments at each end equal to:†

$$M_n' = \phi\left[A_s^* f_{su_1}\left(d_1 - \frac{a_1}{2}\right)\right]$$

When the load is increased to such a degree that a plastic hinge develops at the center of the member, there would be one more hinge than the degree of indeterminacy, and the structure would collapse. The ultimate moment under this condition can be expressed by

$$M_T' = M_n' + M_p' = \phi\left[A_s^* f_{su_1}\left(d_1 - \frac{a_1}{2}\right) + A_s^* f_{su_2}\left(d_2 - \frac{a_2}{2}\right)\right]$$

Assuming the beam of Fig. 8-24 is rectangular, prismatic and $d_1 = d_2$, the ultimate moment the beam can develop is expressed by

$$M_T' = M_n' + M_p' = 2M_n' = \frac{w'l^2}{8}$$

in which w' is the ultimate load the beam can withstand.

In the elastic analysis of such a beam, the moment at the support is twice the moment at the center, and the sum of the moments on the beam is equal to

$$M_T = \frac{wl^2}{8} = 1.5M_n$$

† This is the general equation for ultimate moment.

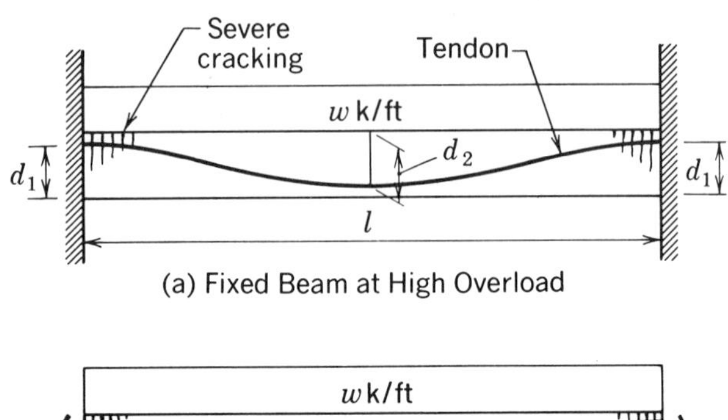

(a) Fixed Beam at High Overload

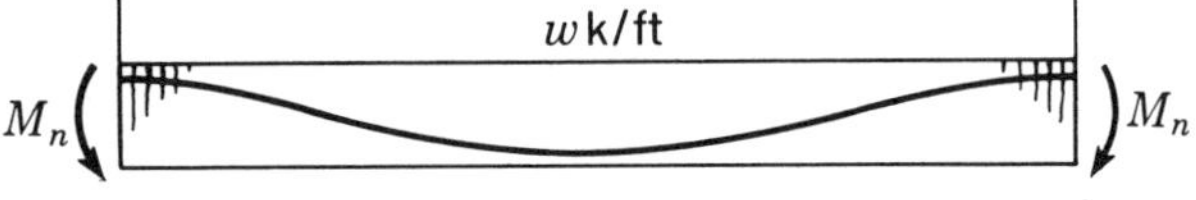

(b) Equivalent Simple-Beam Loading near Ultimate

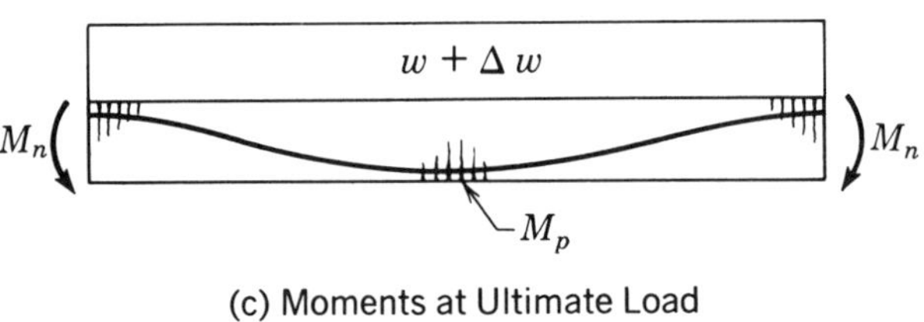

(c) Moments at Ultimate Load

Fig. 8-24 Illustration of the development of plastic hinges in a fixed beam.

Therefore, the ratio between the ultimate resiting moment and the elastic moment is equal to

$$\frac{M_T'}{M_T} = \frac{w'l^2/8}{wl^2/8} = \frac{2M_n'}{1.5M_n}$$

Since the ratio of M_n'/M_n varies for normal values of steel index from approximately 2 to 4, the factor of safety of such a beam will be from 2.67 to 5.33. In beams with variable moments of inertia, the factors of safety indicated by the limit analysis are frequently higher than this. Load factors of the order of 6 to 8 have been found rather frequently in testing continuous beams.

In investigating the safety of a multispan continuous beam under ultimate loading, all conditions of loading must be considered. The loading arrangement that would result in the collapse of the member due to negative moment may

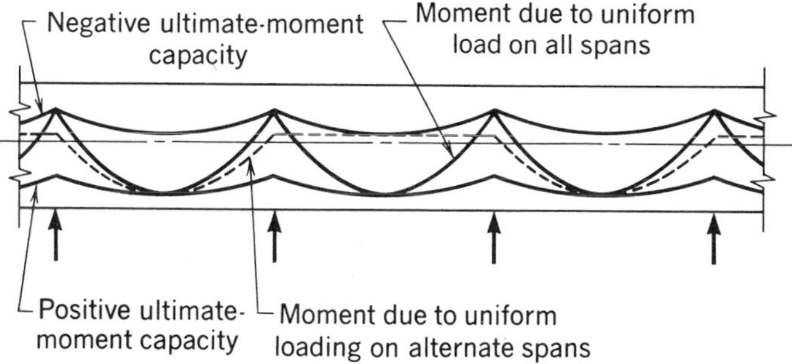

Fig. 8-25 Moment diagrams for uniform loads on all spans and on alternate spans of a multispan beam superimposed on envelope of ultimate moment capacity.

be quite different from the loading that would result in collapse due to positive moment or from positive and negative movement, simultaneously, in the same span. In addition, since reversal of moments occur in continuous structures, it is possible to have critical positive moments develop at ultimate in areas of the beams which are normally stressed by negative moments. The opposite condition is also possible. Therefore, to facilitate the determination of the ultimate strength of a member, an envelope of the maximum and minimum ultimate moments that can be developed at various sections in the member are computed and plotted as shown in the sketch of Fig. 8-25. The moment diagrams due to the various conditions of loading (with the appropriate load factors) are computed and plotted in the envelope. The moments can be distorted from those obtained by the elastic analysis as much as that given in the excerpt from the ACI Building Code given in Sec. 8-9, on projects which are governed by the Code. If the moment diagrams so computed fall within the limits of the ultimate moment envelopes, the design is satisfactory. If not, the design must be altered.

The secondary moments and differential settlement of the supports can be ignored in limit analysis. In addition, linear transformation (which means secondary moment) normally has no effect on the ultimate moment a member can resist.

ILLUSTRATIVE PROBLEM 8-5 Compute the ultimate moment for the beam of Prob. 8-4, using the non-concordant tendon trajectory shown in Fig. 8-22. Assume $f_c' = 4500$ psi, $f_s' = 250{,}000$ psi, $A_s = 7.50$ in.^{2}.

SOLUTION: The ultimate moment capacity for both positive and negative moment are computed using the principles of Sec. 5-4. The computations ignore the existence of the secondary moment.

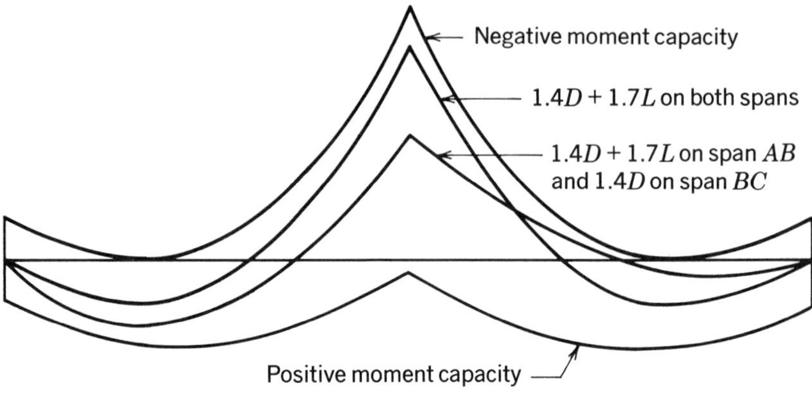

Fig. 8-26 Graphical representation of the ultimate moment capacities and design moments for the beam of Prob. 8-5.

The values of M_u obtained from the above computation are plotted in Fig. 8-26, and the moment curves that result from design (ultimate) uniform loading on both spans and span *AB* only are superimposed on the plot. It will be seen that the design moment curves fall within the envelope of the moment capacity curves. Hence, the beam affords the necessary capacity without redistribution of moments.

It should be recognized that in continuous beams, the steel indices can be substantially higher than is normally experienced in simple beams; in the latter, steel indices of 0.10 to 0.40 are commonly found. In continuous beams, the steel indices may be as high as 2.00, since the tendon may be very near the flange while the section may be subject to moment reversal. Such highly reinforced areas would certainly fail suddenly, and for this reason, critical areas of this type should be investigated with great care and modified so that adequate safety of the structure is assured.

The beam which was selected for use in Probs. 8-4 and 8-5 is certainly not the best solution for the loading conditions specified. The beam does, however, allow illustration of the difficulties that may be encountered with continuous beam design.

In addition, although it may not be apparent due to the omission of much of the routine computations in order to avoid confusion in presenting the examples, the average concrete stress in the straight section of the beam with $P = 1000$ kips is of the order of 1350 psi, which is considerably higher than is normally found to be economical and practical. Furthermore, the 1000-kip force is over 26% larger than the minimum possible required force.

8-11 Additional Considerations

The effect of tendon friction during stressing has not been discussed in the preceding articles on the design of continuous beams. The fundamental principles that govern friction loss of post-tensioned tendons are the same for simple and continuous beams and, since it is more of a practical than theoretical consideration, the discussion of tendon friction is included in the discussion of post-tensioning methods (Chapter 15).

Tendon friction can result in substantial losses of stress, and for this reason, particularly in the design of continuous structures, the designer must select the tendon trajectory that will result in lower friction losses. The designer should estimate the friction loss as various trajectories are studied and, when high losses are of significance and are unavoidable, the design should be made on the basis of a variable prestressing force.

The use of sharp bends or abrupt changes of slope in the tendon is generally avoided, since such sharp bends can result in significant secondary bending stresses in the tendons. When it is desired to change the slope of the tendon rather abruptly, the tendon should be curved over a distance of several feet rather than deflecting it over a small pin or roller.

The angular change through which the tangent to the tendon passes has important influence on the friction loss which results during stressing, with larger losses resulting from the larger curvatures. From this standpoint, a tendon trajectory composed of two chords is preferable to a parabolic trajectory, since the angular change in a parabolic trajectory is twice that of the angular change obtained with chords.

In short-span, continuous prestressed structures, the dead load of the structure is small in comparison to the live load and, if the structure is subjected to a moving live load, the critical sum of the maximum and minimum moments is nearly the same as the simple beam moment for the same span. Therefore, the prestressing force required for such a structure is not significantly less than is required for the simple span. This greatly reduces the economy of materials that one would normally expect to result from the use of continuity. The advantages of less deflection, great resistance to lateral and longitudinal loads, among other things, are still attained through the introduction of continuity.

When the dead-load moment is a large portion of the total moment, the variation in moment in a continuous beam is less than is found in simple beams and, for this reason, the prestressing force is less for the continuous structure. This accounts for the fact that continuity is used on long-span structures to a much greater degree than in short-span structures.

As was shown in Prob. 8-4, the minimum prestressing force required by theoretical considerations may not result in a practical solution, since it may be impossible to find an adequate trajectory for a tendon of such capacity. Therefore, some of the theoretical economy of continuity may be lost in long-span beams in order to increase the limits in which the prestressing force can be placed and to reduce the eccentricity required. Additional loss in the theoretical economy of materials, derived from an elastic analysis, may result from the reversals of moment at over-loads and the necessity of supplying extra prestressing or non-prestressed reinforcement for such conditions of loading.

In spite of these restrictions and disadvantages for the use of continuous prestressed structures, continuity in long-span, prestressed, building and bridge construction is practical and economical under specific conditions.

The same general procedures of design and analysis that have been presented in this chapter are used in designing prestressed rigid frames. In such construction, special attention must be given to the moments that result from the axial shortening of the members due to prestressing.

8-12 Continuous Beams Utilizing Prestressed Beam Soffits

Simple flexural members composed of prestressed components and plain or reinforced-concrete components have been employed in a number of applications domestically and abroad. The most common application of this type of construction is in the use of prestressed bridge stringers used in combination with a cast-in-place slab to form a T beam in the completed structure (*see* Sec. 6-4). Other types of composite simple beams include those with cross sections, as illustrated in Fig. 8-27.

Composite beams composed of a precast, prestressed component and a cast-in-place component can be made continuous at nominal cost under some conditions, by including normal reinforcing steel as negative-moment

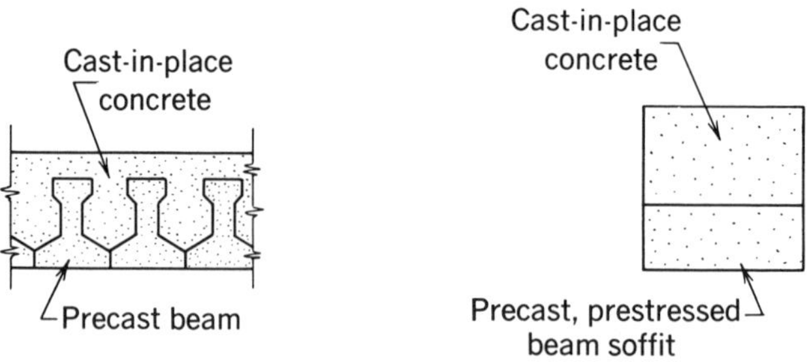

Fig. 8-27 Typical sections of two types of composite beams.

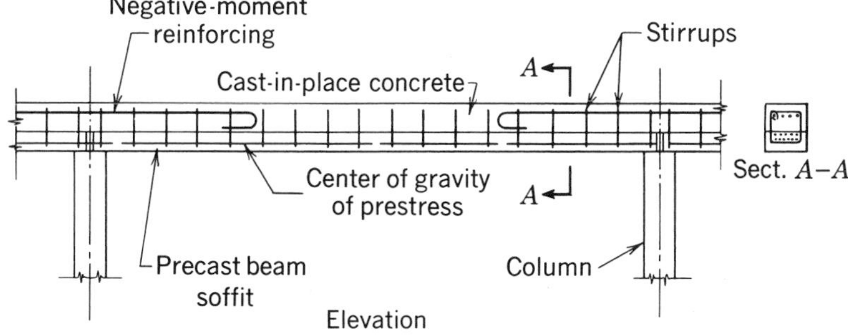

Fig. 8-28 Composite beams utilizing a precast soffit.

reinforcing, as is illustrated in Fig. 8-28. This type of beam may be designed in such a manner that it is necessary to shore the precast soffit in place after erection, during placing, and curing of the cast-in-place concrete. The result of this procedure is that the entire dead load, as well as the live load, is carried by the composite section of the continuous beam. On the other hand, the precast component may be proportioned in such a manner that the entire dead load is resisted by the precast section alone and the composite beam is continuous for live loads alone.

8-13 Continuous Beams Constructed in Cantilever

In Sec. 8-5, in which the assumptions generally made in the analysis of continuous prestressed beams are listed, it is assumed that effects of each cause of moments can be calculated independently and superimposed to attain the result of the combined effect of the several causes. This is the principle of superposition. This assumption is valid for a continuous member that is stressed after the complete structural scheme has been constructed. The assumption is not true for a structure constructed in increments with stresses imposed upon the structure during each construction phase.

In considering this effect, the variation in strain the concrete undergoes during the lifetime of the structure must be considered. As was pointed out in Secs. 6-2 and 6-3, the more sophisticated methods of analyzing the losses of prestress and the deflection of beams must take into account the initial strains to which the structure is subjected, as well as the time dependent changes in strain due to shrinkage, creep, and relaxation. For a continuous member that is cast-in-place in one operation and stressed in its final structural form, the effects of shrinkage and relaxation can reasonably be assumed to be the same at every point in the structure and, since the effect of creep is assumed to be directly proportional to the stress level in the concrete, the creep strain at every section in the structure at any point in time can be assumed to be in direct proportion to the distribution of moments in the structure at the time of stressing. This is not precisely true, since there will be some difference in the relaxation loss along the length of a tendon due to the variation in initial stress in the tendon. This variation results from friction during stressing and other causes.

In the case of members cast in segments, subjected to stresses while still in the form of determinate structural elements, and later rendered continuous and indeterminate by cast-in-place closure joints and additional prestressing, the instantaneous strains in the structure can still be considered to be in proportion to the distribution of moments in the structure at the time the structure was rendered continuous. (An example of this type of structure is shown in Fig. 8-3.) However, the time dependent changes in strain which tend to take place at each section cannot be taken as proportional to the distribution of moments in the structure at the time continuity was established, due to the variations in free strain changes that would take place at various locations along the structure. The variation in strains along the length of the structure results in a redistribution of moments, which can be calculated using the numerical integration methods described in Sec. 6-3 with the additional requirement that at each time increment the conditions of continuity must be met (i.e. deflections at the supports are zero and the slopes of the tangents to the elastic curves are equal at each interior support).

A method of approximating the effect of the creep upon the redistribution of moments has been proposed by Jean Muller. This expression is

$$M_f = M_1 + (M_2 - M_1)\left(1 - \frac{E_f}{E_i}\right) \tag{8-12}$$

in which M_f = the net final moment (effect of dead load and prestressing) after creep. M_1 = net moment (effect of dead load and prestressing) calculated by superimposing the various stresses computed for each step of the construction. M_2 = net moment (effect of dead load and creep) in the structure if constructed and loaded at once (as if it were cast-in-place). E_i = instantaneous modulus of elasticity of the concrete. E_f = final modulus of elasticity of the concrete after creep. In Eq. 8-12, the ratio Ef/Ei is often taken to be 0.33. Due to the fact that some creep would generally take place before continuity is established, the ratio of Ef/Ei may be greater than 0.33 and should be evaluated for each individual case.

This phenomenon can affect the moments computed using the principle of superposition (M_1 above) as much as 20%. (Ref. 4.)

REFERENCES

1. Hognestad, Eivind, Mattock, Alan H. and Kaar, Paul H., "Composite Construction for Continuity," *Journal of the Prestressed Concrete Institute*, **5**, No. 1, 59–72 (Mar. 1969).

2. Foderberg, Dennis L. and Branson, Dan E., "Secondary Moments in Single-Span Prestressed Concrete Beams and Frames Determined by Column Analogy," *Journal of the Prestressed Concrete Institute*, **13**, No. 1, 32–58 (Feb. 1968).

3. Scordelis, A. C., Lin, T. Y. and Itaya, R., "Behavior of a Continuous Slab Prestressed in Two Directions," *Journal of the American Concrete Institute*, **31**, No. 6, 441–459 (Dec. 1959).

4. Muller, Jean, "Long-Span Precast Prestressed Concrete Bridges Built in Cantilever," pp. 705–740, Concrete Bridge Design, ACI Publication SP-23, Detroit, Michigan, 1969.

5. Muller, Jean, "Continuous Prestressed Concrete Structural Design," *Proc. Western Conf. on Prestressed Concrete*, 109–132 (Nov. 1952).

6. Saeed-Un-Din, K., "The Effect of Creep Upon Redundant Reactions in Continuous Prestressed Concrete Beams," *Mag. of Concrete Research*, **10**, 109 (Nov. 1958).

7. Parme, A. L. and Paris, G. H. "Designing for Continuity in Prestressed Concrete Structures," *Journal of the American Concrete Institute*, **23**, 45–64 (Sept. 1951).

8. Muller, Jean, "Flexural Strength of Prestressed Concrete Continuous Structures," pp. 9–19. Paper presented at the Knoxville Convention of the A.S.C.E. (January 1956).

9. Guyon, Y., "The Strength of Statically Indeterminate Prestressed Concrete Structures," *Proc. of a Symposium on the Strength of Concrete Structures*, 305 (May 1956).

10. ACI Committee 318, "Proposed Revision of ACI 318–63: Building Code Requirements for Reinforced Concrete," *Journal of the American Concrete Institute*, **67**, No. 2, 77–186 (Feb. 1970).

9 Direct Stress Members, Temperature and Fatigue

9-1 Introduction

The first portion of this chapter is devoted to a discussion of prestressed members subject to direct stress. With the possible exception of piles, prestressed concrete is not used extensively for members that are to resist direct stress.

The second portion of the chapter is devoted to the consideration of fire resistance, the effect of nominal temperature variations, and fatigue of prestressed concrete.

9-2 Tension Members or Ties

In the application of rigid frames, trusses, certain types of continuous beam framing, and in some water-front structures, among other types of structures, it is necessary to include structural components subject to direct tensile stress alone. Such members are referred to as ties. It may be desirable to provide ties of prestressed concrete rather than of steel or reinforced-concrete for one or more of the following reasons:

(1) Prestressed ties generally deform less under load than ties of reinforced concrete or steel. In addition, the deformation of the ties can be controlled by the designer. For example, a steel or reinforced-concrete tie would normally be designed in such a manner that the stress in the steel would be of the order of 20,000 psi under full load and, if the modulus of elasticity of the steel is assumed to be 29×10^6 psi, the deformation of such a tie 1000 in. long would be

$$\text{Deformation} = \frac{20{,}000 \times 1000}{29 \times 10^6} = 0.69 \text{ in.}$$

A prestressed tie can be proportioned in such a manner that the concrete stress is confined within any desired limits, without affecting the amount of steel required and the total force the tie will develop, simply by varying the area of the concrete section. If the modulus of elasticity of the concrete is 4×10^6 psi and the concrete stress due to effective prestress alone is 2000 psi, the deformation of a tie 1000 in. long under full load (concrete stress = zero under full load) would be

$$\text{Deformation} = \frac{2000 \times 1000}{4 \times 10^6} = 0.50 \text{ in.}$$

By confining the concrete stress due to prestressing to a lower limit, the deformation can be reduced in direct proportion to the stress.

(2) The use of prestressed ties in roof trusses may be preferred over the use of steel ties as a result of the greater fire resistance inherent in concrete members.

(3) Prestressed ties, due to their lack of cracks, are less subject to deterioration by corrosion, and hence, may be preferred over steel and reinforced concrete ties in certain applications.

In using prestressed ties, the designer must consider the effects of creep on the deformation of the tie. If the tie is prestressed and immediately thereafter put into service, assuming the superimposed load and the prestressing force are equal, the new concrete stress in the tie would be near zero and there would be no deferred deformation (creep).

If the tie were prestressed and stored for a period of a year or more before being put into service, a substantial amount of creep deformation would have taken place, and for service loads of short duration which were applied subsequently, the tie would elongate in proportion to the instantaneous modulus of elasticity of the concrete. For service loads of constant duration and of the same magnitude as the effective prestressing force, the concrete stress would be reduced to zero upon the application of the service load. The tie would instantly elongate by an amount determined by the instantaneous modulus of elasticity. The deformation (elongation) would continue until partial recovery of the original creep deformation had been obtained.

Accurate prediction of the amount of deferred strain that would be recovered upon removal of the load can only be made if the properties of the concrete are known. The plastic strain is generally much lower for unloading than for loading and a residual strain remains in the concrete. This is illustrated in the strain-time diagram in Fig. 9-1 (*see* Ref. 1).

The total strains occurring in concrete subjected to constant sustained stresses applied at various ages are as illustrated in Fig. 9-2 (Ref. 2). Using the principles of superposition described in Sec. 6-2, one can employ curves such as shown in Fig. 9-2 to estimate the creep effects due to variation in loading. For example, assume a unit stress of 1000 psi is applied at the age of 28 days and held constant until the age of 91 days, at which time, it is completely removed. The strain vs time diagram would be as in Fig. 9-3. The curve of Fig. 9-3 is constructed using the curves from Fig. 9-2 with the assumption that the strain deformations upon loading or unloading at any particular age of the concrete take place at the same rate and are of the same magnitude, but of opposite sign.

It should be apparent that as the load is increased in a prestressed tie, the concrete stress reduces and the steel stress increases in direct proportion to the strain change in the concrete. Since the elastic modulus of concrete in tension is virtually the same as in compression, the action will continue until the tensile strength of the concrete is reached, at which time, the concrete will crack and the entire load must thereafter be carried by the steel alone. If the

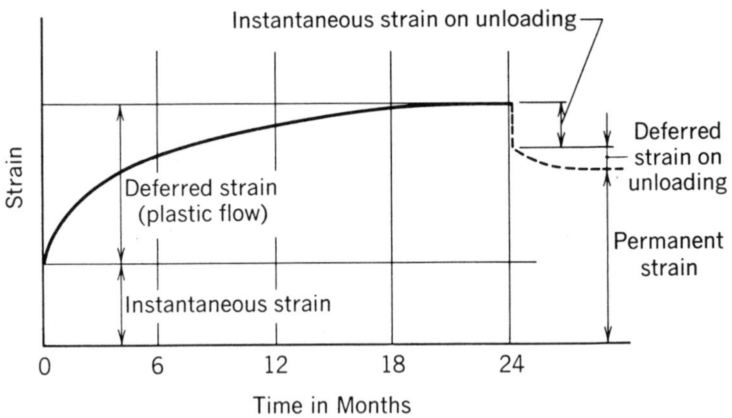

Fig. 9-1 Concrete strains under long-term loading and unloading.

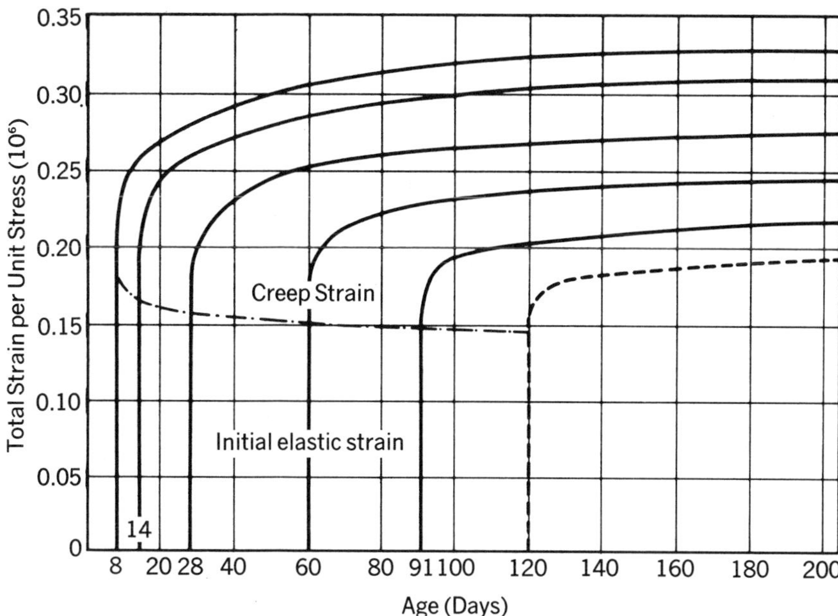

Fig. 9-2 Total strain due to constant sustained stress applied at various ages to a high-strength concrete stored at high humidity.

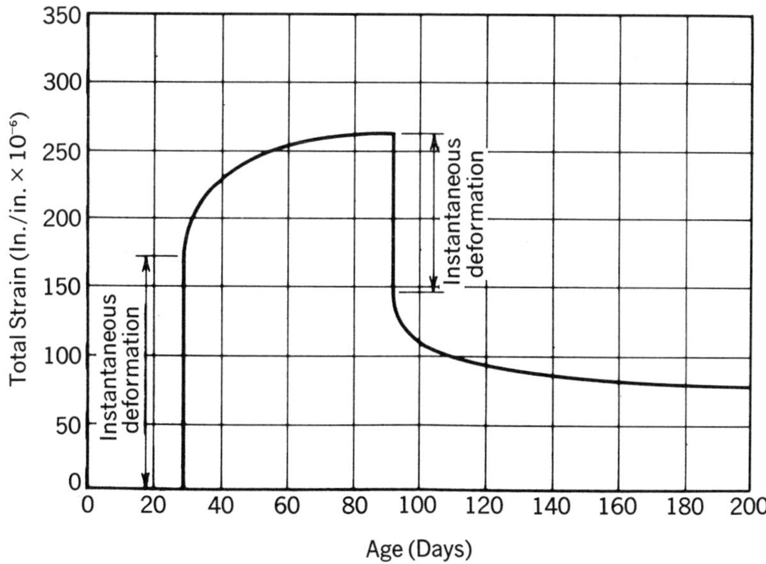

Fig. 9-3 Strain vs time diagram for concrete having creep characteristics of that in Fig. 9-2. For 1000 psi applied at 28 days and removed at 91 days.

load is then reduced below the value of the effective prestress, the cracks will close and there will be a compressive force in the concrete which is equal to the difference between the effective prestressing force and the service load.

If the load resisted by the steel and concrete just prior to cracking is greater than the ultimate strength of the steel, the cracking load will be the ultimate load. This condition can exist in members prestressed with a very small percentage of reinforcing.

Consideration of the action of prestressed ties and the elastic-plastic nature of concrete leads one to the conclusion that the designer must consider the stress level in the concrete and the duration of the applied loads in estimating the deformations that will result in the ties. Obviously, such deformation will result in deflection of a structure and may result in significant secondary stresses.

9-3 Columns and Piles

Prestressed concrete members that are axially loaded in compression are designed in the same manner as are reinforced concrete members. ACI 318-71 provides that members with an effective prestress of less than 225 psi must have the minimum amount of non-prestressed reinforcement that is specified for reinforced concrete columns. There is no minimum amount of non-prestressed reinforcement required for columns with an effective prestress that is 225 psi or more.

The prestressing steel of axially loaded members is required by ACI 318 to be enclosed either by ties that are spaced at the least dimension of the column, but not more than 48 tie diameters, or by spirals that conform to the requirements for reinforced concrete columns.

Prestressed concrete columns are not widely used. Reinforced concrete columns are more efficient and more economical in most applications.

Prestressed-concrete piles have been used to a very notable degree in the United States. The types of piles used can be divided into the three following classifications: (1) cylinder piles, (2) pre-tensioned, precast piles, and (3) Pre-tensioned spun piles.

Post-tensioned, multi-element cylinder piles are made by precasting hollow cylinders of concrete in sections about 16 ft long. Each section has a wall thickness of from 5 to 6 in. and holes are formed longitudinally through the walls at the time the sections are cast. After the precast sections have cured, they are aligned and the post-tensioning tendons are threaded through the holes in the walls and stressed and grouted in place. In this manner, piles up to 150 ft can be made.

Cylinder piles are also made pre-tensioned. In this process, the piles are either cast in conventional molds or may be cast in traveling molds which "extrude" the hollow sections.

The cylinder piles are normally made in diameters from 3 ft to 4 ft 6 in. and have been used with design loads up to 550 tons. The hollow shape is an efficient one for resisting axial loads, as well as for resisting bending moments that may be applied from any direction. Typical details and dimensions for cylinder piles are given in Fig. 9-4.

Pre-tensioned, precast piles have been made with square, triangular, octagonal, and round cross sections, both hollow and solid. Pre-tensioned piles have been used a great deal in the construction of waterfront structures and is fabricated in the normal manner used in pre-tensioned construction. This type of pile has been used more than cylinder or spun piles. Typical details and dimensions of square and octagonal prestressed piles are shown in Figs. 9-5(a) and 9-5(b).

Pre-tensioned spun piles are made in individual molds designed to resist the pre-tensioning force during the casting and curing of the pile. The manufacturing procedure consists of placing the tendons and reinforcing cage in steel molds, stressing the tendons, and placing the mold on revolving wheels that turn the mold as the concrete is placed. The centrifugal force compacts the concrete and forces the excess water from the plastic concrete. The pile is then cured and stripped from the mold.

Prestressed piles generally have better driving characteristics than reinforced-concrete piles. The prestressed piles seem to penetrate better and with less effort. In addition, prestressed piles can be made longer than is practical with reinforced-concrete piles, due to their lower dead weight and higher resistance to bending moments. Prestressed piles will also stand up well under adverse driving conditions, if they are properly designed and fabricated.

Pre-tensioned sheet piles have also been used on a number of projects in this country. Such piles are generally solid and rectangular in cross section, but have a tongue and groove to interlock them with adjacent piles in the completed structure.

The procedure used in the design of prestressed piles is no different than that employed in the design of columns that have axial load or combined axial load and bending. Experience has shown, however, that a minimum prestressing stress of from 700 to 900 psi is required in order to prevent the piles from cracking during driving. Cracking during driving has occurred on many projects and is believed to be the result of tensile stresses in the piles, due to the piles rebounding elastically from the driving hammer. This type of cracking is more apt to occur when driving is commenced (particularly in soft materials) and little tip resistance has been developed.

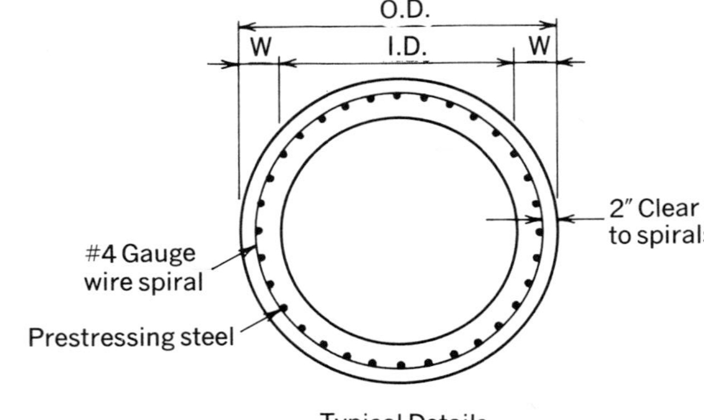

Typical Details

PILE PROPERTIES

Pile Size OD	ID	W	Area A_c	Approx. Weight Per lf (1)	Minimum Prestress Force (2)	Strands Per Pile Diameter (3) 7/16″	1/2″	I	Section Modulus	Perimeter	Design Bearing Capacity Concrete Strength (4) 5000 psi	6000 psi
36 in.	26 in.	5 in.	487 sq in.	508#	414 kips	24	18	60,000 in.⁴	3334 in.³	113 in.	242 tons	292 tons
	24 ,,	6 ,,	565 ,, ,,	590#	481 ,,	28	21	66,100 ,,	3676 ,,	113 ,,	282 ,,	339 ,,
48 in.	38 ,,	5 ,,	675 ,, ,,	703#	574 ,,	33	25	158,200 ,,	6593 ,,	151 ,,	337 ,,	405 ,,
	36 ,,	6 ,,	792 ,, ,,	826#	674 ,,	39	29	178,100 ,,	7422 ,,	151 ,,	396 ,,	475 ,,
54 in.	44 ,,	5 ,,	770 ,, ,,	802#	655 ,,	38	28	233,400 ,,	8645 ,,	170 ,,	385 ,,	462 ,,
	42 ,,	6 ,,	904 ,, ,,	940#	769 ,,	44	33	264,600 ,,	9802 ,,	170 ,,	452 ,,	542 ,,

Fig. 9-4 Pre-tensioned cylinder pile cross section.

NOTES

(1). Weights are based on 150 lbs per cubic foot.

(2). Minimum prestressed force based on unit prestress of 850 psi after losses.

31,000 lbs and 41,300 lbs respectively.

(4). Design bearing capacity based on 5000 psi and 6000 psi concrete and an allowable unit stress on the tip of the pile of $.2f'_cA_c$.

SQUARE PRESTRESSED PILES

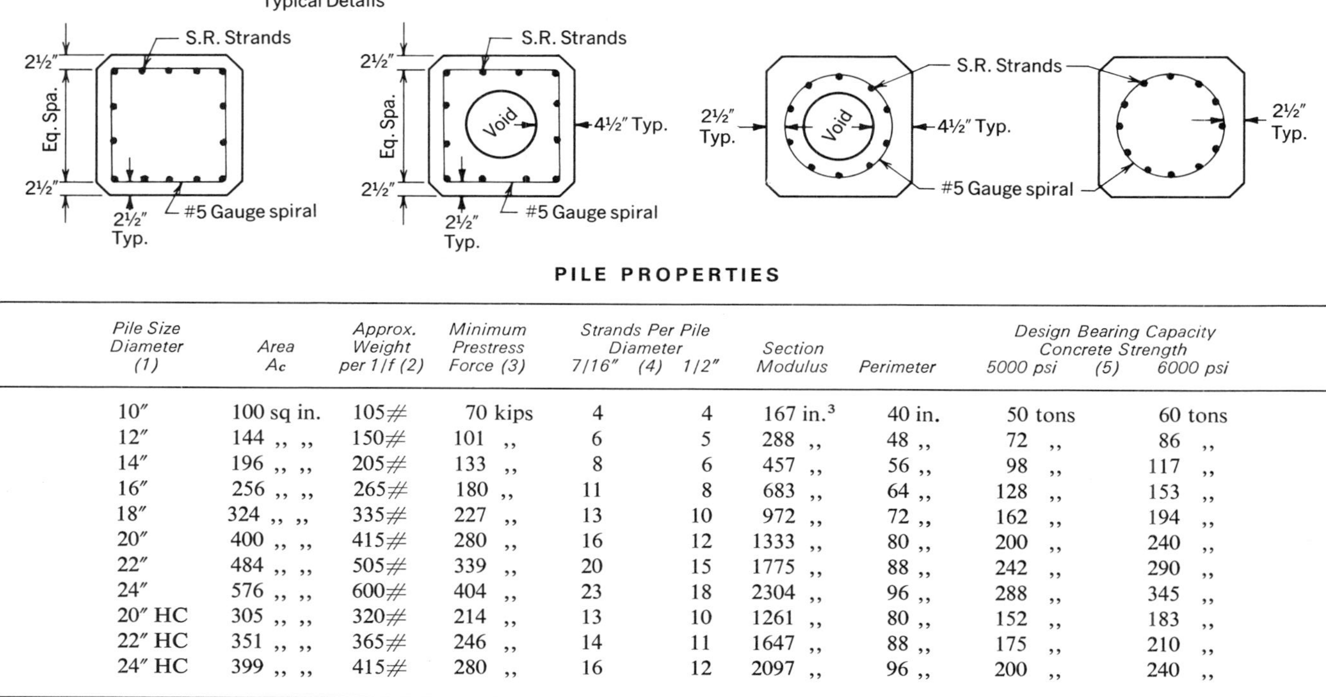

PILE PROPERTIES

Pile Size Diameter (1)	Area A_c	Approx. Weight per 1/f (2)	Minimum Prestress Force (3)	Strands Per Pile Diameter (4) 7/16"	1/2"	Section Modulus	Perimeter	Design Bearing Capacity Concrete Strength (5) 5000 psi	6000 psi
10"	100 sq in.	105#	70 kips	4	4	167 in.^{3}	40 in.	50 tons	60 tons
12"	144 ,, ,,	150#	101 ,,	6	5	288 ,,	48 ,,	72 ,,	86 ,,
14"	196 ,, ,,	205#	133 ,,	8	6	457 ,,	56 ,,	98 ,,	117 ,,
16"	256 ,, ,,	265#	180 ,,	11	8	683 ,,	64 ,,	128 ,,	153 ,,
18"	324 ,, ,,	335#	227 ,,	13	10	972 ,,	72 ,,	162 ,,	194 ,,
20"	400 ,, ,,	415#	280 ,,	16	12	1333 ,,	80 ,,	200 ,,	240 ,,
22"	484 ,, ,,	505#	339 ,,	20	15	1775 ,,	88 ,,	242 ,,	290 ,,
24"	576 ,, ,,	600#	404 ,,	23	18	2304 ,,	96 ,,	288 ,,	345 ,,
20" HC	305 ,, ,,	320#	214 ,,	13	10	1261 ,,	80 ,,	152 ,,	183 ,,
22" HC	351 ,, ,,	365#	246 ,,	14	11	1647 ,,	88 ,,	175 ,,	210 ,,
24" HC	399 ,, ,,	415#	280 ,,	16	12	2097 ,,	96 ,,	200 ,,	240 ,,

Fig. 9-5(a) Typical pre-tensioned pile cross section. (See notes on p. 339).

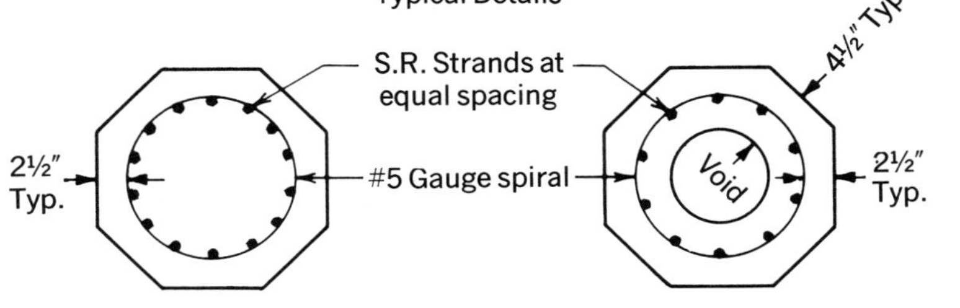

PILE PROPERTIES

Pile Size Diameter (1)	Area A_c	Approx. Weight per 1/f (2)	Minimum Prestress Force (3)	Strands Per Pile Diameter (4) 7/16"	1/2"	Section Modulus	Perimeter	Design Bearing Capacity Concrete Strength (5) 5000 psi	6000 psi
10"	83 sq in.	85#	59 kips	4	4	111 in.^{3}	34 in.	41 tons	50 tons
12"	119 ,, ,,	125#	84 ,,	5	4	189 ,,	40 ,,	59 ,,	71 ,,
14"	162 ,, ,,	170#	114 ,,	7	5	301 ,,	46 ,,	81 ,,	97 ,,
16"	212 ,, ,,	220#	149 ,,	9	7	449 ,,	54 ,,	106 ,,	127 ,,
18"	268 ,, ,,	280#	188 ,,	11	8	639 ,,	60 ,,	134 ,,	160 ,,
20"	331 ,, ,,	345#	232 ,,	14	10	877 ,,	66 ,,	165 ,,	198 ,,
22"	401 ,, ,,	420#	281 ,,	16	12	1167 ,,	72 ,,	200 ,,	240 ,,
24"	477 ,, ,,	495#	334 ,,	19	15	1515 ,,	80 ,,	238 ,,	286 ,,
20" HC	236 ,, ,,	245#	166 ,,	10	8	805 ,,	66 ,,	118 ,,	141 ,,
22" HC	268 ,, ,,	280#	188 ,,	11	8	1040 ,,	72 ,,	134 ,,	160 ,,
24" HC	300 ,, ,,	315#	210 ,,	12	9	1308 ,,	80 ,,	150 ,,	180 ,,

Fig. 9-5(b) Typical pre-tensioned pile cross section.

NOTES

(For both square and octagonal piles)

(1) Voids in 20″, 22″ and 24″ diameter hollow-core (HC) piles are 11″, 13″ and 15″ diameter, respectively, providing a minimum 4-1/2″ wall thickness. If a greater wall thickness is desired, properties should be increased accordingly.

(2) Weights based on 150 lb per cubic foot of regular concrete.

(3) Minimum prestress force based on unit prestress of 700 psi after losses.

(4) Bases on 7/16″ and 1/2″ high strength strand with an ultimate strength of 31,000 lbs and 41,300 lbs respectively. If regular strength strand is used, the number of strands per pile should be increased accordingly in conformance with strand manufacturer's tables.

(5) Design bearing capacity based on 5000 psi and 6000 psi concrete and an allowable unit stress on the tip of the pile of $.2f'_c A_c$. These bearing capacity values may be increased if higher strength concrete is used.

9-4 Fire Resistance

Fire resistance, as determined by standard tests, is a measure of the ability of a structural element to prevent the spread of fire and to retain the necessary structural strength at elevated temperatures. The fire resistance of a member is generally expressed in hours and is indicative of the length of time the member can be subjected to a standard fire without failing. Failure may result from the inability of the member to adequately perform any one of the following functions:

(1) Walls, floors, and roof elements must not allow a temperature rise, on the side opposite to the fire, which would be sufficient to allow inflammable material to ignite, either by conduction through the member or as a result of holes and cracks forming in the member, which would allow flames, hot air, or gases to pass through the member.

(2) All structural members must retain sufficient strength at elevated temperatures to assure safety to the occupants as well as the firemen who are fighting the fire. Severe structural deflection could result in cracking of the supported slabs or panels, which would cause them to fail according to the requirement described above.

(3) The elements must also retain their structural integrity under the action of a stream of water, during or immediately after exposure to the standard fire.

The effects of elevated temperatures on the principal constituents of reinforced and prestressed concrete are interesting from an academic standpoint and can be generalized as follows:

Concrete: The coefficient of thermal conductibility is often assumed to be greater for high-strength concrete, such as is used in prestressed concrete, than for low-strength concrete, although the variation in this coefficient is large. The compressive strength of concrete can be assumed to be approximately 67% and 33% of the strength at room temperature when the temperature is raised to 750°F and 1450°F, respectively. The modulus of elasticity, in terms of the modulus at room temperature, is reduced to approximately 50% at 750°F and 20% at 1450°F. Decomposition of the concrete is evident at about 1300°F (Ref. 6).

High-tensile steel wire: The strength of high-tensile steel generally varies as follows at elevated temperatures: (1) Slight increase in strength up to temperatures of 300°F; and (2) The ultimate tensile strength at 750°F is approximately 50% of the original ultimate tensile strength. The reduction in the elastic modulus of high-tensile steel is of the order of 6% at 400°F and 20% at 600°F. In addition, elevated temperatures result in a significant increase in the relaxation of the steel. The thermal coefficient of linear expansion is not a constant value for cold drawn wire at high temperatures. The expansion of wire heated to temperatures above 300°F is not entirely recovered upon cooling. Virtually all of the strength of wire is regained upon cooling from temperatures as high as 600°F, even though the strain is not. The strength upon cooling for wire heated as high as 750°F may be a considerable percentage of the original strength (Ref. 7).

Reinforcing steel: The strength of normal reinforcing steel is reduced to 50% of its original strength at temperatures of 950 to 1100°F.

In view of the above general properties, at elevated temperatures and after cooling from elevated temperatures, a prestressed structure that was exposed to a fire and did not fail could be expected to possess the characteristics indicated for the following maximum temperature conditions.

Above 400°F, but less than 600°F, the reduction in the effective prestress may reduce the resistance of the cooled structure to cracking, although the ultimate strength may not be materially affected, due to the recovery of the steel strength upon cooling.

Above 600°F, but less than 750°F, the structure would be expected to be badly cracked and have permanent deflection. The ultimate strength of the structure after cooling may still be adequate, since the regain of steel strength would be considerable and the maximum reduction in concrete strength would be of the order of one-third.

Above 750°F, the reduced steel strength and the loss of concrete compressive strength would very likely render the cooled structure unsafe.

In the past several years, there has been considerable research into the ability of prestressed concrete to resist the effects of fire. This research has resulted in the requirements summarized in Table 9-1. The requirements were adopted in the 1967 Edition of the Uniform Building Code (Ref. 8). Grade A concrete is defined as concrete made with aggregates such as limestone, calcareous gravel, trap rock, slag, expanded clay, shale, slate, or any other aggregates possessing equivalent fire resistive properties. Grade B concrete is all concrete other than Grade A concrete and includes concrete made with aggregates containing more than 40% quartz, chert, or flint. In addition, the Uniform Building Code lists the following requirements and interpretations:

Bonded prestressed concrete tendons For members having a single tendon or more than one tendon installed with equal concrete cover measured from the nearest surface, the cover shall be not less than that set forth on Table No. 43-A.

For members having multiple tendons installed with variable concrete cover, the average tendon cover shall be not less than that set forth in Table No. 43-A provided:

TABLE 9-1 Requirements of the 1967 Edition of the Uniform Building Code Stipulated in Table No. 43-A

Structural Parts to be Protected	*Item Number*	*Insulating Material Used*		*Minimum Thickness of Insulating Material for Following Fire-Resistive Periods (In Inches)*			
				4 hr	3 hr	2 hr	1 hr
Bonded Tendons in Prestressed Concrete[5]	30	Grade A[6] Concrete	Beams or girders	4[7]	3[7]	2-1/2[7]	1-1/2
			Solid slabs[8]		2	1-1/2	1

[5] Cover for end anchorages shall be twice that shown for the respective ratings. Where lightweight Grade A concrete aggregates producing structural concrete having an oven-dried weight of 110 lb per cubic foot or less are used, the tabulated minimum cover may be reduced 25%.

[6] For Grade B concrete increase tendon cover 20%.

[7] Adequate provisions against spalling shall be provided by U-shaped or hooped stirrups spaced not to exceed the depth of the member with a clear cover of one inch (1″).

[8] Prestressed slabs shall have a thickness not less than that required in Table 43-C for the respective fire-resistive time period.

(1) The clearance from each tendon to the nearest exposed surface is used to determine the average cover.

(2) In no case can the clear cover for individual tendons be less than one-half of that set forth in Table No. 43-A. A minimum cover of 3/4 in. for slabs and 1 in. for beams is required for any aggregate concrete.

(3) For the purpose of establishing a fire-resistive rating, tendons having a clear cover less than that set forth in Table No. 43-A shall not contribute more than 50% of the required ultimate moment capacity of the member. For structural design purposes, however, tendons having a reduced cover are assumed to be fully effective.

Many types of prestressed concrete standard products, such as double T slabs, cored slabs, etc., have been tested for fire resistance and carry the approval of the Underwriters' Laboratories, Inc. (Ref. 17). In addition, a number of prestressed concrete structures have been subjected to actual fires. The performance of these structures has almost invariably been good.

Additional information on fire resistance of prestressed concrete can be found in the references.

9-5 Normal Temperature Variation

The effect of nominal atmospheric temperature variations on the performance of prestressed structures and on the magnitude of the effective prestress is occasionally questioned by persons who are unfamiliar with prestressed concrete. Because the thermal coefficients of linear expansion for steel and concrete are of the same order (6.5×10^{-6} ft/ft/°F ±), if the steel and concrete have the same temperatures at all times, there is no significant effect on the effective prestress for normal changes in temperature.

If the tendons are not bonded to the member, but are exposed to the atmosphere, it would be possible for the temperature of the tendons to be different from that of the concrete section, due to their difference in mass and exposure to heat sources. Under such conditions, the effect of temperature variations should be studied on the basis of estimated maximum temperature variations and the effect of these variations on the effective prestress.

Atmospheric temperature variations can result in significant stresses in structures prestressed by jacks rather than by tendons (*see* Sec. 1-3). The effect of temperature variations must be given careful consideration in this type of structure.

A prestressed member with either bonded or unbonded tendons will, of course, expand or contract with temperature variations. Provision should be made for thermal expansion and contraction in prestressed construction, just as it is in other types of construction, unless computations indicate the effect can be reasonably neglected. It should be kept in mind that temperature changes can only cause stresses in a structure if the deformation of the structure due to temperature changes is restrained.

9-6 Fatigue

Fatigue strength of structural elements is important if the elements are to be subjected to frequent reversals of stress or variations in stress. Resistance to fatigue is, therefore, an essential property for bridge members, but is not normally an important consideration in building construction.

Although the effect of fatigue on prestressed-concrete beams is not completely understood, there has been considerable research into the causes and types of fatigue failures which may result under the action of stress variations of various magnitudes.

Fatigue failures could be expected to occur in any of the following modes:

(1) Failure of the concrete due to flexural compression.
(2) Failure of the concrete due to diagonal tension or shear.
(3) Failure of the prestressing steel due to flexural, tensile-stress variations.
(4) Failure of pre-tensioned beams due to loss of bond stress.
(5) Failure of the end anchorages of post-tensioned beams.

Although the fatigue limit for concrete in compression alone is generally thought to be from 50% to 55% of the static compressive strength, no failures of prestressed flexural members due to compressive flexural stresses have been reported in the literature. Apparently the restruction of the concrete compressive stress in complete bridge structures to $0.40f_c'$ results in adequate safety

(Ref. 9). It should be pointed out that the fatigue limit mentioned above is for the load alternating from zero to the maximum value, which is a larger variation than is normally experienced in prestressed flexural members in actual service.

In composite bridge stringers ($f'_c = 5000$ psi) of moderately long span, the compressive stress in the top fibers of the precast section may vary from 1500 psi under the loading condition of dead load alone to 2000 psi under the loading condition of dead load plus live load and impact. In a similar manner the compressive stress in the composite flange (cast-in-place deck—$f'_c = 3000$ psi $\pm$) may vary from zero to 900 psi. It is apparent that these ranges of stress variations and maximum values are considerably below the fatigue limit of the concrete in compression (0 to $0.50f'_c$) and for this reason, fatigue of the concrete should not be a problem under these conditions. The variation of stress in the compressive flange from zero to $0.40f'_c$, due to the application of live load plus impact alone, would only be expected to occur in short-span structures in which the dead load of the structure itself is relatively unimportant.

Fatigue failures due to diagonal tension or shear alone have apparently not been observed in prestressed concrete research. Prestressed railroad ties, which are subjected to high-shear stresses, have shown distress when tested under repeated loads, as well as in service, but failure usually is caused by lack of adequate bond rather than insufficient shear or diagonal tensile strength. Railroad ties are very special elements, due to their high-shear loads and short-shear spans.

The majority of fatigue failures that have been found in testing prestressed beams have resulted from fatigue of the tendons. It appears that after the cracking load is reached, concentrations of stress or other phenomena related to the cracks develop in the tendons in the vicinity of the cracks and failure results. It must be emphasized, however, that most tests reveal that the fatigue resistance of prestressed-concrete beams is high and generally superior to conventionally reinforced concrete.

In several fatigue tests conducted abroad it was found that individual wires contained in the prestressing tendons failed by fatigue at points where they passed over spacers which were used to hold the wires in position (Ref. 9). This would lead one to suspect that the secondary stress that develops in the wire due to its passing over a spacer (usually a small diameter wire or bar) can be sufficiently severe to cause a point of fatigue weakness and possible fatigue failure. No failures in actual structures due to this effect are known, but it is believed that designers should avoid the use of spacers and should avoid the use of sharp bends in the tendons when possible, if fatigue is a consideration in the design.

Bond failures have been found in testing very short-span members, as was mentioned above. It appears that cracking of the beam sets up conditions which result in deterioration of the flexural bond between the tendon and the concrete as additional variations in the load are applied. When the flexural bond is destroyed from the point of cracking to the vicinity of the support, in the region where transfer bond is developed, failure ensues. (*See* Sec. 5-7.)

The available experimental data leads one to conclude that the types of tendons normally used domestically in pre-tensioned work provide adequate safety against bond failure for members of usual proportions.

There are indications that a light, hard coating of normal oxidation on the surface of the tendons improves the dynamic-bond properties, just as it improves the static-bond properties (*see* Sec. 5-7 for other factors that affect bond stresses).

No reports of fatigue failures in the steel at the anchorages of post-tensioned members are to be found in the literature. This type of failure is extremely unlikely in bonded construction, since the grout is very effective in developing flexural bond stresses. This was demonstrated in one test in which the end anchorages were removed from the tendons of a grouted beam and the member was then subjected to a fatigue test. The results were satisfactory and failure resulted from fatigue of the tendons, and not from lack of bond.

In unbonded post-tensioned construction, the end anchorages could be subjected to some variation in stress under the action of variation in external load. This type of construction is not generally used in members to be subjected to frequent variations in stress; however, there are very little experimental data available on the performance of this type of construction under repeated loads.

REFERENCES

1. Guyon, Y., *Prestressed Concrete*, p. 58, John Wiley and Sons, Inc., New York, 1953.
2. Ross, A. D., "Creep of Concrete Under Variable Stress," *ACI Journal*, **29**, No. 9, 739–758 (Mar. 1958).
3. ACI Committee 318, "Proposed Revision of ACI 318-63: Building Code Requirements for Reinforced Concrete," *Journal of the American Concrete Institute*, **67**, No. 2, 77–186 (Feb. 1970).
4. PCI Committee on Prestressed Concrete Columns, "Tentative Recommendations for the Design of Prestressed Concrete Columns," *Journal of the Prestressed Concrete Institute*, **13**, No. 5 (Oct. 1968).
5. Zia, Paul. Private communication. (Letter dated Nov. 26, 1969).

6. Guyon, op. cit. p. 101.

7. Hill, A. W. and Ashton, L. A., "The Fire Resistance of Prestressed Concrete," *Proc. World Conference on Prestressed Concrete*, A20-1—A20-8 (July, 1957).

8. "Uniform Building Code," 1967 Editon, Volume I. International Conference of Building Officials, 1967.

9. Nordby, Gene M., "Fatigue of Concrete—A Review of Research," *Journal of the American Concrete Institute*, **30**, 210–215 (Aug. 1958).

10. Sawko, F. and Saha, G. P., "Fatigue of Concrete and Its Effect Upon Prestressed Concrete Beams." *Magazine of Concrete Research*, **20**, No. 62, 21–30 (Mar. 1968).

11. Smith, E. A. L., "Tension in Concrete Piles During Driving," *Journal of the Prestressed Concrete Institute*, **5**, No. 1, 35–40 (Mar. 1960).

12. Gerwick, Ben. C., Jr., "Prestressed Concrete Piles," *Journal of the Prestressed Concrete Institute*, **13**, No. 5, 66–93 (Oct. 1968).

13. Fisher, J. W., "Behavior of AASHO Road Test Prestressed Concrete Bridge Structures," *Journal of the Prestressed Concrete Institute*, **8**, No. 1, 14–38 (Feb. 1963).

14. PCI Committee on Prestressed Concrete Piling for Buildings, "Recommended Practices for Driving Prestressed Concrete Piling," *Journal of the Prestressed Concrete Institute*, **11**, No. 4, 18–27 (Aug. 1966).

15. Heerema, P. S., "Cylindrical Hollow Prestressed Concrete Piles," *Journal of the Prestressed Concrete Institute*, **5**, No. 4, 41–47 (Dec. 1960).

16. Li, Shu-T'ien and Chen-Yeh Liu, Tony, "Prestressed Concrete Piling—Contemporary Design Practice and Recommendations," *Journal of the American Concrete Institute*, **67**, No. 3, 201-220 (Mar. 1970).

17. "Building Materials List, January, 1968," Underwriters' Laboratories, Inc., Chicago, 1968.

10 Cracking and Other Defects-Their Cause and Remedy

10-1 Introduction

Defects, generally in the form of cracks, honeycombing, excessive camber or deflection, often occur in what are otherwise well designed and properly fabricated prestressed-concrete members. The purpose of this chapter is to describe the most common defects and suggest means of preventing their occurrence.

Flexural cracks that are fine and closely spaced are to be expected in prestressed-concrete flexural members in which the tensile stresses exceed the modulus of rupture, just as they are in reinforced concrete. Cracks of this type are considered to be normal and are not considered to be defects.

Prestressed concrete members that have minor defects can be repaired with the modern materials available with the expectation that the member will give many useful years of service. Recommended methods of repair are described in the following articles when applicable.

10-2 Cracking

Undesirable cracking occurs in prestressed members due to a variety of causes. Most cracks that occur in precast elements are not structurally important, but cracks should be avoided whenever possible. The following is a description of cracks which are commonly found in precast prestressed concrete members, as well as an explanation of their cause and methods of avoiding them.

Large Flexural Cracks

Most precast prestressed members are simple beams and are designed to be supported at their ends only. If members of this type are supported or have loads applied to them other than in the direction or at the locations intended by the designer, large flexural cracks may occur. Precast members with overhangs sometimes become supported at locations that are unintended, due to the deflection of the member at the time of stressing. This is illustrated in Fig. 10-1, in which a member with overhangs at each end is shown before and after prestressing. The upward deflection of the relatively long interior span may result in rotations at the intended point of support. These rotations cause the member to be supported at the extreme ends. The application of the reactions at the end of the member results in flexural cracking, as is shown in Fig. 10-1. This type of cracking can be avoided if the designer properly

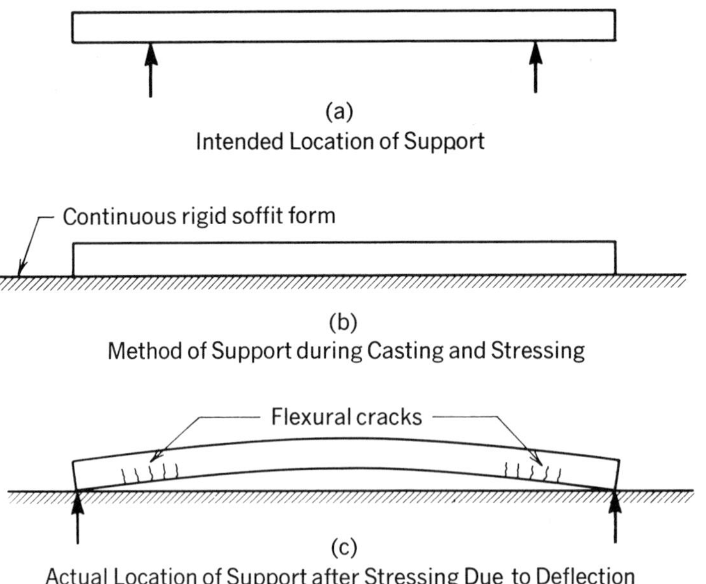

Fig. 10-1 Example of unintended mode of support of a beam with overhangs resulting from deflection due to prestressing.

investigates all facets of the design of each individual member and properly specifies the permissible locations of supports during fabrication and erection, as well as the precautions to be taken to avoid difficulties of this type.

Restraint at Time of Prestressing

At the time of prestressing, the ends of simple precast elements normally rotate and the member shortens as a result of the deformations caused by prestressing. If the soffit form at the time of stressing is rigid and the member becomes supported on the extreme end, the bottom corner of the member frequently cracks, as is shown in Fig. 10-2. This type of cracking is frequently aggravated by the additional localized tensile stresses resulting from the anchorage of the tendons. This type of cracking can be avoided by removing a portion of the soffit form prior to stressing and by providing reinforcing in the ends of the member to strengthen the corners.

Longitudinal Temperature and Shrinkage Strains

In the manufacture of steam cured pre-tensioned concrete products, if the steam curing is stopped without releasing the pre-tensioning force, the concrete members and the tendons exposed between the members cool and contract. The contraction results in an increase in the tension in the exposed tendons

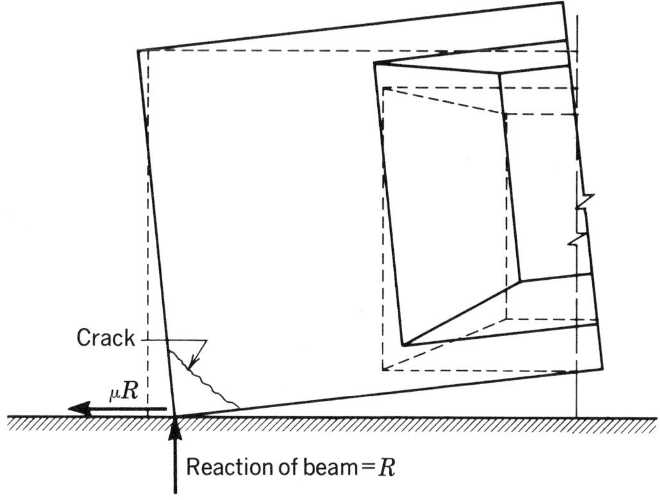

Fig. 10-2 Cracking of the end of a simple beam due to rotation and shortening. The position of the beam before stressing is shown by the broken lines.

between the members, as a result of the ends of the tendons being anchored to the abutments of the pre-tensioning bench and being unable to yield. This tensile stress may be further aggravated by shrinkage of the concrete. The combined effect of these phenomena may result in the concrete members becoming cracked at intervals of 15 to 25 ft, as illustrated in Fig. 10-3, or may result in the tendons breaking at the anchorages where they are subjected to localized stress concentrations. For this reason, steam curing should never be discontinued for any length of time, unless the stress in the tendons is at least partially released. If it is desired to discontinue steam curing after a period of 24 or 30 hr of steaming, and if the concrete cylinders have not gained sufficient strength to allow the full release of the prestress, the tendons should be partially released by an amount which is about 25% of the anticipated contraction of the tendons, which normally would take place if the entire prestress were released. It is best, although not always essential, that the forms be loosened before the partial release is made. In a similar manner, a breakdown in the steam curing facility may result in the cracking of the members, and in such an eventuality, under average conditions of design, a partial release should be made if the concrete strength exceeds 3000 psi.

It is also possible to have similar cracks form in post-tensioned members, due to similar causes (temperature change, shrinkage or both). This is particularly true in long, large members which are lightly reinforced. The cracks can be avoided by keeping the concrete continually wet until prestressing, since by so doing, shrinkage strains will not occur. Another method is to stress the member, either partially or completely, soon after curing is completed. In other words, long periods of drying or large temperature variations on unstressed members should be avoided.

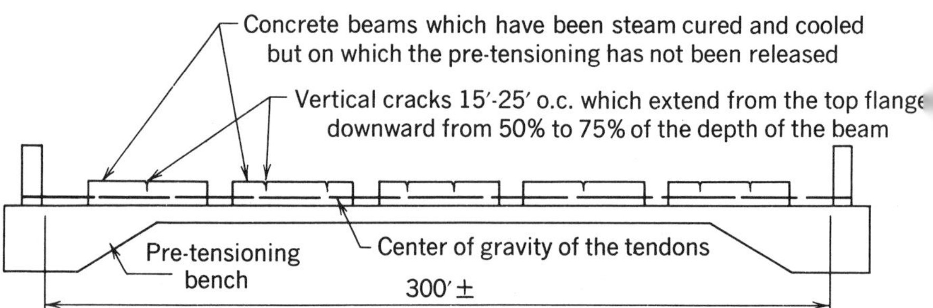

Fig. 10-3 Cracking in pre-tensioned beams due to shrinkage and temperature strains aggravated by steam curing.

Uneven End Bearing Cracking

This phenomenon can occur at the ends of precast members due to stress concentrations resulting from non-uniform bearing. The non-uniform bearing may be the result of the mating planes having been cast incorrectly, due to rotations at the ends of the members as a result of prestressing or applied loads, due to lateral (torsional) rotations resulting from superimposed loads being applied eccentrically, or due to the surfaces not being smooth or being dirty.

The best means of avoiding cracking from these causes is by providing careful attention to erection details. Surfaces intended to be in contact with each other should be observed during erection to be sure they are clean and will fit properly. Beds of mortar or flexible bearing pads are frequently used to insure uniform bearing.

The effects of superimposed loads should be considered and provisions made for them. L-shaped or inverted T-shaped beams that are to receive superimposed loads on one side only will tend to rotate towards the loaded side. Thus, the end of the beam or the supporting column may be subjected to high concentrated stresses and tend to crack. The torsional rotations may be avoided by using temporary shoring. Longitudinal rotations due to superimposed dead and live loads should also be considered at the ends of unrestrained beams. Bearing details should be such that edge loading is avoided.

Form Settlement

If the soffit form of a member settles after the concrete has been placed and has attained its initial set, cracks similar to structural flexural cracks can occur. This type of cracking can be prevented by constructing the soffit forms on a suitable foundation.

Flange-web Cracks

Cracks sometimes occur at the junction of the web and the flanges of precast members that are I or T shaped in cross section. The cracks are more common at the junction of the web and top flange than at the junction of the web and bottom flange. There are two probable causes for this. First, the settlement of the plastic concrete, which sometimes accounts for this type of cracking, is greater and tends to cause this type of cracking near the top flange where the top flange concrete tends to span across the settled web concrete. This effect does not exist near the bottom flange. Second, the slope of the fillet between the top flange and the web is generally flatter than the slope of the fillet at the bottom flange. This type of crack, which is illustrated in Fig. 10-4, is generally thought to not extend completely through the web

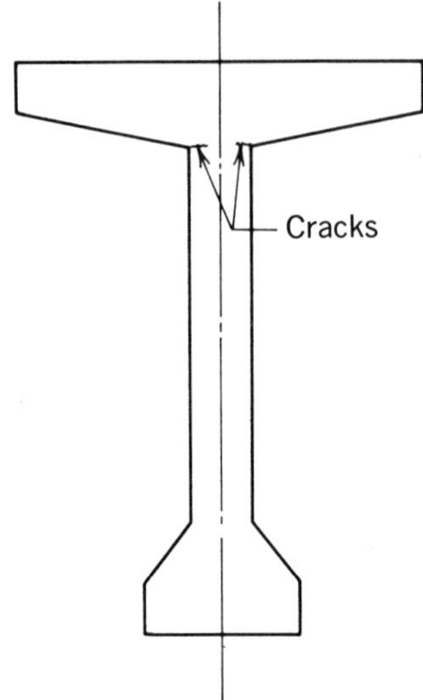

Fig. 10-4 Web-flange cracks due to concrete settlement.

of the member when it is the result of concrete setting in the form, since the plastic top flange concrete will sag down and become reunited with the web concrete, except at the outside surfaces. If this is the cause, the possibility of these cracks occurring can be reduced by allowing the web concrete to settle before the top flange concrete is placed, using concrete of lower settlement and shrinkage characteristics and possible by revibrating the concrete in the vicinity of the potential cracks. These cracks might also be caused by form expansion, in which case, the cause of the form expansion should be eliminated or reduced as much as possible. Wood forms expand when they absorb water or are exposed to steam curing conditions. Steel forms will expand more rapidly than the concrete they contain if they are subjected to a rapid temperature increase.

End Block Cracks

Cracks in the end blocks may occur in the loaded face due to deep beam action. Cracks may also occur along the trajectory of the tendons due either to the tensile stresses resulting from the anchorage forces or to the edge distance not being sufficiently large. Control of end block cracking is best accomplished by providing reinforcing steel in accordance with the methods of Sec. 6-8 and by providing reasonable edge distance for the tendons. Concrete quality in the end blocks is sometimes poor, due to inadequate compaction in these areas. The poor compaction may be due to the end block being congested with reinforcing. As was pointed out in Sec. 6-8, it is generally preferred to obtain a high-quality, well compacted concrete in the end block, even if it must be under-reinforced, than to have an over-reinforced end block in which the concrete can not be well compacted and, hence, will be of low quality.

Web Cracks Following Post-tensioning Tendons

Cracks of this type have been experienced from five different causes. If post-tensioning tendons are grouted during sub-freezing temperatures, the grout may freeze and the resulting increase in volume can cause tensile forces that crack the web. In European practice, some of the mixing water in the grout has been occasionally replaced with alcohol in order to lower the freezing point of the grout and eliminate this problem, although it is recognized that the quality of the grout suffers from this method. Some Canadian producers have adopted the practice of steam curing their beams for two or three days after grouting to eliminate the problem. Cracks following the trajectory of the tendons can also result from settlement of the concrete below the ducts in thin webbed I or T shaped members, which is similar to one of the causes of web-flange cracks. This is best prevented by using members that do not have excessively thin webs and by using concrete mixes that have low settlement characteristics. Cracks following tendons of large dimension in rectangular ducts have been known to be the result of using high pressures during the injection of the grout. This cause of cracking can be eliminated by using lower pressures or by using smaller round ducts. Beams with over hanging ends, as are shown in Fig. 10-5, have been known to develop cracks along the tendons as a result of high tensile stresses acting along the reduced concrete section that follows the tendon. This situation is critical at the time of stressing when the large future reaction at the end of the overhang is not yet in position and the large shear force due to prestressing is acting without the reaction to counteract it. The shear force due to prestressing acts in the direction opposite to that which results from the applied reaction, and hence, the principal tensile stresses are more or less acting on the plane of the tendons anchored in the bottom flange. This condition is best controlled by stressing the tendons in stages with some of the tendons being stressed after the suspended span has been erected. Another cause of this type of cracking is illustrated in Fig. 10-6, in which it will be seen that the beams are prestressed with a combination of pre-tensioned and post-tensioned tendons, the later being straight and near the bottom of the beam for a substantial portion of the length of the beam. These beams were reinforced with stirrups oriented in a plane

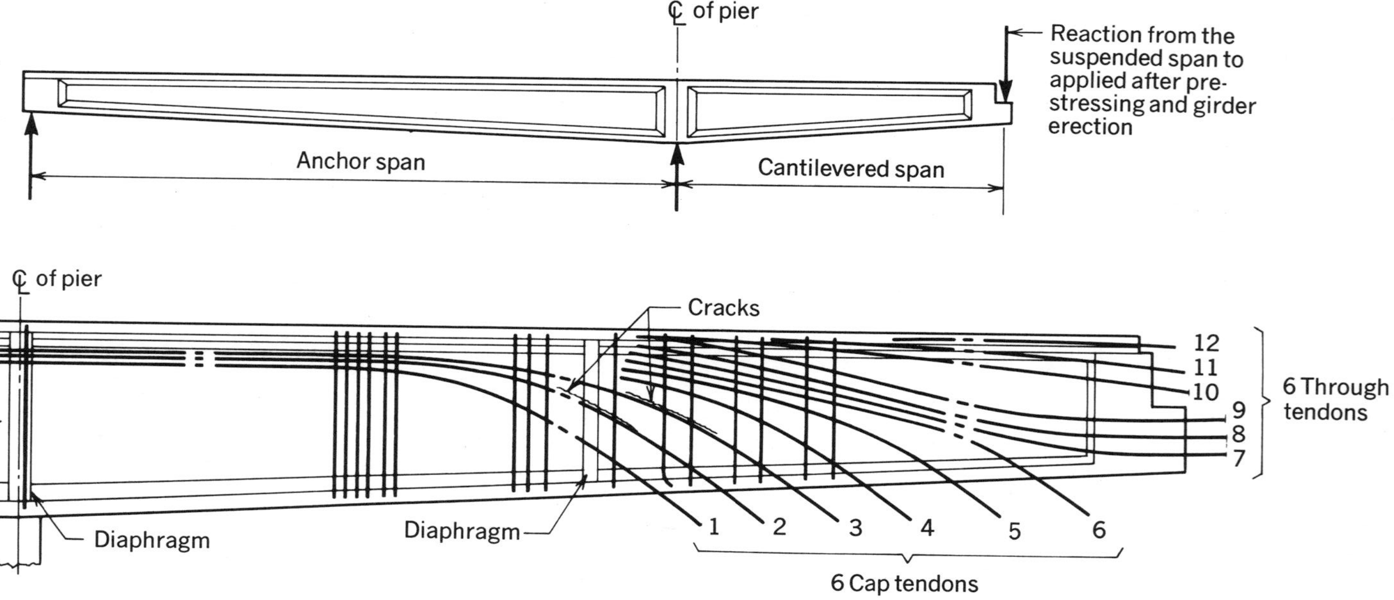

Cantilever Elevation (showing longitudinal tendons)

Fig. 10-5 Beam with overhanging end.

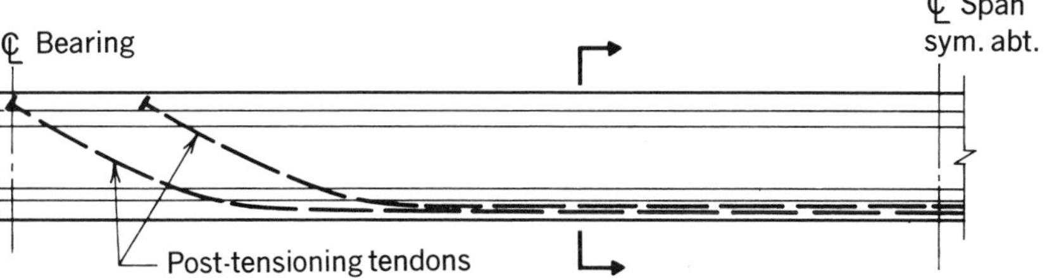

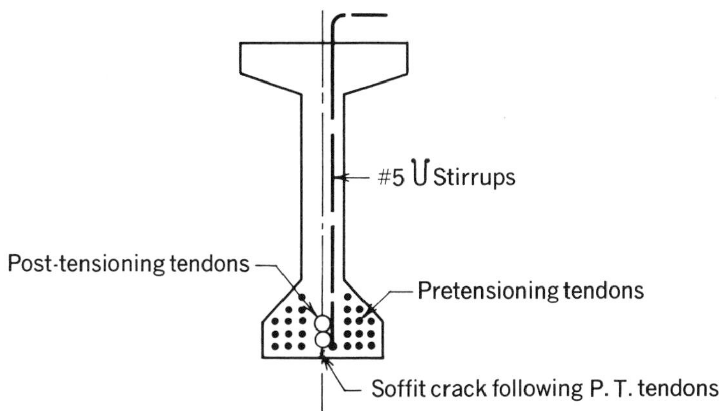

Fig. 10-6 Combined pre-tensioned and post-tensioned girder with soffit cracks.

parrallel to the longitudinal axis of the beams. The stirrups did not extend transversely across the bottom flange of the beam. This type of crack can be controlled by having reasonably closely spaced reinforcing steel extend across the bottom flange near the soffit, as is shown in Fig. 10-7.

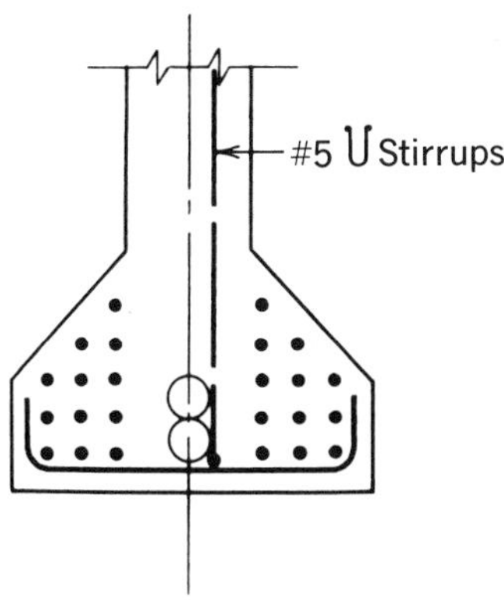

Fig. 10-7 Bottom flange transverse reinforcing.

Restraint of Reinforcing Bars

Large diameter reinforcing bars placed in the concrete immediately und post-tensioning anchorages can result in cracking. This is the result of t large bar being rigid and unable to deform in a manner that is compatib with the highly stressed concrete. The concrete immediately under t anchorages is expected to undergo some localized plastic deformations. T use of large bars in the highly compressed anchorage areas should be avoide

Most of the cracks described above, although not desirable, are not serio structurally. Difficulties due to deterioration are to be expected for norm conditions of service only if the cracks exceed 0.01 in. in width. Cracks whi are 0.01 in. in width or more should be sealed to prevent the intrusion moisture and prevent possible oxidation, loss of steel area, and possib spalling. Cracks in structures exposed to especially adverse conditions shou be sealed, even if they are less than 0.01 in. in width. Cracks which show ru stains should be sealed.

The best method of sealing non-working cracks is to inject the cracks wi an epoxy resin of low viscosity. This is done in such a manner that the cra is filled with the resin and the concrete on each side of the crack is reunited the "glueing" action of the resin. Another method is to rout a groove alo the crack throughout its entire length and fill the groove with an epoxy co pound.

Cracks which are "working" (i.e. opening and closing as a result of loac temperature, etc.) cannot normally be successfully sealed with epoxy co pounds, but must be sealed with flexible sealants that can withstand t movements to which the cracks are subject, without failing.

10-3 Restraint of Volume Changes

If a structural member can deform in accordance with natural laws witho restraint, such deformations do not result in stresses. This applies equal to strain changes resulting from temperature variations, elastic deformatio shrinkage, and creep. If fully restrained, the forces developed by strain chang due to these effects can be enormous and are generally only limited by t ultimate capacity of the weakest portion of the structural member or t restraining elements.

Prestressed concrete members are considerably more critical with respe to restraint than is reinforced concrete, due to the fact the sections are gene ally crack free and creep deformations tend to shorten the length of t members, as does shrinkage and the elastic shortening due to prestressin The creep deformation tends to change the deflection of normal reinforc concrete flexural members, but does not tend to shorten the member. F this reason, the designer must give particular attention to the problems of restraint when designing prestressed-concrete structures.

In Fig. 10-8, plan views showing the structural framing of two buildings are shown. In one plan the shear walls, which are provided to give stability against lateral loads, are at the corners of the building with a maximum distance of 150 ft between the walls. Assuming the prestressed roof members are pre-cast, much of the shortening due to shrinkage and creep may take place before the members are erected and all of the elastic shortening will take place before erection.

For the purposes of this illustration, assume the deferred deformation for the members after they are erected is 600×10^{-6} in./in. Deferred deformation is defined as the deformation that would take place, due to volume changes, if the concrete were unrestrained. If the elastic modulus of the concrete were 4×10^{6} psi, the deferred deformation would result in a unit stress of 2400 psi if the concrete were fully restrained. It is obvious that unit stresses of this order would develop tremendous forces which could easily be expected to exceed the strength of some portion of the structure between the shear walls. The unrestrained deformation would be $600 \times 10^{-6} \times 150 \times 12 = 1.08$ in. and, if the stress in the prestressed member due to the restraint is to be avoided, a movement of this amount must take place somewhere between the shear walls.

In the other plan of Fig. 10-8, the shear walls are located near the center of each wall rather than near the corners. With this layout, the deformation of the prestressed concrete for all practical purposes is not restrained by the walls and the forces due to creep and shrinkage are avoided.

The basic principal one must keep in mind is that one must avoid having shear walls on line with each other that are some distance apart and which are rigidly connected to each other by prestressed concrete. The author has observed buildings in which pilasters have been pulled away as much as one-half inch from the shear walls to which they were attached as a result of volume changes with framing similar to that shown in Framing Plan No. 1 of Fig. 10-8.

Failure of double-T roof slabs which have had each end embedded in cast-in-place beams that were unyielding, due to the presence of shear walls, have been observed. The failures were either in the form of embedded ends pulled from the cast-in-place beam, in which case a large piece of the beam spalled out with the T leg, or the T legs had cracked completely through. The T legs should have been provided with a reliable expansion bearing detail at one end or the shear walls should have been detailed in such a way as to remove the restraint.

Cantilevered roof spans, particularily those with long slender interior spans, can be subject to large vertical movements due to solar heat. The

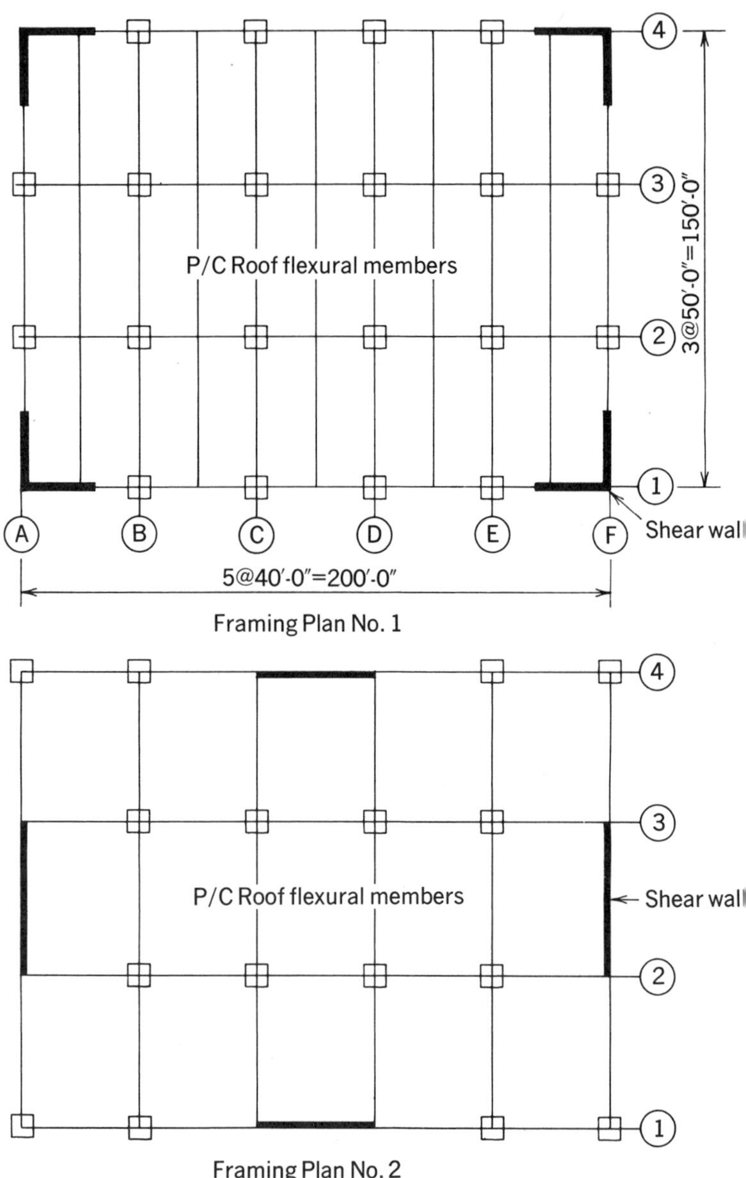

Fig. 10-8 Framing plans with different shear wall layouts.

vertical movements of the cantilever can cause severe damage to walls to which they may be attached or cracking may result in the beam itself. The deflection characteristics of cantilevered spans should be considered carefully before they are used.

Severe movements, some of which can cause serious cracking, can be developed in columns which form a portion of a rigid frame. These movements can result from shortening due to initial prestressing, creep and shrinkage, or from unanticipated movements induced by solar heat. The effect of such movements should be evaluated by the designer and avoided by the provision of hinges and slip joints or adequate reinforcing where required.

10-4 Honeycombing

Honeycombing is defined as voids left in concrete due to the failure of the mortar to effectively fill the spaces among the coarse aggregate particles. Honeycombing is the result of the concrete not being completely compacted, which may be the result of poor vibration. It may also be due to congested reinforcing steel and embedded items which restrict the placing of the concrete. Webs of I-shaped members that are too thin can also cause honeycombing, since thin webs make it difficult, if not impossible, to insert internal vibrators of sufficient size to properly compact the concrete.

Honeycombed areas should be repaired as soon as they are discovered using the methods given in the Concrete Manual published by the Bureau of Reclaimation (Ref. 1). If the curing of the concrete is continued, by keeping it saturated, a good repair can be made that should not adversely affect the strength or durability of the member.

10-5 Buckling

Long narrow beams, which are often used in precast concrete construction in order to minimize their weight, have a tendency to buckle laterally during handling and when subjected to transverse load before they have been incorporated into the final structure. The beams normally become laterally braced by a slab or diaphragms in the completed structure. If a slender beam is loaded, there is a critical value of load that may be much lower than the load causing flexural or shear failure, at which, lateral deflection and rotation start to take place near the center of the span. The lateral deflection and rotation can cause the beam to collapse if it is allowed to develop without restraint. The dead load of the girder alone can be enough to cause this buckling.

Crookedness or lateral deflection due to differential shrinkage or solar-temperature effects can be sufficiently large to result in significant torsional moments in narrow beams. Torsional moments due to these causes can be large enough to render a girder unstable. As an example of this, the girder shown in Illustrative Problem 10-2 is a California Standard I beam. Experience has shown that girders of this type often have lateral deflections initially and generally require lateral bracing during hauling and erection.

The uniformly distributed load, which when applied at the center of gravity of a straight, narrow, slender rectangular beam is critical as far as buckling is concerned, may be written as follows:

$$w_{cr} = \frac{K_1 K_2 M E_c \sqrt{0.4 K_t I_y}}{L^3} \tag{10-1}$$

in which:

M = variable coefficient dependent principally upon the end section conditions with regard to torsion as well as vertical and lateral bending.
L = span length.
I_y = lateral moment of inertia of the concrete section.
E_c = elastic modulus of the concrete.
K_t = torsional modulus.
K_1 = multiplier for point of load application.
K_2 = multiplier for an I section and equals one for a rectangular section.

The value of M for various combinations of end restraint are tabulated in Table 10-1.

TABLE 10-1

End Restraint Conditions			Value of M
for Torsion	*for Vertical Bending*	*for Lateral Bending*	
Fixed	Simply Supported	Hinged	28.3
Fixed	Cantilevered	Hinged	12.8
Fixed	Fixed both ends	Hinged	98
Fixed	Fixed at one end. Simply supported at one end	Hinged	54
Fixed	Simply supported	Fixed	50
Fixed	Fixed at both ends	Fixed	137

For symetrical I beams, additional restraint is offered by the top and bottom flanges and the dimensionless value

$$\beta = \frac{2J}{0.4K_t} \frac{Z^2}{L^2} \tag{10-2}$$

is used. Here, J is the lateral moment of inertia of the top or bottom flange and Z is the moment arm for internal vertical stresses which, for beams having wide flanges, is the distance between the centers of the top and bottom flanges. The factor β from Eq. 10-2 is used to compute the multiplier K_2 that predicts the additional resistance against lateral buckling affected by the flanges. The factor K_2 for beams hinged laterally is:

$$K_2 = \sqrt{1 + \frac{\pi^2 \beta}{4}} \tag{10-3}$$

and for beams which are fixed laterally

$$K_2 = \sqrt{1 + \pi^2 \beta} \tag{10-4}$$

For I beams that are unsymmetrical, Eq. 10-1 applies. But, the value of J in Eq. 10-2 is computed as follows:

$$J = \frac{2}{\dfrac{1}{J_t} + \dfrac{1}{J_b}} \tag{10-5}$$

in which J_t and J_b are the lateral moments of inertia of the top and bottom flanges, respectively. An additional multiplier for cases in which the load is not applied at the centroid of the section is computed from

$$K_1 = 1 - 0.72\,\delta \tag{10-6}$$

in which the dimensionless factor δ is

$$\delta = \frac{2d}{L}\sqrt{\frac{I_y}{0.4K_t}} \tag{10-7}$$

The term d in the Eq. 10-7, is measured from the point of application of the load to the centroid of the section for symmetrical cross sections. For unsymmetrical cross sections, d is measured from the point of application of the load to the center of torsion. The coefficient K_1 is less than unity when the load is applied above the center of torsion.

For average conditions, the product of K_1 and K_2 can be approximated as follows:

$$K_1 K_2 = 1.00 + \frac{6.25JZ^2}{K_t L^2} - \frac{2.27d}{L}\sqrt{\frac{I_y}{K_t}} \tag{10-8}$$

An elastic torsional restraint may be provided by the support end condition rather than being fixed. In this case, for a rotation of Ψ, the torsional moment that is induced is

$$M_x = -R_x \Psi \tag{10-9}$$

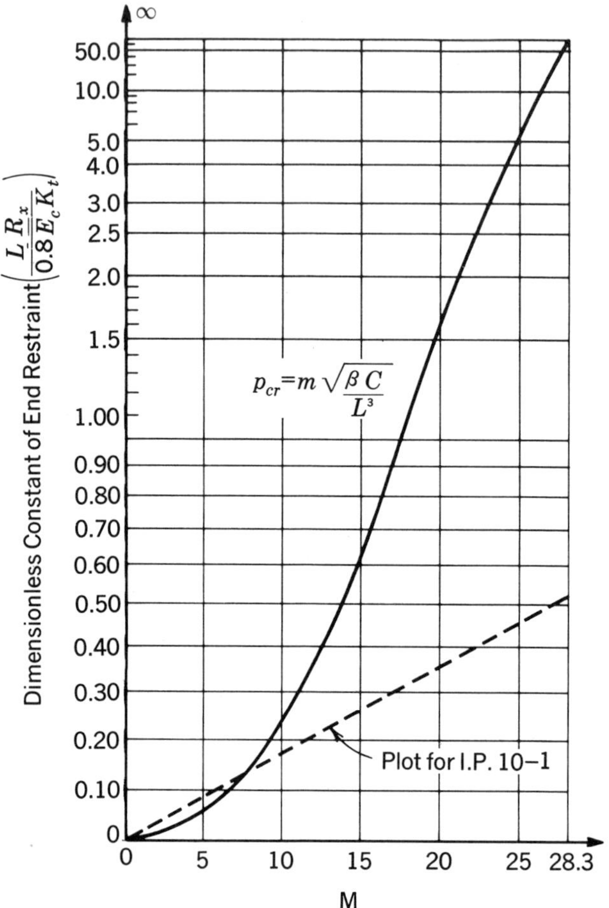

Fig. 10-9 Buckling load of girder with elastic end restraint.

in which R_x is the spring constant for the torsional restraint. Values of M versus a function of the spring constant R_x are shown in Fig. 10-9, in which it will be seen that for high values of R_x, the value of M from Eq. 10-1 approaches 28.3, as is shown in Table 10-1, for a simple beam that is hinged laterally but is torsionally restrained.

For a beam suspended at each end at a distance e above the center of gravity (carrying a uniformly distributed load w) for a small rotation Ψ from the vertical, the torsional moment is equal to

$$-\frac{wLe}{2} \times \Psi \tag{10-10}$$

and the spring constant R_x becomes

$$R_x = \frac{wLe}{2} \tag{10-11}$$

If a beam is lifted by a sling as shown in Fig. 10-10, the value of e is

$$e = y_t' + \frac{H}{1 + K} \tag{10-12}$$

in which:

$$K = \frac{wL \cos \varepsilon}{4E_s A_s \sin^3 \alpha \sin^2 \varepsilon} \tag{10-13}$$

where

w = the uniformly distributed load.
L = the span.
E_s = the elastic modulus of the cables.
A_s = the area of the cables.
α and ε = the angles shown in Fig. 10-10.
y_t = distance between the centrodial axis of the girder and points where lifting cables are connected to the girder.

The value of e computed with Eq. 10-12 is used in Eq. 10-11 to determine the spring constant to be used in analyzing a beam being lifted with a sling.

The critical load increases very rapidly as the points of support of a beam are moved towards the center of the beam. This is illustrated in Fig. 10-11.

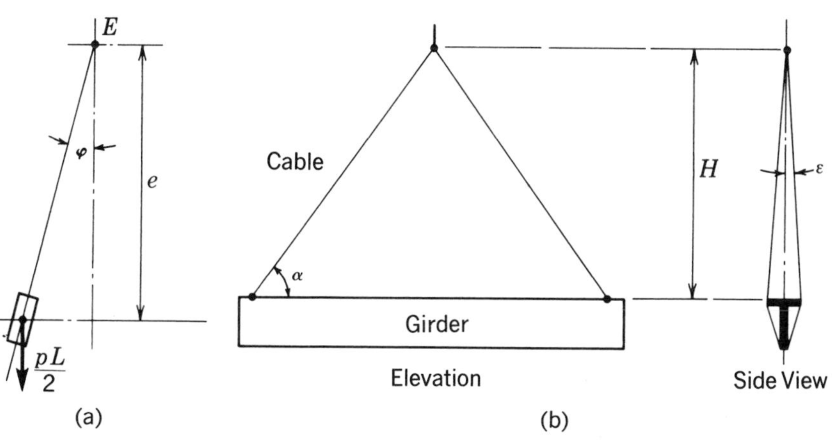

Fig. 10-10

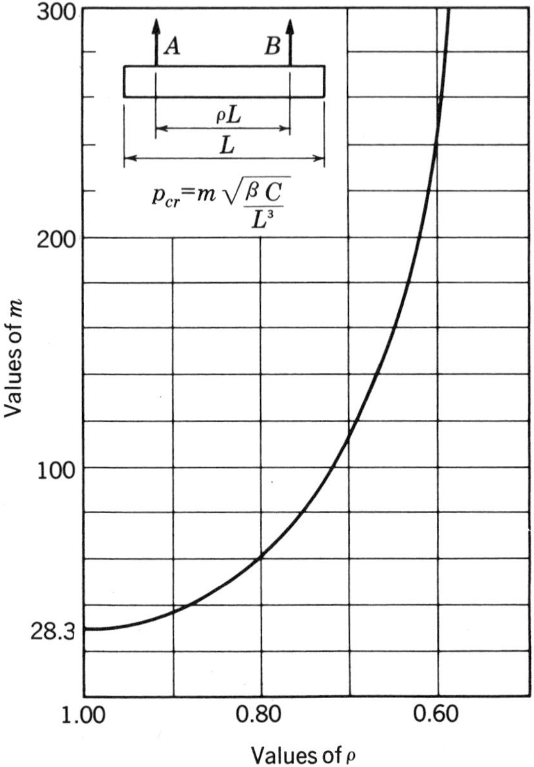

Fig. 10-11 Buckling load vs location of pick-up points.

In a similar manner, if a beam is lifted at its ends but has an additional c centrated load applied to it at the center, as is shown in Fig. 10-12, critical load is markly increased as the concentrated load is increas Neither of these methods of increasing the resistance to buckling is norm practical in prestressed concrete members, due to the fact that precast p stressed members normally do not have significant resistance to loads app upwards at points between the supports.

If lateral loads are applied to a beam, such as due to wind, the tende to buckle is increased and the lateral bending moment is greater than caused by the lateral load alone. Additionally, initial crookedness incre the tendency to buckle and reduces the critical load. The actual lat moment in a beam hinged at both ends with regard to vertical and lat bending, but fixed torsionally, is

$$M_y = \mu \frac{w_1 L^2}{8} \quad (10\text{-}$$

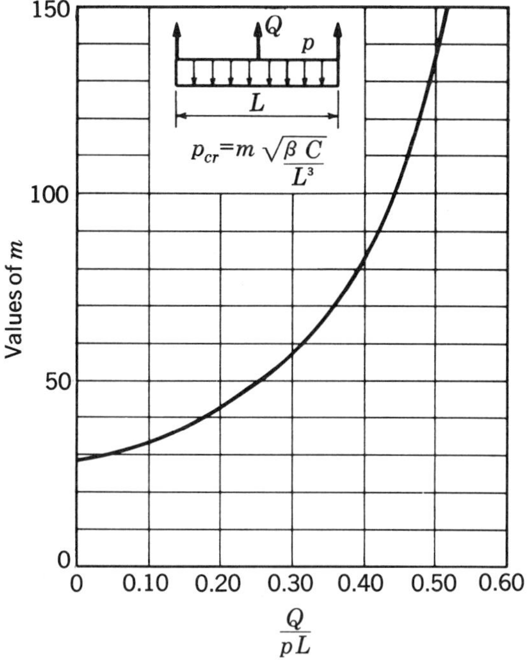

Fig. 10-12 Buckling of beam with uniform and concentrated loads.

in which μ is a factor greater than one and w_1 is the lateral load. Values of μ for a rectangular beam can be obtained from Fig. 10-13. Torsional moments also exist under these conditions and the maximum value near the end of the beam is

$$M_t = \tau \frac{0.8 E_c K_t}{L} \frac{w_1}{w} \tag{10-15}$$

in which w is the vertical load and τ is from Fig. 10-3.

Beams subjected to axial load as well as a vertical load can be analyzed as follows:

$$M_{cr} = M_0 \sqrt{\left(1 - \frac{NL^2}{\pi^2 E_c I_y}\right)\left(1 - \frac{NI_p}{0.4 E_c A_c K_t}\right)} \tag{10-16}$$

in which M_{cr} is the critical moment with regard to lateral buckling under the action of vertical and axial loads, M_0 is the critical moment with regard to lateral buckling for a beam subjected to a uniform bending moment applied

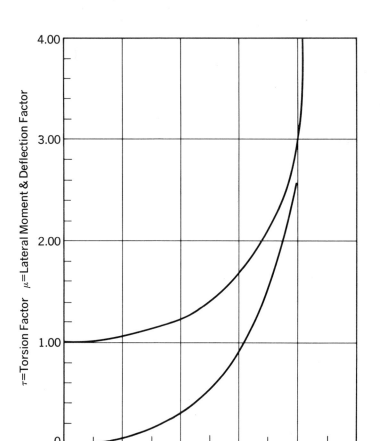

Fig. 10-13 Torsion, moment, and deflection factors vs the ratio of actual to critical load.

in the direction of a vertical load, N is the normal force, and I_p which is the polar moment of inertia, is the sum of I_x and I_y.

Although this relationship is accurate only for beams subjected to a constant moment throughout its length, it can be applied to beams subject to a uniform load when N is small in comparison to the Euler buckling load or to the torsional buckling load. These are

$$\text{Euler buckling load} = \frac{\pi^2 E_c I_y}{L^2} \tag{10-17}$$

Fig. 10-14 Girder with temporary lateral bracing.

and

$$\text{Torsional buckling load} = \frac{0.4E_c A_c K_t}{I_p} \tag{10-18}$$

in which E_c is the elastic modulus of the concrete, A_c is the area of the concrete, and the other terms are as already defined.

In practice it is very common to provide temporary lateral bracing on narrow prestressed concrete beams during periods of handling, transportation, and erection. A frequently used method of providing the lateral bracing is with prestressing tendons placed under nominal stress on each side of the beams in a truss configuration, as is shown in Fig. 10-14.

ILLUSTRATIVE PROBLEM 10-1 Investigate the lateral stability of the girder shown in Fig. 10-15(a) if the section properties are as shown in the figure, and the span is 180 ft. For the purposes of computing the torsion modulus and the location of the torsion center, the cross section is converted to the

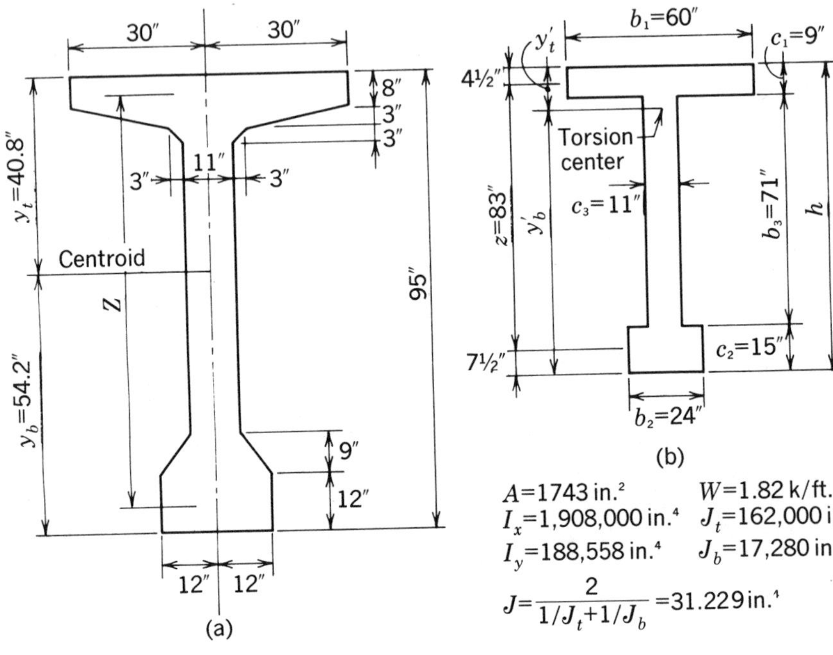

Fig. 10-15 Beam sections for Prob. 10-1.

approximately equivalent section shown in Fig. 10-15(b). The torsion modulus is computed as follows (Ref. 2):

$$K_t = \frac{b_n(c_n)^3}{3} = \frac{60(9)^3 + 24(15)^3 + 71(11)^3}{3} = 73{,}080 \text{ in.}^4$$

The location of the torsion center is computed as follows (Ref. 3):

$$\frac{y_t' - 0.5c_1}{y_b' - 0.5c_2} = \frac{J_b}{J_t}$$

$$y_t' + y_b' = h$$

from which one obtains

$$y_t' = 12.5 \text{ in.}$$

For the condition of the girder erected and laterally braced at the ends only, the critical buckling load is computed as follows from Eq. 10-1:

$$w_{cr} = \frac{K_1 K_2 M E_c \sqrt{0.4 K_t L_y}}{L^3}$$

where $K_1 K_2$ from Eq. 10-8 is

$$K_1 K_2 = 1 + \frac{(6.25)(31{,}229)(83)^2}{(73{,}080)(180^2)(144)} - \frac{(2.27)(-28.3)}{180 \times 12} \sqrt{\frac{188{,}558}{(73{,}080)}}$$

$$= 1 + 0.00394 + 0.0479 = 1.05$$

(It should be noted that in computing $K_1 K_2$, $d = 12.5$ in. $-$ 40.8 in. $= -28.3$)

$$w_{cr} = \frac{(1.05)(28.3)(3 \times 10^6)\sqrt{(0.4)(73{,}080)(188{,}558)}}{(180^3)(1728)}$$

$$= 657 \text{ lb per in.} = 7.88 \text{ kips per foot}$$

and the safety factor is

$$\frac{7.88}{1.82} = 4.3$$

In the above calculation, the elastic modulus of the concrete was taken as 3×10^6 psi, since the beam may be supported in this manner a period of time and some creep will take place.

For the condition of lifting the girder with a crane at each end (vertical lift), assuming the lifting point is 1.5 ft above the top of the girder, the instantaneous elastic modulus is 4×10^3 ksi and the eccentricity to be used in computing the spring constant becomes

$$e = 1.5 + y_t = 4.90 \text{ ft}$$

From Eq. 10-11, the spring constant is

$$R_x = \frac{wLe}{2}$$

To facilitate computations, each side of this relationship is multiplied by $L/0.8E_c K_t$, which gives the dimensionless factors

$$\frac{LR_x}{0.8E_c K_t} = \frac{wL^2 e}{1.6E_c K_t}$$

If U is the factor of safety against buckling; $w_{cr} = U \times 1.82$ k/ft and

$$\frac{LR_x}{0.8E_c K_t} = \frac{wL^2 e}{1.6E_c K_t} = \frac{1.82U \times 180^2 \times 4.90 \times 144}{1.6 \times 4000 \times 73{,}080} = 0.0890U$$

and from Eq. 10-1

$$U = \frac{K_1 K_2 M E_c \sqrt{0.4K_t I_y}}{1.82 \times L^3}$$

$$= \frac{1.05M(4 \times 10^3)\sqrt{(0.4)(73{,}080)(188{,}558)}}{(1.82)(180)^3(144)}$$

$$= 0.204M$$

With this value of U, the expression becomes:

$$\frac{LR_x}{0.8E_c K_t} = (0.0890)(0.204M) = 0.018M$$

This relationship is shown plotted on Fig. 10-9 by the broken line. The intersection with the curve reveals the value of M to be 7.5 and the safety factor is:

$$U = 0.204 \times 7.5 = 1.53$$

ILLUSTRATIVE PROBLEM 10-2 Compute the critical buckling load for the beam shown in Fig. 10-16 for erection with vertical cables at each end, $e = 5.0$ ft, $L = 100$ ft, $E_c = 4 \times 10^3$ ksi, $w = 0.670$ kips per foot, $I_y = 9096$ in.^{4}, $K_t = 11{,}816$ in.^{4}, $J_t = 3430$ in.^{4}, $J_b = 5144$ in.^{4}, $J = 4116$ in.^{4}

$$\frac{LR_x}{0.8E_c K_t} = \frac{wL^2 e}{1.6E_c K_t} = \frac{U \times 0.670 \times 100^2 \times 5 \times 144}{1.6 \times 4 \times 10^3 \times 11{,}816} = 0.0638U$$

and

$$U = \frac{K_1 K_2 M E_c \sqrt{0.4K_t I_y}}{0.670L^3}$$

$$K_1 K_2 = 1 + \frac{6.25 \times 4116 \times 58.5^2}{11{,}816 \times 100^2 \times 144} - \frac{(2.27)(3.70)}{12 \times 100}\sqrt{\frac{9096}{11{,}816}}$$

$$= 100$$

$$U = \frac{4000M\sqrt{0.4 \times 11{,}816 \times 9096}}{0.670 \times 100^3 \times 144} = 0.272M$$

Therefore:

$$\frac{LR_x}{0.8E_c K_t} = 0.0173M$$

From Fig. 10-9, $M = 75$ and $U = 0.272 \times 7.5 = 2.04$.

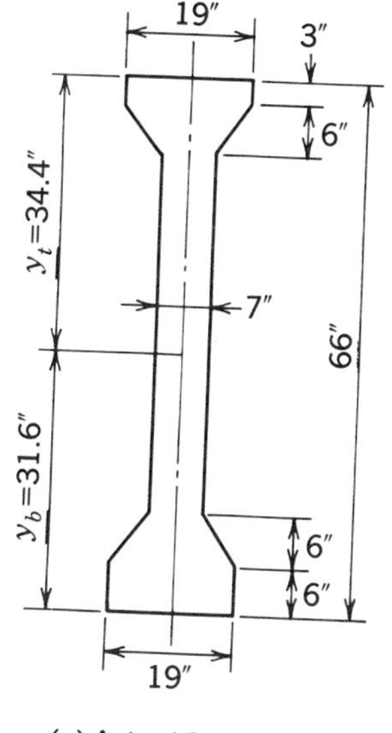

(a) Actual Section

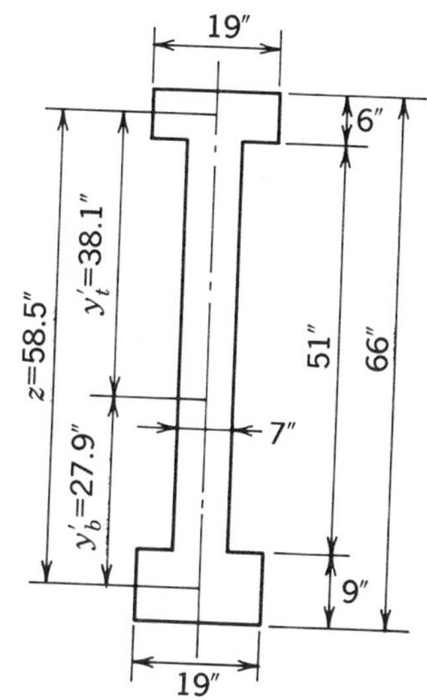

(b) Approximate Section

Fig. 10-16 Beam for Prob. 10-2.

The reader's attention is called to the effect of initial crookedness of narrow girders of this type in the second paragraph of Sec. 10-5.

ILLUSTRATIVE PROBLEM 10-3 For the girders of Illustrative Problems 10-1 and 10-2, compute the effects of a 5 psf wind load. For the girder of Illustrative Problem 10-1:

$$w_1 = \frac{0.005 \times 95}{2} = 0.040 \text{ klf}$$

$$\frac{w}{w_{cr}} = \frac{1}{U} = 0.65 \qquad \text{and from Fig. 10-1} \quad \mu = 1.90, \quad \tau = 1.15$$

$$M_y = \frac{1.90 \times 0.040 \times 180^2}{8} = 308 \text{ ft-kips.}$$

$$f_y = \frac{\pm 308 \times 12 \times 30}{188{,}558} = 0.588 \text{ ksi}$$

$$\delta_y = \frac{1.90 \times 5 \times 0.040 \times 180^4 \times 1728}{384 \times 4000 \times 188{,}558} = 2.38 \text{ in.} \qquad \frac{180 \times 12}{2.38} = 907$$

$$M_t = \frac{1.15 \times 0.8 \times 4000 \times 73{,}080}{180 \times 144} \times \frac{0.040}{1.82} = 228 \text{ ft-kips}$$

For the girder of Illustrative Problem 10-2:

$$w_y = \frac{0.005 \times 66}{12} = 0.0275 \text{ kpf}$$

$$\frac{w}{w_{cr}} = \frac{1}{2.04} = 0.49 \qquad \mu = 1.40 \qquad \tau = 0.51$$

$$M_y = \frac{1.40 \times 0.0275 \times 100^2}{8} = 48.1 \text{ ft-kips}$$

$$f_y = \frac{\pm 48.1 \times 12 \times 9.5}{9096} = \pm 0.603 \text{ ksi}$$

$$\delta_y = \frac{1.40 \times 5 \times 0.0275 \times 100^4 \times 1728}{384 \times 4000 \times 9096} = 2.38 \text{ in.} \qquad \frac{100 \times 12}{2.38} = 504$$

$$M_t = \frac{0.51 \times 0.8 \times 4000 \times 11{,}816}{100 \times 144} \times \frac{0.0275}{0.670} = 55.0 \text{ ft-kips}$$

10-6 Camber-Deflection

The camber of prestressed members can result in construction difficulties in instances when it is significantly greater or less than that which was anticipated. In composite cast-in-place bridge construction, it is customary tc use a detail as shown in Fig. 10-17. This detail anticipates the camber of the girder not being exactly as computed and gives a means of compensating for the deviations from the computed camber as well as for the variations ir camber between adjacent girders. Occasionally, a detail as shown in Fig 10-18 is used, but this detail is not considered as good, since it frequently requires field adjustment of the finished grade.

Variation in camber between adjacent elements in building constructio can present difficulties in constructing structures of precast elements, unles provision is made for the variation. Roofing cannot be applied directly t precast concrete surfaces that have abrupt edges or joints between elements If it is, there is danger that the roofing will become damaged and will leak Therefore, provision should be made to eliminate sharp edges or joints whic could cause roof damage.

Floor construction often consists of precast elements over which a toppin is placed in order to achieve a smooth level wearing surface. Electric conduits and other items are often embedded in the topping and the pruder designer should take this, as well as the estimated camber and variation i camber, into account when specifing the minimum thickness of the toppin

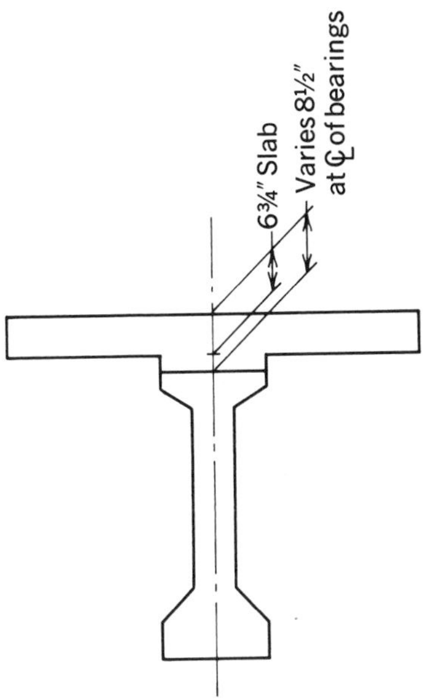

Fig. 10-17 Recommended bridge deck detail.

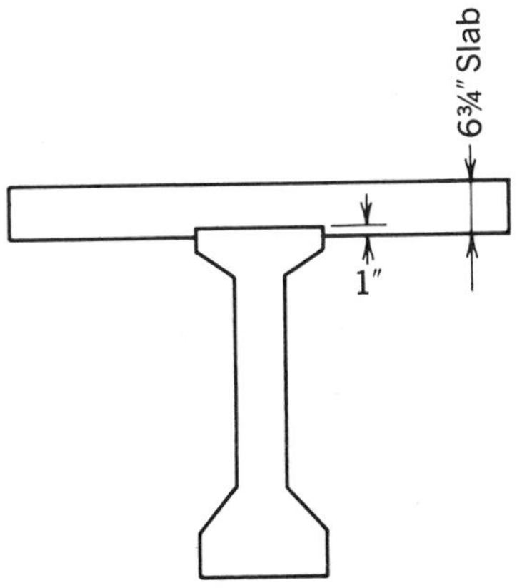

Fig. 10-18 Bridge deck detail not considered as good as that of Fig. 10-17.

10-7 Corrosion of Prestressing Steel

Prestressing steel is generally considered to be somewhat more sensitive corrosion, than ordinary reinforcing steel, due to the fact that the individu wires or strands that are used in prestressing are frequently small in con parison to reinforcing bars. The best protection the steel can have is to b surrounded by cement-rich grout or concrete that is well compacted ar impermeable. Grout and concrete are alkaline and steel cannot corrode whe confined to an alkaline atmosphere.

When the tendons are not protected by sufficient cover of dense concret moisture can reach the steel under some conditions of service, in which cas corrosion may form on the tendons. The products of corrosion may cau cracks, staining of the concrete, and eventually spalling of the concrete.

The prestressing strands shown in Fig. 10-19 were first noticed by a spalle area on the top surface of the concrete. These particular strands were co roded completely through, due to the action of sea water which occasional found its way to the upper surface of the concrete deck during heavy storm The strands were embedded in a joint between precast slabs. The joint ha been filled with a small-aggregate concrete, but it was not sufficiently con pacted. Thus, the strands were not completely covered with cement morta It is interesting to note that on the same project, a post-tensioned pile ha been damaged by a ship which collided with the structure, thus leaving th post-tensioned tendon, which was still encased in its sheath, exposed to th sea in the tidal zone. Several months after the tendon was first exposed to th

Fig. 10-19 Corroded post-tensioned tendon.

sea, the sheath was opened, the grout was removed, and the wires examined. The wires were found to be bright with no signs of oxidation, which clearly demonstrates the ability of grout to protect tendons.

The action of the chloride ion present in some commercially available aggregates for making non-shrink concrete and mortar has been blamed for causing the corrosion at the end anchorages of some post-tensioning tendons. The non-shrink concrete was used to replace the concrete under a post-tensioning anchorage which crushed during stressing. It is believed the non-shrink concrete was porous and permitted rainwater to penetrate to the anchorage and tendon. This, in combination with the chloride ion and perhaps the metalic aggregate, resulted in the wires being severely corroded in the areas adjacent to the repairs. The use of non-shrink aggregates are not recommended in prestressed concrete construction.

Stress corrosion, which has been described in Sec. 2-8, has been blamed for the failure of prestressing wire on two projects in North America. One project consisted of prestressed concrete pipe in which some of the pipe concrete contained calcium chloride and some did not. Stress corrosion was found to varying degrees in all of the pieces of pipe made with concrete containing calcium chloride, while none was found in any of the pipe that did not have calcium chloride. From this experience, it is obvious that calcium chloride (and probably any material containing the chloride ion) should not be used in prestressed concrete. On the second project, some post-tensioning wires were left ungrouted (after they had been stressed for a period of several months) due to certain difficulties that had been experienced in the construction. When the work was resumed, it was found, upon restressing of some of the previously stressed tendons, that many wires had corroded and broken. It is believed this would not have happened if the tendons had been grouted reasonably soon after stressing.

10-8 Concrete Crushing at End Anchorages

Occasionally the concrete in the vicinity of the end anchorages of post-tensioning tendons will not be of proper strength or will not be properly compacted. Thus, the concrete fails by crushing at the time of stressing. This type of failure should not be confused with the splitting type of end block failures or cracking described in Sec. 6-8. The best means of avoiding this type of failure is to take whatever steps are necessary to ensure the concrete in the vicinity of the anchorages is properly compacted and of adequate quality.

If this type of failure occurs, they can generally be satisfactorily repaired be removing all of the fractured concrete and placing new high strength concrete as required to restore the work to the original lines.

The extreme ends of post-tensioned members have been precast in order to facilitate placing concrete in extremely congested end blocks. Using this procedure, the end sections are cast in a horizontal rather than vertical position. This facilitates placing and vibrating of the concrete.

10-9 Deterioration

The concrete used in prestressed construction is subject to deterioration for the same general chemical and mechanical action as any other concrete. Because a higher quality, dense concrete is normally used with prestressed concrete, the resistance to most forms of deterioration can be expected to be greater than for lower quality concrete. In addition, the fact that prestressed concrete can be made crack-free is believed to add to the general resistance to the effects of corrosion and erosion.

When prestressed construction is to be used in applications that will expose the structure to agents known to sometimes promote stress corrosion in high-strength steel, extra precautions should be taken to ensure that the concrete is dense, cement-rich, and that it remains crack free under all conditions of loading. The use of type II (modified) portland cement is common in prestressed members used in structures exposed to sea water.

10-10 Grouting of Post-tensioned Tendons

The equipment that is currently in use in mixing and injecting grout in post-tensioning units is much more efficient than that used when post-tensioning was first introduced in the United States. However, from time-to-time difficulty will still be experienced in grouting, due to such problems as the grout being improperly mixed or proportioned, cement lumps in the grout, improper flushing prior to injection, or unintended obstructions in the post-tensioning ducts. The frequent result is that one or more tendons may become partially grouted.

In order to salvage members with problems such as these, holes must be drilled into the post-tensioning duct with great care in order to not damage the tendons. The extent of the grouting or the location of possible obstructions can thus be observed. After this has been done, the tendons should be flushed with lime water and the tendons grouted, using the drilled holes as ports, with the procedures described in Sec. 15-8.

10-11 Damage Due to Couplers

Post-tensioning tendons are sometimes provided with couplers in order to accommodate a particular construction procedure or in order to make long tendons out of two or more short ones. The couplers are normally considerably larger in diameter than the tendons and they must be provided with a housing to prevent their being bonded to the concrete during concreting, as well as to provide space for the coupler to move during stressing. If, during stressing, the tendon is released due to the tendon breaking or some other mishap, or if the distance provided in the couple housing is not sufficiently large, the coupler may come to bear on the concrete at the end of the housing and may cause spalling of the concrete.

Care must be taken to be sure the housings for the couplers are sufficiently large and that the couplers are properly located in the housing.

If damage occurs as a result of the coupler bearing on the end of the housing, the fractured concrete should be removed and a patch applied to the member. Patches of this type must be applied with care and the use of epoxy mortars would generally be preferred, since this type of damage would normally occur sometime after the curing of the concrete has been ceased.

10-12 Wedge-Type Dead Ends

Wedge type post-tensioning anchorages at the non-jacking end of tendons must move when they are stressed, since the transverse forces that anchor the tendon can only develop as a result of movement of the wedge. Hence, when wedge type anchorages are used as dead end anchorages (non-jacking ends), provision must be made to ensure that the tendon and wedge can move during stressing. If the anchorage is embedded in concrete, provision must be made to prevent concrete from coming in contact with the wedge and tendon. The method employed in one post-tensioning system is illustrated in Fig. 10-20.

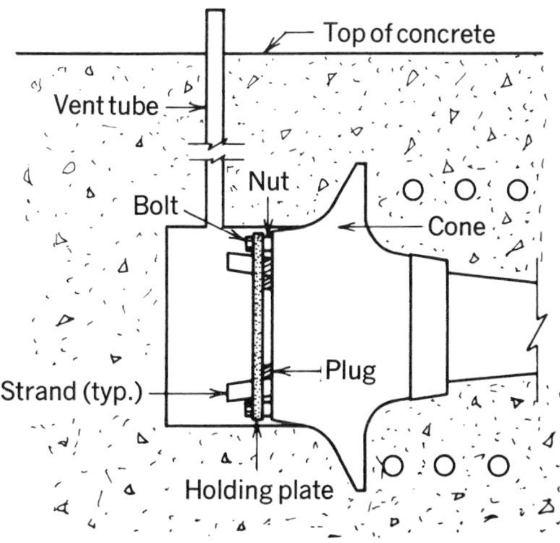

Fig. 10-20 Detail of embedded non-jacking end of a tendon with a wedge-type anchorage.

10-13 Looped or Pig Tail Dead Ends

Dead end anchorages have been made by embedding the ends of tendons either a looped or in a "pig tail" shape, as shown in Fig. 10-21. The meth has the advantage of saving the cost of an anchorage, but it must be do properly or the anchorage will fail as a result of excessive stresses on t concrete. When these methods are used, it is recommended the radius of t bend be computed using the relationship for secondary stresses due to tend curvature (Sec. 6-15). The bearing stress between the tendon and concre should not exceed the cylinder strength of the concrete. In addition, it recommended that the individual wires or strands be spaced apart as sho in Fig. 10-21 Section AA, in order to ensure that the concrete complete surrounds each wire or strand. The wires or strands should not be bund in the curved portion where the anchorage is being developed. At le nominal reinforcing should be provided.

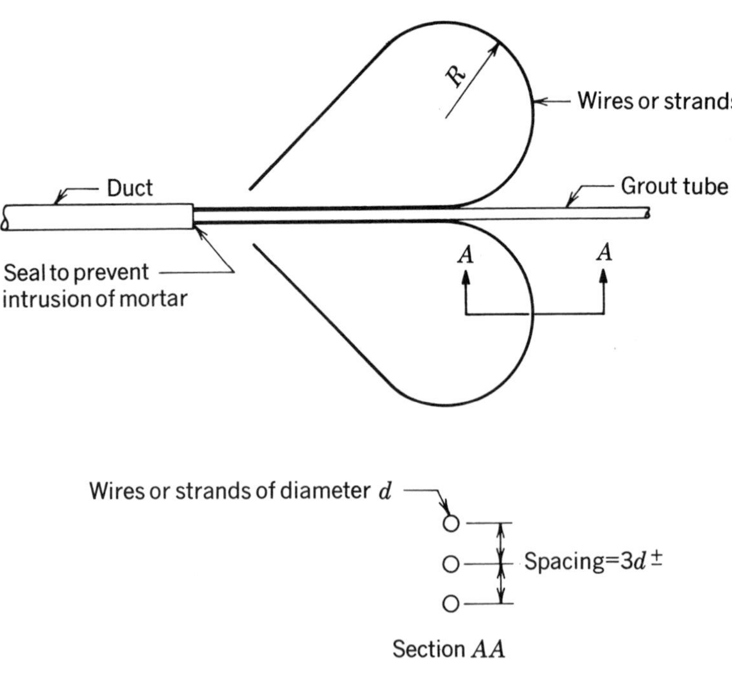

Fig. 10-21 Looped or "pigtailed" tendons.

10-14 Congested Connections

Connections between beams, girders, and columns in precast as well as in cast-in-place construction are frequently detailed with insufficient clearance for construction. The deficiency might be in the amount of space available for a post-tensioning jack that is required to stress tendons at various stages in the construction. The designer may not be aware of the space required for stressing post-tensioned tendons. Congestion may result from the designer overlooking the amount of space required for standard (or even special) bar bends.

The best means of avoiding problems of this type is by drawing details which may be critical to a sufficiently large scale that bar diameters and bends can be shown in the proper perspective, rather than by simple one line drawings. This should be done during the design stage. It is recommended that shop drawings always be made with details drawn to a scale that is sufficiently large to detect possible erection clearance problems before they actually occur.

10-15 Inadequate Welding

The standard specification for deformed billet-steel bars for concrete reinforcement, ASTM designation A615, states in Section 1.3 that the weldability of the steel is not a part of the specification, but that it may be subject to agreement between a particular supplier and user. In addition, under Section 3, Chemical Analysis, the only chemical composition restriction imposed on the bars is with respect to the phosphorous content.

The Recommended Practice for Welding Reinforcing Steel, Metal Inserts, and Connections in Reinforced Concrete Construction of the American Welding Society (AWS D12.1–61) makes various recommendations as to the type and methods of welding reinforcing steel, depending upon the carbon content of the steel as well as upon other considerations. If connections are to be made with reinforcing steel to be welded, it is essential that the carbon content of the steel be known in order to be sure that the proper welding technique will be used. If this is not done, there is no assurance the connections will possess the anticipated or desired strength or ductility.

10-16 Dimensional Tolerances

The designer should give consideration to the dimensional tolerances that might be expected in the construction of prestressed concrete. Special allowances may be required to provide for the fact that concrete members cannot be made to exact dimension, as is the case of members made of other materials.

Cast-in-place prestressed concrete can be expected to be built within t same dimensional tolerances as one would expect for reinforced concr construction. In the case of precast concrete, the designer should spec the maximum dimensional tolerances he is willing to accept. It should recognized that exceptionally small permissible tolerances would be expect to increase the cost of precast members.

The tolerances considered to be standard by the Prestressed Concr Manufacturers Association of California are as follows:

	Tolerances	
Cross Sectional Dimensions:		
Less than 24 in.	±1/4 in.	
24 to 36 in.	±3/8 in.	
Over 36 in.	±1/2 in.	
Length		
Less than 25 ft	±1/2 in.	
25 to 50 ft	±3/4 in.	
Over 50 ft	±1 in.	
Deviation in squareness of ends:		
	Vertical	*Horizontal*
Less than 12 inches	1/32 in. per in.	1/64 per in.
Over 12 inches	3/16 in. + 1/64 per in. Max. ±3/4 in.	1/16 + 1/64 per in. Max. ±1/2 in.
Deviation from straight line (Sweep) 3/16 in. per 10 ft × Total Length		
Deviation from mean camber (as installed) ±1/8 in. per 10 ft × Total Leng		
Prestressing Force:		
Deviation in location from specified C.G.	±3%	
Individual tendon force or elongation	±5%	
Total Prestress, force or elongation	±5%	
Concrete cover over reinforcing	±1/4 in.	

REFERENCES

1. U.S. Department of the Interior, Bureau of Reclamation, "Concrete Manual," U Government Print Office, Washington, D.C., 1966.

2. Timoshenko, S., "Strength of Materials," Part 1, p. 236, D. Van Nostrand Compa Inc., Princeton, New Jersey, 1968.

3. Timoshenko, S., "Strength of Materials," Part II, p. 244, D. Van Nostrand Company, Inc., Princeton, New Jersey, 1968.

4. Timoshenko, Stephen P. and Gere, James M., "Theory of Elastic Stability," McGraw-Hill Book Company, New York, 1961.

5. Muller, Jean, "Lateral Stability of Precast Members During Handling and Placing," *Journal of the Prestressed Concrete Institute*, **7**, No. 1, 20–31 (Feb. 1962).

6. "Prestressed Concrete Inspectors' Manual," Prepared by the Prestressed Concrete Manufacturers Association of California.

7. Monfore, G. E. and Verbeck, G. J., "Corrosion of Prestressed Concrete Wire in Concrete," *Journal of the ACI*, **32**, No. 5, 491–515 (Nov. 1962).

11 Roof and Floor Framing Systems

11-1 Introduction

The subjects of prestressed and reinforced-concrete roof- and floor-framing systems are inseparable because identical concrete sections are used with each mode of reinforcing and because some framing schemes incorporate elements composed of each type of construction. For this reason, each will be considered in this discussion; however, the emphasis will be placed upon the prestressed members, and reinforced-concrete elements will only be discussed when they are used in lieu of or in combination with prestressed elements.

The desirable features of floor and roof systems frequently, but not always, include the following:

(1) Structural integrity at service loads with adequate load factors against ultimate failure.

(2) Economy in first cost and maintenance.

(3) Adaptability to long and short spans with minimum revision to the manufacturing facilities.

(4) Minimum total depth of construction.

(5) Ease in providing vertical openings of various sizes for elevators, stairwells, plumbing, skylights, etc.

(6) High stiffness of individual precast units (low deflection and camber).

(7) Ease in developing diaphragm action of roof or floor structures for resisting horizontal loads (seismic, wind, etc.).

(8) Clean, attractive soffit that is smooth or nearly so and, therefore, can be left exposed.

(9) Stability of precast elements during manufacture, transportation, and erection at the job site.

(10) Low, uniform, stable deflection, or camber, of members.

(11) High fire resistance.

(12) Good thermal and sound insulating qualities.

(13) Large and small daily production possible with minimum capital investment.

(14) Minimum erection time required at the job site.

(15) Large equipment and skilled labor not required for erection.

(16) Minimum additional labor, forms, or welding required for joining the elements at the time of or after erection.

(17) Good acoustical properties if the soffit is to be left exposed.

(18) The precast elements should not be excessively fragile.

It will be seen that no one framing scheme can possibly satisfy all of these requirements. The relative importance of the factors listed above will vary from job to job and in different areas of the country.

It is not possible to discuss each system and method of framing that have been used or produced as a standard product in this country, since there have been many different methods and variations of these different methods. A directory of precast, prestressed-concrete producers and their products has been published by the Prestressed Concrete Institute (Ref. 1). The directory includes thirteen types of beams, girders, and joists; nine types of "stemmed" units (i.e. double T beams, single T beams, etc.); five types of members with continuous internally formed voids; eleven types of piles; and a variety of architectural and miscellaneous units. Therefore, the discussion will be limited to the general schemes which have received reasonably widespread acceptance. Furthermore, the discussion must be confined to general terms, since the many variations that have been made to the several basic schemes render specific limitations for dimensions and other factors nonexistent.

11-2 Double T Slabs

Double T slabs are used extensively in North America for both roof and floor construction. They are made in a variety of widths and depth, as illustrated in Fig. 11-1. When used as floor slabs, a concrete topping from 2 to 4 in.

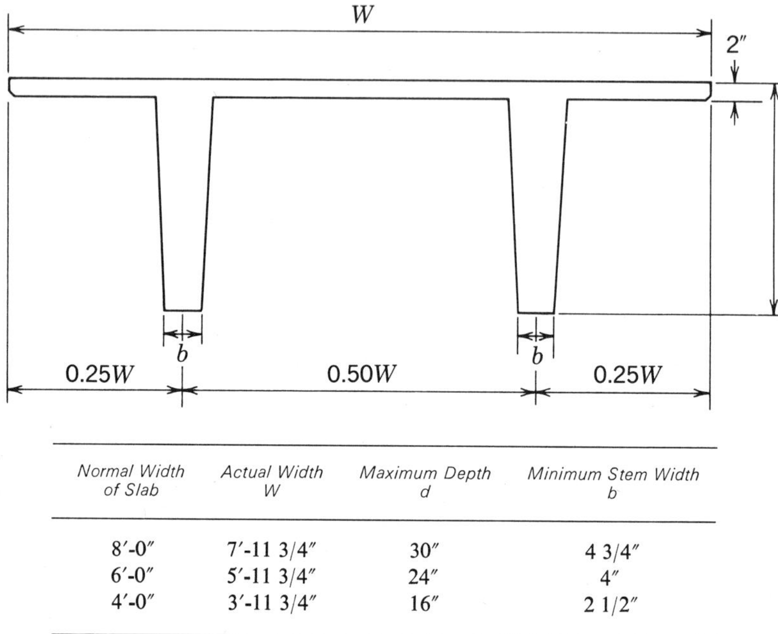

Normal Width of Slab	Actual Width W	Maximum Depth d	Minimum Stem Width b
8′-0″	7′-11 3/4″	30″	4 3/4″
6′-0″	5′-11 3/4″	24″	4″
4′-0″	3′-11 3/4″	16″	2 1/2″

Fig. 11-1 Typical double T slab dimensions.

thick is placed over the top of the slabs. The topping concrete is generally designed to work compositely with the precast section. The topping also provides a means of obtaining a flat wearing surface. It should be recognized that due to casting irregularities, variations in camber or deflection, and other construction inaccuracies, the upper surface of erected double T slabs cannot be expected to be flat and true to line. A concrete topping is generally not used in roof construction and insulation and roofing are applied directly to the precast slabs.

When a concrete topping is not used, it is normal practice to provide weld plates in the edges of the top flanges so that the slabs can develop diaphragm action and thus eliminate differential deflection between the flanges of adjacent double T slabs. Differential deflections may result in the roofing being damaged. Fill material can be provided where necessary between adjacent members in order to eliminate abrupt variations in elevation and thereby avoid possible roof damage.

Double T slabs are efficient structurally. The slab portion is quite thin and has well-balanced negative and positive moments. The legs of the joists or tees are thin and are normally highly stressed. The slabs are relatively light when compared to other types of framing for the longer spans. There is little excess material in the slabs. A significant contribution to the structural efficiency of the slabs results from the fact that a very large portion of the total dead load is acting on the slab at the time of prestressing. Double T slabs can be obtained at relatively low cost in virtually all sections of the country.

When used in applications that do not require beams to support the slabs, such as one-span commercial buildings with bearing walls, the depth of construction that is required for double T slabs is quite low. When it is necessary to support the double T slabs on interior beams and columns, the total depth of construction may become relatively great. This is particularly true if long spans are required in each direction, because, during erection, the slabs must be lowered vertically upon the supporting members and cannot be rotated in a horizontal plane into position, due to the width of the slabs. The inverted T beam, which is illustrated in Fig. 11-2, has been widely used with double T slabs. The inverted T beam may or may not be prestressed and may or may not be made continuous in the completed structure. The inverted T beam does not have an efficient shape for long simple spans. For the longer spans, a large wide top flange is necessary to achieve the required ultimate moment capacity. In applications where a cast-in-place topping will be provided, an adequate top flange may result. An efficient solution for long spans in each direction is the use of double T slabs in combination with beams that have wide flanges, as is illustrated in Fig. 11-3.

The size of vertical openings that can be provided in double T slabs is

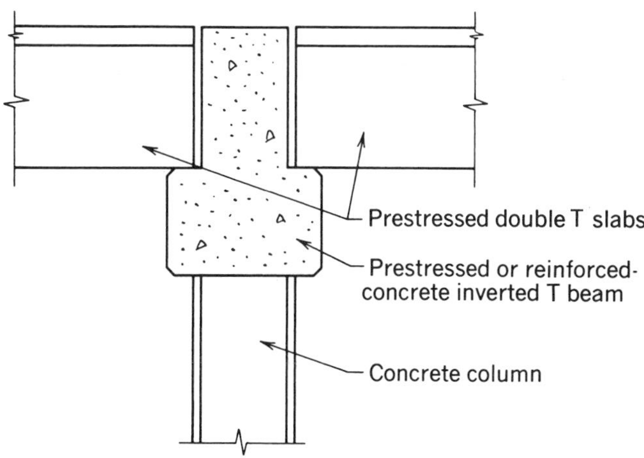

Fig. 11-2 Inverted T beam supporting double T slabs.

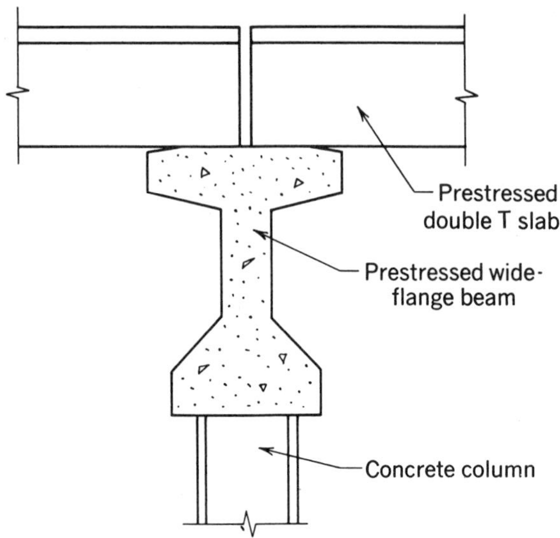

Fig. 11-3 Double T slabs supported by prestressed wide flange beam.

restricted to the clear width of flange between stems, unless special strengthening or intermediate support can be developed. The top flange of the slab is essential in developing ultimate moment capacity and, for this reason, the openings may have to be confined to the areas near the ends of the span where the moments are low. The soffits of double T beams can be left exposed in many industrial and commercial applications.

Double T slabs are stable during manufacture and erection, but they must be handled with reasonable care or the relatively fragile, outstanding flange can be cracked. Double T slabs occasionally have large cambers due to prestressing. The camber is sometimes not uniform from slab to slab and, in an attempt to minimize this undesirable feature, deflected tendons and partial prestressing are frequently used. When fabricated with a top flange of the order of 2 in. thick, as is frequently the case, the fire resistance of double T slabs can be made to have a 2 hr rating if provided with the proper insulation and built-up roofing. When provided with a cast-in-place concrete topping, a fire rating of 2 hr is easily obtained.

Double T slabs are normally made on pre-tensioning benches from 200 to 400 ft long, although some manufacturers use individual molds which resist the pre-tensioning force during concrete placing and curing.

Because of the relatively large size of each double T slab, the erection time normally required with these members is not excessive (on a unit-area basis) and equipment of moderate size is usually adequate. The amount of labor required to complete the structural roof or floor during and after erection varies from job to job and is dependent upon the amount of welding and other tasks required to complete the structure.

Some manufacturers have supplied this type of slab made of ordinary reinforced concrete rather than prestresed. Virtually all double T slabs are currently made pre-tensioned. Adequate results can be obtained with reinforced concrete on short and moderate spans, if camber is provided in the members, so that creep will not result in the members becoming sagged.

Double T slabs fabricated by reputable and skilled manufacturers are efficient in many applications. They not only perform well, but are also attractive in appearance.

11-3 Single T Beams or Joists

Single T beams of two general types are used in roof and floor construction. The first of these is made from the same mold that is used in double T beams and has the general dimensions shown in Fig. 11-4. The advantages and disadvantages of T members of this type are virtually identical to those of the double T slab.

The other type of T beam is made in a mold that allows the dimensions to be varied approximately, as is indicated in Fig. 11-5. This member can be used on roof spans up to 120 ft in length and in bridge spans up to about 60 ft. The web and flange thicknesses of this member are greater than is normally used in the double T slabs and can be varied within limits. The large single T has been used extensively in many areas of the country. The section is one of high structural efficiency.

The single T beam of each type is normally made prismatic. It is essential that some type of temporary support be provided to prevent these members

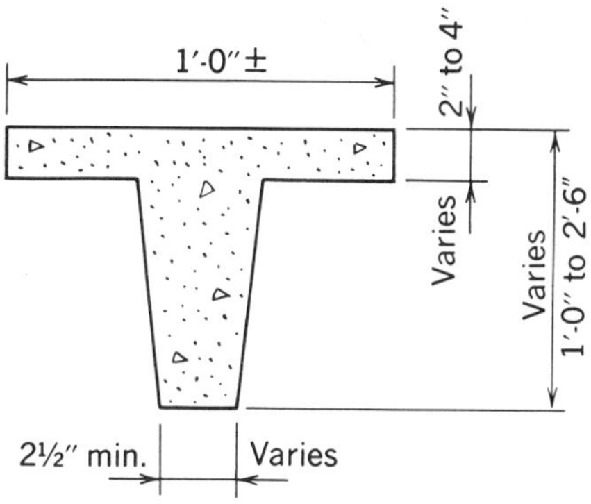

Fig. 11-4 T beam made in double T mold.

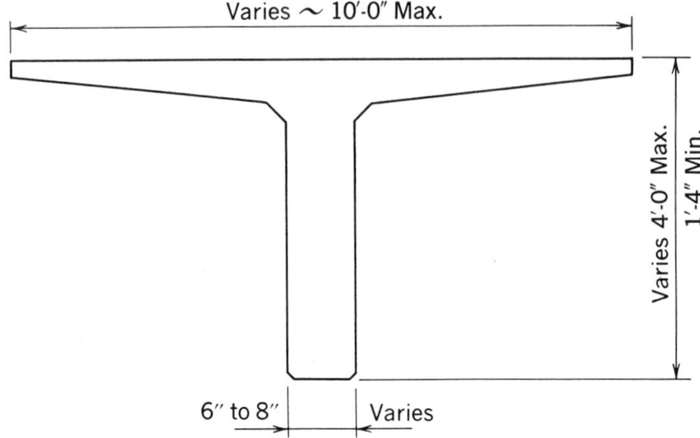

Fig. 11-5 Large T beam made in special mold.

from falling or being accidentally tipped on their side until such time as they are incorporated in the structure.

11-4 Long-Span Channels

Long-span channels are members which, like double T slabs, incorporate a relatively thin slab with a leg in such a manner that the member can be used for relatively long spans without supplementary joists. Therefore, long-span channels are used to span from bearing wall to bearing wall or from beam to beam, just as the double T slabs are. This single factor differentiates long-span channels from short-span channels (normally reinforced concrete) that require joists if the span exceeds 8 to 10 ft. Long-span channels can be used on somewhat greater spans than are possible with standard double T slabs, as a rule, due to their being more narrow and having legs of comparable or larger size than a double T.

Like a single T, some forms of channels are made from the same forms used to make double T slabs, and hence, they have the same general attributes as double T slabs. There is no need to consider this form of channel any further in this discussion, but the general dimensions that are used in some channels of this type are given in Fig. 11-6.

Channels of many other dimensions have been used. When the top flanges and legs are made thicker, the general strength is increased as is the fire resistance, stiffness, and cost. The camber and the variation of camber is less with channels of more ample proportions, such as are illustrated in Fig. 11-7,

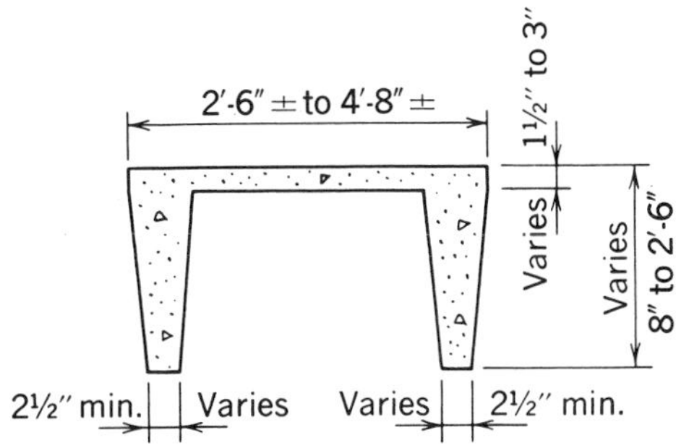

Fig. 11-6 Channel slab made in double T mold.

and with thicker legs, it is possible to develop good shear distribution between the members. In addition, the heavier sections are not as fragile as the lighter double T slabs.

End and intermediate diaphragms or flange stiffeners have been used in some types of prestressed and reinforced-concrete channels, but the provision of such secondary members greatly complicates the manufacturing facilities and techniques that must be employed. This is due to the shrinkage of the concrete. Shrinkage has a tendency to make the precast member cling to the form or mold as does the elastic deformation of the concrete which results from the prestressing. This effect can be minimized by placing a contraction joint in the forms.

The channels have several theoretical or practical advantages over double T slabs as can be seen from the above discussion. The advantages have not proved to be sufficiently important to justify the additional cost that frequently results from the use of channels. Long-span channels are also made in reinforced concrete and such construction is very efficient and economical for moderate spans.

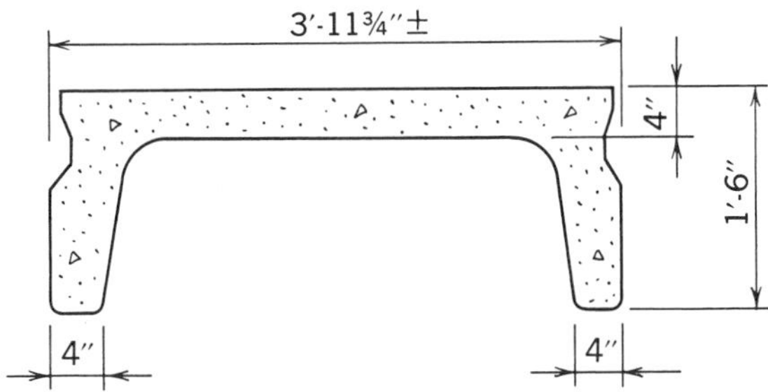

Fig. 11-7 Channel slab of moderate proportion.

11-5 Prestressed Joists

Prestressed joists of various types and sizes are used with many types of de materials, such as short-span channels, concrete plank, cast-in-place concre poured gypsum, and lightweight insulating roof materials composed cement-coated wood fibers. Typical sizes and shapes of joists that have be used are shown in Fig. 11-8. This type of framing is without question the m versatile, since the cross-sectional shape of the joist does not restrict maximum span on which the joist can be used to the same degree as channel slab, double T slab, and T beam construction. This versatility the result of the joist spacing not being a fixed dimension.

The depth of construction required with this type of framing is somew greater than is required in comparable spans of channel slabs and double slabs when each is used in bearing-wall construction, but it may be somew less when interior girders are required. The relatively lower constructi depth required when interior girders are used is a result of using girders t have a flat upper surface on the bottom flange on which the joists are su ported. This type of girder, which is illustrated in Fig. 11-9, can be u

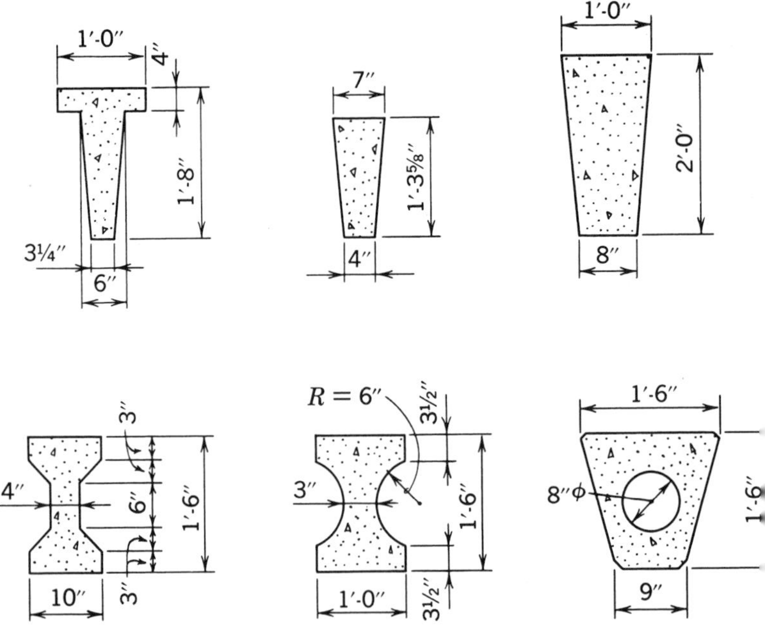

Fig. 11-8 Prestressed joists.

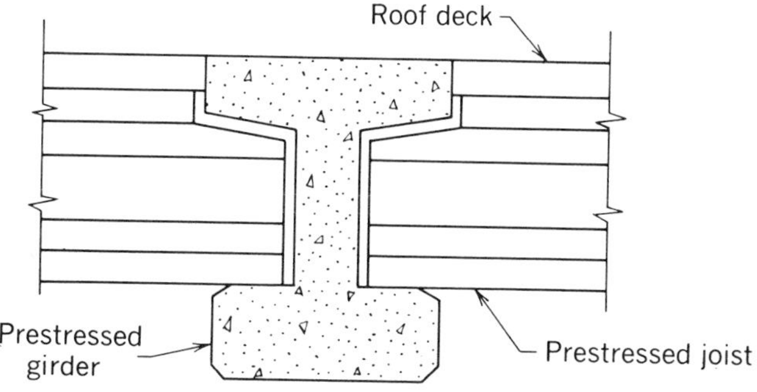

Fig. 11-9 Cross section through girder supporting prestressed joists.

without difficulty in joist construction, since the joists can be rotated into position as a result of their relatively small width. As was explained previously, this cannot be done with double T and channel slabs, due to their width.

Another significant advantage to joist construction is the ease of allowing for vertical openings. Since the joist spacings are normally from 2.5 to 8.0 ft, the size of the vertical openings obtainable without special framing or strengthening is greater than that which is possible with most other types of precast framing.

Many of the other desirable characteristics of the previously listed structural elements are a function of the design of the joists and, therefore, escape generalization. Included in this category are the maximum span, the fire resistance, stability during handling, stiffness, appearance of the joist, and the magnitude and stability of the deflection or camber. Well designed joists, however, will be satisfactory in all of these respects.

The degree of diaphragm action that can be developed, the appearance of the soffit, the fire resistance, the insulating quality, as well as the erection time and labor required, are contingent upon the type of deck that is selected.

11-6 Solid Precast Slabs

Prestressed-concrete solid slabs of two types are possible: small, prestressed planks that are 2 to 4 in. thick, and larger pre-tensioned slabs that are 6 to 10 in. thick. The smaller plank has been used in this country on a few projects, but considerable difficulty was reported to have been experienced in controlling the cambers and straightness of the individual units. It is believed that the large variations in camber are the combined results of the normal variation in the quality of the concrete, the shrinkage of the concrete, and the eccentricity of the tendons not being maintained as precisely as would be

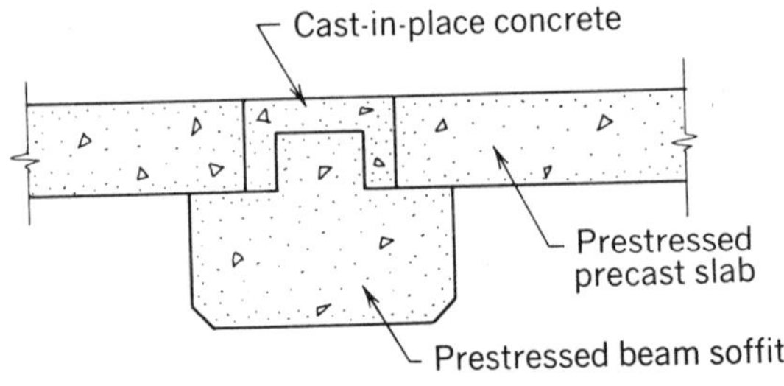

Fig. 11-10 Solid prestressed slabs with prestressed beam soffit.

required in order to obtain a uniform product. Prestressed slabs of the order of 3 in. thick are used in some areas of the country as precast soffits. The precast soffits are erected and shored at the center, after which, a topping is placed. The resulting composite slab can easily be made continuous. The shoring gives a means of equalizing initial differential camber.

Large prestressed slabs capable of carrying roof loads or nominal floo loads on spans to about 30 ft are considered quite practical if made partially prestressed. Although solid pre-tensioned slabs have not been used to a grea extent in this country, hollow slabs and solid, post-tensioned slabs, whic are discussed subsequently, have been used to a very significant degree. Th same general types of structures could be made with each type of construc tion. The principal advantages of pre-tensioned solid slabs are the ease o manufacture, small depth of construction, and the smooth soffit. The latte results in the elimination of the need for a suspended ceiling in some structures The principal disadvantages of solid slabs include the dead weight of th slabs, which can be partially offset by using lightweight concrete, and th difficulty of providing large vertical openings.

Solid slabs can be used with precast, prestressed beam soffits, as is illus trated in Fig. 11-10. The cast-in-place concrete, which is placed between th ends of the slabs, connects the slabs and the beam soffit with the result th beam has a T shape for all subsequently applied loads. In addition, in orde to minimize deflections or to increase the strength, reinforcing can be extende from the slabs into the cast-in-place concrete to develop continuity.

11-7 Precast Hollow Slabs

Hollow slabs or "cored" slabs have been used to a very significant degre The primary advantages and disadvantages of these types of elements a substantially the same as for the solid slabs discussed in Sec. 11-6. Th primary difference is that the hollow slabs are lighter and structurally mo

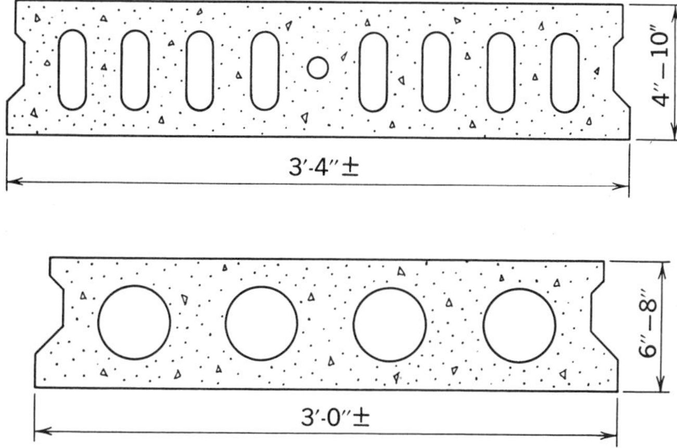

Fig. 11-11 Typical sections of hollow slabs produced domestically.

efficient in the elastic range. Typical cross sections of the hollow slabs currently produced in the United States are illustrated in Fig. 11-11.

One type of hollow slab produced in this country is made by a machine that travels along a pre-tensioning bench, straddling the tensioned wires, and automatically depositing, first, a layer of normal concrete that becomes the soffit of the final slab, next, a layer of lightweight concrete that becomes the web (with holes), and finally, another layer of normal concrete that becomes the top flange. In addition, the machine forms the sides of the slabs and forms the holes through the web. The slabs are made in continuous pieces and cut to the desired length with a saw, after the slabs have been cured. In this process, layers of slabs are cast one on top of the other (over a period of several days) until the stack is several slabs deep. The slabs cannot be cut or removed from the bench until the last slab that was cast has gained sufficient strength to allow the release of stress.

Other types of hollow slabs are made by casting the slabs in the normal manner and forming the voids in the slab by use of paper tubes, which are left in the member, or with metal tubes or inflatable rubber tubes, which are removed from the slab after the slab has hardened.

11-8 Cast-in-Place Prestressed Slabs

Three general types of cast-in-place prestressed slabs are possible in concrete construction. These are one-way slabs, two-way slabs, and flat slabs (or flat plates). These different types of slabs are characterized by the following:

(1) *One-way slabs.* A one-way slab is supported by continuous supports

extending across the entire width of the slabs. The supports may be beams, bearing walls, piers, or abutments. They may be positioned perpendicular to the longitudinal axis of the span or at an angle from the perpendicular to the longitudinal axis of the span (skew angle). One-way slabs may be simple spans or may be continuous over one or more supports.

(2) *Two-way slabs.* A two-way slab is generally square or rectangular in plan and is supported on all four sides by continuous supports. The supports are normally beams or bearing walls, and the slabs may or may not be continuous over the supports.

(3) *Flat slabs.* A flat slab is supported by a series of columns, which are normally positioned on a square or rectangular pattern, and there are no beams spanning from column to column, in either direction. Flat slabs may have drop panels or capitals at the columns and may be statically determinate or continuous over several column rows, and hence, statically indeterminate. Prestressed flat slabs do not normally have drop panels or column capitals. This type of construction has been made using the lift-slab technique, which lends itself to flat-plate framing without drop panels.

The design of one-way slabs, either simple or continuous spans, is done by employing the principles set forth for simple and continuous flexural members in Chapters 4 through 8. In order to simplify the design calculations, slabs are normally analyzed for a 1-ft width rather than working with the actual width. After the amount of prestressing steel and reinforcing steel required for a width of 1 ft has been determined, it is a simple matter to extend these data to the actual slab width.

It should be apparent that transverse bending moments exist in one-way slabs subjected to concentrated loads. The transverse reinforcing required for a specific condition may be dictated by the applicable design criteria or, in special cases, may be determined by the use of an elastic analysis. The elastic analysis of transverse bending moments in one-way slabs is beyond the scope of this book, and the interested reader is referred to the Refs. 2 through 4.

The design of two-way slabs should be done following the procedures given in the "Building Code Requirements for Reinforced Concrete," ACI 318-71. The effects of the transverse loads and the prestressing should be analyzed separately and the effects superimposed. The equivalent loading due to the curvature of the prestressing should be computed, using the methods described in Chapter 8, and these loads used to compute the effects of prestressing. An ultimate moment analysis should be made.

The design of prestressed flat slabs is done by analyzing the slab in each of the two principal directions, as if it were a one-way slab. Because of the difference in the stiffness in the slab along the column lines (column strips) and the slab located between column lines (middle strips), it is customary to place a greater percentage of the prestressing tendons in the column strips than in the middle strips. The proportioning of the tendons between the column strips and the middle strips must be left to the judgment of the designer.

The spacing and distribution of the tendons specified in the "Tentative Recommendations for Concrete Members Prestressed with Unbonded Tendons" (Ref. 6) are as follows:

3.2.4.2.—For panels with length/width ratios not exceeding 1.33, the following approximate distribution may be used: simple spans; 55 to 60% of the tendons are placed in the column strip, with the remainder in the middle strip; continuous spans; 60 to 70% of the tendons are placed in the column strip. When length/width ratio exceeds 1.33, a moment analysis should be made to guide the distribution of tendons. For high values of this length/width ratio, only 50% of the tendons along the long direction shall be placed in the column strip, while 100% of the tendons along the short direction may be placed in column strip. Tests indicate that the ultimate strength is controlled primarily by the total amount of tendons in each direction, rather than by the tendon distribution. Some tendons should be passed through the columns or at least around their edges.

3.2.4.3—The maximum spacing of tendons in column strips should not exceed four times the slab thickness, nor 36 in. (91 cm), whichever is less. Maximum spacing of tendons in the middle strips should not exceed six times the thickness of the slab, nor 42 in. (107 cm), whichever is less.

It should be emphasized that no new theoretical considerations in addition to those previously presented are required in the analysis of prestressed flat slabs. The primary novel features in this type of design include the decisions as to what portion of the tendons will be placed in the middle strips and column strips, as well as the determination of tendon trajectories for the tendons in each direction.

The tendon trajectories generally conflict where the tendons cross near the columns and near the center of the slab panels. The designer normally specified that all the tendons which run in the same direction are on the same trajectory, as far as possible. It is customary to adjust the tendon locations slightly during the preparation of the shop drawings, in order to avoid conflict between tendons and in order to establish a workable sequence for placing the tendons.

Deflection or camber of prestressed flat slabs may be a significant factor in the design of this type of construction, just as it is in other types of framing. The deflection of flat slabs can be determined by using the principles outlined in Sec. 6-3. It should be pointed out that the camber due to prestressing in each direction (span) is additive as is the deflection due to the applied loads.

11-9 Other Types of Framing

Precast and cast-in-place, prestressed, thin-shell or folded-plate construction has been used in the United States to a fairly significant degree. In Southern California, the cast-in-place type of framing has been used in the construction of many churches and in one large bakery. This type of framing is illustrated in Fig. 11-12.

Precast thin shells or folded plates have been used to a less significant degree. This type of element is considered to be very practical for long-span roofs and floors as well as for bridge members, which are discussed in the next chapter. Thin-shell sections, which can be made from one mold with only minor adjustments, are illustrated in Fig. 11-13. Members of this cross section have been made with a vibrating screed that rides on the mold and screeds the top surface to the desired profile and thickness. The precast elements can be made pre-tensioned for lengths up to about 100 ft or even longer, if they can be handled and transported. They could also be precast in short sections and post-tensioned together at the job site when very long members, which cannot be conveniently transported, are required.

Fig. 11-12 Prestressed concrete folded-plate roof.

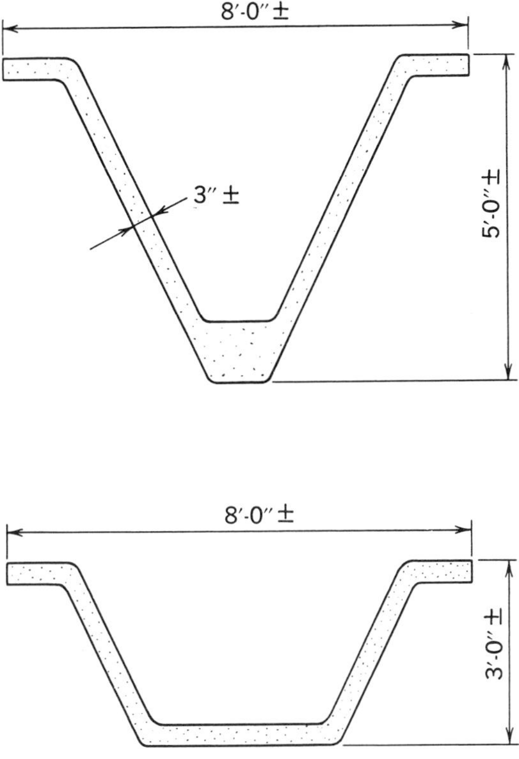

Fig. 11-13 Thin-shell folded-plate roof sections.

Cast-in-place, post-tensioned concrete has proved to be an economical mode of construction for parking structures in many parts of the country. The construction frequently consists of post-tensioned beams spanning 52 to 72 ft, spaced from 18 to 27 ft on centers. The slab between the beams is also frequently post-tensioned. The construction is often done with lightweight concrete. Frequently, the beams are made continuous with the building columns, this results in a frame for vertical loads. Lateral loads are normally resisted by shear walls, although the frames could be proportioned to resist the lateral loads too.

Prestressed-concrete space frames have been used abroad in the construction of exceptionally long spans. These frames are very light and offer a good solution for long-span structures. Although the details of such structures are complicated, it is believed that space frames in prestressed concrete are practical in this country. An interesting view of a space frame is shown in Fig. 11-14. There are many other cross-sectional shapes and types

Fig. 11-14 Prestressed-concrete space frame. (*Courtesy Cement and Concrete Associa-tion, London.*)

of framing that can be used in prestressed construction with good economy. The variety of schemes is limited only by the imagination of the designer.

11-10 Continuity in Precast Construction

By and large, the use of precast, prestressed-concrete building products in the United States is confined to simple beam construction. This may be due to the fact that the use of continuity complicates the design of a structure and renders the use of load tables impractical. Most manufacturers prefer to produce a product that can be advertised with a load table, because the selling of the products is facilitated by the use of such advertising.

Some types of precast, prestressed structures lend themselves to partially continuous designs. One of these is the use of prestressed joists and beam soffits made continuous by the placing of a composite reinforced-concrete deck. Construction of this type is more complicated to design than simple beam construction, but it has good characteristics from the standpoints of fire resistance, deflection, and resistance to vertical and horizontal loads.

REFERENCES

1. "Precast, Prestressed Concrete Producers and Products," Prestressed Concrete Institute, 1969.
2. Kist, H. J. and Bouma, A. L., "An Experimental Investigation of Slabs, Subjected to Concentrated Loads," *Publications, International Association for Bridge and Structural Engineering*, **14**, 85–110 (1954).
3. Kawai, T., "Influence Surfaces for Moments in Slabs Continuous over Flexible Cross Beams," *Publications, International Association for Bridge and Structural Engineering*, **16**, 117–138 (1957).
4. Westergaard, "Computation of Stresses in Bridge Slabs Due to Wheel Loads," *Public Roads* (Mar. 1930).
5. ACI Committee 318, "Proposed Revision of ACI 318-63: Building Code Requirements for Reinforced Concrete," *Journal of the American Concrete Institute*, **67**, No. 2, 77–186 (Feb. 1970).
6. ACI-ASCE Committee 423, "Tentative Recommendations for Concrete Members Prestressed with Unbonded Tendons," *Journal of the American Concrete Institute*, **66**, No. 2, 81–86 (Feb. 1969).

12 Bridge Construction

12-1 Introduction

The discussion of the various factors that influence the design of bridges will be made with respect to simple bridge spans. The same general principles apply in the design of continuous spans. In addition, this discussion is limited to highway bridges designed for normal truck loadings. The same principles apply to bridges designed for other purposes and types of loading. However, the span range in which each basic type of framing is most efficient may be altered if the ratio of the dead load to live load is appreciably different with the other types of live loading.

The basic configuration of the most efficient and economical structural elements in prestressed-concrete bridges for any specific structure is a function of the following:

(1) Magnitude of the span.

(2) Live and impact loads to be carried by the bridge.

(3) Type of structure to be used (simple beams, continuous beams, etc.).

(4) Allowable stresses.

(5) Size of the structure.

(6) Feasibility of various construction types and procedures, as controlled by the requirements of the bridge site.

The effect of each of these factors is discussed below, as a means of introducing the reader to the basic problems. In subsequent articles, the various basic types of simple, highway-bridge constructions commonly used, as well as some less common yet economical modes of prestressed bridge construction, are discussed and the limitations of each are pointed out.

The magnitude of the bridge span affects the design in three ways:

(1) The dead load of a bridge member, in proportion to the live and total loads, increases as the span is increased. Therefore, for very short spans, the live load for which the bridge must be designed is very nearly the total load and, for very long spans, the dead load is of much greater significance than the live load.

(2) The moment for which a flexural member must be designed is a function of the square of the span, while the shear is a direct function of the shear span and the load. Thus, for moving live loads, the shear force in short spans is very large in proportion to the bending moment and, in long spans, the bending moment is of much greater importance than the shear forces.

(3) The impact loads which must be included in the design, and which are usually considered as a function of the live load, are relatively smaller for long spans than for short spans.

These three factors obviously have great influence on the optimum cross section for bridge members to be used on different spans. Bearing in mind that only nominal tensile stresses are allowed in the precompressed tensile zone under the effect of the total load, it is apparent that the principles of elastic design are of somewhat greater significance in bridge design than in building design, where higher tensile stresses can normally be used. The ultimate moment a bridge structure can resist is also very important and must be sufficiently large to ensure that the safety factor is greater than a specific minimum value.

For the purposes of this discussion, it will be assumed that no tensile stresses are to be permitted under total load. Considering the maximum total moment relationship

$$M_t = M_d + M_l = P(e + r^2/y_b) \tag{12-1}$$

it becomes apparent that, since the dead load is very small for short span bridges, the relationship approaches

$$M_t = M_l = P(e + r^2/y_b) \tag{12-2}$$

From this, it follows that when the short span structure is acted upon by dead load alone, the pressure line is located at a distance nearly equal to e below the center of gravity of the section. Therefore, the cross section of the member must provide a relatively large bottom flange to resist the prestressing force during the periods when the live load is not being applied. Solid slabs, hollow slabs, and beams with large top and bottom flanges, but with webs of good proportions, all satisfy these conditions. Bridge cross sections of this type are illustrated in Fig. 12-1.

When the span is very large, shear is less important and dead load is a large percentage of the total load. Consideration of the total moment relationship given in Eq. 12-1 and the location of the pressure line under the action of prestressing plus dead load will reveal that, under the condition of no live load, the pressure line will be some distance above the center of gravity of the tendons (the distance equals M_d/P). Because the dead load is acting at the time of stressing, a large, bottom flange is not required to resist the prestressing force during the periods in which the intermittent live load is not applied. In precast construction, unless the stressing is done in two or more stages, all of the dead load of the structure will not be acting at the time of prestressing, and a bottom flange of moderate size may be required to

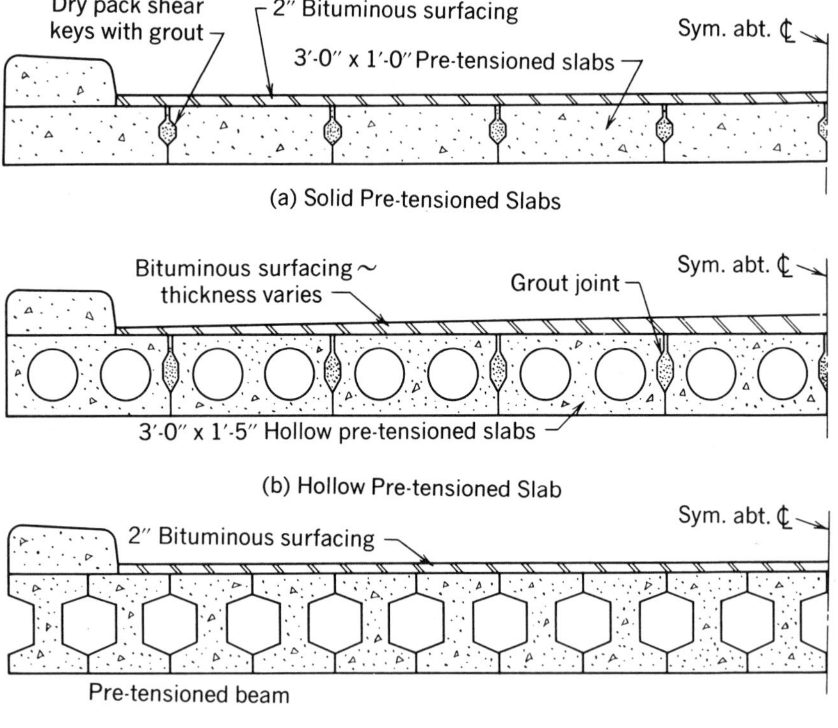

Fig. 12-1 Typical half-sections of short-span bridges.

resist the prestressing force temporarily during construction, even for longer spans. In cast-in-place construction, on the other hand, T shaped beams are often satisfactory for the long spans, since the entire dead load, with the possible exception of sidewalks and wearing surfaces, is acting at the time of stressing. Efficient cross sections for simple prestressed bridges with relatively long spans approach those shown in Fig. 12-2 for cast-in place and precast construction.

It should be apparent that top flanges of large size are required or desirable in both long and short spans to develop adequate ultimate moments and, in long spans, where bending moment is of greater importance than shear, a large top flange is necessary, due to elastic-design considerations.

The optimum cross sections for spans of moderate length are between the short-span solid or hollow slabs and the T beams which are optimum for very long spans. Therefore, either I beams or hollow boxes, such as are illustrated in Fig. 12-3, are most efficient for bridges of moderate spans.

It should be apparent that very heavy, live loads, such as are encountered in railroad bridges, would render the use of solid and hollow slabs practical

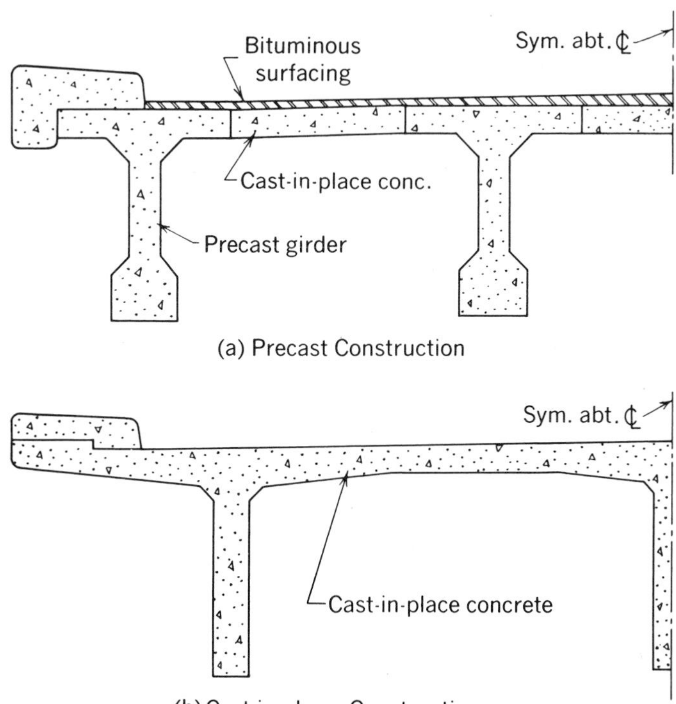

Fig. 12-2 Typical half-sections of long-span bridges.

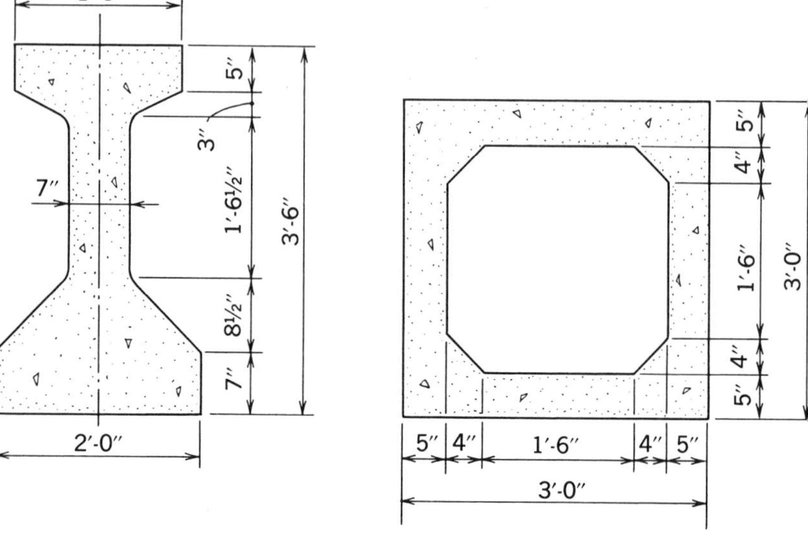

Fig. 12-3 Typical sections of prestressed beams used on moderate spans.

for spans of moderate length, whereas very light, live loads, such as ar encountered in the design of pedestrian bridges, would permit the use of T beams, even for relatively short spans.

Short span continuous structures often have significant reversals o moment due to live load. Also, there is normally a negative moment of grea magnitude at the interior support sections and a significant positive momen at or near the center of the span between supports. Because of these factors short-span continuous structures often require cross sections that are approx imately symmetrical at each section, such as hollow boxes or I shaped ele ments. The structural depth is frequently increased near the support sections but the basic cross-sectional shape is usually maintained throughout th entire length of the beam. Continuous bridges of long spans are not subjec to moment reversals and, because of the greater importance of dead loads larger top rather than bottom flanges are more efficient in areas of positiv moment, while the opposite is true at locations of negative moment.

The allowable stresses in the concrete section and in the prestressing stee will obviously have an influence upon the cross-sectional shape of prestresse elements that are feasible under specific conditions. The allowable concret stress in the top fiber of precast sections has significant influence on th amount of prestressing steel required in pre-tensioned construction. The 196 AASHO bridge design criteria allows small temporary tensile stresses in th top fibers of precast elements, but does not allow tensile stresses in the top fibers of the structure after loss of prestress, even though the application of the live load in simple structures does not increase these tensile stresses. This would seem to be too conservative. It would seem that nominal tensile stresses should be permitted in the top fibers after loss of prestress, without non-prestressed reinforcement, and that higher tensile stresses should be permitted when non-prestressed reinforcement is provided. In continuous structures, where moment reversals are a consideration, it is apparent that tensile stresses in any fiber must be studied with care.

It has been customary to limit the compressive stress to $0.40\, f'_c$ in bridge design. This limitation of compressive stress, for beams of good cross-sectional proportions, is only a problem for girders of very long span. It is the author's opinion that in the future higher compressive stresses will be permitted in members of very long span, since there should be little risk of fatigue in such structures due to the fact that the variation in stress due to the live loads is so small.

On large bridges the designer can use precast or cast-in-place elements and techniques which are designed for the specific conditions and, hence, are the most economical in labor and materials. This is feasible on large jobs, since the cost of the forms and plant required to produce the special members may be less than the savings of materials and labor that results from the special design.

In designing small jobs, the designer must use elements that require simple forms or members that are standard products for the prestressing plants and contractors located in the vicinity of the job site. This is done so that the high cost of elaborate forms and plant facilities are not entirely amortized on the one small job.

Each bridge site is characterized by certain conditions that may dictate which of the design or construction procedures are feasible. Among these are the purpose of the bridge (i.e., grade separation, river crossing, railroad crossing, etc.), accessibility of the site from existing precasting plants, the required skew, the quality of labor and materials available near the site, etc. Any of these factors can obviously govern the feasibility of various types of framing or construction procedure.

The reader will readily understand that the factors briefly discussed above vary considerably from job to job and from one locality to another. The types of framing discussed in Secs. 12-2 and 12-3 are used almost without deviation for bridges of short and moderate spans in the United States, since the construction procedures that are feasible for structures of these types are less apt to be controlling factors in the design. When designing long-span bridges, the designer may be compelled to use cast-in-place or precast construction either to facilitate construction or to maintain minimum horizontal and vertical clearances during construction. The type of framing that may be economical for a bridge or grade separation structure in the United States may be far different than the type of framing that may be economical in some remote country where suitable labor and equipment may not be as available as in this country. In view of this, further attempts will not be made to generalize upon the influence of the various factors that control the selection of the bridge framing.

12-2 Short-Span Bridges

Short-span bridges, for the purposes of this discussion, will be assumed to have a maximum span of 45 ft. It should be understood that this is an arbitrary figure and that there is no definite line of demarcation between short, moderate, and long spans in highway bridges. As has been mentioned above, short-span bridges are most efficiently made of solid slabs, hollow slabs, or I beams, which are of generous proportions.

The solid slabs are most economical when used on short spans of the order of 20 ft or less. The slabs are usually precast 3 or 4 ft wide and have shear keys cast in the sides. After the members are erected and the joints between the slabs are grouted, the keys will transfer shear from one member to another.

Solid slabs may be of the type shown in Fig. 12-1, in which case, from $1\frac{1}{2}$ to 2 in. of asphaltic concrete, portland cement concrete, or bituminous paving material is applied to serve as a wearing surface and leveling course. Solid slabs of the composite type, such as is illustrated in Fig. 12-4, can also be used efficiently. In the composite type of slab, the cast-in-place concrete serves both as the leveling course, the wearing surface, and the structural top flange. It can be used to develop continuity for live loads, if desired, by placing reinforcing steel in the cast-in-place concrete at the locations of negative moment. Because of the relatively high live-load moment and due to moment reversals, which are characteristic of short-span continuous construction, little is to be gained through the use of continuity in short-span bridges, with the exception of increased resistance to lateral loads, reduced deflections, and the elimination of deck joints.

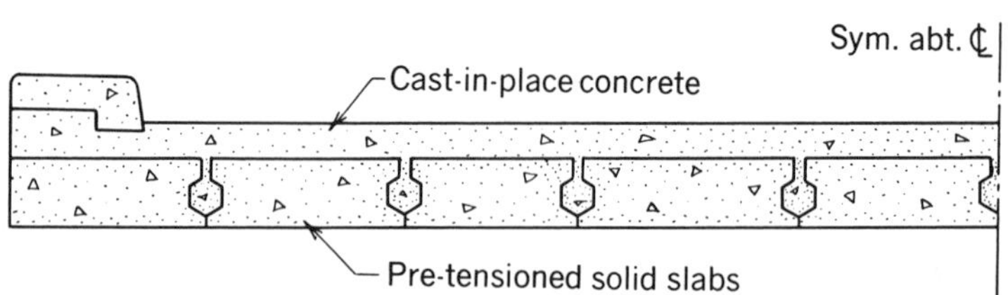

Fig. 12-4 Composite precast-slab construction.

The hollow slabs, which may have round or square voids and which are used on short spans, are generally made in units 3 ft wide and in depths from 18 to 27 in., although they can be made in any convenient width and depth. The hollow slabs are frequently used in bridges with spans from 20 to 50 ft. Longitudinal shear keys are used with the hollow slabs, just as with the solid slabs. Hollow slabs are not frequently used with composite, cast-in-place concrete toppings, but the use of some type of levelling course is normally required.

Some form of transverse tie is required in all types of bridges, in order to ensure that the members will not be spread apart by the application of the live load and that the proper distribution of the live load will be developed between the members. In slab construction, this tie most frequently consists of threaded steel bars placed through small holes which are formed transversely through the member during their fabrication. Nuts are then placed and tightened at each end of the bar. In some instances, the transverse tie consists of a post-tensioning tendon that is placed, stressed, and grouted after the slabs are erected. The traverse tie bar or post-tensioning tendon usually extends from one side of the bridge to the other and is placed along the skew. When large skew angles are encountered, the slabs are frequently tied together by connecting each unit to the next with short, tie bars that extend between two units only and are placed in holes perpendicular to the axis of the slabs. The short bars are offset as required by the skew. This detail is illustrated in Fig. 12-5.

Channel sections of various depths have also been used in several areas of the country. The channels are used in the same general manner that the solid and hollow slabs are used, with the single exception that a transverse tie of the elements in channel-slab bridges is often developed by bolting the legs of the channels together, rather than by using a tie bar that extends across the entire width of the bridge. This is illustrated in Fig. 12-6.

The crown or superelevation of the roadway, when using any of the short-span bridges of the types discussed above, is developed in one of three ways. These are: (1) constructing the bearing seats on the abutments and piers on straight slopes, so that the slabs will be placed on a slope; (2) constructing the abutment and pier bearing seats level and developing the required crown or superelevation with the leveling course; and (3) constructing a level bearing seat for each slab unit, in a series of steps which result in the required superelevation or crown. Each of these methods has been illustrated in Figs. 12-1 through 12-6, and each has merits which require no further elaboration.

Short-span bridges are also made using composite-stringer construction, such as is illustrated in Fig. 12-7, in which the AASHO-PCI, type 1, bridge stringer is shown. There is little advantage in using composite construction for short spans, from the standpoint of flexural stresses, since the flexural

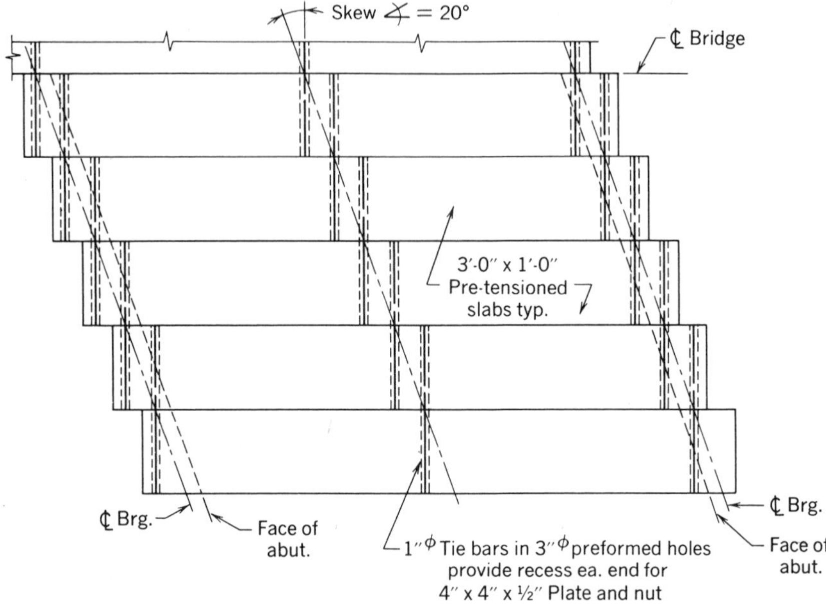

Fig. 12-5 Half-plan of skewed bridge showing staggered transverse tie-bar layout.

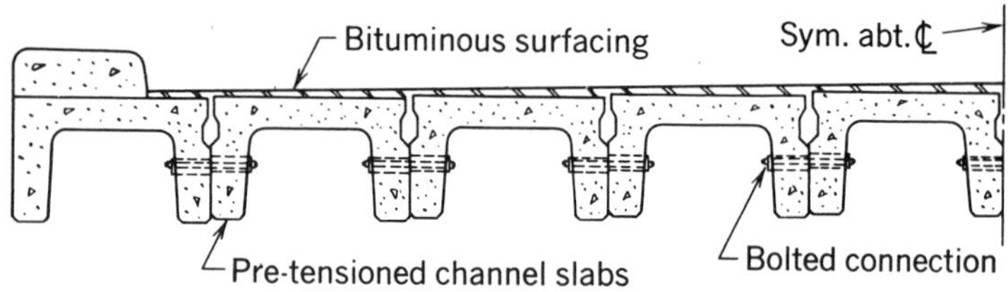

Fig. 12-6 Half-section channel-slab bridge.

stresses are not normally critical. Furthermore, the precast stringer must have at least a nominal top flange in order to provide lateral stiffness during handling, transporting, and erecting the units. The nominal top flange is usually adequate for the small bending stresses, without composite action.

The shear stresses in short-span bridges of composite-stringer construction are frequently very high, and large quantities of web reinforcing may be required in order to ensure that adequate factors of safety against ultimate shear failure is provided in the structure.

When stringer construction is used for spans of between 30 and 45 ft, it is generally considered better practice to use stringers with web thicknesses

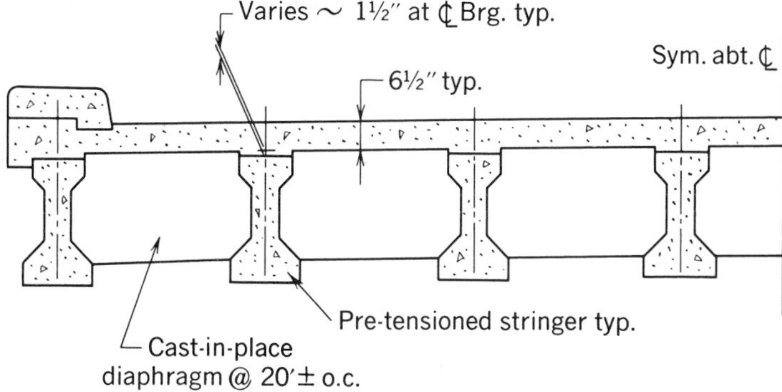

Fig. 12-7 Half-section of short-span composite-stringer construction.

of from 7 to 10 in. in order to reduce the unit shear stresses and the required amount of web reinforcing. In addition, the stringer spacings used in this type of construction are generally restricted to from 4 to 6 ft for short spans. When larger spacings are used, the shear stresses become excessive.

Diaphragms are used in stringer-type construction in order to ensure lateral distribution of the live load. The diaphragms are made post-tensioned or of reinforced concrete. One diaphragm is usually placed at each end and at the center of the shorter spans. The diaphragms are normally placed along the skew, unless the skew angle is very great, in which case, the diaphragms may be staggered across the bridge in much the same manner as in steel bridges or in the hollow-slab bridge illustrated in Fig. 12-5.

No generalities will be given pertaining to the relative cost of each of the above types of construction, since cost is a function of many variables. However, it should be pointed out that the stringer type of construction requires a considerably greater depth of construction when compared with solid, hollow, or channel slabs. The stringer type of construction does not require a separate wearing surface, as do the slab types of construction, unless precast slabs are used to span between the stringers in lieu of a cast-in-place deck. The stringer construction generally results in the minimum total quantity of superstructure materials. The construction time required to complete a bridge after the precast members have been erected is greater with stringer framing than with the slab type of framing.

12-3 Bridges of Moderate Span

Again for the purposes of this discussion only, moderate spans for bridges of prestressed concrete shall be defined as being from 45 to 80 ft. Bridges in this span range can generally be divided into two types—stringer type bridges and slab type bridges. By far the greatest percentage of bridges encountered in American practice has been in the former category.

Stringer type bridges, which employ a composite, cast-in-place deck slab, have been used in virtually all areas of the country. For moderate spans, the AASHO-PCI Stringers, types 2 and 3, are sometimes used. These stringers are normally used at spacings of the order of 5 to 6 ft. The cast-in-place deck is generally of the order of 6.5 in. thick. This type of framing is virtually the same as that used on composite-stringer construction for short-span bridges (illustrated in Fig. 12-7).

It should be pointed out that the AASHO-PCI stringer types 1 through 4 have relatively small flanges and the top flange is smaller than the bottom flange. The dimensions of the AASHO-PCI stringers are given in Fig. 12-8.

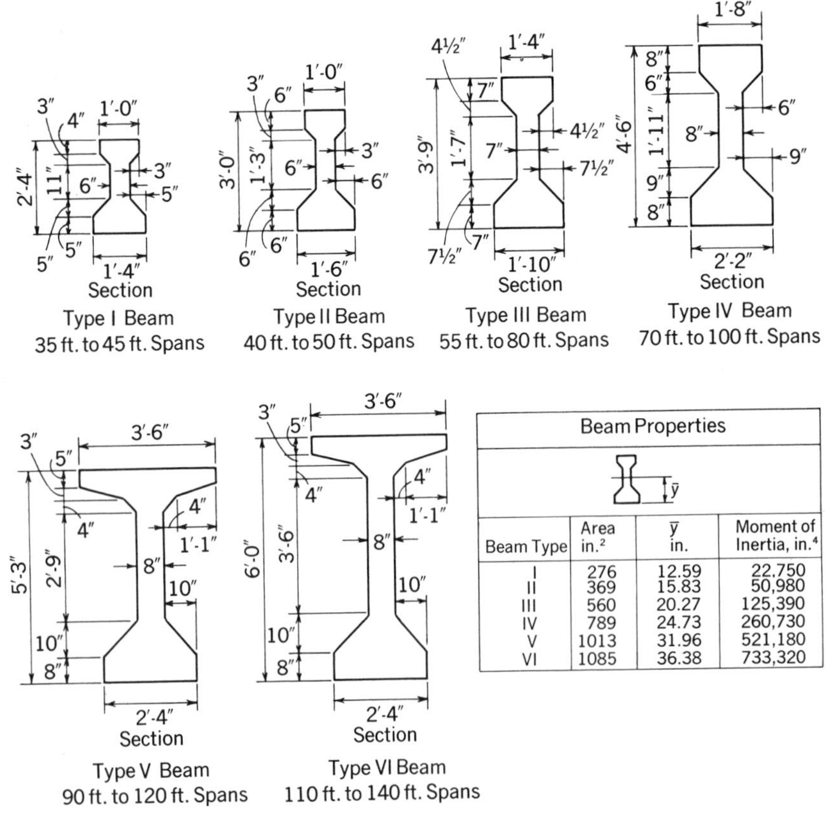

Beam Properties

Beam Type	Area in.²	$\bar{y}$ in.	Moment of Inertia, in.⁴
I	276	12.59	22,750
II	369	15.83	50,980
III	560	20.27	125,390
IV	789	24.73	260,730
V	1013	31.96	521,180
VI	1085	36.38	733,320

Fig. 12-8 AASHO-PCI standard bridge stringers.

Because of the small flanges, a relatively large depth of construction is required for any specific span when the AASHO-PCI Stringers are used. This is not a disadvantage in some applications. A large stringer depth does result in a small prestressing force being required for a specific stringer spacing. In some instances, however, bridge construction depth is of prime importance and, in such instances, stringers with larger flanges, such as those illustrated in Fig. 12-9, can be used at spacings of from 6 to 8 ft with a significantly less depth of construction than that which would be required if stringers with small flanges were used.

Another important consideration is the size of the top flange. As has been pointed out, the dead weight of a structure, as well as the stringer alone, becomes greater as the span is increased. The significance of this can be best understood if a stringer with a smaller top flange than bottom flange is analyzed for various stringer spacings on a span of 70 to 80 ft., with composite construction. It will frequently be found that the bottom flange is adequate and that the capacity and spacing of the member is limited by the compressive stresses in the smaller top flange. If the span were only 50 ft and the same procedure were followed, it would be found that the bottom flange limits the design. The difference is due to the difference in the ratio of dead load to live load which occurs as the span is increased. This restriction can be avoided by selecting stringer shapes similar to that shown in Fig. 12-10, when the span is greater than 70 ft.

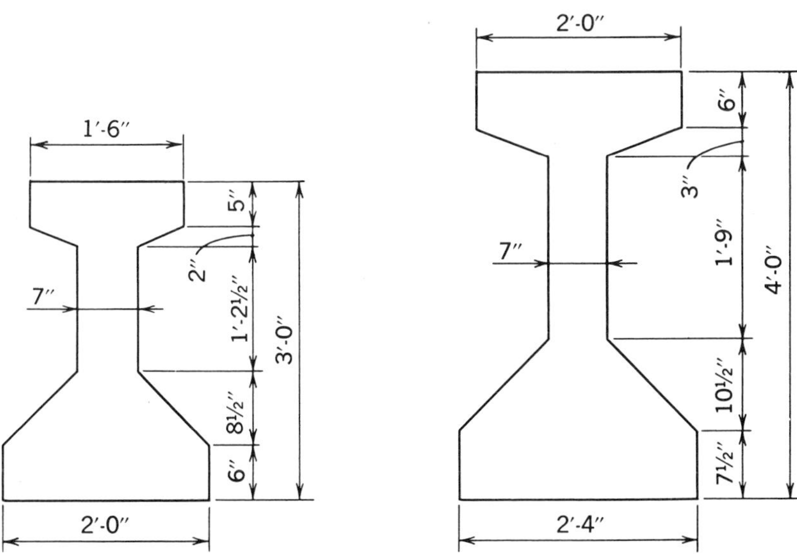

Fig. 12-9 Bridge stringers with large flanges.

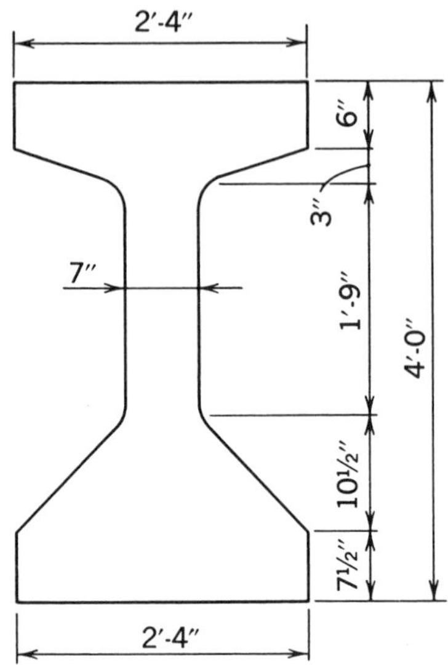

Fig. 12-10 Well proportioned bridge stringer for span of 60-70 ft.

Composite stringer construction is also used with details similar to those shown in Fig. 12-11. In this type of framing, the top flange of the girder is reinforced or post-tensioned traversely and forms a portion of the deck of the completed superstructure. The slab between the flanges of the stringers can be either cast-in-place or precast. Another scheme is to cast the stringers with daps near the top—the daps are used to support precast slabs. This can be

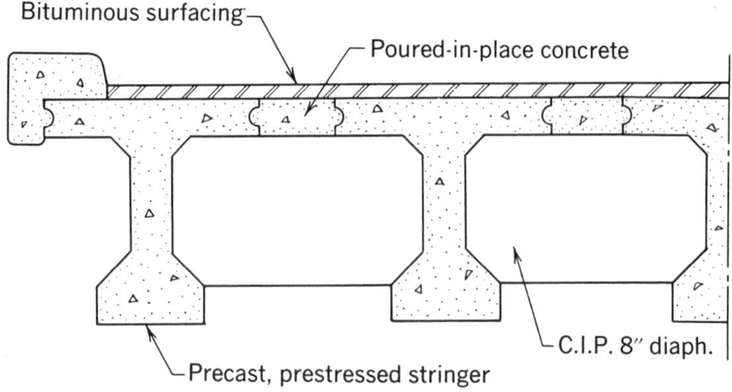

Fig. 12-11 Partial section of a moderate-span bridge.

done with I shaped stringers or with hollow stringers, such as shown in Fig. 12-12.

In order to eliminate the need for large bottom flanges on stringers in the span range of 50 to 70 ft, two-stage post-tensioning has been employed. This procedure consists of using a T-shaped beam in which 50 to 75% of the tendons can be stressed prior to removing the beams from the casting bed. After the deck has been cast, the remaining tendons are stressed. This type of construction is restricted to post-tensioned (or combined pre-tensioned and post-tensioned) construction. The second post-tensioning, which cannot be done until the deck has attained reasonable strength, may prolong the required construction time.

Slab-type bridges in the moderate-span range have been made of solid, cast-in-place, post-tensioned concrete in applications where depth of construction is extremely critical. The depth-to-span ratio for simple spans can be confined from 1 in 26 to 1 in 30. It is more usual in slab-type bridges to use precast, hollow boxes. A typical cross section for an element of this type is illustrated in Fig. 12-3.

Diaphragm details in the moderate-span bridges are generally similar to those of the short spans, with the exception that two or three interior diaphragms are sometimes used, rather than only one at the center line, as in the short-span bridge.

As in the case of short-span bridges, the minimum depth of construction in bridges of moderate span is obtained by using slab construction, which may be either solid- or hollow-box in cross section. Average construction depths are required when stringers with large flanges are used in composite construction. Large construction depths are required when stringers which have small flanges are used. The composite construction may be developed with cast-in-place or precast decks. Less materials are required in the composite type of construction, and the dead weight of the superstructure is less for the stringer construction than for the slab construction.

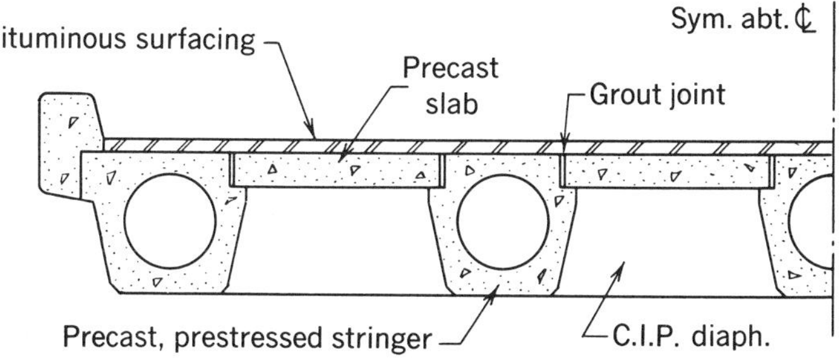

Fig. 12-12 Half-section of a bridge with precast slabs.

12-4 Long-Span Bridges

For bridge spans of the order of 100 ft, the same general types of construction used in bridges of moderate span are frequently used, with the singular exception of solid slabs which would be much too heavy for spans of this magnitude. The stringers are frequently spaced from 7 to 9 ft in bridges of this span range when composite-stringer construction is used. Due to the dead weight of such construction, precast hollow-box construction is generally employed for spans of this order only when the depth of construction must be minimized. Cast-in-place, post-tensioned, hollow-box bridges with simple and continuous spans are frequently used for spans of the order of 100 ft and longer. This has been particularly true in California.

Simple, precast, prestressed stringer construction would be economical in this country in spans up to 300 ft, under some conditions. However, only limited use has been made of this type of construction on spans in excess of 100 ft. For very long simple spans, the advantage of precasting is frequently nullified by the difficulties involved in handling, transporting, and erecting the girders, which may be of the order of 10 ft deep and weigh over 200 tons. The exceptions to this are on large projects on which all of the spans are over water of sufficient depth and character that the precast beams can be handled with floating equipment, when special girder launchers can be used (*see* Chapter 17), and when segmental construction is used (*see* Chapter 15).

Precast, long-span bridges may approach the general shape illustrated in Fig. 12-13. A very long, simple-span bridge of cast-in-place construction will normally approach the cross section shown in Fig. 12-14. These types of

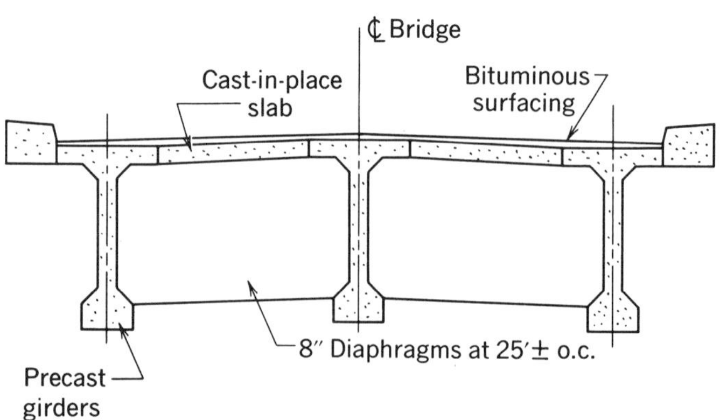

Fig. 12-13 Typical section of a long-span precast bridge.

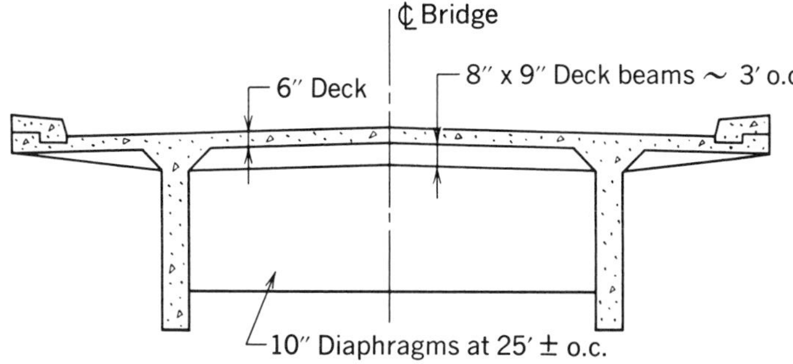

Fig. 12-14 Cross section of a long-span cast-in-place bridge.

bridges may be more accurately described as girder bridges rather than stringer bridges. The general reduction (or elimination) of the size of the bottom flange, as well as the large top flange supplied in the cross-sections of Figs. 12-13 and 12-14, should be noted.

Hollow folded plate beams, such as the one shown in Fig. 12-15, could also be used economically on spans in the range of 80 to 100 ft. As in applying this shape to roof construction, the member would be made with adjustable forms for the bottom and sides, and the internal void would be formed by a traveling vibrating screen. The deck could be precast and set on the plastic concrete of the hollow box immediately after casting the box in the precasting plant or the deck could be cast-in-place at the bridge site. This type of construction would be practical in pre-tensioned and monolithic, post-tensioned construction, as well as in segmental construction.

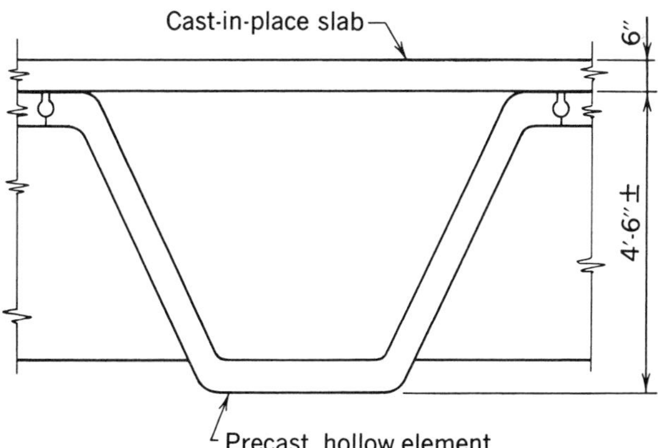

Fig. 12-15 Folded-plate bridge girder.

12-5 Bridges of Special Types

Exceptional procedures and methods may be practical in the construction of bridges that are very large or that have other special and unique requirements.

An illustration of such a project is the Lake Pontchartrain Bridge in Louisiana. This bridge is 24 miles long and consists, in part, of 2235 spans of prestressed concrete. Each span is 56 ft long. The deck and seven stringers which compose the superstructure of each span were cast as one monolith that weighed 180 tons. It is apparent that, with a bridge of such size as the Pontchartrain Bridge, very special methods could be employed. This structure is shown under construction and completed in Figs. 12-16 and 12-17, respectively.

Also illustrative of the elaborate construction methods that can be used under special conditions is the Esbly Bridge in France. This bridge is one of five bridges which were made of the same dimensions and which utilized the

Fig. 12-16 Bridge across Lake Pontchartrain, under construction. (*Courtesy Palmer and Baker, Consulting Engineers, Mobile, Alabama.*)

Fig. 12-17 Completed bridge across Lake Pontchartrain. (*Courtesy Palmer and Baker, Consulting Engineers, Mobile, Alabama.*)

same steel molds. All of the bridges span the River Marne, and, due to the required navigational clearances and the low grades on the roads which approach the bridge, the depth of construction at the center of the span was very restricted. The bridges were formed of precast elements, 6 ft long and were made in elaborate molds by first casting and steam-curing the top and bottom flanges in which the ends of the web reinforcing were embedded. The flanges were then jacked apart, being held apart by the web forms, and the web was cast and cured. Stripping of the web forms resulted in prestressing of the webs. The 6 ft elements were temporarily post-tensioned in the factory into units approximately 40 ft long. The 40 ft units were transported to the bridge site, where they were raised into place and post-tensioned together longitudinally, after which, the temporary post-tensioning was removed. Each span consists of six ribs or beams which were post-tensioned together transversely after they were erected. Hence, the beams are prestressed in all three directions. The completed Esbly Bridge consists of a very flat, two-hinged, prestressed arch with a span of 243 ft and a depth at the center o the span of about 3 ft. The bridge is illustrated in Fig. 12-18.

Cantilevered bridge construction, in which the segments may be cast-in place or precast, has been used extensively in Europe. The first applicatio of cantilevered bridge construction in the United States is expected to b completed in 1974. An example of this type of construction is shown in Fig 12-19. In cantilever bridge construction, the superstructure is constructe segmentally, starting from the piers and working each way until the con struction meets at the center of the spans. Each segment is post-tensioned t the previously constructed portions of the superstructure as the constructio proceeds.

Many other elaborate framing schemes have been used in Europe. Most o these are not considered practical in the American market, due to the relativel high cost of labor in the United States and the normal domestic practice o having separate organizations that are responsible for the engineering an construction. In Europe, where the firm that executes the construction of

Fig. 12-18 The Esbly bridge across the River Marne in France. (*Courtesy Freyssinet Co., Inc., New York.*)

Fig. 12-19 Pierre Bénite Bridge in France. (*Courtesy Freyssinet Co., Inc., New York.*)

project is often responsible for the design, there is less reluctance to use construction methods and procedures that are feasible with only one patented system, if such a system will result in the most economical solution for the particular structure.

13 Connections for Precast Members

13-1 General

The connections between precast members should be designed in such a manner that they are capable of withstanding the ultimate vertical and horizontal loads for which the structure is proportioned, without failure, excessive deformation or rotation. It is normally preferable that the strength of the connections exceed those of the members connected. The details of the connection should be such that they are readily adjusted to accommodate construction tolerances. The tolerances to be considered are not only variations in length, width and elevation, but also possible deviation from the anticipated planes of bearing. Connections must also be detailed to provide for the necessary erection clearances, reinforcing bar bends, and clearances that may be required for special requirements such as post-tensioning tendons after erection.

In the interest of economy, connections should be as simple as possible and should be such that the precast members can be set and disconnected from the erection equipment quickly with the erected members being stable. The connections should be of such a nature that they are easily inspected after they are completed.

The designer should pay attention to the details of the connections in order to be sure the structure will act as has been assumed in the design of the individual elements. If the beams that connect to the opposite sides of a column are designed as simple beams, the connection details should not result in the members being continuous or partially continuous as a result of a cast-in-place reinforced slab or topping, or due to other mechanical fasteners. On the other hand, if the members have been designed as continuous under certain conditions of loading, the connections should be carefully detailed to achieve the required continuity.

Flexural members undergo rotations and deflections due to the application of transverse loads. Rotations and deflection due to other effects such as temperature variation, creep, shrinkage, etc. may be encountered. Connection details, particularly for simple spans, should be made in such a manner the necessary rotations can take place without restraint and the risk of spalling from the member being supported unintentionally at an unreinforced edge, as is shown in Fig. 13-1.

Connections that incorporate welded reinforcing steel should be used with caution, since reinforcing steel frequently contains relatively high amounts of carbon, which necessitates special welding procedures. It is recommended that all welding be done in conformance with the recommendations of the American Welding Society.

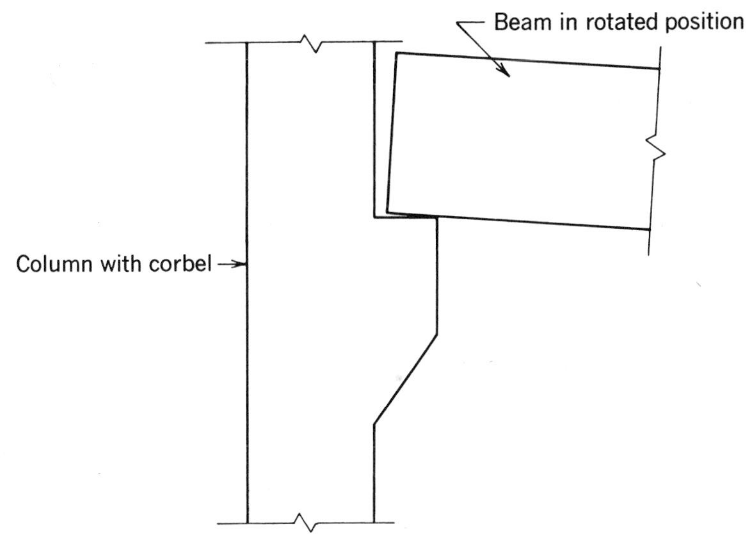

Fig. 13-1 Edge loading due to rotation of a simple beam.

The requirements for fireproofing of connections must be kept in mind Elastomeric bearing pads, which are combustible, and steel bearings wil not be permitted by most building codes to be exposed and unprotected in fire-rated construction. If connections are fireproofed by rigid materials, th fireproofing must not restrain the connection and cause it to act in an unin tended manner.

Booklets showing standard connection details are available from pre stressed-concrete manufacturer's associations.

13-2 Computation of Horizontal Forces

The computation of horizontal forces to be used in the design of connection for restrained members can be done by first determining the unrestrained change in length (Δ) of the member that would be expected to occur afte the member has been erected and the connection effected. The force from th restraint must then be determined.

If a simple beam were fully restrained, such as being attached to two infinitely stiff shear walls as shown in Fig. 13-2, the restraining force R_O would be that which would cause an increase of strain equal to Δ in the bottom fibers and can be computed as follows:

$$R_O = \frac{E_c A \Delta}{L\left(1 + \frac{y_b^2}{r^2}\right)} \tag{13-1}$$

In the case of a single span supported vertically and restrained agains translation, but not rotation, by two columns of equal stiffness (as shown in Fig. 13-3), the value for R_O can be computed as

$$R_O = \frac{3E_c I_c \Delta}{2H^3} \tag{13-2}$$

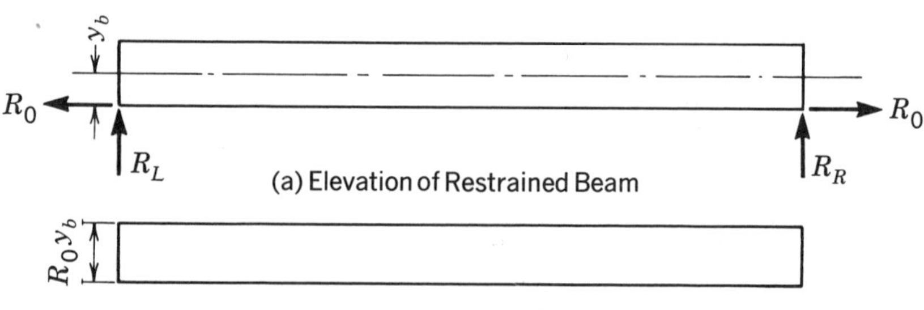

Fig. 13-2 Restrained simple beam.

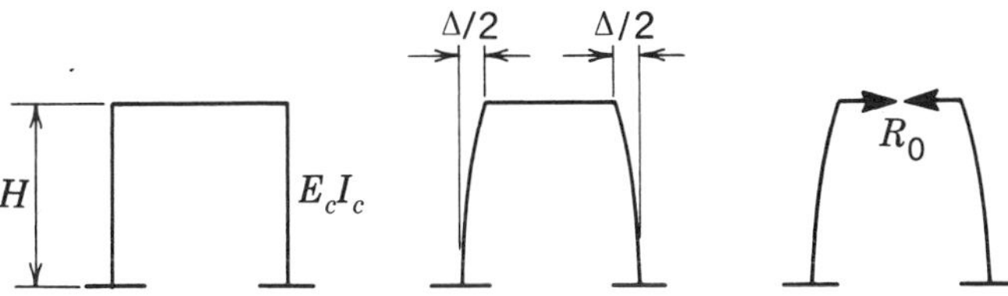

Fig. 13-3 Effect of beam shortening for a beam supported by two columns in which the beam ends are not restrained against rotation.

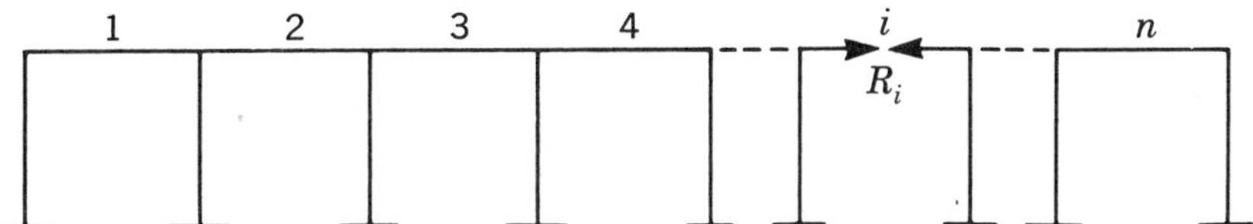

Fig. 13-4 Effect of restraint for a multi-span frame in which beam ends are not restrained against rotation.

in which E_c and I_c are the elastic modulus and the moment of inertia of the column, respectively. For a multi-span frame containing a number of equal spans restrained by columns of equal stiffness, such as shown in Fig. 13-4, the force in the interior spans can be approximated by

$$R_i = R_O\, i(n + 1 - i) \tag{13-3}$$

in which n is the number of spans, i is the number of the spans under consideration reckon from the end, and R_O is computed from Eq. 13-2. In Eqs. 13-1 and 13-2, the effective modulus can be used in computing R_O. (*See* Sec. 3-8.)

In multi-span structures where difficulty is experienced in applying Eq. 13-3, the restraining force in interior spans can be roughly approximated as being equal to 50% of the applied prestressing force (Refs. 3 and 4).

13-3 Corbels

Corbels for the support of precast beams on columns or walls were studied by tests performed at the Portland Cement Association Laboratories (Ref. 5). These tests resulted, in part, in the recommendations shown in Fig. 13-5, which can be summarized as follows:

(1) The minimum distance from the bearing plate to the edge of the corbel should be 2 in.

(2) The tensile reinforcement should be welded to an anchorage bar of the same size as the tensile reinforcement.

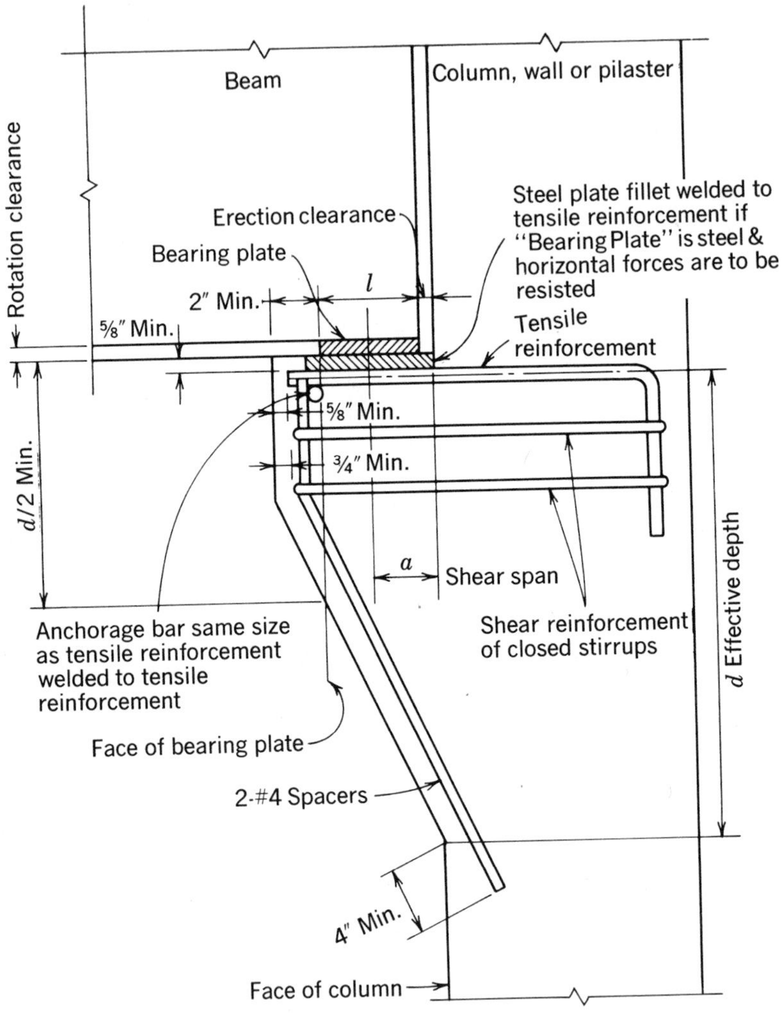

Fig. 13-5 Typical details for corbel design.

(3) The depth of the corbel at the outside face of the bearing plate should be not less than one-half the depth (d) at the face of support.

(4) If a horizontal (tensile) force is to be resisted in addition to the vertical force, a steel plate should be provided welded to the tensile reinforcement and embedded in the top of the corbel.

(5) The minimum cover on the tensile reinforcement and the shear reinforcement should be 5/8 and 3/4 in., respectively.

The shear span, a, and the distance from the extreme fiber to the tensile reinforcement, d, to be used in the design of corbels are illustrated in Fig. 13-5. The provisions of ACI 318-71 relative to the design of corbels include the following:

(1) a/d shall not exceed unity.

(2) For $a/d = 0.50$ or less, the corbel may be designed as for a shear-friction connection (*see* Sec. 13-12), but the quantity and spacing of reinforcing specified under this section shall be met.

(3) The percentage of tensile reinforcement, p, shall not exceed $0.13f'_c/f_y$.

(4) The ratio of the design tensile force on the corbels (T_v) to the design shear load (V_u) shall not be taken as less than 0.20.

(5) If provisions are made to prevent tension from restrained shrinkage and creep strains ($T_v = 0$), the corbel will be subjected to shear and moment only and the combined percentage of reinforcement (p_v) shall not exceed $0.20f'_c/f_y$. The area of the shear reinforcement parallel to the tensile reinforcement A_{vh} shall not exceed A_s. The combined percentage of reinforcement shall be taken as $p_v = (A_s + A_{vh})/bd$.

(6) The minimum area of closed stirrups, A_{vh} shall be 0.50 A_s and the stirrups shall be uniformly distributed within a distance of $2d/3$ adjacent to the tensile reinforcement.

(7) The minimum percentage of tensile reinforcement shall be $0.04f'_c/f_y$.

With these qualifications, the shear stress shall not exceed

$$v_u = [6.5 - 5.1\sqrt{(T_v/V_u)}][1 - 0.5(a/d)] \times \{1 + [64 + 160\sqrt{(T_v/V_u)^3}]p\}\sqrt{f'_c} \tag{13-4}$$

or, for the special condition where there is no creep or shrinkage restraint

$$v_u = 6.5[1 - 0.5(a/d)][1 + 64p_v]\sqrt{f'_c} \tag{13-5}$$

The bearing stress due to V_u shall not exceed 0.85 $\phi f'_c$, in which $\phi = 0.70$, based upon the contact area when the area of load is equal to the support area. When the area of load is less than the area of support and extends beyond the loaded area on all sides, the value of V_u shall not exceed 0.85 $\phi\, b'l'f'_c$ nor 1.5 ϕ 0.85 blf'_c in which b, l, b' and l' are as shown in Fig. 13-6.

ILLUSTRATIVE PROBLEM 13-1 Determine the required dimensions and reinforcing for a corbel that is to be provided on a 14 in. square column. The corbel is to be designed for an ultimate vertical load of 140 kips. There is no restraint for possible creep and shrinkage strains. The concrete strength is 4500 psi and $f_y = 60{,}000$ psi.

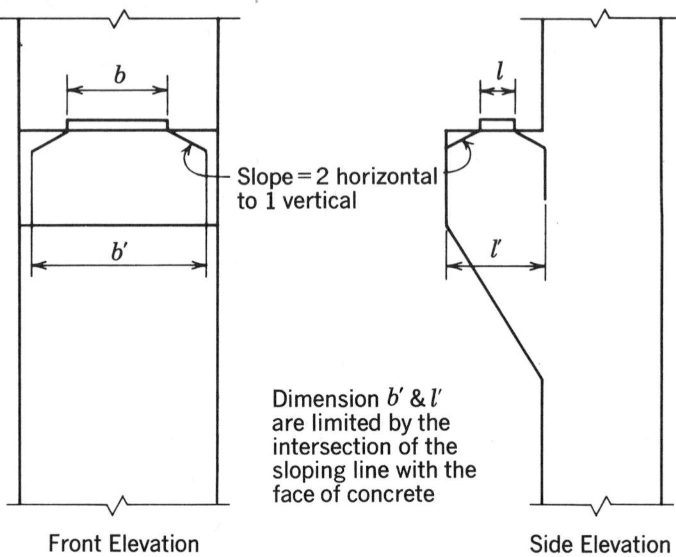

Fig. 13-6 Limits for computation of bearing stresses in accordance with ACI 318-71 for the condition of loaded area less than supporting area.

SOLUTION:

The minimum bearing plate length for a plate that extends the width of the corbel can be computed as follows:

$$l = \frac{V_u}{0.85\phi b f_c'} = \frac{140{,}000}{0.85 \times 0.70 \times 14 \times 4500} = 3.73 \text{ in. Use } 4.5 \text{ in.}$$

Providing for an erection gap of 1 in. $\pm$ 0.75, the maximum possible shear span a becomes

$$a = 1.75 + \frac{4.5}{2} = 4.00 \text{ in.}$$

The maximum combined reinforcement is

$$p_v = \frac{A_s + A_{vh}}{bd} = 0.20\left(\frac{4{,}500}{60{,}000}\right) = 0.015$$

Adopting a trial dimension of 18 in. for d, $v_u = \dfrac{140{,}000}{14 \times 18} = 556$ psi and the minimum combined area of steel is:

$$A_s + A_{vh} = 0.015 \times 14 \times 18 = 1.78 \text{ sq in.}$$

Try 5 No. 5 bars for A_s and 2 No. 4 closed stirrups for A_{vh}

$$A_s = 1.55 \text{ in.}^2$$

$$A_{vh} = 0.80 \text{ in.}^2 > 0.5\, A_s$$

$$p_v = \frac{2.35}{14 \times 18} = 0.0093$$

and the allowable shear stress from Eq. 13-5 is

$$v_u = 6.5\left[1 - \frac{0.5a}{d}\right][1 + 64\, p_v]\sqrt{f'_c}$$

$$v_u = 6.5\left[1 - \frac{0.5 \times 4.00}{18}\right][1 + 64 \times 0.0093]\sqrt{4500}$$

$$= 618 \text{ psi} > 556 \text{ psi}$$

The detailed corbel is shown in Fig. 13-7. Note that the stirrups are within the upper two-thirds of the effective depth.

ILLUSTRATIVE PROBLEM 13-2 Design the corbel of Illustrative Problem 13-1 assuming $V_u = 140$ k and $T_u = 100$ k, all other details remain the same.

$$a = 4.0 \text{ in.}$$

$$\frac{T_v}{V_u} = 0.714$$

Assuming a value of d of 24 in., the shear stress due to V_u is $v_u = \dfrac{140{,}000}{14 \times 24} =$ 417 psi and the maximum amount of tensile reinforcing is

$$A_s = \frac{0.13 f'_c}{f_y} bd = \frac{0.13 \times 4500 \times 14 \times 24}{60{,}000} = 3.28 \text{ in.}^2$$

Adopting 4 No. 8, $A_s = 3.16$ sq in., $p = 0.0094$, $a/d = 0.167$, from Eq. 13-4, the allowable shear stress is

$$v_u = [6.5 - 5.1\sqrt{0.714}][1 - 0.5 \times 0.167][1 + (64 + 160\sqrt{0.714^3})0.0094][67.08]$$

$$= [2.19][0.917][2.51][67.08] = 338 \text{ psi}$$

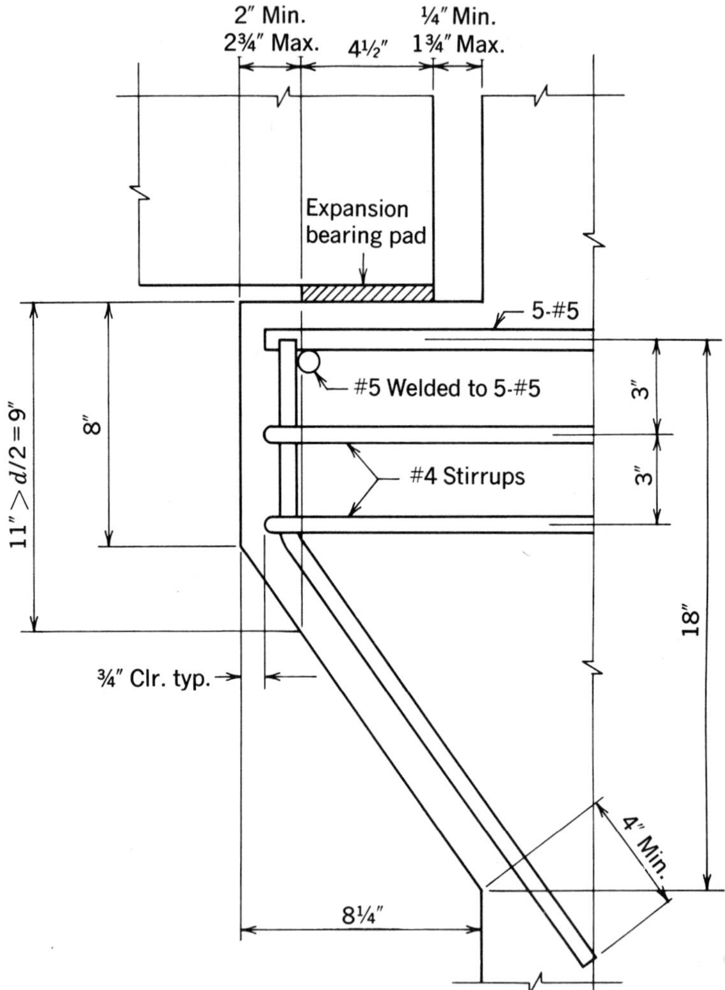

Fig. 13-7 Corbel for Prob. 13-1.

Thus, the depth must be increased. For a depth of 30 in.,

$$v_u = \frac{140{,}000}{14 \times 30} = 333 \text{ psi}$$

Therefore, use a depth of 30 in., 4 No. 8 bars for tensile reinforcing and 4 No. 4 closed stirrups, $A_{vh} = 1.60 \text{ in.}^2 > 0.5A_s$, spaced at 4 in. o.c.

13-4 Column Heads

The provisions of the building codes relative to bearing stresses can be a critical limitation in the design of column-beam connections in precast construction. A study of the ultimate strength of column heads with various reinforcing details, loaded in various manners, was conducted by the Portland Cement Association in order to determine relationships for the accurate determination of the strength of these connections.

The study revealed that without restricting the bearing stress to conventional levels, column heads that are adequately reinforced with lateral reinforcing (when loaded to failure) would be expected to fail by crushing. Without lateral reinforcing, they would be expected to fail in shear if the load is applied at a distance of 1.5 in. or less from the edge of the column and by splitting when loaded at a distance greater than 1.5 in. from the face of the column. This is illustrated in Fig. 13-8. Lateral reinforcement should be anchored by welding cross bars close to the ends of the lateral reinforcement or by welding bearing plates or angles to the lateral reinforcing, as is shown in Fig. 13-9. Horizontal loads, as in the case of corbels, have a significant effect on the strength of column heads and should be eliminated when possible. The uniformity of the bearing stress between the beam and the column head is also very significant in affecting the strength of a column head.

The prudent designer will take the tolerances of casting precast members, as well as the effects of end rotation, into account when designing column

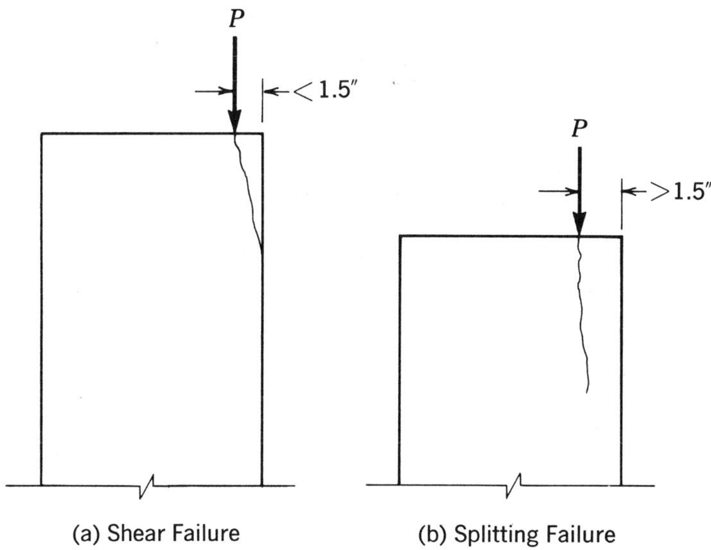

Fig.13-8 Modes of failure for properly reinforced column heads.

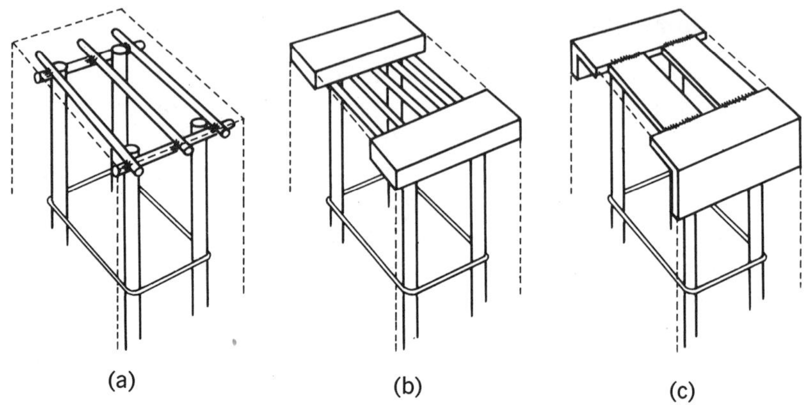

Fig. 13-9 Modes of welding lateral reinforcement in column heads.

head connections and provide elastomeric or other types of bearings to eliminate the adverse effects of irregularities and rotations.

The ultimate bearing strength, of a laterally reinforced column head that is uniformly loaded in bearing, in which the horizontal forces are either eliminated or accurately determined, and in which the product of sl is two or more can be computed by

$$f_b' = (A)(B)(C) \tag{13-6}$$

in which

$$A = \phi\left(69\sqrt{f_c'}\sqrt[3]{\frac{s}{l}}\right) = 58.7\sqrt{f_c'}\sqrt[3]{\frac{s}{l}} \tag{13-7}$$

$$B = \left(1 + C_1\sqrt{\frac{A_{sl}}{b}}\right) \qquad \text{(Maximum value = 2)} \tag{13-8}$$

$$C = \left(\frac{1}{16}\right)\frac{H}{V} \qquad \text{when the lateral reinforcing is welded to transverse bars of equal size.} \tag{13-9}$$

$$C = \frac{1}{9}\frac{H}{V} \qquad \text{when the lateral reinforcing is welded to embedded steel bearing plates or steel angles.} \tag{13-10}$$

The value of ϕ has been taken as equal to 0.85 as specified in ACI 318-71 for shear stresses, and:

A_{sl} = area of the lateral reinforcing (minimum yield strength = 40,000 psi) with a maximum value of 0.16 sq in. per inch of bearing plate width.
b = width of bearing plate.

C_1 = a constant which equals 0 when s is less than 2 in. and which is equal to 2.5 when s is 2 in. or more.
f'_b = concrete bearing strength in psi.
f'_c = specified cylinder strength in psi.
H/V = ratio of horizontal and vertical ultimate design loads.
s = distance from the edge of the column to the center line of the bearing plate in inches.
l = length of the bearing plate in inches.

It should be noted that very little increase in strength (if any) is achieved by using lateral reinforcing having a yield point greater than 40,000 psi.

In Eq. 13-6, the term A represents the strength of a column head that has no lateral reinforcing and which is subjected to vertical loads only. The term B is a strength increase factor that reflects the effects of properly anchored lateral reinforcing when the distance s exceeds 2 in. The term B should not exceed 2, since an increase in the amount of lateral reinforcing above that which results in B being equal to 2 causes little increase in the bearing strength. The term C is a strength reduction factor that accounts for the effects of horizontal loads acting in combination with the vertical loads.

The lateral reinforcing should be placed near the top of the column with a concrete cover of 5/8 in. or the minimum allowed by the applicable code. The lateral reinforcing can be placed in two layers when necessary to facilitate concrete placing and compaction.

ILLUSTRATIVE PROBLEM 13-3 Design the column head for a 12 in. × 12 in. column that is to support two beams symmetrically placed about the center line of the column. Horizontal loads are eliminated by elastomeric bearing pads. The ultimate design loads for each beam is 200,000 lb, $f'_c = 4000$ psi, and the bearing pad width is 3 in.

SOLUTION: Provide a 1 in. gap between girders and use a distance s of 2.75 in. in order to provide equal edge distances for the beam and column without lateral reinforcing.

$$f'_b = 58.7\sqrt{4000} \times \sqrt[3]{\frac{2.75}{3.00}} = 3600 \text{ psi}$$

$$P_b = 3600 \times 3 \text{ in.} \times 12 \text{ in.} = 129{,}600 \text{ lb} < 200{,}000 \text{ lb}$$

Therefore, lateral reinforcing is required and the term B from Eq. 13-8 must equal

$$B = \frac{200{,}000}{129{,}600} = 1.54 = 1 + 2.5\sqrt{\frac{A_{sl}}{12}}$$

since $s > 2.0$ inches and

$$A_{sl} = 12\left(\frac{1.54 - 1.0}{2.5}\right)^2 = 12\left(\frac{0.54}{2.5}\right)^2 = 0.560 \text{ in.}^2$$

Three No. 4 bars, which are provided with a No. 4 anchorage bar welded at each end, give a lateral reinforcing area of 0.60 sq in. and provide a good solution. Note that the actual ultimate bearing stress on the concrete with this solution would be 5560 psi.

13-5 Post-tensioned Connection

Continuity of beam-column connections can be obtained by utilizing the principle of segmental construction shown in Fig. 13-10. This connection has the advantage that continuity can be developed without the use of embedded metal bearing plates and field welded connections. The principal disadvantage of this detail is that it is necessary to use a mortar or concrete joint, unless the beams and columns can be cast one against the other, match-marked, and later erected in the same relative position. This type of joint may result

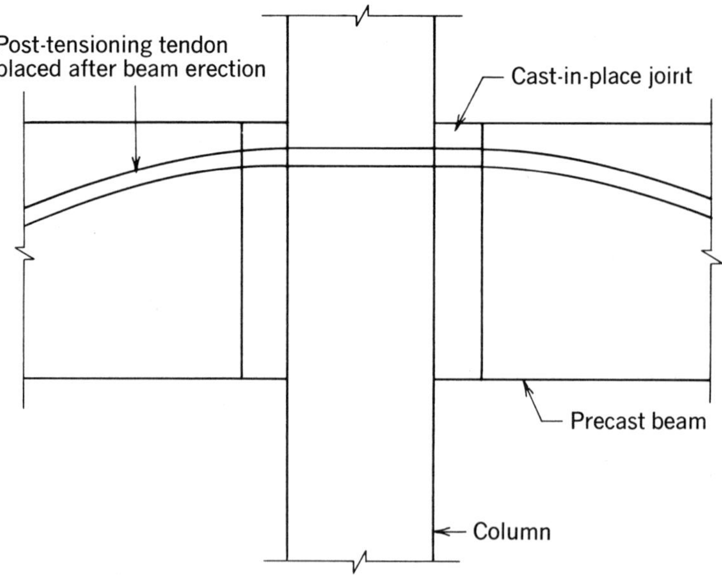

Fig. 13-10 Post-tensioned beam-column connection.

in a slightly longer construction time than would be experienced with other types of connections and may induce slightly larger moments in the columns, due to beam shortening.

13-6 Other Beam Connections

A type of beam connection designed for the transfer of shear and axial load has been proposed by Loov (Ref. 7). This connection is shown in Fig. 13-11 and the forces which act on the connection are shown in Fig. 13-12. Advantages of the connection are claimed to be low cost and, because the connection is in the top of the beam, the beam is stable as soon as it is placed on its supports. This expedites the erection.

It is recommended that the design of this connector be based upon the ultimate strength method, using the relationship of Kriz and Raths (Ref. 6), for column heads without lateral reinforcing in computing the compressive strength of the concrete. The designer using this connection must include the effect of the horizontal force due to shrinkage and creep when determining the size of the horizontal bar.

ILLUSTRATIVE PROBLEM 13-4 Determine the capacity of a connector using 2 No. 6 inclined bars welded to a 1 in. thick plate. Proportion the horizontal bar and plate. The angle of inclination is 30°, $f_c' = 5000$ psi $f_y = 40{,}000$ psi, and the width of the beam is 10 in. The vertical reaction is applied 3 in. from the end of the beam. The plate is to be located in such a manner that the

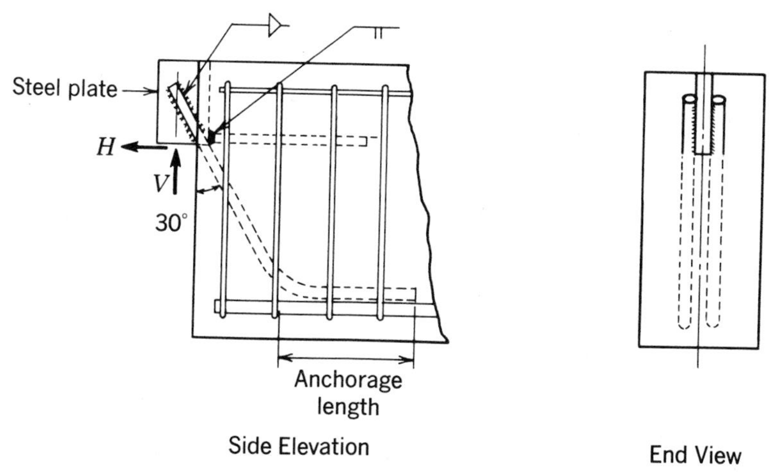

Fig. 13-11 Beam connection proposed by Loov (Ref. 7).

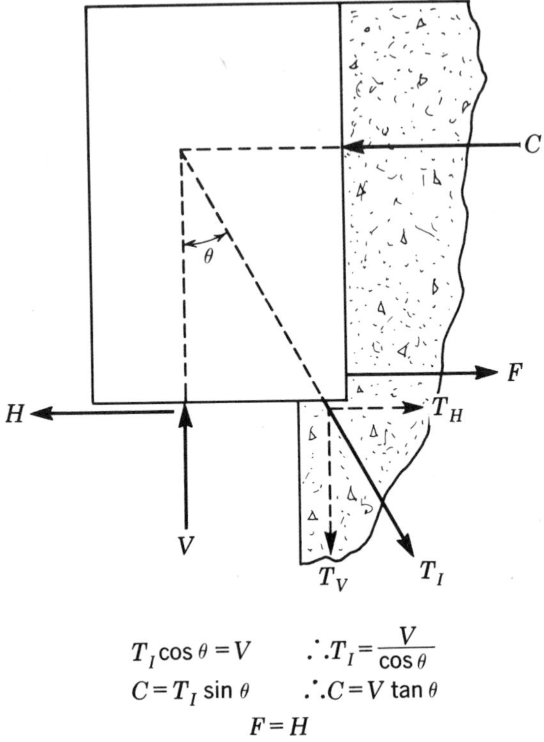

$$T_I \cos\theta = V \qquad \therefore T_I = \frac{V}{\cos\theta}$$
$$C = T_I \sin\theta \qquad \therefore C = V\tan\theta$$
$$F = H$$

Fig. 13-12 Forces acting on the connection plate.

compressive force from the plate is applied at a distance (s) 4 in. from the top surface of the beam.

SOLUTION:

$$T_I = 2 \times 0.44 \times 40 = 35.2^{k}$$
$$V = T_I \text{ Cos } 30^\circ = 35.2^{k} \times 0.866 = 30.5^{k}$$
$$T_H = T_I \sin 30^\circ = 35.2^{k} \times 0.500 = 17.6^{k}$$
$$f_b' = 58.7\sqrt{f_c'}\sqrt[3]{\frac{s}{l}} = 58.7\sqrt{5000}\sqrt[3]{\frac{4}{1}} = 6600 \text{ psi}$$
$$\text{Depth of compression block} = \frac{17{,}600}{6600 \times 1} = 2.67 \text{ in.}$$
$$\text{Plate depth} = \frac{3.0}{\tan 30^\circ} + \frac{2.67}{2} = 6.53 \text{ in say } 6.50 \text{ in.}$$

The required horizontal bar is proportioned as follows

H/V	H	$A_s = H/f_y$	*Horizontal Bar Size*
0	0	0	None
0.2	6.1	0.15	~1 No. 4
0.4	12.2	0.31	~1 No. 5
0.6	18.3	0.48	~1 No. 6
0.8	24.4	0.61	~1 No. 7
1.0	30.5	0.76	~1 No. 8

13-7 Column Base Connections

Column base connections shown in Fig. 13-13 are commonly used in precast concrete structures. Each detail has certain advantages relative to fabrication of the precast columns, but there is little difference in the structural performance of the types of details. The recessed column detail requires careful filling of the recesses with high quality concrete after erection if the full strength capacity of the connection is to be achieved. The welding of the reinforcing to the bearing plates and the grouting of the plates after erection must be done properly in order to obtain optimum results. It should be recognized that the capacity of this type of connection is generally considered to be limited by the allowable bearing stress in the weaker of the two concretes. If a compression force is considered to be transferred from the column reinforcing to the anchor bolts through the bearing plate, it may be rationalized

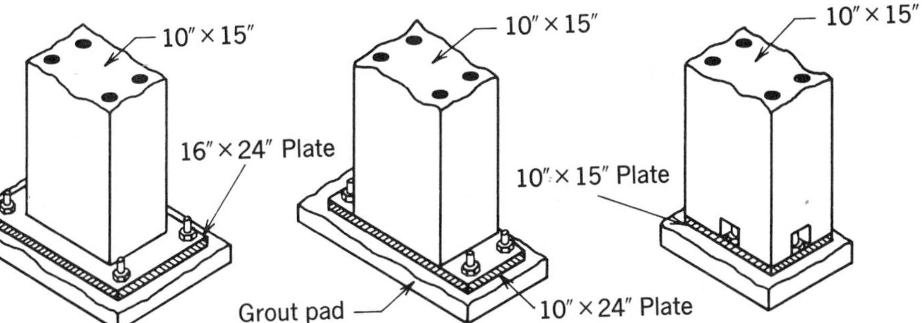

Fig. 13-13 Typical column base connections for precast columns.

that a force larger than that controlled by the concrete bearing stress may be allowed.

The provision of jam nuts gives an easy means of plumbing and adjusting the elevation of the columns at the time of erection.

Column connections of this type should be designed using the capacity reduction factor $\phi = 0.70$. The critical section in design of the extended base plate can be taken as the plane tangent to the column reinforcing steel nearest the anchor bolts that are loaded in tension.

The stiffness of the connection must include the effects of bolt elongation, plate deflection, as well as foundation rotations. The bolt design should be based upon the area at the root of the threads and not the nominal diameter. If shear forces are to be transmitted by the connection, special consideration should be given the combined shear and tensile stresses in the anchor bolts.

13-8 Elastomeric Bearing Pads

Elastomeric bearing pads are widely used in both concrete and steel structures. They provide a means of accommodating vertical forces in combination with horizontal displacements and rotations in all directions. This type of bearing is particularly interesting, since it functions in proportion to the shear modulus of the elastomer and to a shape factor and is not affected by friction, dirt or oxidation.

Some elastomeric bearing pads are solid elastomer, while others are formed of laminations of steel plates and elastomer. The provision of bonded steel plates between layers of elastomer reduces the bulging of the elastomer and, since shear stresses result from the bulges, higher loads can be carried on laminated pads. Plain elastomer pads are suitable for use where low stresses, due to vertical loads, can be accepted and where larger horizontal movements are not needed.

The "Standard Specifications for Highway Bridges" Tenth Edition, published by the American Association of State Highway Officials includes a specification for the design of elastomeric bearing pads. This procedure is based upon a shape factor that is equal to the area of the loaded face divided by the side area that is free to bulge. The shape factor is used with curves available from the manufacturers of this type of bearing. The AASHO specifications make the following additional limitations:

(1) T (the total effective elastomer thickness) = the summation of thickness of individual layers of laminated bearings.

(2) Maximum allowable compressive deflection under dead load due to nonparallel surfaces = $0.06\ T$.

		Plain Bearing	*Laminated Bearing*
(3) Minimum length of rectangular bearings (parallel to direction of translation)	=	$5T$	$3T$
(4) Minimum width of rectangular bearings (perpendicular to direction of translation)	=	$5T$	$2T$
(5) Minimum radius for circular bearings	=	$5T$	$3T$
(6) Total displacement due to temperature (sum of positive and negative displacements)	=	$0.5T$	
(7) Maximum allowable unit stress due to vertical load:			
Total load (without impact)		800 psi	
Dead load only		500 psi	

(8) Maximum allowable compressive deflection in a plain bearing or in any layer of a laminated bearing under dead plus live loads (without impact) is $0.07T$.

A typical plot of the relationship between the shape factor and compression strain for an elastomer having a hardness of 52 durometer is shown in Fig. 13-14.

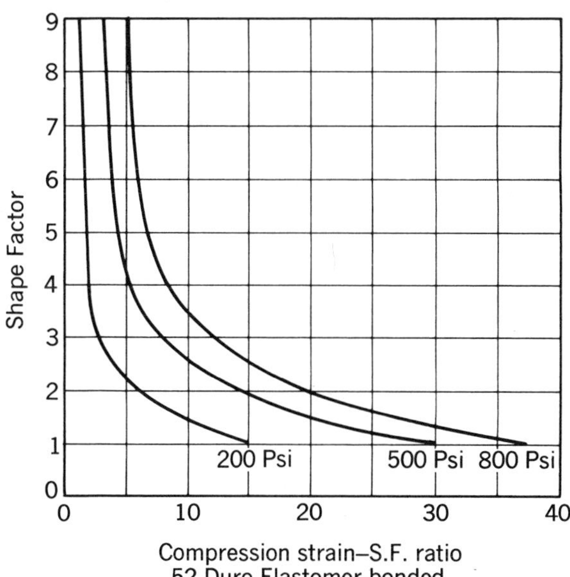

Fig. 13-14 Shape factor vs compression strain for elastomer of 52 durometer hardness loaded to various stress levels.

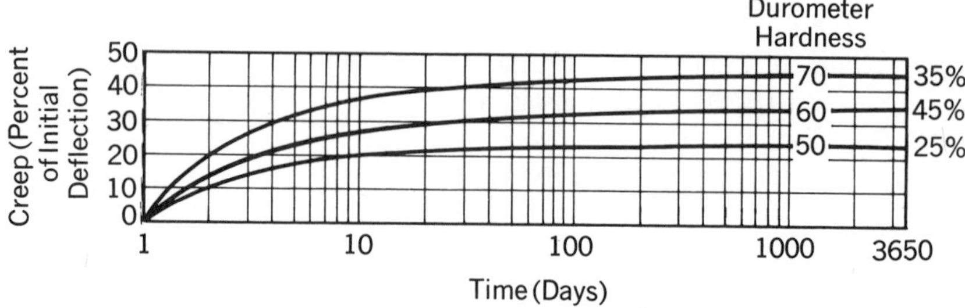

Fig. 13-15 Creep in compression of typical elastomers.

The variation in shear modulus for elastomers of different hardnesses is given in Table 13-1 and the creep in compression is shown in Fig. 13-15.

The design of an elastomeric bearing pad for a bridge girder is given in the following illustrative problem.

TABLE 13-1 Modulus of Elasticity in Shear

50 Durometer Hardness	*60 Durometer Hardness*	*70 Durometer Hardness*
110 psi at 70°F	160 psi at 70°F	215 psi at 70°F
1.1 × 110 psi at 20°F	1.1 × 160 psi at 20°F	1.1 × 215 psi at 20°F
1.25 × 110 psi at 0°F	1.25 × 160 psi at 0°F	1.25 × 215 psi at 0°F
1.9 × 110 psi at −20°F	1.9 × 160 psi at −20°F	1.9 × 215 psi at −20°F

ILLUSTRATIVE PROBLEM 13-5 Design an elastomeric bearing pad that is 24 in. wide (the width of the girder) for a dead load of 110 kips and a total load of 145 kips. Use the AASHO specifications and a movement of 1.25 in. Assume that the shear modulus of the elastomer is 120 psi at the minimum service temperature (from the manufacturer's data).

SOLUTION:

The bearing length by dead load is

$$L = \frac{110{,}000}{24 \times 500} = 9.16 \text{ in.}$$

and by total load:

$$L = \frac{145{,}000}{800} = 7.55 \text{ in.} \qquad \text{Use } 9.25 \text{ in.}$$

Refering to Fig. 13-14, assume a shape factor of 6 in order to keep the compressive strain low.

$$\text{Shape factor} = 6 = \frac{LW}{2t(L + W)} = \frac{9.25 \times 24}{2t(9.25 + 24)}$$

$$t = \frac{9.25 \times 24}{2 \times 6(9.25 + 24)} = 0.556 \text{ in.} \qquad \text{Say } 9/16 \text{ in.}$$

The maximum allowable horizontal movement is equal to one-half the height, therefore, each layer of 9/16 in. can provide a movement of 9/32 in. Thus, for the 1.25 in. of total movement, assume 4 interior layers of 9/16 in. and 1/4 in. layers at the top and bottom. The total allowable movement is

$$\Delta t = \frac{2 \times 0.25 + 4 \times 0.563}{2} = 1.375 \text{ in.}$$

which is greater than 1.25 in.

From Fig. 13-15 it will be seen the compression strain at 500 psi (dead load) is about 4% and under total load (653 psi) is of the order of $4\frac{1}{2}\%$, which is less than the 7% allowed. The force resulting from the movement of 1.25 in. is computed as follows:

$$F = \frac{(\text{Shear Modulus})(\text{Area})(\text{Movement})}{\text{Height of elastomer}}$$

$$= \frac{120 \times 9.25 \times 24 \times 1.25}{2.75} = 12{,}100 \text{ lb}$$

13-9 Other Expansion Bearing Pads

Other materials such as steel plates coated with low friction coatings, graphite impregnated materials, steel rollers and bronze plates containing grooves filled with lubricants have been used with varying degrees of success in prestressed concrete structures. The newer low-friction plastic-coated steels would appear to have good promise. The designer should investigate any proposed bearing material well before use, since the horizontal forces which can develop when bearings do not perform as expected can be very large and can result in serious structural problems.

In addition, the rotations resulting from the applied loads should be considered in detailing the bearings.

13-10 Fixed Steel Bearings

Bearing details similar to those shown in Fig. 13-16 are frequently recommended and have been used in many instances without difficulty. If the steel plates are welded together or if they become fixed to each other by the

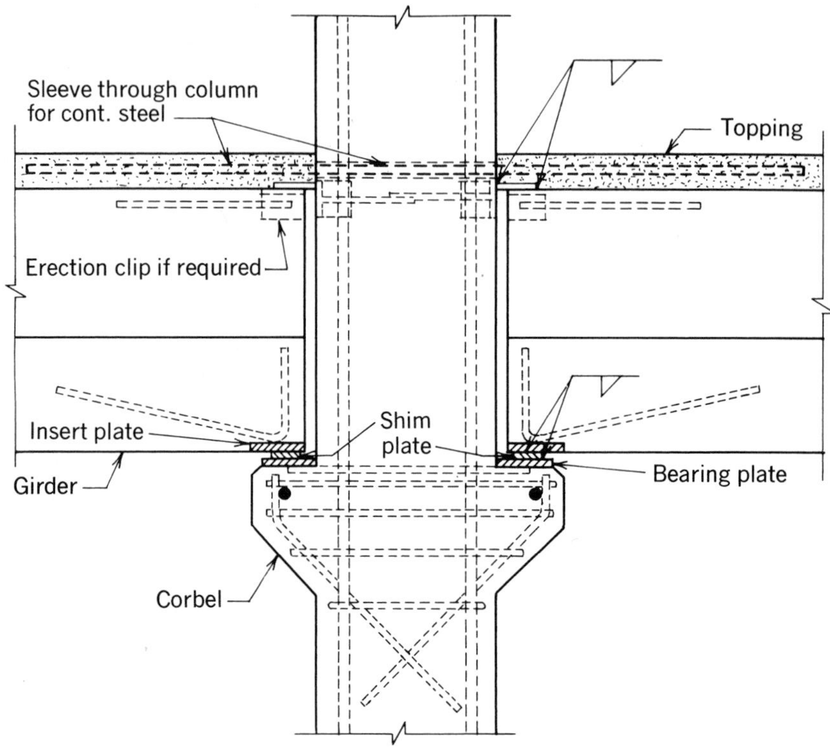

Fig. 13-16 Fixed bearing.

products of corrosion, they can transmit large forces that may cause difficulties. If the plates are not welded together and the center plate width is small as compared to the upper and lower plates, the action of the bearing could approximate that of a rocker.

If top reinforcing is provided at a connection and the bottom bearings are welded, the beam will be continuous and the reinforcing, as well as bearings, should be designed for the forces resulting from the continuity and from restrained volume and temperature strains.

13-11 Wind/Seismic Connections

It is often necessary to provide a means of transferring shear forces between precast concrete structural elements in order to develop diaphragm action for the resistance of lateral loads. If a cast-in-place concrete topping is used, it is generally structurally adequate to develop the required shear transfer.

When a concrete topping is not provided, shear connections between the precast units must be provided by concrete shear keys, adhesives, or metal inserts that are connected by welding or bolting. The use of concrete shear keys for this purpose is not common, since in order to be effective, the units must be restrained against movements that tend to separate the units. Although this is sometimes possible through the inclusion of continuous reinforcing or prestressing in the end joints (over the supporting beams or girders), the details can be involved and not straight forward. The use of adhesives is not common for providing shear strength, since they are not resistant to heat, and hence, cannot be used structurally in fire-rated construction. A commonly used shear connector detail for double T, single T and channel slabs are shown in Fig. 13-17. It should be noted that these details frequently include the welding of reinforcing steel, and hence, caution is required (*see* Sec. 10-16).

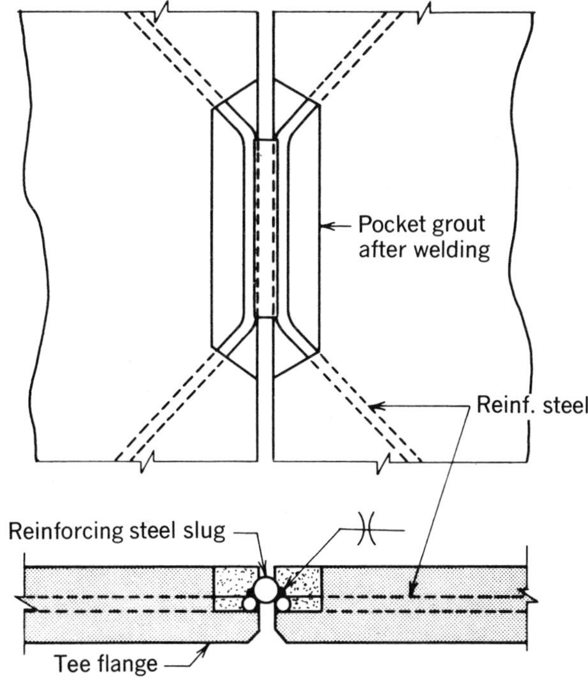

Fig. 13-17 T flange seismic connection.

Hollow core slabs (for building construction) made with some processes can be provided with metallic inserts that can be used to develop the required shear forces. Other manufacturing processes do not permit the inclusion of metal inserts and must normally be used with a structural topping in areas where lateral loads must be resisted.

Concrete shear keys can be used efficiently in bridges to provide the required lateral and longitudinal shear resistance for horizontal forces. The shear keys are provided between the end diaphragms and the bent or pier caps or at the abutments. The keys can be lined with thin (0.25 in.) expansion joint material (to prevent bond) where the keys are designed to be fixed against displacements, but free to rotate. Thicker expansion joint material or expanded polystyrene is frequently used to line the keys, which are designed to provide for displacements (expansion bearings) as well as rotations.

13-12 Shear-friction Connections

When it is inappropriate to consider shear as a measure of diagonal tension, ACI 318-71 provides the shear-friction concept as a means of designing. The method can be used in the design of corbels when the ratio of the shear span to the effective depth is 0.50 or less. There are many areas where the concept will find application in precast structural elements.

In the shear-friction concept, a crack is assumed to exist in the member in the location where one might expect the element to fail in shear. Reinforcing steel that crosses the crack is assumed to provide a force (normal to the crack) which develops a frictional resistance along the crack. If the reinforcing is present in sufficient quantity, the frictional resistance will be greater than the shear stress. The reinforcing should be approximately perpendicular to the crack.

The shear stress v_u is not permitted to exceed 800 psi or $0.2f_c'$ and the area of steel required is computed from

$$A_{vf} = \frac{V_u}{\phi f_y \mu} \tag{13-11}$$

in which the yield strength, f_y, cannot exceed 60,000 psi, $\phi = 0.85$ and, μ is equal to 1.4 for concrete cast monolithically, 1.0 for concrete placed against a clean, artificially roughened, hardened concrete (construction joints) surface, and 0.70 for concrete plated against clean, unpainted as-rolled steel.

If tensile stresses exist across the crack, additional reinforcement must be provided to resist them. In any case, the reinforcement must be placed in such a manner that it is adequately anchored on each side of the assumed crack and must be well distributed over the area.

REFERENCES

1. "Recommended Practices for Welding Reinforcing Steel, Metal Inserts and Connections in Reinforced Concrete Construction," AWS D12, 1-61, American Welding Society, New York, New York, 1961.

2. "PCMAC Connections Manual," prepared by the Prestressed Concrete Manufacturers Association of California, Inc. (Oct. 1969).

3. "Connection Details for Precast Prestressed Concrete Buildings," Prestressed Concrete Institute, Chicago, Illinois, 1963.

4. Burton, K. T., Corley, W. G. and Hognestad, E., "Connections in Precast Concrete Structures—Effect of Restrained Creep and Shrinkage," *Journal of the Prestressed Concrete Institute*, **12**, No. 2, 18–37 (April 1967).

5. Kriz, L. B. and Raths, C. H., "Connections in Precast Concrete Structures—Strength of Corbels," *Journal of the Prestressed Concrete Institute*, **10**, No. 1, 16–61 (Feb. 1965).

6. Kriz, L. B. and Raths, C. H., "Connections in Precast Concrete Structures—Strength of Column Heads," *Journal of the Prestressed Concrete Institute*, **8**, No. 6, 45–75 (Dec. 1963).

7. Loov, Robert, "A Precast Beam Connection Designed for Shear and Axial Load," *Journal of the Prestressed Concrete Institute*, **13**, No. 3, 12–27 (June 1968).

8. Ife, J. S., Vzumeri, S. M. and Huggins, M. W., "Behavior of the 'Cazaly Hanger' Subject to Vertical Loading," *Journal of the Prestressed Concrete Institute*, **13**, No. 6, 48–66 (Dec. 1968).

9. La Fraugh, R. W. and Magura, D. D., "Connections in Precast Concrete Structures —Column Base Plates," *Journal of the Prestressed Concrete Institute*, **11**, No. 6, 18–35 (Dec. 1966).

10. Rejcha, C., "Design of Elastomer Bearings," *Journal of the Prestressed Concrete Institute*, **9**, No. 5, 62–78 (Oct. 1964).

11. "Sky Span Elastomeric Bridge Bearings Engineering and Standards Handbook," General Tire and Rubber Company, Wabash, Indiana.

12. "Design of Neoprene Bearing Pads," E. I. du Pont de Nemours and Company, Inc., Wilmington, Delaware.

14 Pre-tensioning Equipment and Procedures

14-1 Introduction

Pre-tensioning equipment and methods which are characteristic of American practice vary in many respects from the equipment and methods employed abroad. The basic principles in the design of prestressing facilities here and abroad are identical, but, because the normal type of products produced in domestic prestressing plants are larger than the average pre-tensioned products produced abroad, the prestressing facilities required domestically are also larger. No attempt will be made to describe the prestressing facilities which are typical of foreign practice, since a considerable amount of information on this subject is available in the literature. The scope of this chapter is confined to the discussion of equipment and procedures that are peculiar to pre-tensioning as practiced in the United States.

The stress in the pre-tensioning tendons must be maintained as nearly constant as possible during placing and curing of the concrete. This can be accomplished in two ways.

(1) The tendons are stressed and anchored to individual steel molds designed to withstand the prestressing force as well as the stresses which result from the plastic concrete.

(2) The tendons are restrained by a special device, called a pre-tensioning bed or bench, which restrains the pre-tensioning force and provides a level surface on which the forms are placed.

In addition to these devices, other equipment peculiar to pre-tensioned construction, including the mechanism used to stress and release the prestressing tendons, the forms, vibrators, and, tendon deflectors, are discussed in this chapter.

14-2 Pre-tensioning with Individual Molds

With the exception of a few firms which employ stress-resisting molds in the manufacture of double T roof slabs and pre-tensioned spun piles, this technique is not considered practical in this country. In Europe, where pretensioned railroad ties and small joists for residential construction are economical in pre-tensioned concrete, this method has the advantage of allowing individual units to be mass produced with the products (and molds) moving through the plant in a production cycle, rather than requiring that the materials and plant be brought to the molds or forms, as is required with a pre-tensioning bench. In manufacturing spun pre-tensioned piles, this technique must be used, since there is no other practical means of spinning the mold and pre-tensioned tendons as a unit.

Another advantage of this method, when employed on small pre-tensioned products, is that the prestressing plant need not be as large and as elongated as is required in using a bench, since small pre-tensioned products in their individual molds can be stacked and need not be arranged in long rows. This advantage applies, but to a lesser degree, with large products that can only be handled with very large cranes and that cannot be stacked very high, if at all.

14-3 Pre-tensioning Benches

Pre-tensioning benches are normally designed to withstand a specific maximum force applied at a specific maximum eccentricity. Therefore, it is customary, when stipulating the capacity of a pre-tensioning bench, to give the maximum permissible force (shear) and maximum permissible moment the bench can safely withstand. The maximum moment is normally expressed in terms of the bench proper (slab portion of the bench which extends between the uprights at the abutments) and not necessarily in terms of the top surface of the abutments, which may be recessed to accommodate the stressing mechanism.

Pre-tensioning benches are generally one of the following types:

(1) Column-type bench, which may serve as the mold or form, as well as the device which restrains the tendons.

(2) Independent abutment-type bench in which the independent abutments rely upon soil pressure, piling, or rock foundations for stability.

(3) Strut-and-tie-type bench.

(4) Abutment-and-strut-type bench.

(5) Tendon-deflecting-type bench.

(6) Portable benches.

Each of these types of benches has specific areas of application, and each will be discussed separately.

One additional definition that must be given before discussing the types of pre-tensioning benches pertains to the use to which the bench is to be subject. Some benches are designed to produce a specific product and may be termed a fixed bench. Other benches are designed to produce any type of product that is normally encountered in practice and are termed universal benches.

Column Benches

Column benches rely upon the column action of the bench alone to resist the prestressing force. The eccentricity of the prestressing force, with respect to the bench, must be confined to relatively low values in order to achieve economy with this type of bench. For this reason, the use of column benches is generally restricted to fixed benches designed to produce a single T, double T or pile. An example of a column-type bench, which is designed to produce double T slabs, is shown in Fig. 14-1.

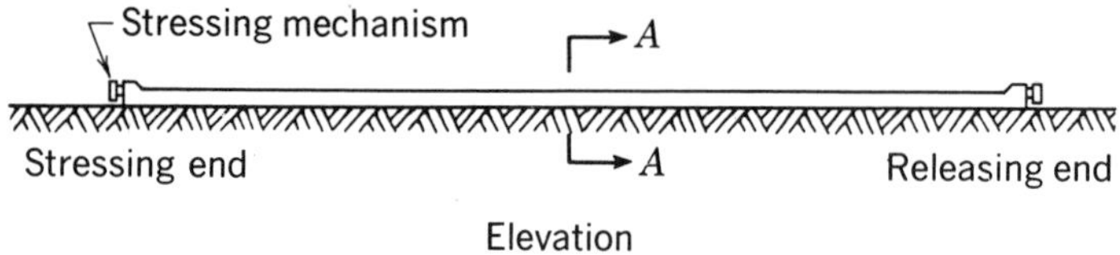

Elevation

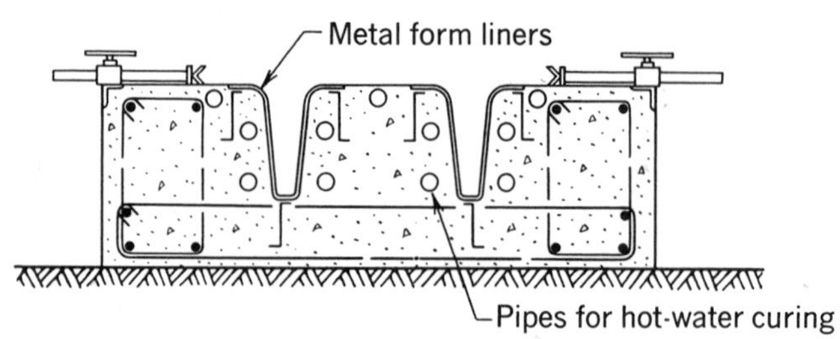

Section *A-A*

Fig. 14-1 Column-type double T bench.

The column-type benches are generally designed using the Euler formula to compute the critical buckling stress. Adequate safety factors against both crushing of the concrete and buckling of the column must be allowed in the design. The dead weight of the bench alone is relied upon to prevent the column from buckling. The buckling of the bench could occur at the center of the bench by buckling upward; at the ends of the bench which could buckle upward; or by a combination of both. These are illustrated in Fig. 14-2.

Column-type benches are not normally used for universal benches or in benches in which a relatively large eccentricity of the prestressing force must be accommodated.

Independent Abutment Benches

This type of pre-tensioning bench is composed of two large abutments that are structurally independent of each other as well as of the paving material used as a casting surface between the abutments. The abutments, when embedded in soil, may rely exclusively upon the weight of the abutment and passive soil pressure for stability, but it is more common to incorporate piling in the abutments, in order to increase the stability. Abutments of these two types are shown in Fig. 14-3.

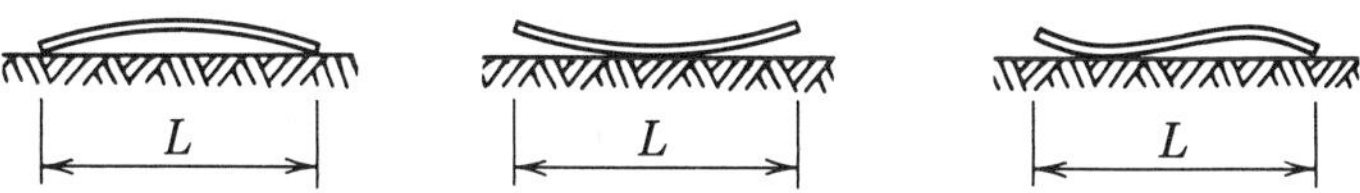

Fig. 14-2 Possible forms of column buckling.

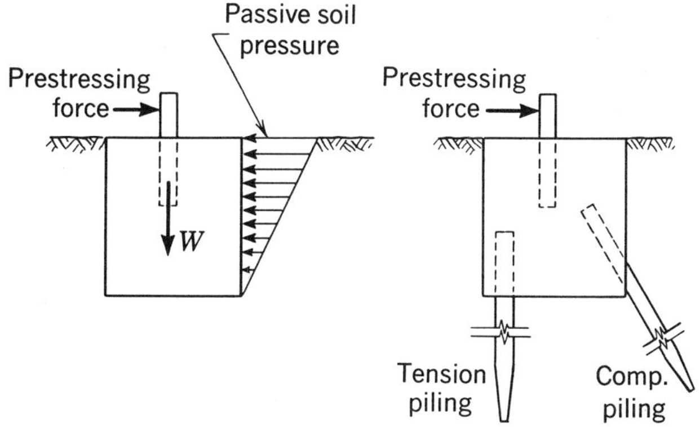

Fig. 14-3 Independent abutments in soil.

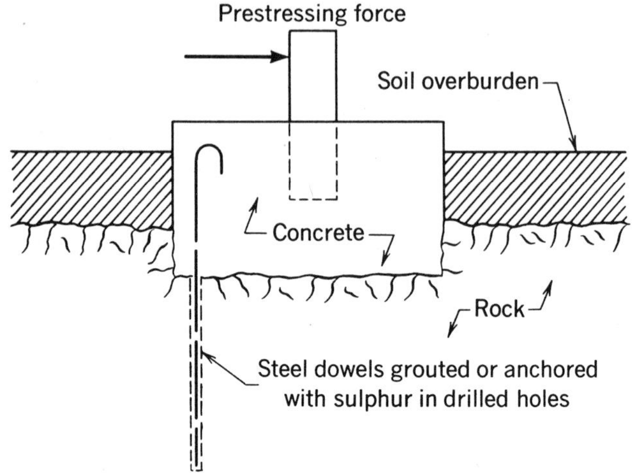

Fig. 14-4 Independent abutments on rock.

When founded on sound rock, independent abutments can be formed by keying the abutments into the rock and, if necessary, increasing the resistance of the abutments to overturning by providing anchors that are embedded in the rock. This is illustrated in Fig. 14-4.

The design of independent abutments requires accurate knowledge of the character of the soil or rock on which the abutments are to be located. The effects of long-term loading, variation of loading, and variation of the moisture content on the mechanical properties of the foundation material are also important and must be known. Due to the difficulty in accurately determining the mechanical properties of the foundation material, this type of pre-tensioning facility is not frequently used. When the foundation material is adequate, this type of pre-tensioning bench is economical for long benches, since the casting surface is not a structural component and hence can be considerably lighter than is required for other types of benches.

Strut and Tie Bench

The principle of the strut and tie pre-tensioning bench is illustrated in Fig. 14-5. Examination of this illustration will reveal that the prestressing force (P) results in a tensile force in the tie (T) and a compressive force in the strut (C). The uprights are relied upon to distribute the three forces.

This type of bench can be used for large eccentricities and is adaptable to universal pre-tensioning benches for this reason. One serious objection to this type of bench, when large prestressing forces are used, is that the compressive

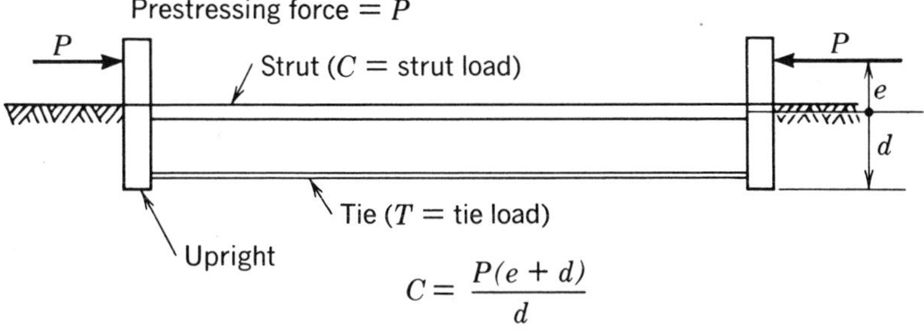

Fig. 14-5 Elevation of strut and tie bench.

force in the strut is larger than the prestressing force ($C = P + T$) and the dimensions of the strut may have to be large in order to prevent buckling. Another objection to the use of this principle on long benches is that the effect of the deformation of the bench during stressing can be large, because the tie becomes longer and the strut becomes shorter as the prestressing force is applied—these deformations are amplified at the level of the prestressing tendons by the lever action of the uprights. For these reasons, strut and tie benches are generally used only on short benches, such as universal benches used in laboratories.

Abutment and Strut Bench

This is the most frequently used type of pre-tensioning bench. The structural principle of this type of bench is illustrated in Fig. 14-6, in which it will be seen that the prestressing force applied on uprights embedded in each abutment has the tendency to overturn the abutments about the concrete hinges, as well as to force the two abutments to slide toward each other. The overturning of the abutment is prevented by the weight of the abutment, and sliding of the abutments is prevented by the slab or strut which separates the two abutments. The provision of the hinge between the abutments and the slab insures that the slab section is subjected to a direct axial force (alone) which is equal to the prestressing force in magnitude.

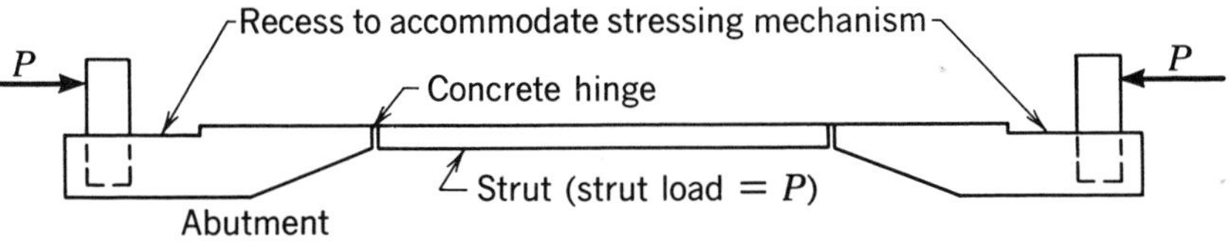

Fig. 14-6 Elevation of abutment and strut bench.

The design of this type of bench consists of determining the amount and shape of the concrete abutment that will provide an adequate safety factor against overturning, the reinforcing of the abutment, and the proportioning of the slab, which is generally designed as a plain concrete column. The abutment of this type of bench is usually made of heavily reinforced concrete or of post-tensioned concrete. The slab portion is generally reinforced with a welded wire fabric in order to help control cracking due to shrinkage.

The approximate quantity of concrete required for the two abutments of abutment- and strut-type benches can be computed from

$$Q = 35 + 0.06M \tag{14-1}$$

in which Q is the approximate quantity of concrete in cubic yards and M is the moment due to prestressing in kip-feet for which the bench is designed. The reinforcing required can be taken at 75 lb per cubic yard. The approximate weight of the structural steel uprights required for the two ends can be computed from

$$W = 3500 + 6M \tag{14-2}$$

in which W is the approximate weight of the uprights in pounds. The quantity computed from Eq. 14-2 does not include the cross-beams, templates, pull rods or other components of the stressing mechanism that may be required.

Tendon-Deflecting Benches

Most of the methods used in deflecting the tendons in pre-tensioned beams and girders require that the tendons be held in the lower position by devices attached to the slab portion of the pre-tensioning bench. The tendons are held in the upper position by devices which bear on the slab portion of the bench. The slab is thus subjected to a series of vertical loads, as well as to the axial load associated with abutment and strut benches. This type of loading and the bench shape usually employed under such conditions are illustrated in Fig. 14-7.

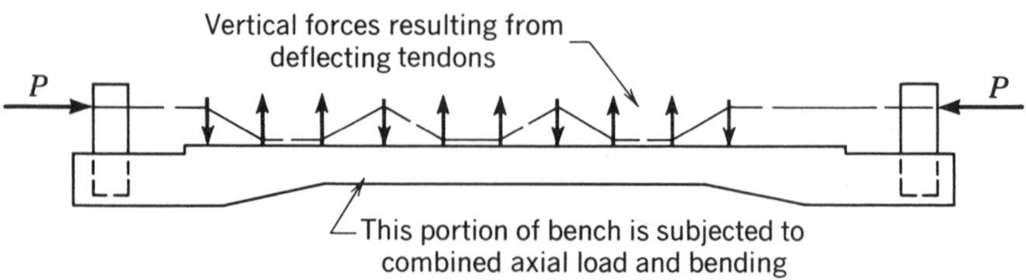

Fig. 14-7 Elevation of tendon-deflecting bench.

Due to the large vertical forces, which may have to be applied at virtually any point along the bench in order to achieve the desired results in prestressed products produced with deflected tendons, the use of the concrete hinge, which is characteristic of the abutment and strut bench, is not feasible. Furthermore, since the slab portion of the bench is subject to combined bending and direct stress, the slab must be made of reinforced or prestressed concrete rather than the plain concrete used in abutment- and strut-type of benches. The cost of this type of bench is substantially greater than abutment and strut benches of equal capacity.

The quantity of materials required for the abutments of a tendon-deflecting bench can be approximated from Eqs. 14-1 and 14-2.

Portable Benches

Pre-tensioning benches that can be moved from job site to job site have been used to a limited degree. The portable benches may be of any of the above types and may be entirely or only partially portable. An example of a partially portable bench is a bench of the abutment- and strut-type in which the principal components of the abutment are portable and the strut and counterweight portions of the abutments are not moved.

A universal, portable pre-tensioning bench of the abutment- and strut-type is a very practical piece of equipment for general contractors engaged in large highway and marine projects on which large quantities of pre-tensioned concrete may be required in areas where there are no permanent pre-tensioning plants. In a similar manner, portable benches of the column type are occasionally feasible for job site fabrication of double T and single T roof members for large building projects.

In the design of universal pre-tensioning benches, it is often desired to provide a means for adjusting the length of the bench so that the waste of the pre-tensioning tendons can be minimized for different production problems and maximum of economy can be achieved in operating the bench. This has been accomplished by the use of pull-rod extensions in the stressing mechanisms, the mechanics of which will be apparent to the reader after considering the discussion of stressing mechanisms which follows this section. Another method used quite extensively is to provide a long dead-end abutment with several alternate positions for the dead-end uprights, as is illustrated in Fig. 14-8. The provision of an intermediate abutment with removable uprights has also been used successfully. Intermediate abutments designed to withstand the prestressing load from either direction, as illustrated in Fig. 14-9, have been used.

The maximum force used in the design of the pre-tensioning bench should be about 15% greater than the total initial force applied to the maximum

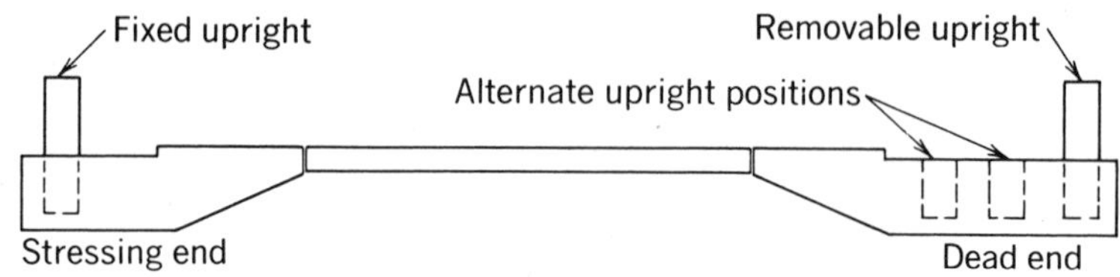

Fig. 14-8 Bench with adjustable length.

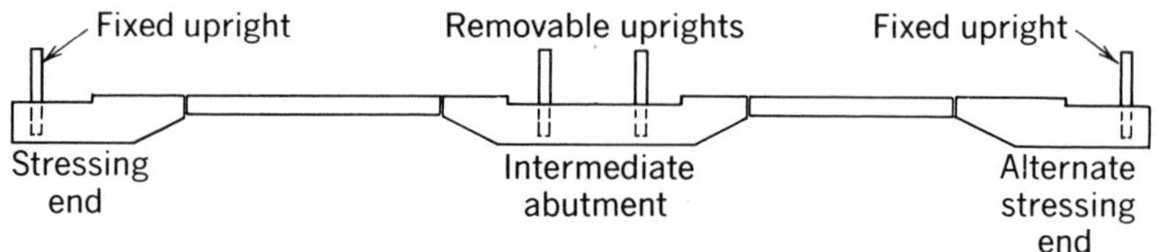

Fig. 14-9 Bench with intermediate abutment.

number of tendons used with the bench. Experience has shown that the shrinkage of the concrete and the effect of temperature variations, which take place during curing and stripping of the concrete, result in the prestressing force being larger at the time of release than it is immediately after stressing. The increase in the stress is a function of: (1) the ratio of the length of the tendon, which is embedded in concrete, to the total length of the tendon between the uprights of the bench; (2) the type of cement; (3) the type of curing used; (4) the air temperature during stripping; and (5) other factors. The gain in stress in the tendons increases with the ratio of length of the embedded tendon to total tendon length. The gain is also affected by the curing time and is more severe when the air temperature during stripping is low.

14-4 Stressing Mechanisms and Related Devices

The tensioning of the pre-tensioning tendons can be done by stressing each tendon individually or by stressing all of the tendons at one time. Each method has advantages, and the method used most frequently has significant influence on the design of the stressing mechanism. It is possible to design the stressing mechanism so that the tendons can be stressed with either procedure. In the majority of the plants the tendons are stressed individually. In addition, the stressing mechanism may be designed so that the pre-tensioning tendons can all be released simultaneously with hydraulic jacks, rather than by cutting the tendons.

If the stressing is to be done by stressing one strand at a time, the jack used for the stressing must normally have a stroke of 30 to 48 in. and a working capacity of from 10 to 15 tons, depending upon the length and size of the tendons to be stressed. If the tendons are to be released simultaneously, large jacks, the capacity of which depends upon the maximum prestressing force for which the bench is designed, must be provided for this purpose. The jacks which are used to release the bench do not normally require a stroke in excess of 6 in.

Releasing the tendons hydraulically all at one time has the advantage that the force can be transmitted to the concrete products slowly, and hence, shock or impact loading is avoided. The result is that the transmission length is minimized. In addition, with this type of equipment, the products can be partially released if this becomes necessary or desirable. The principal disadvantage of releasing the tendons with hydraulic jacks is that all of the strain is released at one end of the production line. The result is that the products at the releasing end tend to move away from the releasing end. The amount of movement is a direct function of the elastic deformation of the concrete products and unembedded tendons. When deflected tendons are used, releasing the tendons simultaneously may not be possible, since it may not be possible to release the tendon deflecting devices before the prestressing force is released, and hence, the necessary movement of the products can not take place freely. Another advantage of releasing all of the tendons at one time is that with this procedure the cutting of the tendons between precast units can be done without adhering to a strict schedule.

Releasing the tendons individually is generally done by cutting them one at a time with an acetylene torch. The tendons must be cut according to a strict sequence in order to avoid eccentric loading in the products and in order to prevent too many tendons being cut at one location, which will result in failure of the remaining tendons. In other words, the tendons at each end of each product in the line must be cut at approximately the same time.

When all of the tendons are to be stressed simultaneously, the jacks used to stress the tendons must be of large capacity. Unless the stressing is done in several increments, which is a procedure that is not recommended for normal operation, the stroke of the jacks must be of the order of 30 to 48 in. The same jacks used to stress the tendons can be used to release the pretensioning force, but, during the releasing operation, the stroke required is again generally less than 6 in.

It should also be mentioned here that, when the tendons are stressed individually, it is not necessary to apply an initial load to each tendon in order to equalize the length of the tendons, as is often required (by construction specifications) when all of the tendons are stressed simultaneously.

Furthermore, if an anchor slips on one tendon during or after the stressing of the tendons, the tendon that has slipped can be simply restressed, if the tendons are being stressed individually. If, on the other hand, the tendons are all being stressed simultaneously, it is necessary to release all of the tendons and restress after reanchoring the slipped tendon to the stressing mechanism.

It will be found that the cost of the hydraulic jacks is greater in installations designed to stress all of the tendons simultaneously than in installations designed for individual stressing of the strands. The total labor cost of stressing is about equal with both methods, if an initial force must be applied to the tendons when stressed simultaneously, but more time may be required for stressing the tendons individually in members in which there are a large number of tendons. The time factor may be important in some instances, since in the interest of safety, all work in the vicinity of a pre-tensioning bench must be stopped during the stressing operation.

The stroke of the jacks specified for any particular installation should be based upon the anticipated elongation of the steel during stressing plus an allowance for slack in the tendon and slack in the anchorages. The normal theoretical elongation for the pre-tensioning tendons is 7 to 8 in. per 100 ft of length. The slack and anchorage take-up, for which jacking stroke must be provided, is much larger in the case of single tendon stressing than in stressing all the tendons simultaneously, assuming an initial load is applied to the tendons individually, in the latter case, in order to equalize the length of the tendons. It is recommended that the jacks used for stressing tendons individually be 18 in. longer than is required for the theoretical, elastic elongation of the tendon. An extra 10 in. of stroke is recommended for jacks used for the simultaneous stressing of tendons which have been equalized in length by the application of a force equal to about 5% of the initial prestressing force. These recommendations apply for pre-tensioning application in which the maximum length of the tendons to be stressed is of the order of 300 ft.

Experience has shown that for the capacity, stroke, and use to which the hydraulic jacks are subjected in pre-tensioning installations, the jacks should be double acting rather than relying upon springs to return the pistons to the closed (or open) position. Furthermore, it is recommended that the jacks be designed to develop the maximum normal load that will be used with the jacks at a pressure of the order of 5000 to 6000 psi and that the jacks be guaranteed for intermittent service for pressures up to 10,000 psi. The piston rods of the jacks to be exposed during the stressing and releasing operations should be hard-chrome plated in order to protect the jack against corrosion and subsequent damage to the hydraulic seals in the jack, which often results from corrosion on the piston rods.

The hydraulic pumping unit used to operate the jacks should also be designed for a maximum intermittent operating pressure of 10,000 psi and a minimum continuous operating pressure of the order of 5000 to 6000 psi. The pumping unit should be designed as simply as possible, in order to avoid confusing the workmen, and at least one extra pressure gauge that is not used under normal operating conditions should be provided in the system for use in checking the calibration of the other gauges in the system. In addition, the unit should be made in such a manner that damaged pressure gauges can be easily removed and replaced.

The structural frames to which the tensioned tendons are attached and which project above the top surface of the abutments generally consist of uprights, pull rods, cross-beams, and templates. The purpose of these frames is to transfer the load that is in the pre-tensioning tendons to the abutments, as well as to provide a means of applying and releasing the prestress. Unless screw jacks or hydraulic jacks with threaded piston shafts are used to maintain the prestressing force during curing, the structural frames must have a positive means of connecting the tendons to the uprights, since simple hydraulic jacks cannot be relied upon to maintain a constant strain on the tendons over a long period of time.

There are several types of structural arrangements that can be used for stressing mechanisms, and the type best suited for a specific situation is dependent upon a number of factors including the capacity of the bench, the type products to be made on the bench (i.e., fixed or universal bench), and the method of stressing to be used. Rather than discuss each of these factors separately, three types of structural stressing and releasing systems are described. The designer must determine which type of arrangement is most suited to the needs of any particular installation, after carefully weighing the advantages and disadvantages of each system.

The first system to be considered is that which is shown in Fig. 14-10. It consists of cross-beams, templates, fixed uprights, and, at the stressing end, vertical beams and pull rods. The vertical beam at the stressing end is held away from the fixed uprights by a strut at the top and the concrete abutment at the bottom. The clear space which results between the vertical beams and the fixed uprights is used to accommodate nuts and plates that hold the force in the pull rods when the jacks are not in use. This system is shown set up with four, small-stroke, high-capacity jacks (100 ton) which have a hole through their center so that the pull rods can be placed through the jacks. This mechanism can be used in stressing the tendons individually. With the 15-ton jack, all of the tendons can be stressed simultaneously, either by using the four, small-stroke high-capacity jacks, as shown, and taking several increments of loading in order to obtain the required elongation of the tendons, or by using four, long-stroke high-capacity jacks and obtaining the entire elongation in one stressing increment. It must be pointed out that unless the center of gravity of the prestressing force is midway between the planes

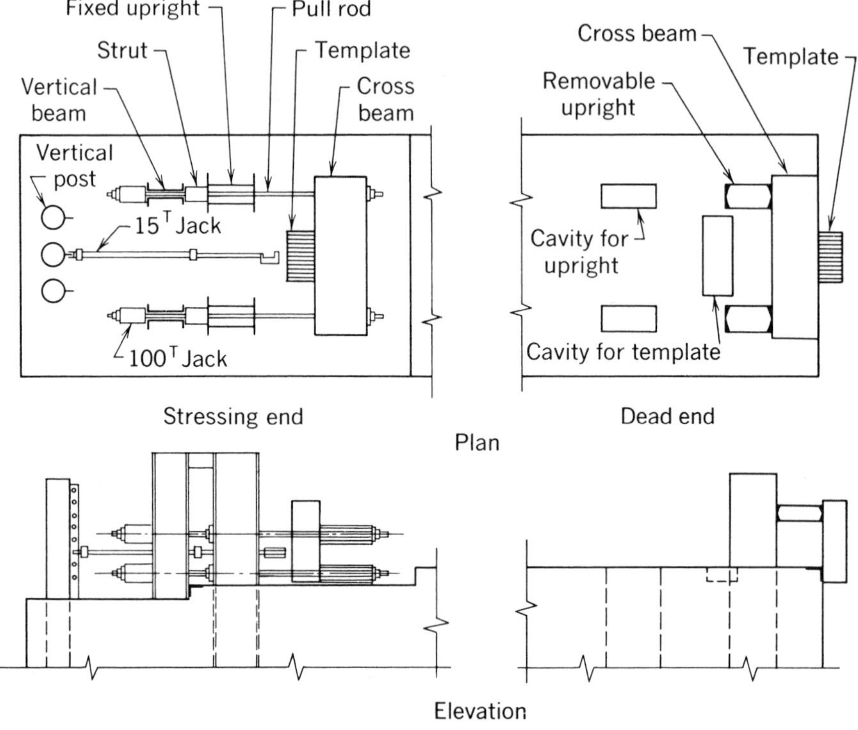

Fig. 14-10 Stressing mechanism, type 1.

of the pull rods in each direction, the forces in the pull rods, and hence the pressures in the four jacks, will not be equal during stressing (if all tendons are stressed simultaneously) and releasing. This can result in the operation of the hydraulic system being somewhat complicated and dangerous if not controlled properly. From the plan view in Fig. 14-10, it will be seen that the area between the uprights is free of obstructions, which is essential when the tendons are to be stressed individually.

The system shown in Fig. 14-11 is designed for tensioning the tendons individually at one end and releasing them simultaneously at the other end. The stressing end consists of uprights, which support the template, and has provision for a long-stroke small-capacity stressing jack that may be used with an individual abutment or that may be of the type which bears directly on the template or anchorage device during stressing. The releasing end in this bench consists of a template, cross-beams, uprights, jacking yokes, and pull rods. The releasing is done with two jacks. If desired, the tendons

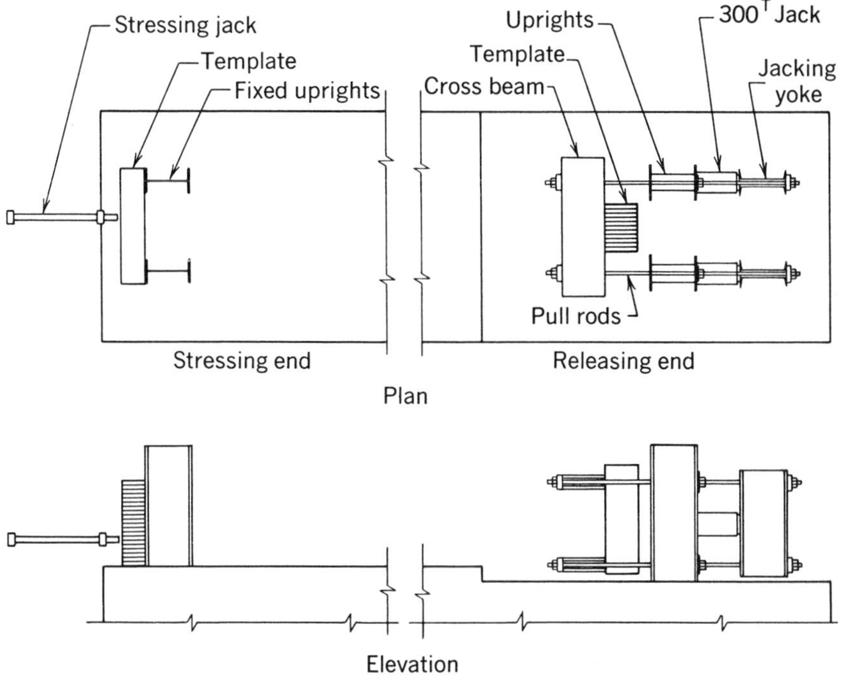

Fig. 14-11 Stressing mechanism, type 2.

can be stressed simultaneously with this bench by stressing at what is normally the releasing end. This system is less difficult to operate than the system discussed above, since the large-capacity jacks have been reduced to two. By using a releasing mechanism that is a combination of the one shown in Fig. 14-11 and the one shown in Fig. 14-12, the number of releasing jacks can be reduced to one. It must be emphasized that the center of gravity of the prestressing force and the axis of the releasing jacks must be coincident in this scheme, in order to maintain stability of the mechanism.

The system shown in Fig. 14-12 is designed for stressing all of the tendons simultaneously. Since the ends of the tendons are not accessible to a long-stroke, low-capacity jack in this mechanism, single tendon stressing cannot be used. This device utilizes only one large-capacity jack which would normally have a long stroke. The jack must be placed at a location that renders the axis of the jack coincident with the center of gravity of the prestressing force. If this is not done, the mechanism will not be stable during stressing and releasing.

In order to simplify the illustrations of the mechanisms in Figs. 14-10,

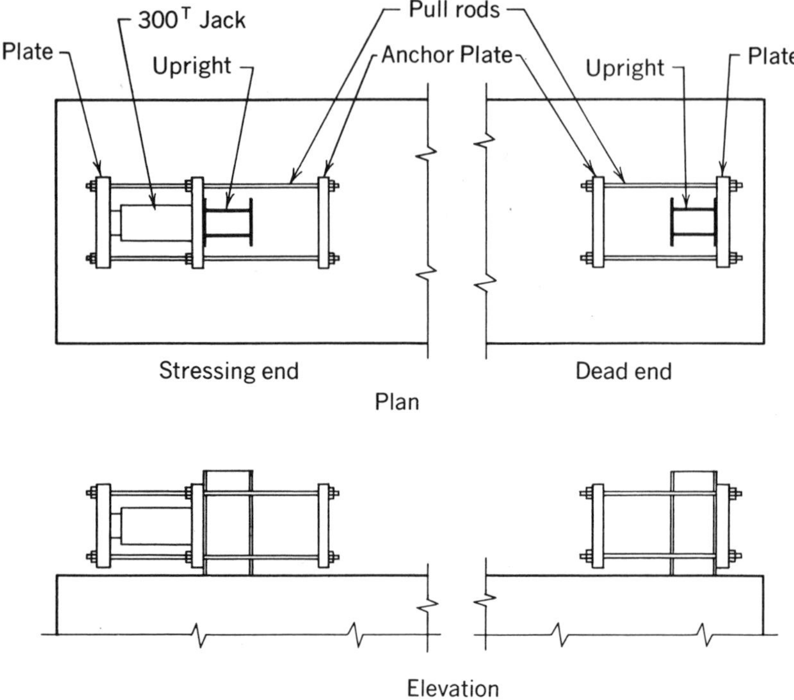

Fig. 14-12 Prestressing mechanism, type 3.

14-11, and 14-12, rollers or other devices that support the members (which move during stressing and releasing) were not shown nor were the devices used to adjust the vertical location of the cross-beams and releasing jacks. Accessories of this nature should be provided with each type of mechanism in order to facilitate the stressing and releasing and in order to reduce the friction in the system during these operations.

Anchorage devices, which are placed on the tendons to hold them to the templates during stressing of the tendons and during curing of the concrete, are also a part of the stressing mechanism. There are several satisfactory types of devices available for this purpose, and the selection of the device to use does not materially affect the operation of the stressing mechanism.

A dynamometer is frequently used to calibrate the pressure gauge used with the long-stroke jack in stressing the tendons individually. The dynamometer is not ordinarily used to measure the stress in the tendons during normal production, due to the risk of breaking the device if an accident occurs.

14-5 Forms for Pre-tensioned Concrete

Stress-resisting forms or molds used in the manufacture of pre-tensioned concrete are special structural elements not normally encountered in practice. For this reason, this discussion will be confined to the forms or molds used on pre-tensioning benches.

The desirable features of forms to be used in pre-tensioned concrete are varied, and the form requirements vary for different products. In general, the desirable characteristics can be summarized as follows:

(1) High resistance to damage due to rough handling and to the high humidity associated with steam curing. This requirement normally eliminates the use of wood forms, which do not perform well under repeated use and, particularly, when exposed to steam curing. Although concrete forms have been used successfully, the lighter steel forms are generally preferred.

(2) Precision of the form units and dimensions. Since the forms are generally made in panels which can be connected together to form a large member, it is essential that the panels fit together precisely.

(3) Ease of handling. When erecting or stripping the form, it is essential that the individual pieces of the form that must be handled are not awkward to handle and remain in a generally upright position. This characteristic facilitates laying the forms on their backs so that they may be cleaned easily and it facilitates setting and adjusting the form in the precise position required during assembly.

(4) The form should be designed in such a manner that one side may be erected in the final position independently of the opposite side. This facilitates the layout of the member being made as well as the forming of special block-outs and transverse holes through the member, thus securing the web reinforcing and post-tensioning units, if any, in the proper location.

(5) Adjustability of the forms. The forms or components of the forms should be adjustable in such a manner that members of several shapes can be made from the form or form components.

(6) Form vibration. The forms should be sufficiently strong to withstand the effects of form vibration. Brackets or rails to facilitate placing form vibrators should be supplied with the forms.

(7) Rigid, structural soffit form. The soffit form must be rigid and must not deform during use, since such deformation results in the soffits of the products being curved and uneven. In addition, the soffit form should be a structural element to which the side form can be securely attached (anchored against lateral and uplift loads) in order to prevent the side forms from moving during placing of the concrete. This latter requirement is particularly significant in the manufacture of I shaped beams that have large, bottom flanges, since the uplift may be very large under such conditions.

(8) The forms should be made with a minimum of joints, and all joints should be as tight as possible in order to minimize leakage and bleeding.

Standard forms that satisfy the above requirements are produced by several firms. Custom-made forms that incorporate the above characteristics can be made by many fabricators.

An illustration of a type of custom-made form that incorporates the above characteristics is shown in Fig. 14-13. It should be noted from this illustration that the side-form units are narrow and, when stripped, can be easily turned on their backs to facilitate cleaning of the forms. Each side form can be set up independently of the other by attaching the form to the concrete soffit and plumbing the side by use of the adjustable brace. The removable web forms can be replaced by web forms of various shapes, and the waste concrete soffit can be adjusted in height. In this manner, the side forms can be utilized in making a variety of beams. The continuous rail on each side form allows heavy form vibrators to be rolled to any position along the form where they can be firmly clamped to the form. This avoids the necessity of the workmen carrying the vibrator from location to location.

The form illustrated in Fig. 14-13 is only one of many possible solutions. The special conditions that exist in a particular prestressing plant or on a specific project may require other form details.

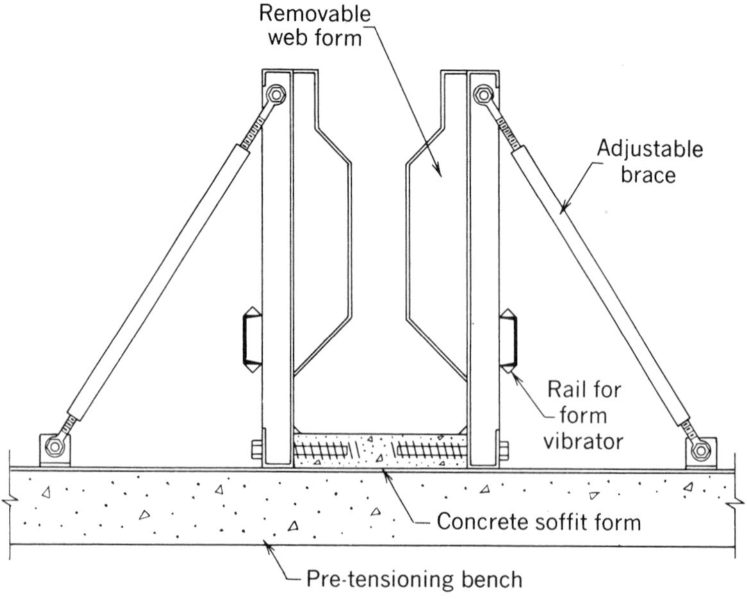

Fig. 14-13 Cross section of an adjustable steel form suitable for the manufacture of pre-tensioned concrete.

The end forms or bulkheads must have provision for the tendons to pass through them, and yet, the holes, slots, or grooves through which the tendons pass must not be so large (if left unplugged during placing of the concrete) to allow mortar to bleed out of the form. One obvious method of making the end bulkheads is to cut the bulkhead from a steel plate through which holes are drilled for each tendon. This type of bulkhead has two serious disadvantages: the tendons must be threaded through the holes when they are being placed in the bench; and, if the holes are small enough to prevent excessive leakage of the mortar during placing of the concrete, the threading and moving of the bulkheads into position along the bench is very difficult.

Bulkheads with large holes for the tendons have been used successfully by placing corks, which have holes through their centers for the tendons, in the holes to prevent mortar leakage and to form a recess in the end of the members to allow the patching of the ends of the members and prevent corrosion of the ends of the tendons. The placing of the corks in such bulkheads can be very difficult if there are many tendons in the members and if the tendons are closely spaced.

Bulkheads with slots in them for the tendons have also been used, but in such cases, it is usually necessary to tape the slots to prevent mortar leakage. The taping of the slots is often a slow and laborious operation.

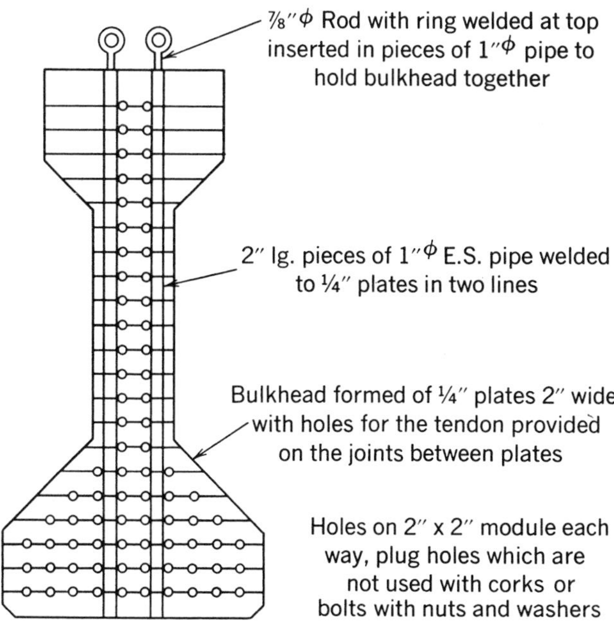

Fig. 14-14 Bulkhead for steel forms used in pre-tensioned construction.

Another type of bulkhead that has been used successfully is composed of a series of plates which can be connected with a removable bar assembly and which have holes provided at the joints between the plates. This type of bulkhead, which is illustrated in Fig. 14-14, does not allow significant mortar leakage and can be installed after the tendons have been placed and stressed.

14-6 Tendon-Deflecting Mechanisms

As has been stated previously, it is often desirable in long-span pre-tensioned members to have the tendons more eccentric near the center of the beam than at the ends of the beam. In manufacturing pre-tensioned products, the normal procedure is to stress the tendons from one abutment to the other, which results in the tendons being very straight. If the pre-tensioned tendons are to follow a path other than a straight one, additional forces must be made to act on the tendons to cause the trajectory of the tendons to be other than straight. The devices used to deflect the tendons from the straight line are called "tendon deflectors" or "tendon-deflecting mechanisms." Tendon-deflecting mechanisms frequently consist of devices which support the tendons in the high position and devices which hold the tendons in the lower position. These components are generally referred to as "hold-up devices" and "hold-down devices," respectively.

In applying the principle of deflected tendons to double T roof and floor slabs, it is customary to deflect the tendons at one or two points near the center of the slab in such a manner that the lowest tendon is straight through the entire length of the member and the remaining tendons are spaced out at the end of the slab, but are touching each other near the center of the slab. This is illustrated in Fig. 14-15. In this type of member, the steel forms or bulkheads that form the ends of the members and that space the tendons at the ends also serve as the hold-up devices. The hold-down devices, which hold the tendons down from the top, are similar to the one shown in Fig. 14-16.

Because in double T construction the sloping distance from the end bulkhead to the lowest point of the tendon trajectory is rarely of significantly greater length than the horizontal distance between the end bulkhead and the

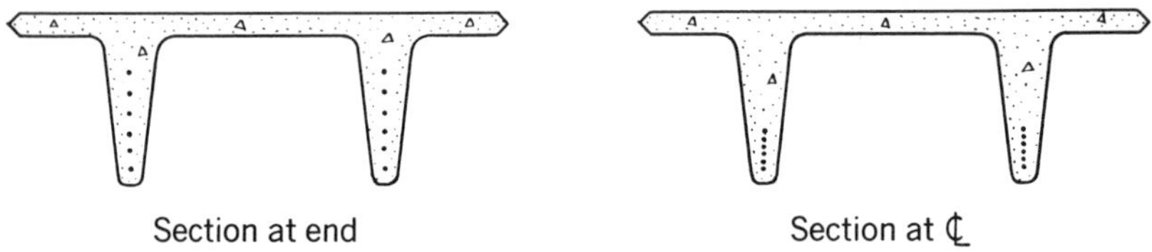

Fig. 14-15 Double T slab with deflected tendons.

Fig. 14-16 Hold-down device shown spanning a double T form. Workmen are in process of deflecting the tendons with a hydraulic jack. (*Courtesy of Food Machinery and Chemical Corp., Lakeland, Florida.*)

hold-down device, the usual construction procedure is to stress all of the tendons to the desired initial stress in the straight trajectory, install the end bulkheads, place the hold-down devices at the proper locations, and push the tendons down to the lower position with a hydraulic jack that is temporarily installed in the hold-down device. The tendons are held in the lower position by nuts or pins and the hydraulic jack is moved to the next hold-down device. The tendons are then jacked down at this location. The procedure is continued until the tendons are jacked down at all the hold-down devices.

After the concrete has gained sufficient strength to allow the prestress to be released, the hold-down devices are removed, after which, the pre-tensioning force is released. If released with jacks, the slabs which are nearest the releasing end move a few inches at the time of release. It is for this reason that it is necessary to remove the hold-down devices before releasing the pre-tensioning tendons. If this were not done, the hold-down devices would restrain the movement of the slabs, with the possibility of cracking the slab or damaging the hold-down devices.

After the slabs have been removed from the pre-tensioning bench, the metal rods or pipes, which extend into the slab to hold the tendons in the deflected position, are removed and the holes which are formed by these rods are filled with grout.

When the length of the tendon from the hold-up device to the hold-down device is significantly larger than the horizontal distance between these two devices, the initial stress applied to the tendons in the straight position should be lower than the stress that is desired initially in the deflected position. In this manner, the deflecting of the tendons raises the stresses in the tendons to the desired values.

In the case of bridge girders, because the vertical depth or amplitude of the deflected tendon is generally large in proportion to the horizontal dimension of the sloping portion of the tendon, the additional stress which results in the deflected tendons (due to pushing the tendons into the deflected trajectory) is generally quite important and must be taken into account when calculating the required initial stress that must be applied to the tendons.

At the present time, there is no standard method used to deflect the pre-tensioning tendons in bridge-girder construction. Several methods have been used or have been proposed for deflecting the tendons in bridge girders, and these methods are described in the following paragraphs.

Jacking Down at the Hold-Down Points

The method is shown schematically in Fig. 14-17. The procedure followed in this system is substantially the same as that used in deflecting the tendons in double T slabs, except the length of the deflected-tendon path may be quite significantly greater than the length in the straight path, and hence, the initial prestress applied to the tendon must be adjusted accordingly. In addition, the end bulkheads are not normally sufficiently strong to act as the hold-up device and, as a result, special devices must be supplied for this purpose.

One means of deflecting the tendons in this manner is shown in Fig. 14-18.

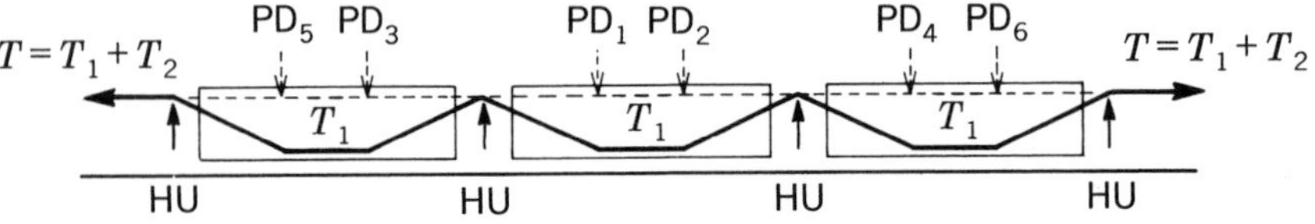

Initial tension T_1, in "up" position. Strand profile and final tension T by push down (PD), which increases strand tension by T_2.

Fig. 14-17 Jacking down at the hold-down points.

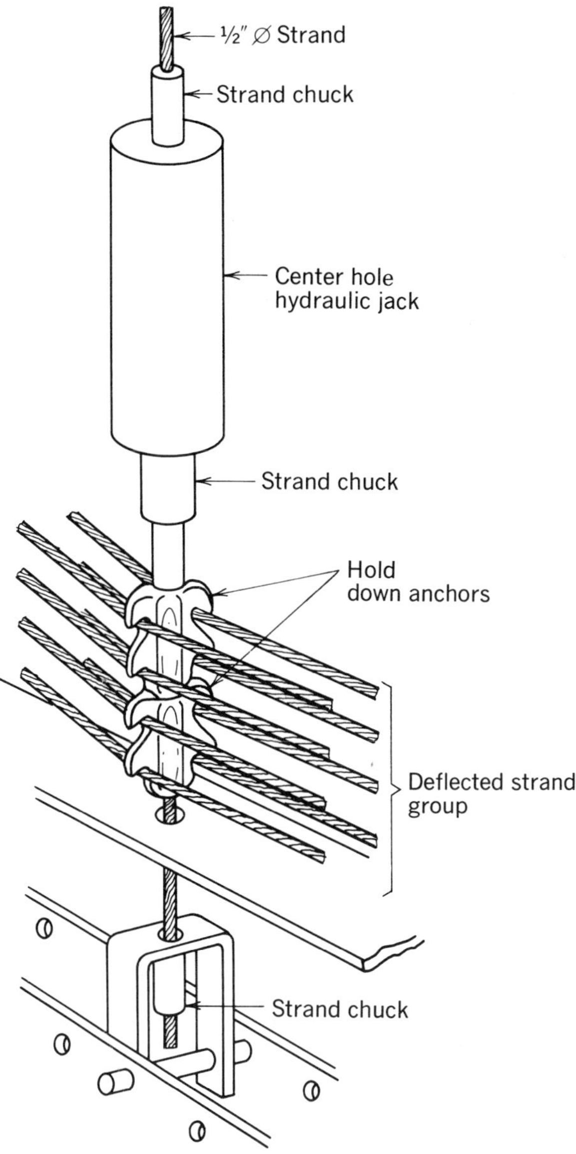

Fig. 14-18 Tendon hold-down device for use with jacking down at the hold-down points.

In this method, metal anchors provided in the upper surface of the prestressing bench are used to anchor the bottom end of the hold-down device and a center-hole jack is used to jack the tendons down. A strand chuck anchors the tendons in the deflected position. The strand chuck and the hold-down anchors are expended with this procedure. The jacking must be done according to a predetermined sequence in order to equalize the effects of friction along the bench.

Jacking Up at the Hold-Up Points

This procedure, which is shown schematically in Fig. 14-19, is virtually the same as that used in jacking down at the hold-down points. The principal difference is that the tendons are stressed in the lower position, attached securely to the pre-tensioning bench at the hold-down points by a device which extends through the soffit form, and the tendons are jacked up at the hold-up points (*see* Fig. 14-20).

After the members are removed from the bench, the cavity at the bottom of the beam at each hold-down point must be patched.

Stressing the Deflected Tendons Individually

Satisfactory results have been obtained in some instances by placing and stressing the tendons in the deflected trajectory. When this procedure is used, some means of reducing the effect of the friction which occurs between the tendons and the hold-down and hold-up devices must be employed. The methods that have been used to reduce the effect of the friction include stressing the tendons from each end, supplying rollers with needle bearing at each hold-down and each hold-up point during the stressing (the low friction rollers are removed after stressing is completed so they can be re-used many times), and applying a vibration to the tendons as they are stressed, which reduces the coefficient of friction at the hold-down and hold-up points. Again

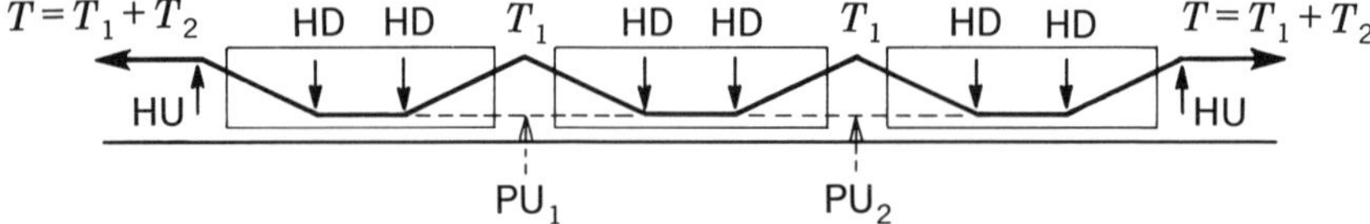

Initial tension T_1, in "down" position. Strand profile and final tension T by push up (PU), which increases strand tension by T_2.

Fig. 14-19 Jacking up at the hold-up points.

(a)

(b)

Fig. 14-20 (a) The expendable hold-down device attached to a removable portion of the soffit form, which is anchored to the pre-tensioning bench. (b) A general view of the deflecting mechanism showing the hold-down devices, deflected tendons, and the tower used to deflect the tendons.

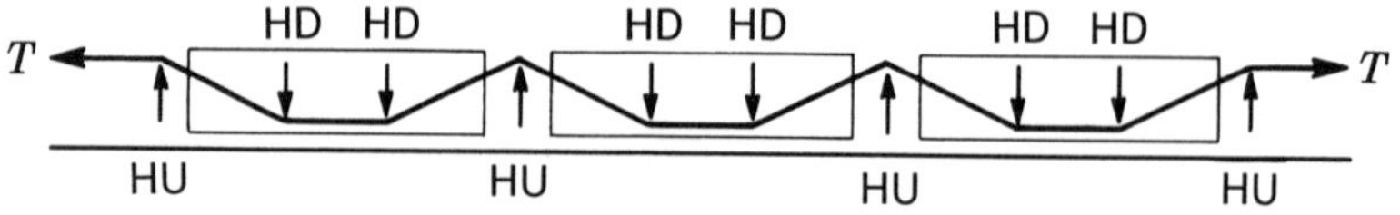

Single or multiple strand tensioning in "draped" pattern. Strand profile established by hold downs (HD) and hold ups (HU).

Fig. 14-21 Deflected or draped tendons stressed in the deflected position.

in this method, the hold-down devices must extend through the soffit form, which results in a cavity that must be patched after the members are removed from the pre-tensioning bench. The scheme is illustrated in Fig. 14-21.

Each of the methods discussed above have advantages and disadvantages, and the method to be used on any particular project must be selected on the basis of the available equipment and the results that are desired. Significant disadvantages of the use of deflected tendons include the large capital investment in a stressing bed which is capable of deflecting tendons and the deflecting mechanisms which are required, the friction losses which occur along the deflected tendons, the secondary stresses in the tendons which result, from bending the tendons over small diameter pins and rollers (does not apply to all methods), and the extra labor required to deflect the strands.

15 Post-tensioning Systems and Procedures

15-1 Introduction

Post-tensioning materials and equipment are available from a number of firms in the United States. These firms sell the necessary prestressing steel, ducts, anchorages and accessories, as well as either rent or sell the necessary stressing and grouting equipment. As a part of their service, the firms normally provide shop drawings for specific projects with the sale of their materials. In addition, when requested, they can normally supply field technicians who are qualified to instruct workmen in the proper methods of installing, stressing, and grouting their materials. Sometimes, the suppliers of post-tensioning materials sell the materials completely installed, stressed, and grouted.

The details of the systems available in the United States are continually changing. For this reason, no attempt will be made in this book to describe the precise details of the systems available at the time of this writing. The reader who is interested in learning the details of the individual systems is advised to contact the Prestressed Concrete Institute.* They will be able to supply a current list of suppliers of post-tensioning materials who can be contacted for specific details.

* Prestressed Concrete Institute, 20 N. Wacker Drive, Chicago, Illinois, 60606.

The primary differences between post-tensioning and pre-tensioning, from the construction viewpoint, can be enumerated as follows:

(1) Post-tensioning offers a means of making prestressed concrete, either in a plant or at the job site, without requiring a large capital investment in pre-tensioning facilities.

(2) Post-tensioned tendons allow the construction of cast-in-place structures which would not be feasible in pre-tensioned concrete.

(3) Post-tensioning tendons can be placed on curved trajectories easily and without large, special deflecting equipment.

(4) Friction losses during stressing of post-tensioned tendons must be considered in the design of such construction, since the friction losses may have significant influence on the performance.

(5) Each post-tensioned tendon must be stressed individually and this, in combination with the cost of the end anchorages, sheath, special equipment, and grouting of the tendons, results in the unit cost (cost per pound of effective prestressing force) of post-tensioned tendons being substantially greater than the cost of pre-tensioned tendons, in most instances.

(6) Post-tensioning anchorage devices and stressing equipment are often protected by patents and frequently require manufacturing techniques and facilities that are quite elaborate. As a result, each system is generally available from only one supplier or representative of the manufacturer. (One system that is covered by a basic patent is marketed by several firms, many of whom have developed end-anchorage details which are unique to the materials they supply).

In addition to the brief description of the general types of post-tensioning systems available in this country, the general construction procedures recommended in post-tensioned construction are discussed in this chapter, as are the computation of friction losses, gauge pressures, and elongations to be attained during stressing. Special construction methods and devices unique to post-tensioned construction are also considered.

15-2 Description of Post-tensioning Systems

The post-tensioning systems commonly used in this country can be separated into three general categories, which are:

(1) Parallel-wire systems.

(2) Strand-tendon systems.

(3) High-tensile bar systems.

The steel that comprises the tendons of these various systems has the general characteristics discussed previously in Chapter 2.

Parallel-wire systems generally employ a number of wires, 0.250 in. in diameter, which are bundled or grouped into a unit referred to as a tendon or a cable. The term "parallel wire" may not be precisely accurate in describing the relative position of the individual wires in the tendons, since in some instances, the individual wires are bundled or grouped together without spacers and with no real attempt being made to ensure that the individual wires in a tendon are parallel. Tests that have been conducted by the manufacturers and suppliers of the components used in these systems have proved that the spacers used abroad to maintain the wires in a parallel position (to facilitate grouting and to prevent the sheath from collapsing during handling in similar tendons) are unnecessary, due to the size and types of sheath used domestically. Hence, the term parallel wire serves as a means of differentiating systems that employ tendons composed of a number of wires bundled into a group from the "strand tendon" in which a number of factory-twisted strands are bundled together to form a tendon.

The original Freyssinet system is one of the first parallel-wire post-tensioning systems. It received its name from the eminent French engineer Eugène Freyssinet who invented the anchorage device used in this system. This system was introduced in this country when the use of linear prestressed concrete was just beginning to gain impetus. The anchorage device used in this system consists of two parts, called the "female cone" and the "male plug." An illustration of early anchorage cones is given in Fig. 15-1. Anchorage cones anchor the wires by the friction which results from the wedge shape of the male plug and the hole in the female cone.

The original Freyssinet system, which utilized the concrete female cone and male plug together with tendons composed of wire, is no longer marketed in the United States. A more recent evolution in the Freyssinet system utilizes

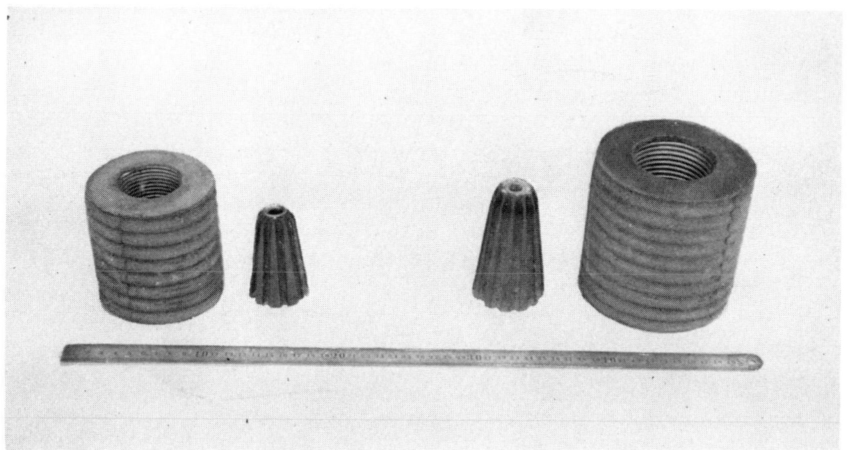

Fig. 15-1 Freyssinet anchorage cones. (*Courtesy Freyssinet Co., Inc., New York.*)

a cast steel anchorage that employs the same wedge principle to anchor tendons composed of a number of seven-wire strands. This anchorage is illustrated in Fig. 15-2.

Special jacks are required when using the Freyssinet system. These jacks actually consist of two rams connected in series. Steel wedges on the periphery of the main piston are used to attach the wires of the parallel wire tendons. Strand anchors are used in a similar manner with the strand tendons. After the wires or strands have been stressed by the introduction of hydraulic pressure to the main piston, the pressure is held in the main piston while pressure is applied to the secondary or plugging piston that forces the male plug into the cavity of the female cone. The stressing is completed by releasing the pressure in the plugging piston and then in the main piston. The jack is then disconnected from the tendon by removing the steel wedges. After stressing, the excess wire at the ends of the tendons are cut off and the tendon is grouted in place. A Freyssinet jack, connected to a wire tendon in stressing position as well as a tendon that has been stressed and one that has not been stressed, are shown in Fig. 15-3. The mechanics of the stressing sequence are shown in Fig. 15-4.

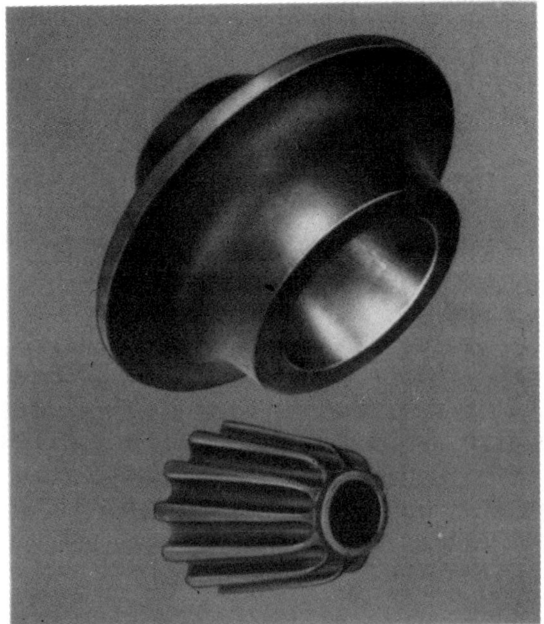

Fig. 15-2 Freyssinet anchorage of cast steel for use with tendons composed of 12 seven-wire strands of $\frac{1}{2}$ in. diameter. (*Courtesy Freyssinet Co., Inc., New York.*)

Fig. 15-3 Freyssinet jack in stressing position. (*Courtesy Freyssinet Co., Inc., New York.*)

The Freyssinet anchorage cone is provided with a hole through the male plug for the introduction of grout after stressing. Before grouting, it is necessary to plug the openings between the wires, the male plug, and the female cone, since large quantities of grout escape through these openings and reduce the effectiveness of the grouting apparatus if this is not done.

Due to the necessity of a wedge moving longitudinally to develop lateral forces, which is essential if a wedge or conical plug is to anchor a tendon by friction, a portion of the elongation of the tendon that results from stressing is lost when the prestressing force is transferred from the jack to the anchorage cone. This loss of elongation is increased in the case of a Freyssinet cone by the elastic expansion of the female cone which takes place upon anchoring the tendon. The entire loss of elongation which results from these actions is referred to as the *loss due to the elastic deformation* of the anchorage, or the *loss due to seating the anchorage*. The loss due to the elastic deformation of the anchorage can be of significance under certain conditions. This subject is discussed further in Sec. 15-6.

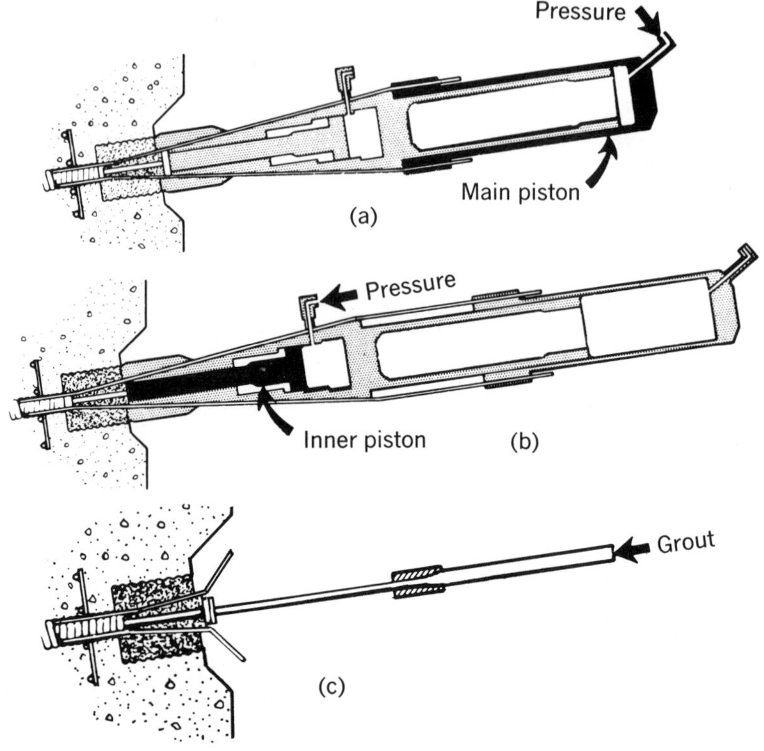

Fig. 15-4 Stressing sequence in the Freyssinet system. Showing (a) the stressing, (b) the plugging, and (c) the grouting of the tendon.

Several other systems employing from one to 48 seven-wire strands in each tendon are available in the United States. All of these systems rely upon wedges for the anchorage of the strands. The details of the anchorages, grouting procedure, and stressing equipment must be obtained from the individual supplier.

End anchorage of parallel-wire post-tensioning tendons is also achieved by cold-formed heads on the ends of 0.25 in. diameter wire. Several systems that utilize this principle are available in the United States. The system is used with a number of variations, but in all of them, the button heads bear on a stressing ring or stressing nut to which a hydraulic jack is attached for stressing. After stressing, the stressing ring is locked in the stressed position either with shims or with threaded nuts. This is illustrated in Fig. 15-5.

Another type of post-tensioning tendon commonly used in this country is the high-tensile steel bar. Tendons of this type were first developed in England where they are called "Lee McCall bars." In this country, this type of tendon is available smooth or with deformations and with threaded ends,

Fig. 15-5 End anchorages of tendons with cold-formed button heads. (*Courtesy Prescon Corp., Chorpus Christi, Texas.*)

nuts, and couplers or with wedge-type nuts and couplers. High-tensile bars have been widely used in recent years. The anchors used with one of the bar systems are illustrated in Fig. 15-6.

It should be pointed out that couplers are required to join individual bars together at the job site to form the longer bar tendons. This restriction affects the detailing and construction of longer members designed to be stressed with bar tendons.

The jack required to stress the button-head tendons and the high-tensile bar tendons is a simple center-hole ram or jack. This type of equipment is readily available in the United States in various sizes. Center-hole rams are much less expensive and lighter than a Freyssinet jack of equivalent capacity.

Each system of post-tensioning has certain advantages and disadvantages and no attempt is made here to discuss all of the systems or factors that may influence the designer or contractor in selecting a system for a particular project. The designer should select and specify the systems which will best meet the requirements of each individual job. In most applications, however, all of the systems will work equally well and the controlling factor is the cost of the tendons and the labor required to install, stress, and grout the tendons.

15-3 Sheaths and Ducts for Post-tensioning Tendons

In prestressed construction, it is essential that the tendons do not become bonded to the concrete until they are stressed. In post-tensioned construction, since the tendon cannot be stressed until the concrete is hardened, it is necessary to ensure that the concrete is not in contact with the tendon during placing

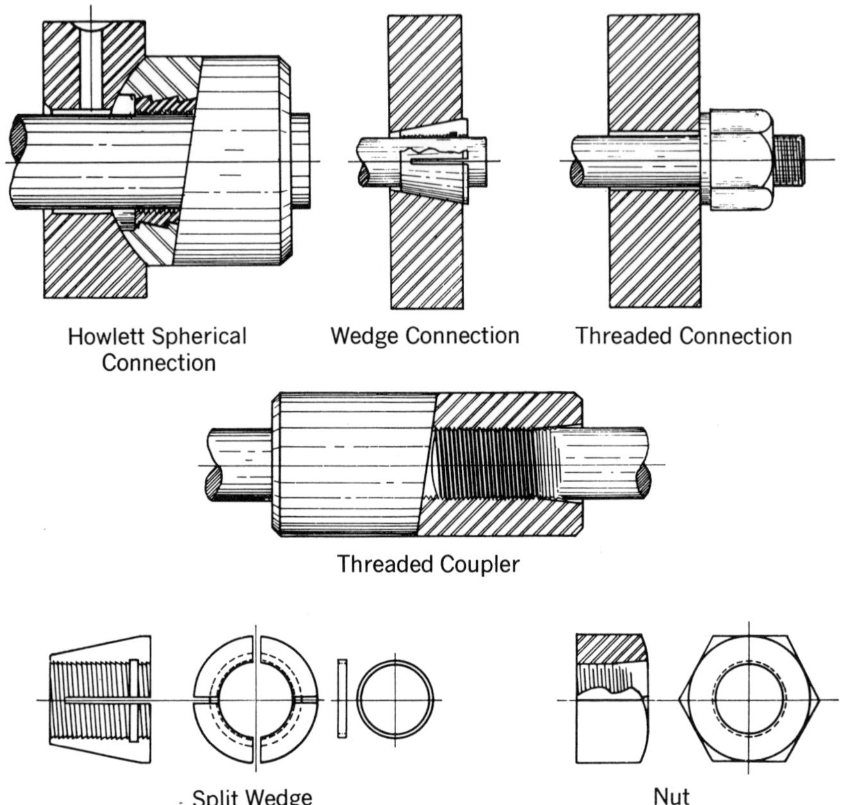

Fig. 15-6 Some of the types of anchorages and couplers used with the bar systems. (*Courtesy Rods, Inc., Berkeley, California.*)

and curing of the concrete. This can be done by placing the tendon in a tube or covering it with a material that will prevent the concrete from coming in contact with the tendon. In addition, a hole can be formed through the concrete section with a form which may or may not be removable. The tendon can thus be inserted in the preformed hole after the concrete has hardened.

In virtually all cast-in-place post-tensioned building construction in the United States, the tendons are placed in sheaths and the complete unit is embedded in the concrete at the time the concrete is placed. A large portion of the precast post-tensioned concrete is constructed by performing holes for the tendons through the members with rigid tubing. After the concrete has gained sufficient strength for stressing, the tendons are pulled into the preformed holes, stressed, and grouted. Preformed holes can also be made by placing rubber tubes, which are either inflated with water or air or which are stiffened with metal rods, in the member and withdrawing them after the concrete has set. When the tendons are to be factory-fabricated and are to be grouted after stressing, the sheath most commonly used is a flexible metal hose of interlocking construction. When the tendons are to be left unbonded after stressing, the usual practice has been to coat the tendons with a rust inhibitor and wrap the tendon with paper or plastic tubing.

When the wire or strand tendons are factory made, it is necessary that the sheath be of a construction that will allow the tendons to be coiled to facilitate handling and shipping. The sheath must be strong so that it will not become damaged in transit and so that it will retain its shape during the placing and vibrating of the concrete. If the sheath does not retain its shape during placing and vibrating of the concrete, a large friction loss will develop during stressing. In addition, the coefficient of friction between the sheath and the tendon should be low and the sheath must be impervious to cement, although it is not essential that the sheath be impervious to water.

All of these properties can be obtained with an interlocking, flexible, metal sheath. This type of sheath is available bright and galvanized. The former is adequate for most tendons in which friction is not expected to be a serious problem. A galvanized sheath is recommended for longer tendons, due to its lower coefficient of friction, and for tendons which may be subjected to a long exposure to moist air before stressing and grouting. The interlocking sheath is also available with an asbestos packing in the interlock which reduces the intrusion of cement and water during placing of the concrete.

Rigid metal tubing should be used for sheath on long tendons, since the friction characteristics of this type of sheath have proved to be very favorable.

The use of paper or plastic tubes for sheaths is limited to applications in which the tendons are to be left unbonded. When using paper and plastic tubes for a sheath, the tendons can be coiled for ease of handling and shipping. Normal handling will not damage the sheath. Care must be exercised, however, to ensure that adequate coatings of rust inhibitors are applied to the tendons before the paper or plastic is applied. The rust inhibitor is essential to protect the tendon, as well as to ensure a low coefficient of friction between the tendon and sheath during stressing.

The use of preformed holes with the high-tensile bar tendons is not recommended. Due to the stiffness of the bars, preformed holes must be very straight, with no secondary curvature of significance, in order to allow the bar to be inserted.

As stated above, preformed holes can be made with rubber tubes stiffened by inflating the tubes with air or water. Holes can also be preformed with rubber hoses that are stiffened with steel rods that are removed before the hoses are pulled out. The use of neoprene is preferred to natural rubber, due to the high resistance to deterioration when exposed to petroleum derivatives. When air or water is used in the tubes, extra care must be exercised in tying the tubes in place, since the inflated tubes are lighter than the plastic concrete and they tend to float when the concrete is placed and vibrated.

When using steel rods to stiffen neoprene hoses, the rods should be smooth (not deformed bars) and lubricated to facilitate inserting and removing them from the hoses. When this procedure is used, the steel rods are frequently made in two pieces—and one piece is removed from each end. Hoses that are in two pieces and that are taped together near the center have also been used, but there is a chance that the tape will loosen during placing of the concrete, in which case, the hole through the tube may not be continuous and the beam may have to be rejected. Furthermore, when the hose is in two pieces, each half must be pulled out simultaneously. If this is not done, the taped joint may hold just long enough for the pulling at one end to draw the half of hose at the other end into the beam.

15-4 Forms for Post-tensioned Members

Post-tensioning is frequently used on projects in which only a few prestressed members are required and in which the cost of setting up an efficient pre-tensioning operation is greater than the savings that can be made by the use of pre-tensioning in lieu of post-tensioning. This situation frequently exists, even though the basic pre-tensioning facilities may be available. In post-tensioning, one set of side forms may be used with two or three soffit forms to obtain an efficient and economical production cycle. In pre-tensioning, if the same amount of forms are used, by moving them along the bench as the members are produced, the entire pre-tensioning bench (or line) will be tied up and the production will not be greater than is possible with post-tensioning. In addition, post-tensioning has been used frequently for the construction of "custom" beams for bridges in which only 10 or 12 beams are required. In such cases, the beams have often been made on the job site by the general contractor.

Wooden forms have been used in many post-tensioning applications, because the cost of steel forms is prohibitive unless they can be re-used many times. The wooden forms are not usually made adjustable, but are built to produce members of only one size. Well built wooden forms that are properly oiled can frequently be used ten or more times before they must be completely rebuilt or discarded. Lining the inside of wood forms with light-gauge metal reduces the amount of moisture wood absorbs from the plastic concrete and increases the life expectancy of the forms.

When using wood forms, the method of curing the concrete must be such that expansion of the wood due to the absorption of moisture will not crack the concrete. This is particularly dangerous with I shaped beams and necessitates protecting the backs of the forms, as well as the front surfaces which are in contact with the concrete, when water curing is applied before stripping of the form. Steam curing is not recommended for members made in wood forms.

Steel forms are used in the construction of post-tensioned members when the number of re-uses which can be obtained from the forms justifies the higher form cost. In this case, the forms may be made adjustable or for only one shape of member, depending upon the circumstances.

The same basic characteristics that are desirable for forms used in the manufacture of pre-tensioning are desirable in the manufacture of post-tensioned concrete, with the possible exception of the adjustability. In manufacturing pre-tensioned concrete, it is frequently necessary to keep the forms as close to the top surface of the pre-tensioning bench as possible, in order to reduce the stress in the uprights and the tendency toward overturning the abutments. This condition does not exist in post-tensioned construction and the forms can be built elevated on small columns which have rubber vibration insulators that will allow maximum use and benefit from form vibration. This scheme is illustrated in Fig. 15-7. It will be seen that the form vibrator

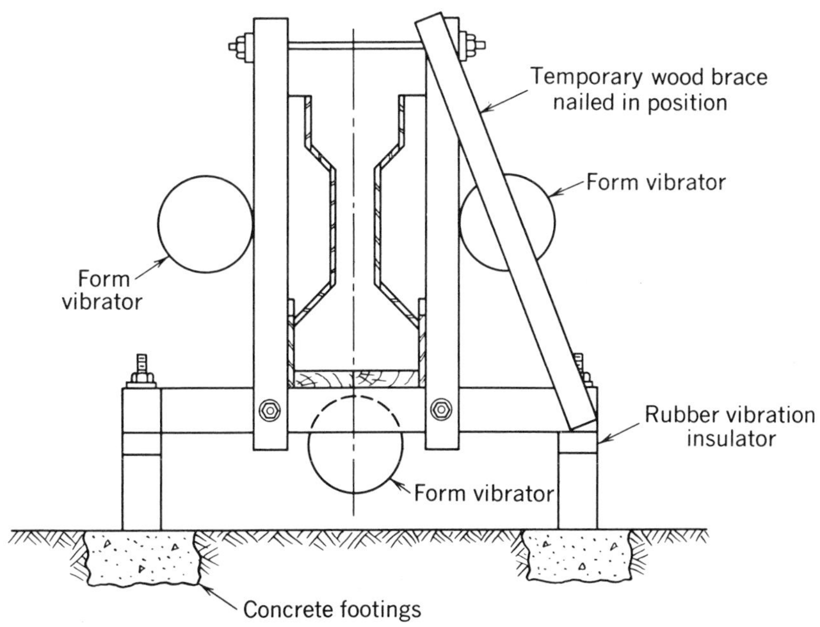

Fig. 15-7 Form designed for use with external vibrators.

can be attached to the soffit form as well as the side forms. The vibrators may be moved along the length of the form as the member is cast or, if desired, vibrators may be positioned at various locations along the form rather than moving the vibrators during the placing of the concrete. When this type of vibration is used, it is normally necessary to supplement the form vibration with internal vibration. This method, which for all practical purposes makes the form into a combination form and vibrating table, such as is used in the manufacture of small concrete products, is only feasible in precasting plants and on large casting operations conducted at the job site, due to the relatively high cost of the equipment.

15-5 Effect of Friction During Stressing

The variation in stress in post-tensioned tendons due to friction between the tendon and the duct during stressing is considered to be the function of two effects. These are the primary curvature or *draping* of the tendon, which is intentional, and the *secondary curvature* or wobble which is the unavoidable minor horizontal and vertical deviations of the tendon from the theoretical position. Coefficients for each of these effects, which are commonly specified by post-tensioning design criteria in this country, are listed in Table 15-1.

TABLE 15-1

Type of Tendon	*Type of Duct*	*Design Values* μ	K
Uncoated wire or large diameter strands	Bright flexible metal sheath	0.30	0.0020
	Galvanized flexible metal sheath	0.25	0.0015
	Galvanized rigid metal sheath	0.25	0.0002
	Mastic coated, paper or plastic wrapped	0.05	0.0015
Uncoated seven-wire strand	Bright flexible metal sheath	0.30	0.0020
	Galvanized flexible metal sheath	0.25	0.0015
	Galvanized rigid metal sheath	0.25	0.0002
	Mastic coated, paper or plastic wrapped	0.08	0.0014
Bright metal bars	Bright flexible metal sheath	0.20	0.0003
	Galvanized flexible metal sheath	0.15	0.0002
	Galvanized rigid metal sheath	0.15	0.0002
	Mastic coated, paper or plastic wrapped	0.05	0.0002

The stress at any point in a tendon can be determined by substituting the proper coefficients in the relationship.*

$$T_O = T_x e^{(\mu\alpha + KX)} \tag{15-1}$$

* This relationship can be written for unit stresses as follows:

$$f_O = f_x e^{(\mu\alpha + KX)}$$

in which: T_O = the force at the jacking end, T_x = the force at point X feet from the jacking end, e = base of the Naperian logarithm, K = secondary curvature coefficient, μ = primary curvature coefficient, α = total angle change between the tangents to the tendon at the end and at point X (sum of the horizontal and vertical angles) in radians, and X = distance in feet from the jacking end to the point under consideration.

For low values of $\mu + KX$, Eq. 15-1 is approximately equal to:

$$T_O = T_x(1 + \mu\alpha + KX) \tag{15-2}$$

Eq. 15-2 is permitted for use with values of $\mu\alpha + KX$ as high as 0.3 by ACI 318-71.

The variation in stress in a post-tensioned tendon stressed from one end, according to the above relationship, will be as is illustrated in Fig. 15-8. The maximum value of stress in the tendon is at the jacking end and the minimum value is at the dead end. If the tendon were stressed from each end simultaneously, the curve would be symmetrical about the center line and would have the shape of AB on each side.

The effects of friction during post-tensioning should be investigated by the designer at the time the trajectory of the post-tensioning tendons is selected. The computations will reveal the magnitude of the friction loss that can be expected and will allow the designer to determine if special precautions should be required in the specifications in order to reduce the friction loss. Special procedures to reduce the effects of friction include using galvanized sheath in lieu of bright sheath, using rigid sheath in lieu of flexible, reducing the curvature of the tendons, or using a water-soluble oil on the tendons. The oil is removed by flushing the ducts with water before the tendon is grouted. The use of water-soluble oil to reduce the friction is generally employed only as a last resort when serious friction is encountered on the job site. It is not recommended that this procedure be relied upon during design.

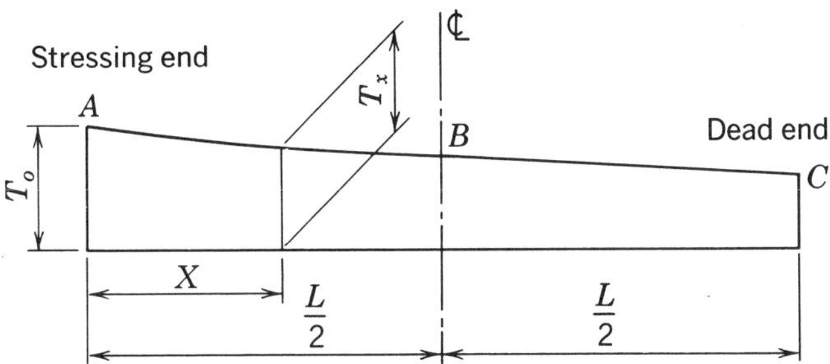

Fig. 15-8 Variation in stress due to friction in a tendon that is post-tensioned from one end.

ILLUSTRATIVE PROBLEM 15-1 Compute the force which would be expected to result at midspan of a tendon that is 100 ft long and which is on a parabolic curve having an ordinate of 3 ft at the center line. Assume the sheath is to be a bright flexible metal hose and that the tendons are to be composed of parallel wires.

SOLUTION: Since the tangent of the angle between the tangents to the tendon can be assumed to be numerically equal to the value of the angle expressed in radians, the value of α is found by

$$\alpha = \tan \alpha = \frac{4e}{L} = \frac{4 \times 3 \text{ ft}}{100 \text{ ft}} = 0.12$$

Using the coefficients from Table 15-1

$$= 0.30 \times 0.12 = 0.036$$

$$KX = 0.002 \times 50 \text{ ft} = \underline{0.100}$$

$$0.136$$

$$T_O = T_x e^{0.136} = 1.15 T_x$$

$$T_x = 0.873 T_O$$

The loss of stress due to friction is 13%.

With galvanized sheath rather than bright sheath, the loss of prestress would be computed as follows:

$$= 0.25 \times 0.12 = 0.030$$

$$KX = 0.0015 \times 50 = \underline{0.075}$$

$$0.105$$

$$T_O = 1.111 T_x$$

$$T_x = 0.900 T_O$$

15-6 Elastic Deformation of Post-tensioning Anchorages

As explained in Sec. 15-5, the variation in stress along a post-tensioned tendon at the time of stressing is assumed to follow a curve such as *ABC* in Fig. 15-9. The curve *EDBC* in this figure indicates the assumed variation in stress after the tendon has been stressed and anchored. It will be noted that a reduction in the stress at the end of the tendon resulted from the anchoring procedure. The reduction in stress occurs when the prestressing force is transferred from the jack to the anchorage device, at which time, a portion

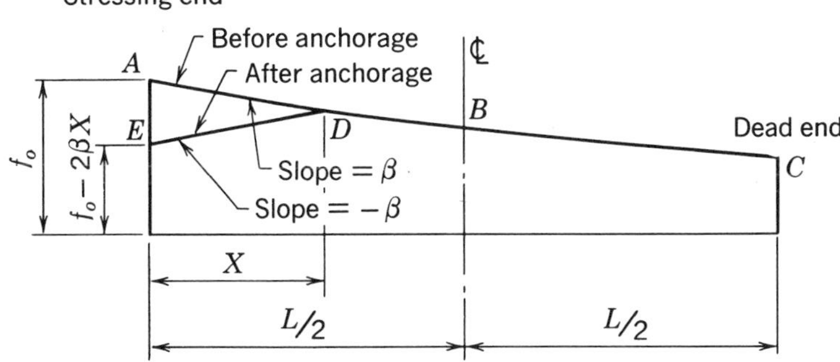

Fig. 15-9 Variation in stress in a post-tensioned tendon before and after anchorage. Condition I.

of the elongation of the tendon obtained during stressing is lost due to the deformation of the anchorage device. Although some *positive type* anchorage devices, such as the button-head systems, when properly applied, have no appreciable deformation of anchorage, the wedge or cone-type anchorages often deform significantly as the load is applied to them. Other types of anchorage which have components stressed in the plastic range may require several hours to reach the limiting value of anchorage deformation. The manufacturers of the anchorage devices that are to be allowed on any project should be consulted in order to determine the limits of the anchorage deformation that may be encountered in the field. If the computations indicate that the deformation may result in a significant reduction in stress of the tendons, the deformation should be measured in the field as a means of ensuring that the desired results are obtained.

In order to simplify the computation of the effect of the anchorage deformation on the stress in the tendon, the curves *AB*, *AD*, *BC* and *DE* in Fig. 15-9 are assumed to be straight lines. The slopes of lines *AB* and *DE* are assumed to be of equal magnitude, but of opposite sign. Tests have shown that this assumption is approximately correct.

The state of stress indicated in Fig. 15-9 will be referred to as condition I. This condition is characterized by the fact that the stress at the center of the tendon is not affected by the deformation of the anchorage, which means that the length X is less than one-half of the length of the tendon. This condition generally occurs in long tendons or in members that have high friction, such as those with high primary curvatures. In some instances, it is desirable to stress tendons having stress distributions of this type from one end only, but, since this condition generally exists only in tendons of considerable length, they are more frequently stressed from both end simultaneously.

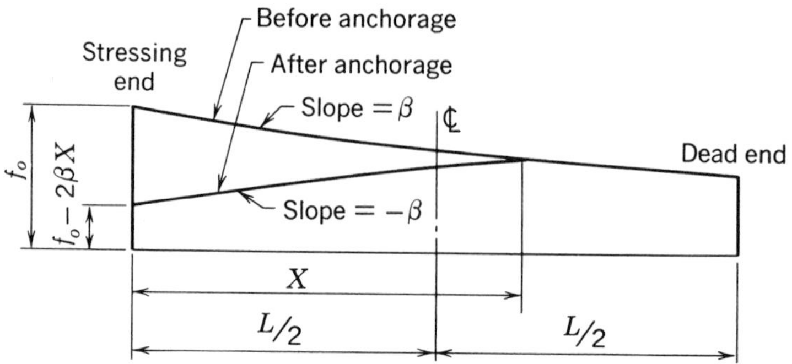

Fig. 15-10 Variation in stress in a post-tensioned tendon before and after anchorage. Condition II.

The assumed variation in stress referred to as condition II is shown in Fig. 15-10. In this case, the stress at the center line is affected by the deformation of the anchorage, since the distance X is greater than one-half the length of the tendon. There is generally no advantage to be gained in stressing a tendon having this type of stress distribution from each end simultaneously.

The most severe effects from the deformation of the anchorage are found in short cables and those with very low friction. In this case, the stress at the dead end is reduced by the deformation of the end anchorage. This condition is characterized by the computed length of distance X being greater than the length of the tendon. The assumed distribution of stress for condition III is illustrated in Fig. 15-11.

Determination of the effect of anchorage deformation on the stress at midspan, which is generally the most critical section from the designer's viewpoint, will be developed. A similar procedure is used in structures having the critical sections at other locations. The procedure is as follows:

(1) Determine the ratio between the stress at the end and midspan

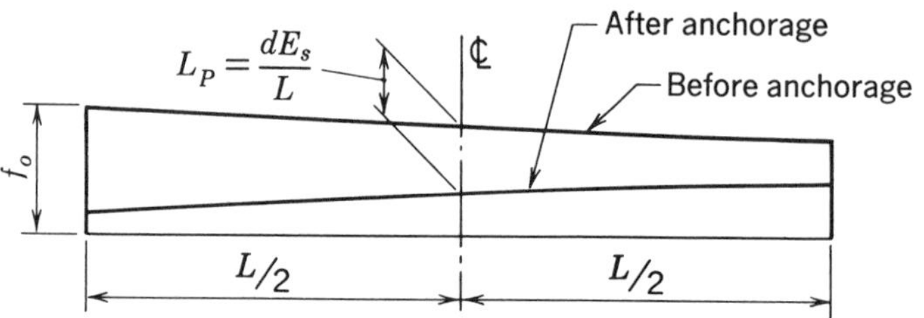

Fig. 15-11 Variation in stress in a post-tensioned tendon before and after anchorage. Condition III.

resulting from the friction, as explained in Sec. 15-5. This can be expressed as

$$f_O = f_{\mathbb{C}} e^{\phi}$$

where

$$\phi = \mu\alpha + \frac{KL}{2}$$

(2) Assume the deformation of anchorage does not affect the stress at midspan of the beam (condition I) and compute the slope of the curve between the center line and stressing end as follows:

$$\text{Slope} = \beta = \frac{2f_{\mathbb{C}}(e^{\phi} - 1)}{L} \quad (15\text{-}3)$$

(3) Compute the length of the tendon on which the stress is reduced by the anchorage deformation

$$X = \sqrt{\frac{dE_s}{\beta}} \quad (15\text{-}4)$$

where X = length of tendon which is affected in in. β = slope in psi/in., E_s = modulus of elasticity of the steel in psi., d = deformation of the anchorage in in.

4. If X is less than $L/2$, condition I exists, the stress at the center line is not affected, and the stress at each end is readily computed if desired. If X is greater than $L/2$, but less than L, condition II exists. The stress loss at midspan resulting from the anchorage deformation is equal to

$$L_p = 2\beta(X - L/2) \quad (15\text{-}5)$$

If this loss is too great to be tolerated, higher initial stresses should be investigated in a trial and error procedure until a satisfactory solution is found. If X is found to be greater than L, condition III exists and the loss at the center of the tendon is equal to

$$L_p = \frac{dE_s}{L} \quad (15\text{-}6)$$

and the stress after anchorage at the center of the tendon can be computed directly.

It must be emphasized again that the anchorage deformation for some post-tension systems is very small and can be reasonable neglected. Anchorage deformations as high as 1 in. have been observed with wedge-type anchors under very unusual conditions. It is important to recognize that this phenomenon exists and that it must be taken into consideration during the design. This is particularly true when short tendons are to be used.

ILLUSTRATIVE PROBLEM 15-2 Determine the effect of the elastic deformation of the anchorage cone on the stress at midspan of a tendon 100 ft long with a maximum ordinate of 3 ft, if the desired stress at the center line is 165,000 psi, the deformation of the anchorage is 0.50 in., the modulus of elasticity of the steel is 28,000,000 psi, and a bright metal sheath is used. Also determine the effect if the sheath is galvanized rather than bright.

SOLUTION: The effect of friction for these conditions were determined in Prob. 15-1. Using the calculated value for e^{ϕ}, the computation of β and X becomes:

$$\beta = \frac{2 \times 165{,}000 \times 0.15}{1200 \text{ in.}} = 41.2 \text{ psi/in.}$$

$$X = \sqrt{\frac{0.50 \times 28 \times 10^6}{41.2}} = 584 \text{ in.} < 600 \text{ in.} = \frac{L}{2}$$

Therefore, condition I exists and the deformation of the anchorage of 0.50 in. does not reduce the stress in the tendon at the center line.

Using galvanized sheath:

$$\beta = 41.2 \times \frac{0.11}{0.15} = 30.2 \text{ psi/in.}$$

$$X = \sqrt{\frac{0.50 \times 28 \times 10^6}{}} = 680 \text{ in.} > 600 \text{ in.} = \frac{L}{2}$$

The reduction in stress at midspan due to the deformation of the anchorage is:

$$L_p = 2\beta\left(X - \frac{L}{2}\right) = 2 \times 30.2 \times 80 = 4830 \text{ psi}$$

Therefore, the stress in the tendon at the center after the tendon is anchored is

$$f = 165{,}000 - 4800 = 160{,}200 \text{ psi}$$

If the stress at midspan is increased to 169,000 psi at the time of stressing, the values of β and X and the loss of the prestressing stress in the tendon at midspan become:

$$\beta = \frac{2 \times 169{,}000 \times 0.11}{1200} = 31.0 \text{ psi}$$

$$X = \sqrt{\frac{0.50 \times 28 \times 10^6}{31.0}} = 672 \text{ in.}$$

$$L_p = 2 \times 31.0 \times 72 = 4500 \text{ psi}$$

Therefore, if the stress at midspan is increased to 169,500 psi at the time of stressing, the deformation of the anchorage of 0.50 in. will reduce the stress to approximately 165,000 psi, which is the desired value.

It should be noted that the stress at the end of the tendon which is required to obtain 165,000 psi in the tendon at the center is equal to 165,000 × 1.15 = 190,000 psi when a bright sheath is used. When galvanized sheath is used, the stress at the end of the tendon which is required to obtain the desired stress at the center line after anchoring is equal to 169,500 × 1.11 = 188,000 psi. It will be seen that the required tendon stress at the end of the tendon is not materially reduced by the use of galvanized sheath in this case.

15-7 Computation of Gauge Pressures and Elongations

At the time the design of post-tensioned members is made, the computation of gauge pressures and elongations for post-tensioning should be made and summarized on the drawings, if the post-tensioning system is specified by the designer. In following this procedure, the designer is assured that the stressing can be achieved as intended. When the system of post-tensioning is not specified, the computation of gauge pressures and elongations must be supplied by the contractor for review and approval by the engineer. In this case, it is recommended that the summary of the stressing data be placed on the shop drawings.

The computation of gauge pressures and elongations are conveniently made and summarized in tabular form as illustrated in Table 15-2. Some systems have peculiarities of their own which are not included in the list of items in Table 15-2. These must be covered by special notes or instructions and will not be discussed here.

The length of the prestressing tendon used in the computations is normally taken as the horizontal projection of the tendon. For more precise computations or where there are large curvatures of the tendons, the length can be computed by:

$$L' = \left(1 + \frac{8}{3}\frac{a^2}{L^2}\right) L \qquad (15\text{-}7)$$

in which L is the horizontal projection of the tendon, a is the vertical deflection of the tendon and L' is the correct length. Provision is made in the table for the length to be recorded in both feet and inches, since the former length is required in the computation of the effect of friction and the latter is required in the computation of elongations.

Under item 5, the distance from the anchorage to the mark or marks on the tendon which are used in measuring the elongation during stressing is recorded. Although this may at first appear to be an unnecessary refinement, it

TABLE 15-2 Typical Calculation Sheet for Use in Computing the Gauge Pressures and Elongations in Post-tensioning

Item Number	*Tendon Numbers* 1	2	3
1. Type of tendon			
2. Area of steel (in.^{2})			
3. Length of tendon (ft)			
4. Length of tendon (in.)			
5. Distance from anchorage to marks (in.). Note: For one-end stressing, use only the length at stressing end. For two-end stressing use the length at both ends			
6. Total length of unit to be stressed (in.)			
7. Desired effective stress (psi)			
8. Losses of prestress.			
a. Elastic shortening*			
b. Relaxation, creep and shrinkage			
9. Desired stress before anchorage deformation (psi)			
10. Effect of anchorage deformation (psi)			
11. Desired jacking stress at center (psi)			
12. Anticipated jacking stress at jack (psi)			
13. Anticipated gauge pressure (psi)			
14. Maximum gauge pressure (psi)			
15. Elongation to be obtained (in.)			
a. Total			
b. Net			
16. Special instructions			

* The losses of prestress should include all the effects listed in Sec. 6-2. The loss due to elastic shortening is often left to be included in the stressing computations. This should be investigated for each job.

must be pointed out that in some instances, the distance from the anchorage to the marks on the wires may be as great as 18 in., which can be a significant portion of the total length of the tendon. In such a case, the elongation of the tendon in this length only during stressing may be as high as 1/8 in. for one-end stressing and 1/4 in. for two-end stressing. It is apparent that elongations of this magnitude should be included in the computations.

If the designer has specified the final tendon stress, the computation of the losses of prestress must be made as described in Art. 6-2. If the initial tendon stress has been specified, it can be entered directly in item 9. The effect of the deformation of the anchorage (Sec. 15-6), if any, is then computed and entered in item 10. The desired, jacking stresses at the center and at the end, which are found in computing the effects of the anchorage deformation, are entered as items 11 and 12, respectively. The value of the jacking stress at the end is not normally allowed to exceed 80% of the guaranteed, ultimate tensile strength of the tendon.

The gauge pressure to be anticipated during the stressing of a tendon is higher than the theoretical gauge pressure, due to internal friction of the jack and, in some systems, due to the friction of the tendons which must slide over the anchorage and the jack during stressing. This friction ranges from 2% to 15% and a reasonable estimate of the friction can only be obtained by calibrating the jack when all of the contributing factors are present. The data obtained in calibrating a jack can be plotted in a curve that shows the relationship between the gauge pressure and the steel stress. The gauge pressure that would be anticipated in obtaining the desired initial stress at the end of the tendon can be taken directly from such a calibration curve. This gauge pressure is termed anticipated, since it is computed on the basis of assumed friction coefficients and is subject to error. For this reason, the elongation of the tendon is generally used as the controlling measurement during stressing and the normal procedure is to apply the anticipated pressure and then check the elongation. If the elongation is not satisfactory, the pressure is increased until a satisfactory elongation is obtained. In order to ensure that the tendon will not be overstressed, a maximum permissible gauge pressure (corresponding to the maximum allowable jacking stress) is taken from the calibration curve. The maximum permissible gauge pressure is entered in item 14.

Before the reference marks used in measuring the elongation of the tendons are placed on the tendons, it is normal procedure to apply a small pressure of about 10% of the anticipated gauge pressure to the tendon in order to tighten up the jack and eliminate any slack that may be in the tendon. This value should be constant on the entire job in order to eliminate the possibility of the workmen using the incorrect value when changing from tendons of different size.

Since the initial pressure applied to the tendon results in a small stress in the tendon, the amount of elongation that is obtained during the application of the remaining force is the result of the deformation of the steel between the lower value of stress and the higher value. In addition, because there is a variation in stress in the tendon due to friction, an average value of stress must be used in determining the elongation that is to be obtained in the stressing. In the case of stressing from one end only, the average stress would be the stress at the center of the tendon in most instances, while, in the case of stressing from each end simultaneously, the average stress would occur approximately midway between the end and the center of the tendon. After the appropriate average stress is determined, the total elongation that would result from this average stress can be calculated from the stress-strain diagram for the steel that is to be used. The net elongation is determined by taking the increment of strain from the lower stress to the higher stress, for the steel at the average location in the tendon, and multiplying this value by the length of the tendon.

15-8 Construction Procedure in Post-tensioned Concrete

Although the construction details and the system of post-tensioning to be used will influence the construction procedure to be followed, general statements and precautions are considered of value. Careful planning and attention to details of construction in post-tensioned elements is important and can have material influence on the cost and performance of post-tensioned construction.

The normal procedure followed in the construction of a post-tensioned beam consists of erecting the soffit form, and one side form, placing the reinforcing steel, prestressing steel, end anchorages, erecting the remaining side form, placing and curing the concrete, stripping the forms, and finally stressing and grouting the post-tensioning tendons. In large productions of post-tensioned members, the reinforcing and post-tensioning steel may be tied together in a jig and set in the forms as a unit. On smaller jobs, the post-tensioning units are normally tied in place after the reinforcing steel cage is partially or completely assembled in the forms. The latter procedure is much the more common, and it must be emphasized that the erection of one side form, which can be used to facilitate layout and to secure the reinforcing cage, inserts, and blockout forms in place, expedites the assembly of post-tensioned members.

When post-tensioning tendons in sheaths are used, the tendons must be securely tied in place at close intervals and in such a manner that secondary curvatures of the tendons are minimized and the primary curvature of the tendon is close to the curve specified on the drawings. It must be emphasized that it is important for the location of the tendons at the center of simple flexural members to be close to that specified on the drawings. As was shown in Sec. 4-7, the tendon does not normally need to be placed within precise limits at points between the center of the beam and the ends in order to achieve a satisfactory post-tensioned flexural member. Hence, it is generally important that the tendon be placed on smooth curves that will minimize friction losses during stressing, but it is not of great importance if a tendon is not precisely on the specified trajectory.

No general statement can be made on the maximum spacing that should be permitted for the ties and supports that secure the post-tensioning tendons in place during placing of the concrete. This is a function of the type and size of tendons, as well as the arrangement of the reinforcing steel and required trajectory of the tendons.

Care must be exercised when tying the tendons, since it is possible to damage the sheath during the tying operation. This is particularly true when paper or plastic sheath is used. Damaged sheath may permit grout to enter the sheath and bond the tendon to the beam. This may render the stressing of the tendons impossible.

After the tendons and anchorages have been tied in place, they should be carefully inspected to be sure they are securely tied at all locations and that there is no possibility of mortar leaking into the sheath or anchorage device during placing and vibrating the concrete. Although mortar that leaks into a sheath or anchorage does not always result in the tendon being bonded and impossible to stress, the extra labor required to clean the anchorage before stressing can be started (or to overcome the high friction in the tendon which may be encountered during stressing) usually far exceeds the amount of labor required to properly seal the sheath and anchorage before stressing.

When ducts are to be preformed in the concrete, the procedure is very similar to that followed for tendons which are in sheath. Rubber ducts should not be tied so tightly to the reinforcing that they will be difficult to remove.

The manufacturers of some types of post-tensioning tendons recommend that the tendon be moved back and forth in the sheath during and after the placing of the concrete in order to be certain that the tendon is not bonded to the concrete by any mortar which may have leaked into the sheath. Although there is no technical objection to this procedure, it is not practical for all systems of post-tensioning and is considered unnecessary if sufficient attention is given to the prevention of leaks in the sheaths and anchorages before the concrete is placed in the forms.

After post-tensioned members are cured, the tendons are usually stressed and grouted as soon as possible. The stressing is done according to the procedure recommended by the manufacturers of the post-tensioning anchorages and jacking equipment. In post-tensioning, the elongation of the tendons is generally used as the controlling measurement for ensuring proper stress at the center of a member. As has been explained, the elongation is a better measure of the stress at the center of the tendon than are the gauge pressure and the stress in the tendon at the end of the tendon.

The procedure used in grouting post-tensioning units also varies with the post-tensioning system being used. In general, the grouting ports and sheaths are very small and prohibit the use of large-grain sand in the grout. Very fine sand is used occasionally in order to reduce the quantity of cement required and in an attempt to reduce the shrinkage of the grout. Although the addition of an adequate quantity of very fine sand to mortar may result in the shrinkage of the mortar being only one-half of that which results from the use of neat cement, the addition of fine sand in the amounts which can be added to the grout for post-tensioning tendons will not materially reduce the shrinkage. Pozzolanic material (fly ash) is sometimes specified as an admixture for the grout, but little if any benefits are gained from this procedure, because the shrinkage of concrete is not reduced and may be increased by such admixtures. For these reasons, the grout used in domestic post-tensioning operations rarely contain sand.

The type of cement used in the grout for the post-tensioning ducts should be the same type used in the concrete, because a difference in the electrolytic properties of the cement in the grout and the basic concrete may result in the deterioration of the prestressing steel. Normally, a grout mixture of 4.0 to 4.5 gallons of water per sack of cement will result in a grout that can be easily injected. It is important that grouting tubes, hoses, ports, and valves do not leak during grouting. Leaks may result in plugging of the system and the trapping of water in the sheath. When small round sheaths are used, the grout can be introduced at pressures up to 100 psi. When large round or rectangular ducts are used, the grout pressure may have to be restricted to atmospheric pressure, in order to avoid cracking of the concrete section along the trajectory of the tendon.

During freezing weather, the grout that is injected into post-tensioning ducts may freeze and cause cracking of the concrete along the trajectory of the post-tensioning tendons. This can be avoided by heating the member after grouting for 24 to 48 hr, in order to allow the grout to harden, or by replacing some of the mixing water in the grout mixture with alcohol in order to reduce the freezing point of the grout. The use of one part alcohol to eight parts of water has been recommended by some authorities as a satisfactory method of providing protection against freezing of the grout for temperatures as low as 20°F, without excessive sacrifice in the strength of the grout. The use of more alcohol will render the grout more resistant to freezing, but will also reduce the strength of the grout (Ref. 6).

If post-tensioned members are to be exposed to freezing weather before the tendons are grouted, the ducts or sheaths should be drained of all water that may collect in the sheath during placing and curing of the concrete. If this is not done and the water freezes, the members may be cracked, just as if grout had frozen in the ducts. Water should not be allowed to remain in the ducts of post-tensioning tendons for more than just a few days, even if there is not any danger of freezing, because corrosion of the tendons will generally proceed at a more rapid rate when the tendons are submerged or in a very moist atmosphere.

15-9 Construction of Multi-element Beams

Post-tensioned beams may be made of a number of precast elements that are post-tensioned together to form one structural unit. This procedure is used in order to facilitate handling, transportation, and forming large members. Types of multi-element beams are illustrated in Fig. 15-12. It will be seen that a member may be formed of many elements made in small forms on vibrating tables or of only a few large elements made with the same

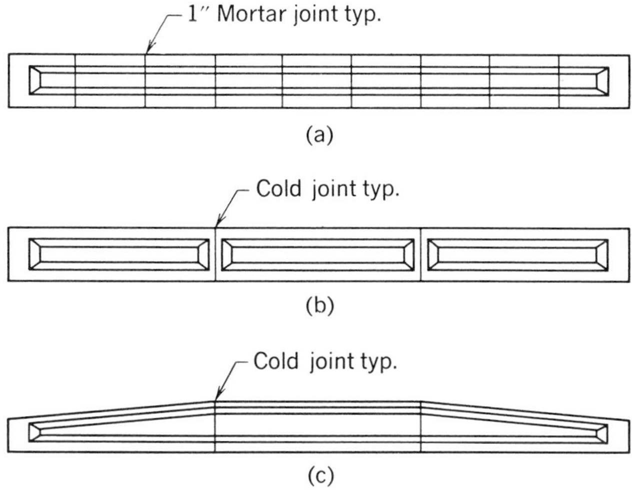

Fig. 15-12 Multi-element post-tension beams.

general procedure used in the construction of monolithic members. Preformed holes are generally used for the tendons in multi-element members. When preformed holes cannot be employed, because of the post-tensioning system to be used, the tendons must be coupled at each joint.

Members composed of many small elements normally must have mortar placed in the joints at the time the elements are aligned for stressing. The mortar is necessary to assure a uniform distribution of stress between the elements. The mortar joints present a difficulty in the assembly of the members, because it is necessary to prevent the mortar from entering the post-tensioning ducts and bonding to the tendons. Additionally, the placing of the mortar requires considerable labor. In many instances, the prestress can be applied to the multi-element beams immediately after the mortar is placed, since the total deformation of the plastic mortar is small, due to the relatively short length of the mortar joints. The thin mortar joints cannot fail due to prestressing, even if the mortar is partially plastic at the time of stressing.

The need of mortar joints in multi-element members can be eliminated by casting, first, one element, and, after the first element has hardened, casting the second element against the end of the first element. In this manner, the cold joint formed between the two elements must be a perfect fit, and yet the elements can be separated for handling and transportation purposes. A light coating of oil is often placed on the end of the elements to reduce the bond between the elements during manufacture. Shear keys, which provide vertical and horizontal projections of the end surfaces, are often provided at the ends of each element to facilitate aligning the elements in the field. Shear keys are not normally required due to structural considerations. The joints between the individual pieces of multi-element beams are frequently coated with epoxy resin at the time of assembly. This glues the pieces together. A bridge on which this procedure was used is shown in Fig. 15-13.

When multi-element beams composed of only a few pieces are assembled in the field, it is not necessary to set the elements close together and in precise alignment with the cranes or other handling equipment being used. It is generally easier and more expeditious to set the elements on wooden blocks, with a space of approximately 1 ft between the elements. This procedure facilitates threading the post-tensioning tendons in the ducts. The elements can be pulled together with the use of the post-tensioning jacks after the tendons have been inserted.

Fig. 15-13 Multi-element post-tensioned construction was used on the bridge between the mainland and the island of Oléron in France. (*Courtesy Freyssinet Co., Inc., New York.*)

REFERENCES

1. Bumanis, Alfreds, "Friction Loss Study of 402-Ft. Tendons," *Journal of the Prestressed Concrete Institute*, **11**, No. 4, 57–63 (Aug. 1966).

2. Polivka, Milos, "Grouts for Post-tensioned Prestressed Concrete Members," *Journal of the Prestressed Concrete Institute*, **6**, No. 2, 28–38 (June 1961).

3. Committee on Grouting of Post-tensioned endons, "A Report of Field Experiences in Grouting Post-tensioned Prestressed Cables." *Journal of the Prestressed Concrete Institute*, **7**, No. 4, 13–17 (Aug. 1962).

4. Committee on Grouting of Post-tensioned Tendons, "Tentative Recommended Practices for Grouting Post-tensioned Prestressed Concrete," *Journal of the Prestressed Concrete Institute*, **5**, No. 2, 78–81 (June 1960).

5. Grouting Subcommittee of the Combined Western Concrete Reinforcing Steel Institute Post-tensioning Committee and the Prestressed Concrete Manufacturers Association of California Task Force on Post-tensioning, "Recommended Practice for Grouting Post-tensioning Tendons," (July 1967)- Tentative.

6. STUVO Committee on Grouting, "Some Facts About Grouting," *Proceedings of the Second Congress of the Federation Internationale de la Precontrainte*, Session 1, Paper No. 3, Amsterdam, Netherlands, 1955.

16 Erection of Precast Members

16-1 General

An important consideration in the use of precast concrete is the methods of erection that is feasible under the jobsite conditions. This obviously has significant effect on the cost of construction. Occasionally, high-rise buildings cannot be made at a reasonable cost with precast construction due to the inability of cranes to lift and place heavy precast members at the heights and reaches that are necessary. The large cranes required to erect the larger precast members are costly and are not readily available in all localities. In bridge construction, the use of cranes to erect precast members may not be feasible due to the reaches that are involved, due to the risk of floods, or due to other considerations. The designer must give careful consideration to all the factors that affect the methods that can be used in the construction of each particular structure and prepare his design to best suit job site conditions.

16-2 Truck Cranes

Truck cranes are often used to erect precast units in building and bridge construction. Although cranes with rated capacities of 100 tons or more are frequently used, the designer should be aware of the fact that the rated capacity is the maximum load the crane can lift with a relatively short boom and with the minimum possible load radius. The load radius is the horizontal distance measured from the vertical axis about which the crane cab rotates to the line of the load. This is illustrated in Fig. 16-1.

Capacities for cranes of 35 tons, 65 tons, and 82 tons are given in tables 16-1, 2, and 3. It should be noted that for large loads on short radii, the capacity of a crane is limited by structural considerations, where as for loads

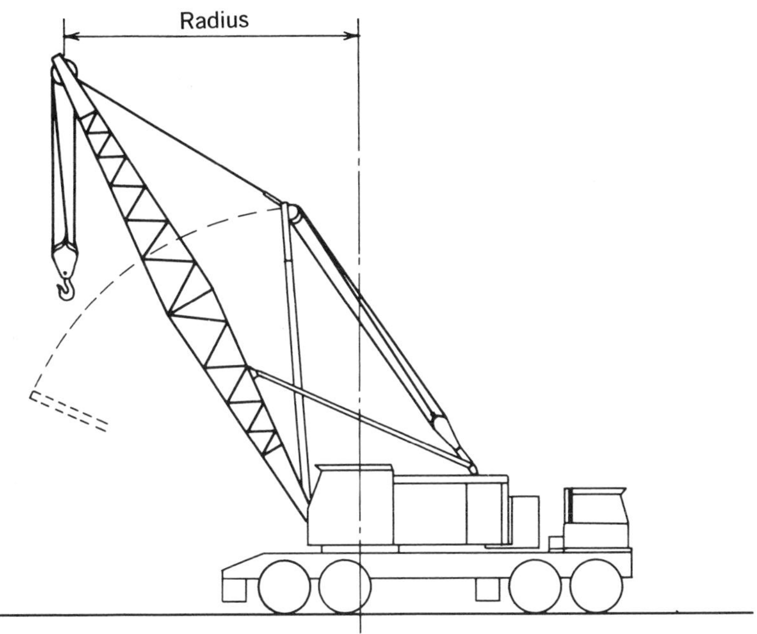

Fig. 16-1 Load radius for a truck crane.

at larger radii, the capacity is limited by tipping. From table 16-3 it will be seen that a 82 ton crane with a 100 ft boom working on a load radius of 40 ft has an approximate capacity of from 21 to 23 tons on outriggers, depending upon whether the load is being lifted over the side or over the end. This clearly illustrates the fact that one must carefully investigate the equipment available in each locality and be certain the capacity will be adequate for the intended use.

It should also be noted that the capacity of a crane is less when it is on its tires and without the benefit of its outriggers. In addition, it is very important that a crane be exactly level when making lifts that are close to its capacity.

TABLE 16-1 Capacities Capacity of a 35 Ton Truck Crane with Angle Boom

* Capacities are based on machine equipped with Retractable High Gantry, 8 × 4 drive Carrier—9′ 0″ wide, 12:00 × 20 14-ply rating tires, Power Hydraulic Outriggers, 13,200# ctwt.

BOOM				On Outriggers		On Tires	
Length	Radius	Angle	Point Ht. W	Rear	Side	Rear	Side
	10′	79°	40′ 9″	70,000*	70,000*	53,100	40,800*
	11′	77°	40′ 7″	70,000*	70,000*	46,300	37,800*
	12′	75°	40′ 3″	69,300*	69,300*	41,000	35,200*
	13′	74°	39′11″	64,500*	64,500*	36,800	32,300
35′	14′	72°	39′ 9″	60,300*	60,300*	33,300	29,200
	15′	70°	39′ 4″	56,600*	56,600*	30,400	26,500
	20′	61°	37′ 1″	42,100*	42,100*	21,000	18,100
	25′	51°	33′ 9″	32,600*	32,600*	15,900	13,600
	30′	40°	28′ 9″	26,500*	26,500*	12,600	10,700
	35′	24°	20′ 9″	22,200*	22,200*	10,400	8,800
	10′	80°	45′10″	70,000*	70,000*	53,000	40,500*
	11′	79°	45′ 8″	70,000*	70,000*	46,200	37,600*
	12′	77°	45′ 5″	69,000*	69,000*	40,900	35,000*
	13′	76°	45′ 2″	64,200*	64,200*	36,600	32,200
	14′	74°	44′11″	60,100*	60,100*	33,100	29,000
40′	15′	73°	44′ 7″	56,400*	56,400*	30,200	26,400
	20′	65°	42′ 8″	41,900*	41,900*	20,800	17,900
	25′	57°	39′10″	32,400*	32,400*	15,700	13,400
	30′	48°	36′ 1″	26,300*	26,300*	12,500	10,600
	35′	37°	30′ 7″	22,000*	22,000*	10,200	8,600
	40′	23°	21′10″	18,900*	18,900	8,600	7,200
	12′	80°	55′ 8″	68,600*	68,600*	40,600	34,500*
	13′	79°	55′ 5″	63,800*	63,800*	36,300	31,900
	14′	77°	55′ 3″	59,600*	59,600*	32,800	28,700
	15′	76°	54′11″	55,900*	55,900*	29,900	26,100
50′	20′	70°	53′ 5″	41,500*	41,500*	20,500	17,600
	25′	64°	51′ 4″	32,000*	32,000*	15,400	13,100
	30′	57°	48′ 7″	25,900*	25,900*	12,200	10,200
	35′	50°	44′11″	21,600*	21,600*	9,900	8,300
	40′	42°	40′ 2″	18,500*	18,500*	8,300	6,900
	50′	20°	23′ 9″	13,900	13,500	6,100	4,900
	13′	81°	64′ 9″	63,300*	63,300*	36,000	31,700
	14′	80°	64′ 6″	59,100*	59,100*	32,600	28,500
	15′	79°	65′ 3″	55,400*	55,400*	29,600	25,800
	20′	74°	63′11″	41,100*	41,100*	20,200	17,400
60′	25′	69°	62′ 3″	31,600*	31,600*	15,100	12,800
	30′	63°	60′ 1″	25,500*	25,500*	11,800	9,900
	35′	58°	57′ 3″	21,200*	21,200*	9,600	8,000
	40′	52°	53′ 9″	18,100*	18,100*	8,000	6,600
	50′	39°	43′10″	13,600*	13,300	5,800	4,600
	60′	18°	25′ 5″	10,500	10,200	4,300	3,400

BOOM				On Outriggers		On Tires	
Length	Radius	Angle	Point Ht. W	Rear	Side	Rear	Side
	15′	80°	75′ 5″	53,100*	53,100*	29,400	25,500
	20′	76°	74′ 4″	40,700*	40,700*	19,900	17,100
	25′	72°	72′11″	31,200*	31,200*	14,800	12,500
	30′	67°	71′ 1″	25,100*	25,100*	11,500	9,600
70′	35′	63°	68′ 9″	20,900*	20,900*	9,300	7,700
	40′	58°	65′11″	17,700*	17,700*	7,700	6,300
	50′	48°	58′ 4″	13,200*	13,000	5,500	4,300
	60′	36°	47′ 2″	10,300	9,900	4,000	3,100
	70′	17°	26′11″	8,100	7,800	3,000	2,200
	20′	78°	84′ 8″	40,400*	40,400*	19,600	16,800
	25′	74°	83′ 4″	30,900*	30,900*	14,500	12,200
	30′	70°	81′ 9″	24,700*	24,700*	11,200	9,300
	35′	67°	79′ 9″	20,500*	20,500*	9,000	7,400
80′	40′	63°	77′ 5″	17,300*	17,300*	7,400	5,900
	50′	54°	71′ 3″	12,900*	12,700	5,100	4,000
	60′	45°	62′ 8″	9,900*	9,600	3,700	2,800
	70′	33°	50′ 3″	7,900	7,500	2,700	1,900
	80′	16°	28′ 5″	6,100*	6,000	1,900	1,200
	20′	79°	94′10″	36,300*	36,300*	19,300	16,500
	25′	76°	93′ 9″	30,500*	30,500*	14,200	11,900
	30′	73°	92′ 4″	24,400*	24,400*	10,900	9,000
	35′	70°	90′ 7″	20,100*	20,100*	8,700	7,100
90′	40′	66°	88′ 7″	16,900*	16,900*	7,000	5,600
	50′	59°	83′ 3″	12,500*	12,500	4,800	3,700
	60′	51°	76′ 2″	9,600*	9,400	3,400	2,400
	70′	42°	66′ 8″	7,500*	7,300	2,400	1,600
	80′	31°	53′ 2″	5,800*	5,700	1,600	900
	90′	15°	29′ 9″	3,900*	3,900*	1,000	400
	20′	80°	105′ 1″	32,400*	32,400*	19,000	16,200
	25′	77°	103′11″	27,000*	27,000*	13,900	11,600
	30′	74°	102′ 9″	22,800*	22,800*	10,600	8,700
	35′	71°	101′ 2″	19,300*	19,300*	8,400	6,800
	40′	68°	99′ 4″	16,300*	16,300*	6,700	5,300
100′	50′	62°	94′ 9″	12,100*	12,100*	4,500	3,400
	60′	55°	88′ 8″	8,800*	8,800*	3,100	2,100
	70′	48°	80′ 9″	6,600*	6,600*	2,000	1,200
	80′	40°	70′ 4″	5,000*	5,000*	1,300	600
	90′	30°	55′10″	3,600*	3,600*	700	—
	100′	14°	31′ 1″	2,300*	2,300*	200	—

(*Courtesy of Link-Belt Speeder.*)

TABLE 16-2 Capacities of a 65 Ton Truck Crane

*Capacities are based on machine equipped with Boom Gantry, 8 × 4 drive Carrier—11′ 0″ wide, 14:00 × 20 18-ply rating tires, Power Hydraulic Outriggers, 18,000 lbs. ctwt and 4,000 lbs. Bumper ctwt.

BOOM				ON OUTRIGGERS		ON TIRES	
Length	Radius	Angle	Point Ht. W.	Rear	Side	Rear	Side
40′	12′	78°	45′ 11″	130,000*	130,000*	76,700*	62,310*
	13′	76°	45′ 7″	128,040*	127,140*	73,870*	58,450*
	14′	74°	45′ 4″	119,790*	119,080*	71,250*	55,010*
	15′	73°	45′ 0″	112,500*	111,960*	68,790*	51,930*
	20′	65°	43′ 1″	86,000*	85,930*	47,760	35,510
	25′	57°	40′ 5″	67,530*	66,680	35,540	26,140
	30′	48°	36′ 6″	54,380*	49,320	28,010	20,370
	35′	37°	31′ 1″	44,890*	38,810	22,880	16,460
	40′	23°	22′ 5″	37,240*	31,730	19,150	13,600
50′	13′	79°	55′ 11″	127,360*	127,110*	73,630*	58,370*
	14′	78°	55′ 8″	119,760*	119,050*	71,010*	54,930*
	15′	76°	55′ 5″	112,480*	111,930*	68,560*	51,850*
	20′	70°	53′ 11″	85,980*	85,900*	47,890	35,650
	25′	64°	51′ 10″	67,600*	66,990	35,650	26,250
	30′	58°	48′ 11″	54,490*	49,560	28,100	20,470
	35′	50°	45′ 5″	45,000*	39,030	22,980	16,560
	40′	43°	40′ 7″	38,150*	31,960	19,270	13,730
	50′	21°	24′ 4″	28,390	23,030	14,220	9,870
60′	15′	79°	65′ 10″	112,630*	112,070*	68,410*	51,920*
	20′	74°	64′ 5″	86,080*	85,990*	48,170	35,930
	25′	69°	62′ 8″	67,770*	67,420	35,850	26,450
	30′	63°	60′ 6″	54,670*	49,880	28,250	20,630
	35′	58°	57′ 8″	45,150*	39,270	23,100	16,680
	40′	52°	54′ 2″	38,270*	32,160	19,380	13,830
	50′	39°	44′ 4″	28,580	23,220	14,330	9,980
	60′	19°	26′ 1″	22,060	17,760	11,020	7,440
70′	20′	76°	74′ 10″	85,930*	85,840*	48,150	35,920
	25′	72°	73′ 4″	67,680*	67,530	35,810	26,410
	30′	68°	71′ 6″	54,600*	49,930	28,200	20,580
	35′	63°	69′ 2″	45,080*	39,300	23,050	16,630
	40′	58°	66′ 5″	38,190*	32,170	19,320	13,780
	50′	48°	58′ 11″	28,580	23,220	14,280	9,930
	60′	36°	47′ 10″	22,100	17,800	11,010	7,430
	70′	17°	27′ 7″	17,710	14,130	8,690	5,650
80′	20′	78°	85′ 0″	85,750*	85,660*	48,100	35,870
	25′	74°	83′ 10″	67,560*	67,560*	35,730	26,340
	30′	70°	82′ 2″	54,500*	49,940	28,120	20,490
	35′	67°	80′ 2″	44,980*	39,280	22,960	16,540
	40′	63°	77′-11″	38,090*	32,140	19,220	13,690
	50′	54°	71′ 10″	28,530	23,170	14,180	9,840
	60′	45°	63′ 1″	22,050	17,760	10,930	7,350
	70′	33°	50′ 10″	17,700	14,120	8,640	5,600
	80′	16°	29′ 1″	14,540	11,470	6,920	4,270
90′	20′	79°	95′ 4″	81,510*	81,510*	48,030	35,800
	25′	76°	94′ 2″	67,420*	67,420*	35,640	26,250
	30′	73°	92′ 8″	54,380*	49,920	28,010	20,390
	35′	69°	91′ 0″	44,840*	39,240	22,840	16,430
	40′	66°	89′ 0″	37,950*	32,080	19,110	13,570
	50′	59°	83′ 2″	28,460	23,090	14,060	9,720
	60′	51°	76′ 7″	21,970	17,670	10,810	7,240
	70′	42°	67′ 1″	17,620	14,040	8,530	5,500
	80′	31°	53′ 8″	14,500	11,420	6,840	4,200
	90′	15°	30′ 6″	12,110	9,420	5,510	3,170
100′	25′	77°	104′ 5″	67,260*	67,260*	35,530	26,140
	30′	74°	103′ 2″	54,240*	49,890	27,890	20,270
	40′	68°	99′ 10″	37,790*	32,010	18,980	13,440
	50′	62°	95′ 2″	28,360	23,000	13,930	9,590
	60′	55°	89′ 1″	21,870	17,570	10,680	7,100
	70′	48°	81′ 2″	17,520	13,940	8,400	5,360
	80′	40°	70′ 10″	14,400	11,320	6,720	4,080
	90′	30°	56′ 5″	12,040	9,350	5,410	3,070
	100′	14°	31′ 10″	10,170	7,780	4,350	2,250
110′	25′	79°	114′ 8″	64,820*	64,820*	35,410	26,030
	30′	76°	113′ 6″	54,090*	49,840	27,760	20,150
	40′	70°	110′ 6″	37,630*	31,920	18,840	13,300
	50′	65°	106′ 5″	28,260	22,890	13,790	9,440
	60′	59°	101′ 0″	21,750	17,460	10,530	6,960
	70′	53°	94′ 2″	17,400	13,820	8,260	5,220
	80′	46°	85′ 7″	14,280	11,200	6,580	3,930
	90′	38°	74′ 5″	11,930	9,240	5,280	2,940
	100′	28°	59′ 0″	10,080	7,690	4,240	2,140
	110′	14°	33′ 0″	8,580	6,430	3,370	1,470
120′	30′	77°	123′ 10″	53,930*	49,780	27,630	20,010
	40′	72°	121′ 1″	37,460*	31,820	18,690	13,160
	50′	67°	117′ 4″	28,140	22,780	13,630	9,290
	60′	62°	112′ 6″	21,620	17,330	10,380	6,800
	70′	56°	106′ 6″	17,270	13,680	8,100	5,060
	80′	50°	99′ 0″	14,140	11,070	6,420	3,780
	90′	44°	89′ 8″	11,790	9,100	5,130	2,790
	100′	36°	77′ 8″	9,960	7,560	4,090	2,000
	110′	27°	61′ 6″	8,470	6,320	3,250	1,350
	120′	13°	34′ 2″	7,240	5,280	2,520	——
130′	30′	78°	134′ 0″	52,380*	49,720	27,490	19,870
	40′	74°	131′ 6″	37,280*	31,720	18,540	13,010
	50′	69°	128′ 1″	27,960*	22,660	13,470	9,130
	60′	64°	123′ 10″	21,490	17,200	10,210	6,640
	70′	59°	118′ 4″	17,130	13,550	7,940	4,900
	80′	54°	111′ 8″	14,000	10,930	6,260	3,620
	90′	48°	103′ 7″	11,650	8,960	4,960	2,630
	100′	42°	93′ 7″	9,810	7,420	3,940	1,840
	110′	35°	80′ 11″	8,340	6,190	3,090	1,200
	120′	26°	63′ 10″	7,120	5,160	2,390	——
	130′	13°	35′ 4″	6,080	4,290	1,770	——
140′	30′	79°	144′ 2″	47,730*	47,730*	27,340	19,730
	40′	75°	141′ 11″	37,090*	31,610	18,380	12,850
	50′	71°	138′ 10″	27,760*	22,530	13,310	8,970
	60′	66°	134′ 10″	21,360	17,060	10,050	6,480
	70′	62°	129′ 11″	16,980	13,400	7,770	4,740
	80′	57°	123′ 11″	31,850	10,780	6,090	3,450
	90′	52°	116′ 8″	11,500	8,810	4,800	2,460
	100′	46°	107′ 11″	9,660	7,270	3,770	1,680
	110′	40°	97′ 4″	8,190	6,040	2,930	1,040
	120′	33°	84′ 0″	6,980	5,020	2,230	——
	130′	25°	66′ 1″	5,960	4,160	1,640	——
	140′	12°	36′ 5″	5,070	3,420	1,110	——
150′	35′	78°	153′ 5″	39,180*	38,750	21,980	16,580
	40′	76°	152′ 4″	34,890*	31,500	18,220	12,690
	50′	72°	149′ 4″	27,320*	22,400	13,150	8,800
	60′	68°	145′ 8″	21,210*	16,920	9,880	6,310
	70′	64°	141′ 1″	16,840*	13,250	7,600	4,560
	80′	59°	135′ 8″	13,700*	10,630	5,920	3,280
	90′	55°	129′ 2″	11,350*	8,660	4,620	2,290
	100′	50°	121′ 5″	9,510*	7,120	3,600	1,510
	110′	45°	112′ 1″	8,030*	5,880	2,760	——
	120′	39°	100′ 11″	6,820*	4,870	2,070	——
	130′	32°	86′ 11″	5,770*	4,010	1,470	——
	140′	24°	68′ 4″	4,810*	3,290	——	——
	150′	12°	37′ 6″	4,060*	2,640	——	——
160′	35′	79°	163′ 7″	36,070*	36,070*	21,830	15,420
	40′	77°	162′ 6″	31,450*	31,380	18,060	12,530
	50′	73°	159′ 10″	25,070*	22,270	12,980	8,640
	60′	69°	156′ 5″	20,110*	16,770	9,710	6,140
	70′	65°	152′ 2″	16,680*	13,100	7,430	4,390
	80′	61°	147′ 2″	12,760*	10,470	5,740	3,100
	90′	57°	141′ 2″	10,520*	8,500	4,450	2,120
	100′	53°	134′ 5″	8,660*	6,960	3,420	1,330
	110′	48°	125′ 11″	7,110*	5,720	2,590	——
	120′	43°	116′ 2″	5,830*	4,710	1,890	——
	130′	38°	104′ 5″	4,770*	3,860	1,300	——
	150′	23°	70′ 5″	3,130*	2,500	——	——
	160′	11°	38′ 6″	2,540*	1,940	——	——

(Courtesy of Link-Belt Speeder.)

TABLE 16-3 Capacities of a 82 Ton Truck Crane With Tubular "Hi-Lite" Boom

BOOM Length	R	A	W Boom Point Height	ON OUTRIGGERS Side	ON OUTRIGGERS Rear	ON TIRES Side	ON TIRES Rear
40'	12'	78°	45' 7"	164,000*	164,000*	77,590*	119,850*
	13'	76°	45' 4"	153,080*	153,380*	72,250*	115,550*
	14'	75°	45' 2"	143,280*	143,560*	67,540*	104,770
	15'	73°	44'11"	134,610*	134,890*	63,400*	95,160
	20'	66°	43' 0"	102,000*	103,260*	45,470	64,770
	25'	57°	40' 4"	80,280*	81,300*	33,570	48,620
	30'	48°	36' 7"	65,000*	65,850*	26,250	38,580
	35'	38°	31' 4"	51,800	55,060*	21,290	31,720
	40'	24°	23' 0"	42,390	45,540*	17,670	26,720
50'	13'	79°	55' 8"	143,850*	143,850*	72,200*	115,270*
	14'	78°	55' 6"	140,420*	140,420*	67,510*	105,000
	15'	77°	55' 4"	134,590*	137,210*	63,360*	95,370
	20'	71°	53'10"	101,890*	103,260*	45,640	64,920
	25'	64°	51' 8"	80,370*	81,400*	33,700	48,740
	30'	58°	49' 0"	65,090*	65,940*	26,370	38,690
	35'	51°	45' 5"	52,060	55,160*	21,410	31,840
	40'	43°	40'10"	42,660	47,210*	17,820	26,860
	45'	34°	34' 6"	35,930	41,100*	15,090	23,070
	50'	22°	25' 0"	30,850	34,680*	12,930	20,070
60'	15'	79°	65' 6"	130,000*	130,000*	63,530*	95,860
	20'	74°	64' 4"	102,090*	103,380*	45,980	65,230
	25'	69°	62' 7"	80,580*	81,610*	33,940	48,960
	30'	64°	60' 5"	65,240*	66,100*	26,560	38,860
	35'	58°	57' 8"	52,360	55,290*	21,560	31,970
	40'	52°	54' 2"	42,910	47,330*	17,940	26,980
	50'	39°	44' 6"	31,060	36,340*	13,050	20,190
	60'	20°	26'10"	23,890	29,090*	9,850	15,750
70'	20'	76°	74' 7"	100,500*	100,500*	45,960	65,200
	25'	72°	73' 2"	80,490*	81,520*	33,900	48,910
	30'	68°	71' 5"	65,150*	65,990*	26,500	38,800
	35'	63°	69' 1"	52,400	55,180*	21,500	31,910
	40'	59°	66' 4"	42,930	47,220*	17,890	26,910
	50'	48°	58'11"	31,070	36,270*	13,000	20,130
	60'	36°	47'11"	23,930	29,060*	9,840	15,740
	70'	18°	28' 6"	19,110	23,920*	7,590	12,610
80'	20'	78°	84'11"	95,000*	95,000*	45,910	65,130
	25'	74°	83' 8"	80,350*	81,380*	33,820	48,820
	30'	71°	82' 1"	65,000*	65,860*	26,410	38,700
	35'	67°	80' 1"	52,380	55,040*	21,400	31,800
	40'	63°	77'10"	42,890	47,080*	17,780	26,800
	50'	55°	71' 8"	31,020	36,120*	12,900	20,030
	60'	45°	63' 2"	23,880	28,940*	9,750	14,640
	70'	34°	51' 1"	19,100	23,850*	7,540	12,550
	80'	17°	30' 0"	15,630	20,000*	5,860	10,230
90'	20'	79°	95' 1"	90,000*	90,000*	45,820	65,040
	25'	76°	94' 0"	77,030*	77,030*	33,720	48,700
	30'	73°	92' 7"	64,840*	65,690*	26,290	38,570
	35'	70°	90'11"	52,340	54,870*	21,270	31,670
	40'	66°	88'11"	42,830	46,910*	17,650	26,670
	50'	59°	83' 8"	30,940	35,950*	12,770	19,890
	60'	51°	76' 8"	23,790	28,760*	9,620	15,510
	70'	42°	67' 2"	19,010	23,690*	7,420	12,430
	80'	32°	54' 0"	15,580	19,890*	5,780	10,150
	90'	16°	31' 6"	12,970	16,920*	4,480	8,350
100'	25'	78°	104' 4"	72,720*	72,720*	33,590	48,570
	30'	75°	103' 0"	64,640*	65,490*	26,160	38,430
	35'	72°	101' 6"	52,270	54,670*	21,130	31,520
	40'	69°	99' 8"	42,750	46,720*	17,500	26,520
	50'	62°	95' 1"	30,830	35,760*	12,620	19,730
	60'	56°	89' 1"	23,680	28,580*	9,470	15,350
	70'	48°	81' 4"	18,900	23,500*	7,270	12,280
	80'	40°	71' 0"	15,470	19,710*	5,640	10,010
	90'	30°	56'10"	12,880	16,770*	4,380	8,240
	100'	15°	32'10"	10,840	14,400*	3,340	6,810

BOOM L	R	A	W Boom Point Height	ON OUTRIGGERS Side	ON OUTRIGGERS Rear	ON TIRES Side	ON TIRES Rear
110'	25'	79°	114' 6"	68,000*	68,000*	33,460	48,430
	30'	76°	113' 5"	62,600*	62,600*	26,010	38,280
	35'	73°	112' 0"	52,190	54,470*	20,980	31,370
	40'	71°	110' 5"	42,650	46,510*	17,340	26,350
	50'	65°	106' 4"	30,710	35,560*	12,450	19,570
	60'	59°	101' 0"	23,550	28,370*	9,300	15,180
	70'	53°	94' 2"	18,760	23,300*	7,100	12,120
	80'	46°	85' 7"	15,340	19,510*	5,480	9,840
	90'	38°	74' 7"	12,760	16,580*	4,220	8,090
	100'	29°	59' 5"	10,740	14,250*	3,220	6,690
	110'	14°	34' 1"	9,090	12,290*	2,370	5,520
120'	30'	77°	123' 7"	59,140*	59,140*	25,860	38,120
	35'	75°	122' 5"	52,100	54,270*	20,810	31,200
	40'	72°	120'11"	42,540	46,290*	17,180	26,180
	50'	67°	117' 2"	30,590	35,320*	12,280	19,390
	60'	62°	112' 6"	23,410	28,150*	9,130	15,010
	70'	56°	106' 6"	18,610	23,080*	6,930	11,940
	80'	50°	99' 1"	15,190	19,300*	5,310	9,670
	90'	44°	89'10"	12,610	16,370*	4,050	7,920
	100'	36°	77'11"	10,590	14,040*	3,050	6,520
	110'	27°	61'11"	8,970	12,130*	2,230	5,380
	120'	14°	35' 5"	7,610	10,510*	1,530	4,410
130'	30'	78°	133'11"	55,590*	55,590*	25,690	37,950
	35'	76°	132' 8"	51,160*	51,160*	20,640	31,020
	40'	74°	131' 5"	42,420	45,400*	17,000	26,000
	50'	69°	128' 0"	30,450	35,100*	12,100	19,210
	60'	64°	123' 8"	23,260	27,930*	8,940	14,820
	70'	59°	118' 4"	18,460	22,850*	6,750	11,750
	80'	54°	111' 8"	15,030	19,070*	5,120	9,480
	90'	48°	103' 7"	12,450	16,160*	3,870	7,730
	100'	42°	93' 8"	10,430	13,830*	2,880	6,340
	110'	35°	81' 2"	8,820	11,920*	2,060	5,210
	120'	26°	64' 4"	7,480	10,320*	----	4,260
	130'	13°	36' 7"	6,350	8,900*	----	3,430
140'	30'	79°	144' 1"	52,170*	52,170*	25,530	37,780
	35'	77°	143' 0"	47,450*	47,450*	20,470	30,850
	40'	75°	141'10"	42,300	42,790*	16,820	25,820
	50'	71°	138' 8"	30,300	34,780*	11,910	19,020
	60'	66°	134' 8"	23,100	27,680*	8,760	14,630
	70'	62°	129'10"	18,290	22,610*	6,560	11,560
	80'	57°	123'11"	14,860	18,840*	4,930	9,290
	90'	52°	116' 8"	12,280	15,930*	3,680	7,540
	100'	46°	108' 0"	10,260	13,600*	2,690	6,150
	110'	41°	97' 6"	8,650	11,700*	1,880	5,020
	120'	34°	84' 2"	7,320	10,100*	----	4,080
	130'	25°	66' 7"	6,200	8,750*	----	3,270
	140'	13°	37' 8"	5,240	7,480*	----	2,570
150'	35'	78°	153' 4"	44,330*	44,330*	20,290	30,660
	40'	76°	152' 1"	39,730*	39,730*	16,640	25,630
	50'	72°	149' 2"	30,160	31,690*	11,720	18,830
	60'	68°	145' 7"	22,940	25,900*	8,560	14,430
	70'	64°	141' 1"	18,130	21,410*	6,360	11,360
	80'	59°	135' 7"	14,680	16,860*	4,740	9,090
	90'	55°	129' 2"	12,100	14,410*	3,490	7,340
	100'	50°	121' 5"	10,090	12,230*	2,490	5,960
	110'	45°	112' 2"	8,470	10,360*	1,680	4,830
	120'	39°	101' 1"	7,150	8,880*	----	3,890
	130'	32°	87' 2"	6,040	7,610*	----	3,090
	140'	24°	68'10"	5,090	6,510*	----	2,400
	150'	12°	38'10"	4,250	5,610*	----	1,780

BOOM L	R	A	W Boom Point Height	ON OUTRIGGERS Side	ON OUTRIGGERS Rear	ON TIRES Side	ON TIRES Rear
160' ②	35'	79°	163' 6"	41,120*	41,120*	20,110	[illegible]
	40'	77°	162' 5"	36,900*	36,900*	16,450	[illegible]
	50'	73°	159' 8"	29,270*	29,270*	11,530	[illegible]
	60'	69°	156' 4"	22,780	24,000*	8,370	[illegible]
	70'	65°	152' 1"	17,960	19,740*	6,160	[illegible]
	80'	61°	147' 1"	14,510	15,660*	4,540	[illegible]
	90'	57°	141' 2"	11,920	13,140*	3,290	[illegible]
	100'	53°	134' 2"	9,900	11,020*	2,290	[illegible]
	110'	48°	126' 0"	8,290	9,280*	1,490	[illegible]
	120'	43°	116' 4"	6,960	7,770*	----	[illegible]
	130'	38°	104' 7"	5,860	6,530*	----	[illegible]
	140'	31°	90' 1"	4,910	5,500*	----	[illegible]
	150'	24°	70'11"	4,100	4,600*	----	[illegible]
	160'	12°	39'11"	3,370	3,920*	----	
170' ②	35'	79°	173' 8"	38,110*	38,110*	19,930	[illegible]
	40'	78°	172' 7"	34,270*	34,270*	16,270	[illegible]
	50'	74°	170' 1"	27,170*	27,170*	11,340	[illegible]
	60'	71°	166'11"	22,100*	22,100*	8,170	[illegible]
	70'	67°	163' 1"	17,780	19,920*	5,960	[illegible]
	80'	63°	158' 5"	14,330	15,770*	4,330	[illegible]
	90'	59°	153' 0"	11,740	13,100*	3,080	[illegible]
	100'	55°	146' 7"	9,720	10,840*	2,090	[illegible]
	110'	51°	139' 1"	8,100	8,980*	----	[illegible]
	120'	47°	130' 5"	6,780	7,460*	----	[illegible]
	130'	42°	120' 2"	5,670	6,240*	----	[illegible]
	140'	37°	107'11"	4,730	5,200*	----	[illegible]
	150'	30°	92'10"	3,920	4,310*	----	
	160'	23°	73' 0"	3,200	3,520*	----	
	170'	12°	40'11"	2,560	2,820*	----	
180' ②	40'	78°	182'11"	31,110*	31,110*	16,070	[illegible]
	50'	75°	180' 6"	25,100*	25,100*	11,140	[illegible]
	60'	72°	177' 6"	20,350*	20,350*	7,970	[illegible]
	70'	68°	173'11"	16,600	19,360*	5,760	[illegible]
	80'	65°	169' 6"	14,140	15,550*	4,130	[illegible]
	90'	61°	164' 6"	11,550	12,710*	2,880	[illegible]
	100'	58°	158' 7"	9,530	10,480*	1,890	[illegible]
	110'	54°	151' 8"	7,910	8,700*	----	[illegible]
	120'	50°	143'10"	6,580	7,240*	----	[illegible]
	130'	45°	134' 8"	5,480	6,030*	----	[illegible]
	140'	41°	123'11"	4,540	4,990*	----	[illegible]
	150'	35°	111' 2"	3,730	4,100*	----	
	160'	30°	95' 6"	3,020	3,320*	----	
	170'	22°	75' 0"	2,400	2,640*	----	
	180'	11°	41'11"	1,830	2,010*	----	

(*Courtesy of Link-Belt Speeder.*)

If it is necessary to move a large precast member some distance with one or more cranes, the ground over which the cranes move must be very level and firm.

Precast building beams and girders can often be efficiently erected using truck cranes in combination with dollies, even when the building dimensions, site conditions or crane capacity preclude the possibility of the crane setting the member into its final location. In this case, one or two cranes are used to hoist the precast member to the proper level and set it on dollies which then are used to move the member horizontally to the proper location. The dollies can be provided with hydraulic jacks to allow the members to be lowered into place after they are correctly positioned horizontally. This method does not work well on sloping floors such as those found in some types of parking garages.

16-3 Crawler Cranes

Crawler cranes having high capacities are occasionally used to erect precast members. Crawler cranes must be moved from site to site on trailers, and for this reason, they are generally more costly and more difficult to use in developed areas. The load radius for a crawler crane is measured in a manner similar to that of a truck crane (*see* Fig. 16-2).

The capacities of a 103.5 ton crawler crane are given in Table 16-4. Crawler cranes working under maximum lifts, like truck cranes, must be level and be on firm ground.

16-4 Floating Cranes

Floating cranes with very large capacities are available in some localities. The girder shown being erected in Fig. 16-3 weighs 230 tons and is approximately 200 ft long. Some floating cranes have capacities as high as 600 tons. The principal limitation on the use of floating cranes is the availability of sufficient depth of relatively calm water on which the crane can work.

16-5 Girder Launchers

Large precast bridge girders are sometimes erected with girder launchers such as the one shown in Fig. 16-4. The launchers are generally steel trusses, although aluminum has been used. For shorter spans with lighter girders, steel beams may be used rather than trusses.

There have been many different variations in the construction of launchers, but they all work on the same basic principle of moving over the already erected girders and cantilevering out to the next pier in order to provide an overhead structure that is sufficiently strong to support the necessary erection loads. The sequences in erecting a bridge with a girder launcher is illustrated in Fig. 16-5.

TABLE 16-4 Capacities of a 103.5 Ton Crawler Crane With Tubular "Hi-Lite" Boom.

BOOM Length	BOOM Radius	BOOM Angle	W Boom Point Height	Ctwt. "A"	Ctwt. "AB"
50'	13'	80°	56' 0"	207,000*	207,000*
	14'	79°	55'10"	185,090*	207,000*
	15'	78°	55' 8"	160,470*	207,000*
	16'	77°	55' 6"	144,660	200,000*
	17'	76°	55' 3"	129,310	187,910
	18'	74°	54'10"	116,830	169,980
	19'	73°	54' 7"	106,490	155,050
	20'	72°	54' 4"	97,780	144,720
	25'	66°	52' 5"	69,010	102,660
	30'	59°	49'10"	52,910	79,140
	35'	53°	46' 5"	42,600	64,090
	40'	45°	42' 1"	35,420	53,620
	50'	25°	27' 8"	26,000	39,940
60'	14'	81°	66' 0"	185,090*	201,160*
	15'	80°	65'10"	160,470*	197,940*
	16'	79°	65' 8"	144,660	197,500*
	17'	78°	65' 5"	129,310	187,910
	18'	77°	65' 2"	116,830	169,980
	19'	76°	64'11"	106,490	155,050
	20'	75°	64' 9"	97,780	144,720
	25'	70°	63' 2"	69,010	102,660
	30'	65°	61' 1"	52,910	79,140
	35'	60°	58' 6"	42,600	64,090
	40'	54°	55' 2"	35,420	53,620
	50'	41°	45'11"	26,000	39,940
	60'	23°	29' 9"	20,410	31,690
70'	16'	81°	75'11"	144,660	179,500*
	17'	80°	75' 8"	129,310	179,000*
	18'	79°	75' 5"	116,830	169,980
	19'	78°	75' 3"	106,490	155,050
	20'	77°	75' 0"	97,780	144,720
	25'	73°	73' 8"	69,010	102,660
	30'	68°	71'11"	52,910	79,140
	35'	64°	69' 9"	42,600	64,090
	40'	60°	67' 1"	35,420	53,620
	50'	50°	60' 0"	26,000	39,940
	60'	38°	49' 6"	20,410	31,690
	70'	21°	31' 8"	16,510	26,000
80'	17'	81°	85' 9"	129,310	164,000*
	18'	80°	85' 6"	116,830	162,000*
	19'	79.5°	85' 5"	106,490	155,050
	20'	79°	85' 3"	97,780	144,720
	25'	75°	84' 1"	69,010	102,660
	30'	72°	82' 7"	52,910	79,140
	35'	68°	80' 9"	42,600	64,090
	40'	64°	78' 6"	35,420	53,620
	50'	55°	72' 7"	26,000	39,940
	60'	46°	64' 5"	20,410	31,690
	70'	35°	52' 9"	16,510	26,000
	80'	20°	33' 5"	13,620	21,800
90'	19'	81°	95' 8"	106,490	147,000*
	20'	80°	95' 5"	97,780	139,290*
	25'	77°	94' 5"	69,010	102,660
	30'	74°	93' 1"	52,910	79,140
	35'	70°	91' 5"	42.600	64,090
	40'	67°	89' 6"	35,420	53,620
	50'	60°	84' 5"	26,000	39,940
	60'	52°	77' 8"	20,410	31,690
	70'	43°	68' 7"	16,510	26,000
	80'	33°	55'10"	13,620	21,800
	90'	18°	35' 1"	11,370	18,560
100'	20'	81°	105' 7"	97,780	127,550*
	25'	78°	104' 8"	69,010	102,660
	30'	75°	103' 6"	52,910	79,140
	35'	72°	102' 0"	42,600	64,090
	40'	69°	100' 3"	35,420	53,620
	50'	63°	95'10"	26,000	39,940
	60'	56°	90' 0"	20,410	31,690
	70'	49°	82' 4"	16,500	25,980
	80'	41°	72' 5"	13,580	21,770
	90'	31°	58' 9"	11,350	18,540
	100'	17°	36' 8"	9,550	15,970
110'	25'	79°	114'10"	69,010	102,660
	30'	77°	113' 9"	52,910	79,140
	35'	74°	112' 5"	42,600	64,090
	40'	71°	110'11"	35,420	53,620
	50'	65°	106'11"	26,000	39,940
	60'	60°	101' 9"	20,410	31,670
	70'	54°	95' 2"	16,390	25,870
	80'	47°	86'10"	13,480	21,660
	90'	39°	76' 1"	11,260	18,460
	100'	30°	61' 6"	9,500	15,920
	110'	17°	38' 2"	8,040	13,830
120'	25'	80°	125' 0"	69,010	101,600*
	30'	78°	124' 0"	52,910	79,140
	35'	75°	122'10"	42,600	64,090
	40'	73°	121' 5"	35,420	53,620
	50'	68°	117'10"	26,000	39,940
	60'	63°	113' 2"	20,250	31,540
	70'	57°	107' 4"	16,260	25,740
	80'	51°	100' 1"	13,350	21,540
	90'	45°	91' 0"	11,140	18,330
	100'	37°	79' 6"	9,400	15,810
	110'	29°	64' 1"	7,970	13,760
	120'	16°	39' 7"	6,760	12,030
130'	25'	81°	135' 2"	69,010	93,280*
	30'	79°	134' 3"	52,910	79,140
	35'	77°	133' 2"	42,600	64,090
	40'	74°	131'10"	35,390	53,620
	50'	70°	128' 7"	25,960	39,900
	60'	65°	124' 4"	20,110	31,400
	70'	60°	119' 1"	16,110	25,600
	80'	55°	112' 8"	13,210	21,390
	90'	49°	104' 9"	11,000	18,190
	100'	43°	95' 0"	9,260	15,670
	110'	36°	82'10"	7,840	13,640
	120'	27°	66' 7"	6,670	11,940
	130'	15°	41' 0"	5,640	10,490
140'	30'	80°	144' 7"	52,910	79,140
	35'	77°	143' 5"	42,540	64,030
	40'	75°	142' 3"	35,260	53,460
	50'	71°	137' 4"	25,820	39,750
	60'	67°	135' 4"	19,950	31,240
	70'	62°	130' 6"	15,960	25,440
	80'	57°	124' 8"	13,050	21,230
	90'	52°	117' 8"	10,840	18,030
	100'	47°	109' 2"	9,100	15,520
	110'	41°	98'11"	7,700	13,490
	120'	35°	86' 0"	6,530	11,800
	130'	26°	68'11"	5,540	10,380
	140'	15°	42' 3"	4,600	9,140
150'	30'	80°	154' 7"	52,900	75,890*
	35'	78°	153' 8"	42,400	63,900
	40'	76°	152' 6"	35,120	53,320
	50'	72°	149' 9"	25,600	39,600
	60'	68°	146' 2"	19,790	31,080
	70'	64°	141' 9"	15,790	25,280
	80'	60°	136' 5"	12,880	21,060
	90'	55°	130' 1"	10,670	17,870
	100'	51°	122' 6"	8,940	15,350
	110'	45°	113' 6"	7,530	13,320
	120'	40°	102' 7"	6,370	11,650
	130'	33°	89' 0"	5,390	10,230
	140'	26°	71' 3"	4,540	9,020
	150'	14°	43' 6"	3,780	7,950
160'	30'	81°	164' 9"	52,810	70,120*
	35'	79°	163'10"	42,290	63,780
	40'	77°	162' 9"	34,990	53,190
	50'	74°	160' 2"	25,520	39,460
	60'	70°	156'10"	19,640	30,930
	70'	66°	152' 9"	15,640	25,120
	80'	62°	147'10"	12,730	20,910
	90'	58°	142' 0"	10,520	17,710
	100'	53°	135' 2"	8,780	15,190
	110'	49°	127' 1"	7,370	13,170
	120'	44°	117' 7"	6,210	11,490
	130'	38°	106' 1"	5,240	10,080
	140'	32°	92' 0"	4,400	8,880
	150'	25°	73' 6"	3,670	7,830
	160'	14°	44' 9"	3,000	6,900
170'	35'	80°	174' 0"	42,170	
	40'	78°	173' 0"	34,860	
	50'	75°	170' 7"	25,380	
	60'	71°	169' 6"	19,490	
	70'	67°	163' 8"	15,480	
	80'	64°	159' 1"	12,570	
	90'	60°	153' 9"	10,360	
	100'	56°	147' 5"	8,620	
	110'	51°	140' 1"	7,220	
	120'	47°	131' 7"	6,060	
	130'	43°	121' 6"	5,090	
	140'	37°	109' 6"	4,250	
	150'	31°	94'10"	3,530	
	160'	24°	75' 8"	2,890	
	170'	13°	46' 0"	2,300	
180'	35'	80°	184' 2"	42,040	
	40'	79°	183' 3"	34,720	
	50'	75°	180'11"	25,230	
	60'	72°	178' 0"	19,340	
	70'	69°	174' 5"	15,320	
	80'	65°	170' 2"	12,410	
	90'	62°	165' 2"	10,200	
	100'	58°	159' 5"	8,460	
	110'	54°	152' 8"	7,060	
	120'	50°	144'10"	5,900	
	130'	46°	135'10"	4,930	
	140'	41°	125' 4"	4,100	
	150'	36°	112'10"	3,380	
	160'	30°	97' 7"	2,750	
	170'	23°	77' 9"	2,180	
	180'	13°	47' 1"	1,660	
190'	35'	80°	194' 4"	41,900	
	40'	79°	193' 5"	34,580	
	50'	76°	191' 3"	25,080	
	60'	73°	188' 7"	19,180	
	70'	70°	185' 1"	15,160	
	80'	67°	181' 2"	12,240	
	90'	63°	176' 6"	10,030	
	100'	60°	171' 0"	8,290	
	110'	56°	164'10"	6,890	
	120'	53°	157' 8"	5,730	
	130'	49°	149' 5"	4,760	
	140'	45°	140' 0"	3,940	
	150'	40°	129' 0"	3,220	
	160'	35°	116' 0"	2,590	
	170'	30°	100' 2"	2,040	
	180'	23°	79' 9"	1,540	
	190'	13°	48' 3"	1,070	
200'	40'	80°	203' 7"	34,430	
	50'	77°	201' 6"	24,920	
	60'	74°	198'11"	19,020	
	70'	71°	195' 9"	14,990	
	80'	68°	192' 0"	12,080	
	90'	65°	187' 7"	9,860	
	100'	61°	182' 6"	8,120	
	110'	58°	176' 8"	6,720	
	120'	55°	170' 1"	5,560	
	130'	51°	162' 6"	4,590	
	140'	47°	153'11"	3,770	
	150'	43°	144' 1"	3,050	
	160'	39°	132' 8"	2,430	
	170'	34°	119' 2"	1,880	
	180'	29°	102'10"	1,380	
	190'	22°	81' 9"	——	
	200'	12°	49' 4"	——	

(*Courtesy of Link-Belt Speeder.*)

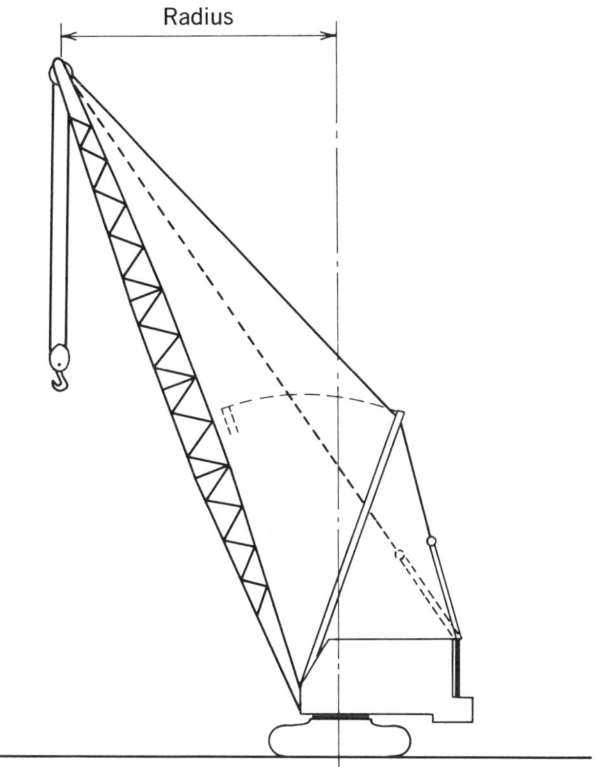

Fig. 16-2 Load radius for a crawler crane.

It should be noted that the launcher shown in Fig. 16-5 is approximately 1.7 times the length of the girder that is to be erected. A girder is used for a counterweight when the launcher is being positioned.

The approximate weight of concrete bridge girders suitable for normal launcher erection (wide top flanges) and that of the launching truss alone is shown in Fig. 16-6. The weight of the dollies, hoists, and other necessary equipment is approximately equal to that of the truss.

Girder launchers capable of erecting girders 220 ft long weighing 270 tons have been designed and proved to be economical. For the longer and heavier girders, guyed launchers having a central tower, as shown in Fig. 16-4 are considered to be most economical. For erecting shorter spans, cantilevered trusses, without the central tower as shown in Fig. 16-7, have proved to be economical and efficient.

Fig. 16-3 Floating crane erecting a 230 ton girder on Crossbay Parkway Bridge in New York City.

Fig. 16-4 Girder launcher erecting a 70 ton, 130-ft long girder.

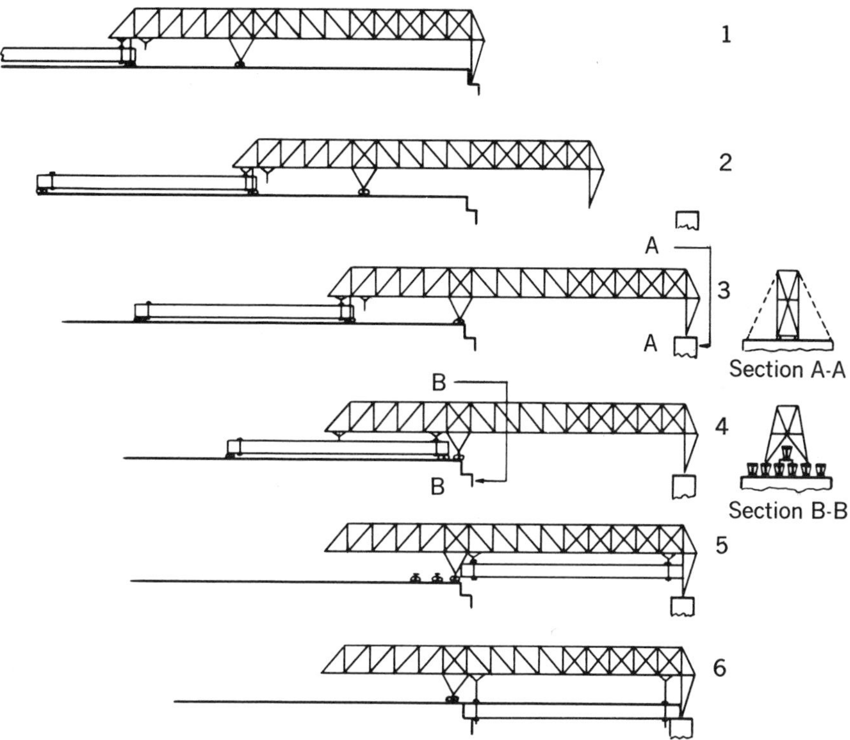

Fig. 16-5 Sequence in using a girder launcher. (1) Launcher assembled with girder connected for a counterweight. (2) Launcher and girder being moved into place for erecting first span. (3) Launcher in place for erecting first span. (4) Girder connected to launcher at outboard end. (5) Girder supported by launcher. (6) Girder erected.

16-6 Falsework

Falsework can be used to erect precast bridge girders economically under certain conditions. The falsework may consist of wood or steel towers which support steel or wood beams. The precast girders are normally rolled on to the beams and then slid transversely into place. Alternately, the girders may be precast in segments, assembled, stressed, and grouted on falsework. This later procedure eliminates the need of moving the assembled girder.

Many methods of sliding girders transversely have been used. The details of the scheme used on the Bamako Bridge are shown in Fig. 16-8 (Ref. 3).

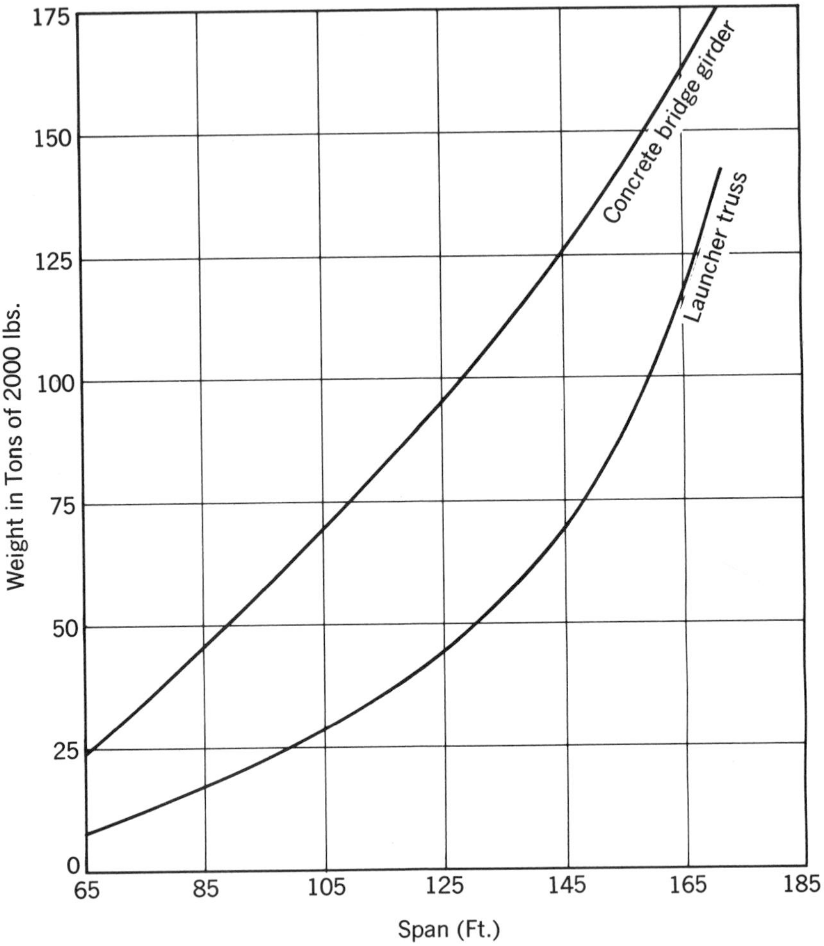

Concrete bridge girder
Launcher truss
Weight in Tons of 2000 lbs.
0 25 50 75 100 125 150 175
65 85 105 125 145 165 185
Span (Ft.)

Fig. 16-6 Approximate bridge girder and launch-truss weights for various spans.

Fig. 16-7 Cantilevered truss launcher.

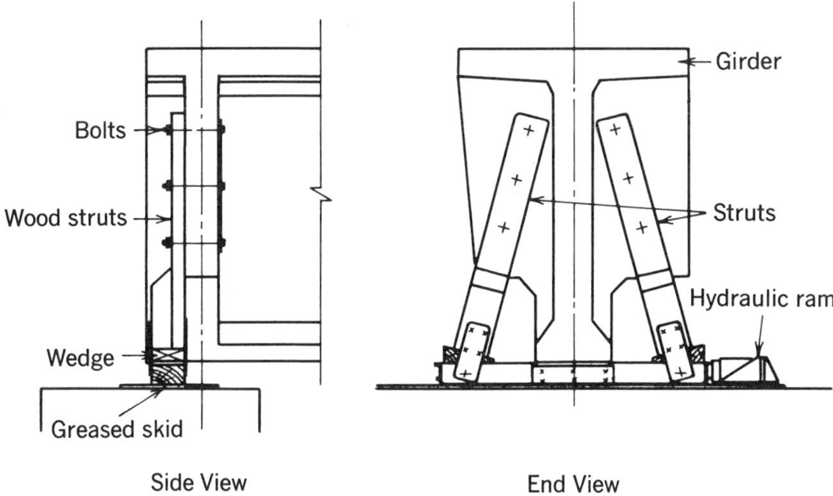

Fig. 16-8 Bamako bridge apparatus for transverse sliding of girder.

16-7 Cable Ways and Highlines

Cable ways and highlines are occasionally considered for use in erecting precast concrete bridge girders. The cost of cableways is very high and they are not a type of device that is moved at low cost.

Cableways find their use on dam construction projects where they can be installed in a single location and used for a period of several months or years in handling loads of nominal magnitudes (say 25 tons maximum). Sites which particularly lend themselves to cableways are normally deep narrow gorges. Rather large sags with relatively small (at least not very large) spans are used.

Cableways could conceivably be used to good advantage to erect precast segments in cantilever construction, but it is doubtful they are practical for use in erecting complete girders of anything but short spans.

16-8 Towers

Rolling towers, as illustrated in Fig. 16-9, offer a means of erecting bridge girders of short to moderate length. The scheme utilizes a girder as counterweight in combination with an additional nominal counterweight. It has the disadvantage that the girders over which the tower moves during erection must bear the weight of two girders, the counterweight, and the tower complete with rigging. The result of this disadvantage is that the method is not feasible for the erection of long span girders where dead load is so important.

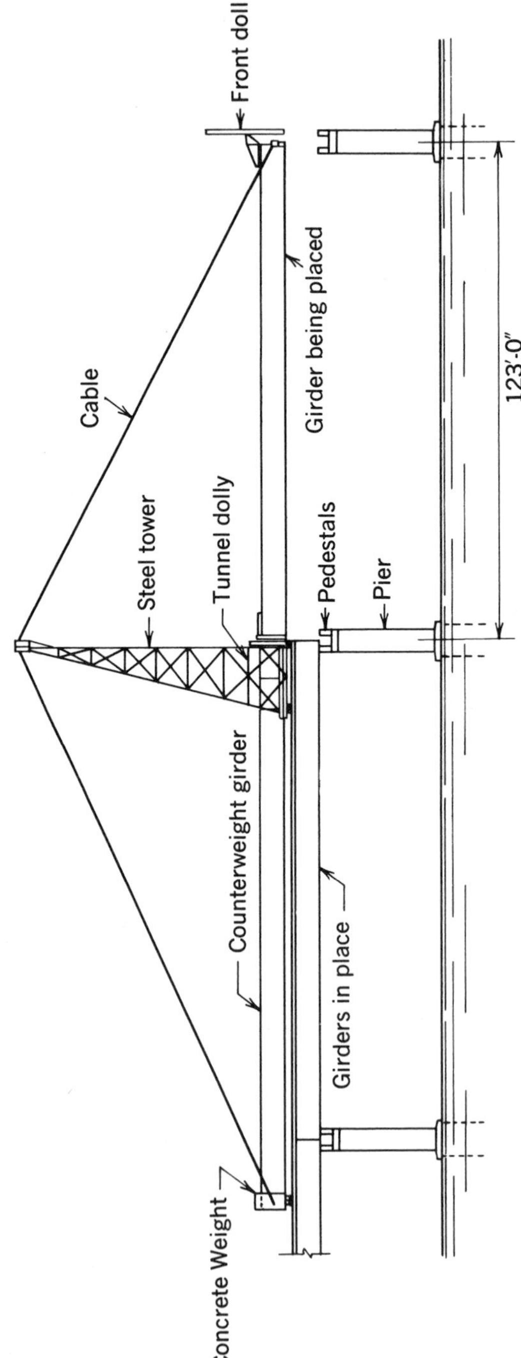

Fig. 16-9 Tower for erecting bridge girders.

REFERENCES

1. Société Technique Pour L'Utilisation de la Precontrainte, Private Communication (July 7, 1966).

2. Stergiou, Paul, "Launcher Erects Post-Tensioned Girders," *Civil Engineering*, pp. 66–67 (Aug. 1967).

3. Muller, Jean, "Engineering Feats With Post-Tensioning," *Journal of the Prestressed Concrete Institute*, **5**, No. 4, 12–40 (Dec. 1960).

Index

Index